Antarctic Marine Geology

The fragile Antarctic environment consists of a closely linked system of the lithosphere, atmosphere, cryosphere, hydrosphere and biosphere. Changes in this system have influenced global climate, oceanography and sea level for most of Cenozoic time. The geological history of this region therefore provides a special record of important interactions between the various components of the Earth System.

Antarctic Marine Geology is the first comprehensive single-authored book to introduce students and researchers to the geological history of the region and the unique processes that occur there. The Antarctic geological literature is widely disseminated, and until now no single reference has existed which provides such a summary. The book is intended as a reference for all scientists working in Antarctica, and will also serve as a textbook for graduate courses in Antarctic marine geology.

John B. Anderson has worked in Antarctica for 29 years, and he has participated in 19 scientific expeditions. He has published over 100 refereed papers, most dealing with Antarctic marine geology. He is a former member of the Polar Research Board and 1997 NSF–OPP External Panel. Anderson received the 1996 Rice University graduate teaching award and the 1992 Gulf Coast Association of Geological Societies Outstanding Educator Award.

Antarctic Marine Geology

JOHN B. ANDERSON
Rice University

CAMBRIDGE
UNIVERSITY PRESS

CAMBRIDGE UNIVERSITY PRESS
Cambridge, New York, Melbourne, Madrid, Cape Town, Singapore,
São Paulo, Delhi, Dubai, Tokyo

Cambridge University Press
The Edinburgh Building, Cambridge CB2 8RU, UK

Published in the United States of America by Cambridge University Press, New York

www.cambridge.org
Information on this title: www.cambridge.org/9780521131681

First published 1999
This digitally printed version 2010

A catalogue record for this publication is available from the British Library

Library of Congress Cataloguing in Publication data
Anderson, John B., 1944–

  Antarctic marine geology / John B. Anderson.
    p.   cm.
  Includes index.
  ISBN 0-521-59317-4 (hardback)

  1. Geology – Antarctica.   2. Submarine geology.   I. Title.
QE350.A49   1999
559.8′9 – dc21   98-37219
                                              CIP

ISBN 978-0-521-59317-5 Hardback
ISBN 978-0-521-13168-1 Paperback

Contents

Acknowledgments

This book draws heavily from the theses and dissertations of my graduate students who have traveled to the Antarctic with me to collect data, have conducted most of the analyses of these data, and have been a constant source of inspiration these past two decades. I owe a special word of thanks to all of them. Thank you Barbara Andrews, Mohammed Baegi, Laura Banfield, Phil Bart, Lou Bartek, Kathy Balshaw Biddle, Chris Brake, Jana DaSilva, Laura De Santis, Eugene Domack, Robyn Wright Dunbar, Pam Fisco, Tom Griffith, Elke Jahns, John Jeffers, Doug Kennedy, Dennis Kurtz, Margaret Herron, Ashley Lowe, Scott MacDonald, Robert Milam, Nathan Myers, Peter Pope, David Reid, Tony Rodriguez, Christina Roqueplo, Stephanie Shipp, Jill Singer, Fernando Siringan, Mike Smith, Mark Thomas and Julia Wellner. I cannot begin to list my colleagues in the field who have contributed through stimulating discussion and interaction.

I owe a special word of thanks to Stephanie Shipp, who edited several versions of the book and helped with its overall organization. She saw this project through to the end. Thank you Stephanie.

Michelle Fassell, Charlotte Kelchner, Ashley Lowe, Tony Rodriguez, Cindy Ross and Julia Wellner provided valuable editorial assistance and David Matty and Robyn Wright Dunbar contributed to the writing of Chapter 2. Stephanie Shipp, Nicki Atkinson, Ashley Lowe and Breezy Long drafted the figures for the book and assisted with other technical aspects of publication.

I also want to express my gratitude to the crews and technical support staff from all my cruises on the USCGC *Glacier,* ARA *Islas Orcadas,* R/V *Polar Duke,* and the R/V *NB Palmer.* Their dedication and highly professional support has made all my ventures to Antarctica rewarding experiences. I appreciate the support of the Antarctic Marine Geology Research Facility, particularly Dennis Cassidy and Tom Janecek, for providing samples and being so helpful during our many visits to the facility to study the Antarctic core collections. My research during the past two decades has been funded by the National Science Foundation–Office of Polar Programs.

Lastly, I must thank my wife Doris and my children John and Heather for their love, encouragement and support. They spent many winters at home while I was in Antarctica. And lastly, thanks to my mother and father, Walter and Lena Anderson, who have always inspired me by their positive outlooks on life.

CHAPTER ONE

Antarctica's Environment

This chapter introduces the reader to the various components of Antarctica's glacial, atmospheric, and oceanographic environment. It is not intended as an in-depth treatment of these subjects. Rather, this chapter is intended to provide marine geologists with sufficient information to focus on and solve problems related to sedimentary processes and to examine the stratigraphic record as it relates to changes in the Antarctic environment. It is also intended to provide the information needed to plan and execute successful marine geologic and geophysical expeditions.

ANTARCTICA'S ROLE IN THE GLOBAL ENVIRONMENT

Isolated from other continents by the Southern Ocean (oceans south of ~60° S), Antarctica is the only continent that experiences a true polar climate. The Antarctic Ice Sheet is almost as large as it was during the last glacial maximum, containing at least 80% of its previous volume (Denton and Hughes, 1981). Thus, Antarctica provides a modern setting in which to study "Ice Age" Earth. Antarctica claims a long history of Cenozoic glaciation and has played a key role in regulating world climate, oceanography, and eustasy for most of this geologic era.

Antarctica's lithosphere, atmosphere, cryosphere, hydrosphere, and biosphere are closely linked and have been for much of the continent's glacial history, but the continent has not always been glaciated. During the Cretaceous, the coastal regions of Antarctica supported lush, temperate forests inhabited by a wide diversity of animals. At that time, the continent was situated at a latitude similar to southern South America. As Antarctica drifted away from the other Gondwana continents and toward the south, its climate cooled and daylight hours became more seasonally distributed. The evolutionary changes of plants and animals trying to adapt to these harsher conditions on the continent are among the most spectacular and poorly understood paleontologic events in Earth's geologic history. There were also significant changes in water mass properties and circulation patterns in the Southern Ocean, and these changes influenced organisms far

from the Antarctic coast. Even early humans may have responded to its influence (Denton, Prentice, and Burckle, 1991). Conversely, there is little doubt that human beings have seriously altered the fragile ecosystem of the Southern Ocean by uncontrolled harvesting of seals, whales, finfish, and krill.

Tectonic activity on the continent, especially the uplift of the Transantarctic Mountains (TAM), strongly influenced the early evolution of the Antarctic Ice Sheet (Fig. 1.1). Eventually, the TAM formed a physiographic boundary between the East and West Antarctic ice sheets. These ice sheets have had a very different evolution and are still quite different in terms of their size, shape, and dynamics: these differences are, in large part, due to their different geologic settings.

The Antarctic atmosphere-cryosphere-lithosphere system constantly changes, continuously influencing the global system. On a short-term scale, Antarctica serves as an atmospheric heat sink causing the temperature gradient that drives atmospheric circulation in the Southern Hemisphere (Mullan and Hickman, 1990). The stable upper-atmospheric air mass that resides over the continent has been seriously altered by manufactured chemicals, resulting in the creation of the well-known Antarctic Ozone Hole (Farman, Gardiner, and Shanklin, 1985; Solomon et al., 1986). In addition, ice core records provide evidence that the quantity of atmospheric greenhouse gases (CO_2 and CH_4) has increased markedly during the past 200 years and that these gases are presently more abundant than at any time in the past 160,000 years (Oeschger and Siegenthaler, 1988; Lorius, Jouzel, and Raynaud, 1993).

Sea-ice growth and decay around Antarctica influence climate and ocean circulation throughout much of the Southern Hemisphere. The temperature gradient associated with Antarctica's sea-ice zone is one of the strongest on Earth. The oscillating sea-ice canopy around Antarctica is intimately associated with surface water masses and is the primary regulator of the ocean's vertical heat flux and stability (Mullan and Hickman, 1990; Martinson and Iannuzzi, 1998). Predicted global warming trends are expected to reduce significantly the areal extent of this

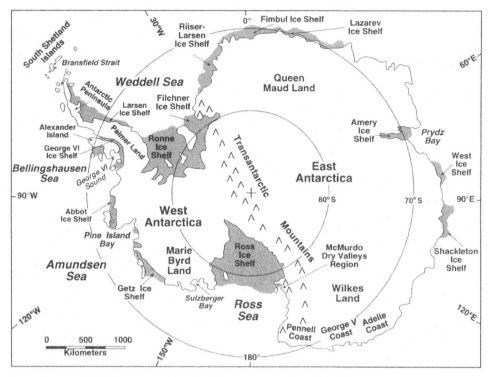

Figure 1.1. Geographic locations on the Antarctic continent and the surrounding Southern Ocean. Shaded areas are ice shelves.

Figure 1.2. Subglacial topography of Antarctica corrected for isostatic uplift. Subglacial basins referred to in the text are labeled: BSB = Byrd Subglacial Basin; WSB = Wilkes Land Subglacial Basin; TT = Thiel Trough (from Denton et al., 1991).

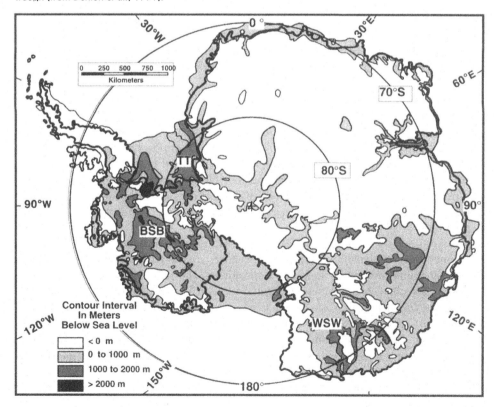

sea-ice zone around the continent (Budd, 1991). Removal of the sea-ice canopy would permit increased heat loss from the ocean to the atmosphere by as much as two orders of magnitude (Budd, 1991). There is a well-documented correspondence between the location of the sea-ice margin, oceanographic fronts, and levels of primary productivity in surface waters of the Southern Ocean. However, predicting the ultimate biologic perturbation of a decrease in the areal extent of sea ice around Antarctica is problematic (Mortlock et al., 1991). It is also known that the extent of sea-ice around Antarctica fluctuates over annual and decade time scales and this variability has been correlated to change in atmospheric circulation, especially wind intensity. There is some evidence that these variations affect production rates of Antarctic Bottom Water, one of the principal deep waters in the world ocean (Comiso and Gordon, 1998).

Another potential impact of global warming is increased melting of Antarctica's ice shelves. Indeed, this effect is perhaps already being observed in the Antarctic Peninsula region. A warming trend in the region during this century coincided with a significant reduction in the size of the Larsen and George VI ice shelves (Potter and Paren, 1985; Rott, Skvarca, and Nagler, 1996) and the disappearance of the smaller Wordie Ice Shelf (Doake and Vaughan, 1991) and Müller Ice Shelf (Domack et al., 1995) in the Antarctic Peninsula region. Is this warming trend due to anthropogenic effects or part of a natural climate cycle? The answer to this important question will perhaps be found by studying ice cores and sediment cores from the region.

Retreat of ice shelves initiates increased discharge of outlet glaciers and ice streams flowing to sustain these ice shelves. Ultimately, this results in a reduction in the size of the ice sheet. Furthermore, there is a strong feedback between ice shelves and oceanographic processes. If ice shelves are bathed in warmer water masses, the net effect is an order of magnitude increase in the basal melting rate of the ice shelves, an impact that is being observed today throughout West Antarctica (Potter and Paren, 1985; Jenkins and Doake, 1991; Jacobs et al., 1992; Jacobs, Hellmer, and Jenkins, 1996; Jenkins et al., 1996). In turn, ice shelves play a key role in the formation of water masses on the Antarctic continental shelf.

Removal of the ice shelves would lead to a decrease in the production and volume of very cold shelf water masses, which are a vital component in Antarctic Bottom Water. This feedback between ice shelves and water masses

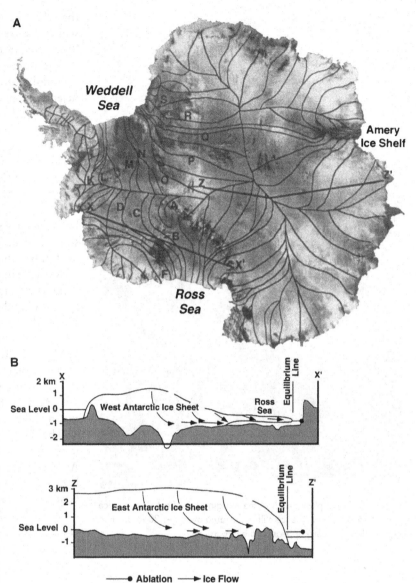

Figure 1.3. (A) Drainage map for Antarctica and (B) profiles of the East and West Antarctic ice sheets showing ice flow trajectories (modified from Bentley, 1964). Letters designate ice streams. Continental flowlines are modified from Hughes (1973) and Drewry (1983). Ice shelf drainage is from Stuiver, Denton, Hughes, and Fastook (1981).

is one of the most delicate features of the Antarctic environment and one that would quickly respond to global warming (Budd, 1991). The impact of changing water mass production around Antarctica would be global, because deep and intermediate water masses formed near the continent play an integral role in ocean circulation across the globe.

Studies of large ice streams flowing into the Ross Sea provide evidence that the West Antarctic Ice Sheet (WAIS) is presently unstable and may experience rapid retreat some time in the next few millennia (Hughes, 1973, 1987; Alley, 1990). This instability is, in part, caused by the interaction between the ice sheet and the bed upon which it rests. The bed has been subjected to such great stress

Figure 1.4. Aerial photograph of the margin of the East Antarctic Ice Sheet off Wilkes Land. The ice cliff in this area averages 20 to 30 m in height. A large iceberg appears in the foreground.

that the basal material acts as a deforming layer along which the ice sheet slides, thereby greatly increasing its flow velocity (Alley et al., 1989). If the flow rate of the ice sheet increases without a corresponding increase in ice accumulation, the ice sheet thins. If the ice sheet thins sufficiently, it may become buoyant and decouple from the seafloor under the influence of tidal pumping. This mechanism may act independently of climate. Decoupling could lead to the collapse of the ice sheet and rapidly accelerate the rate of sea-level rise (Thomas and Bentley, 1978; Hughes, 1987; Alley, 1990; Alley and Whillans, 1991). Anderson and Thomas (1991) suggested that rapid rises in sea level during the Holocene may have been caused by marine ice sheet collapses. These eustatic rises, though small in magnitude, had a profound impact on coastal evolution (Thomas and Anderson, 1994).

Model simulations conducted by MacAyeal (1992) support the idea that the WAIS may have a history of sporadic, and possibly very rapid, collapse. Alley (1990) argues that, given the delayed response of the ice sheet to changes in the bed upon which it rests, rapid retreat of the WAIS may already have been set in motion. There is some evidence that the East Antarctic Ice Sheet (EAIS) may also be decreasing in size, at least locally (Nakawo, 1989). Jacobs (1992), however, argues that the available data on the mass balance of the EAIS are inconclusive.

The linkages between Antarctica and other global systems have changed through time, particularly as Antarctica's ice sheet has fluctuated in size. The ice sheet has deeply eroded the continent and continental shelf. Glacial erosion is most pronounced along geologic boundaries, such as fault zones, resulting in the creation of glacial troughs on the shelf. These troughs are conduits for

Figure 1.5. An ice cliff grounded at sea level. The cliff is ~40 m in height.

ice discharging from the continent (ice streams). As the bed beneath the ice sheet is lowered, the ice sheet becomes increasingly unstable, responding more radically to rising and falling sea level. In this manner, the WAIS has become increasingly sensitive to volumetric changes in the Northern Hemisphere ice sheets. Northern Hemisphere ice sheets are especially sensitive to climatic changes, which occur at high frequencies, because they extend into lower latitudes. Evidence for the linkage between Northern and Southern Hemisphere ice sheets is found in the higher frequency of ice sheet grounding events on the Ross Sea continental shelf (Alonso et al., 1992; Anderson and Bartek, 1992) and Antarctic Peninsula continental shelf (Bart and Anderson, 1995) during Plio-Pleistocene time compared to Miocene time.

Given the strong linkages between Antarctica's lithosphere, atmosphere, cryosphere, and biosphere, as well as the impact of the Antarctic realm on the rest of the globe, we should increase our knowledge of this system. This chapter provides the reader with enough information about the Antarctic cryosphere, atmosphere, and hydrosphere to understand how they function and relate to one another. Chapter 2 provides an overview of the Antarctic lithosphere and geologic history.

Figure 1.6. Aerial photograph showing the edge of an ice stream in the Ross Sea drainage basin. A crevasse pattern, transverse to the flow direction and ~100 m wide, marks the boundary between the fast-flowing ice stream and adjoining slower-moving ice (from Swithinbank, 1988; original source, U.S. Navy trimetrogen aerial photograph no. 10 [can no. 4145] in the Antarctic Map and Photograph Library, U.S. Geological Survey).

Figure 1.7. Aerial photograph of the margin of the West Antarctic Ice Sheet in Sulzberger Bay. Abrupt change in the surface profile coincides with the grounding line of the ice sheet.

Figure 1.8. Aerial photograph of the Ross Ice Shelf. The large iceberg in the foreground is about to calve from the ice shelf.

PRESENT GLACIAL SETTING

Ice Sheets

The Antarctic Ice Sheet presently contains about 30×10^6 km^3 of ice and covers an area of 13.6×10^6 km^2 (Denton et al., 1991). It accounts for 90% of the Earth's total ice volume and, if melted, would raise sea level by ~ 66 m (Denton et al., 1991). One of the most important unsolved problems in Antarctic research concerns the present mass balance and stability of the ice sheet. The mass balance is estimated by differencing the accumulation over the ice sheet and ablation by iceberg calving. Melting of the ice sheet is insignificant in the present polar setting. Jacobs (1992) recently reviewed the literature on this subject and concluded that uncertainties about the mass balance of the ice sheet stem largely from poor estimates of ice sheet ablation. The most reliable estimates for ablation are for West Antarctica, and they suggest that the ice sheet is currently losing mass (Jacobs et al., 1996).

The Antarctic Ice Sheet is divided into two portions, the EAIS and the WAIS. The TAM form the dividing line (Fig. 1.1). The ice sheets covering the East and West Antarctic subcontinents differ in several aspects.

East Antarctic Ice Sheet. The EAIS is a terrestrial ice sheet, meaning that it is grounded predominantly above sea level. It rests on several large interior basins, some over 1 km deep (Fig. 1.2). The EAIS averages just over 3 km in thickness with a steep marginal profile (Figs. 1.3 and 1.4) that corresponds to glaciologic models based on mass balance and plastic flow with no basal sliding (Hughes, 1973). Flow trajectories of the ice sheet are, for the most part, directed downward (Fig. 1.3B).

Figure 1.9. Map of mean annual atmospheric isotherms in the Antarctic Peninsula region (modified from Reynolds, 1981).

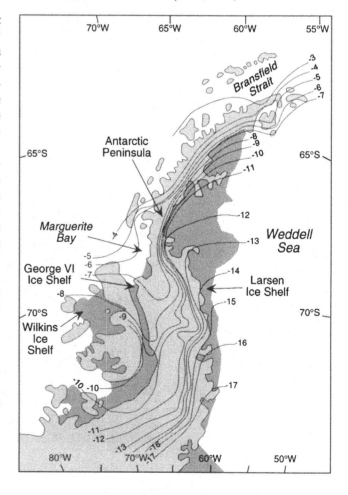

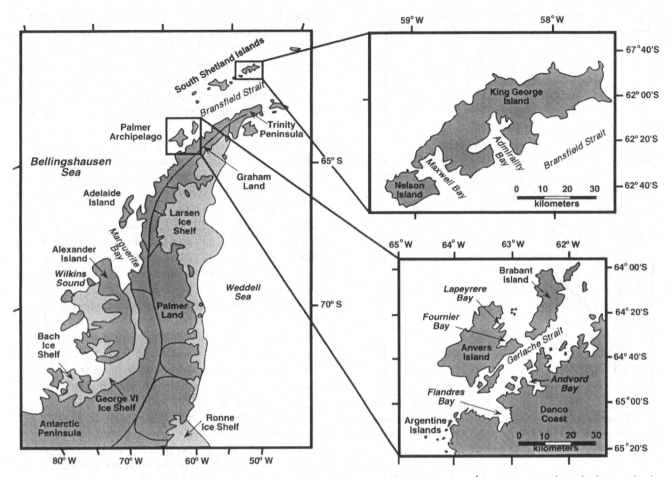

Figure 1.10. Location map of Antarctic Peninsula and adjacent islands.

Drainage from the EAIS is divergent primarily toward the coast. Important exceptions occur in the Amery Ice Shelf region and those portions of the ice sheet that drain through the TAM into the Ross and Ronne–Filchner ice shelves (Fig. 1.3A). Because of this divergent drainage pattern, flow velocities of the ice sheet decrease toward the coast and typically range from a few meters per year to a few tens of meters per year. This indicates that basal meltwater, if present, does not exist in extensive or wide enough layers to enhance flow of the ice sheet by basal sliding. The most notable exception to this is Lake Vostok, situated near the center of East Antarctica (Williams, 1996). It is a freshwater lake of ~14,000 km², or roughly the size of Lake Ontario.

Ice cliffs occur in coastal areas of East Antarctica with slowly advancing ice fronts, because waves erode the ice front as quickly as it advances (Fig. 1.5). Although these ice cliffs are typically grounded at sea level, they may be grounded several hundred meters below sea level. Ice streams occupy areas where glacial drainage converges toward the coast. These streams extend into the sea as either fringing ice shelves or smaller glacier tongues associated with large outlet glaciers. Although these glacier tongues are relatively small, they accommodate significant glacial drainage.

West Antarctic Ice Sheet. The WAIS is a marine ice sheet. Most of the WAIS is grounded at over 1 km below sea

level and at more than 2 km below sea level in the Byrd Subglacial Basin (Fig. 1.2). Drainage from the WAIS is predominantly convergent toward the coast, creating a series of rapidly flowing ice streams (Fig. 1.3A). These ice streams account for most of the ice flowing from the WAIS (Hughes, 1973; Goldstein, Engelhardt, Kamb, and Frolich, 1993; Whillans, Jackson, and Tseng, 1993). The rapid flow of ice streams implies that all three ice sheet flow processes – internal deformation, basal sliding, and deformation of the bed on which the ice sheet rests – are active.

Studies in the Ross Sea region have shown that different ice streams have different flow rates, mass balances, and physical behavior (Shabtaie and Bentley, 1987; Alley and Whillans, 1991; Jacobel and Bindschadler, 1993). Ice Stream C (Fig. 1.3A) is presently inactive (Shabtaie and Bentley, 1987) whereas the adjacent Ice Stream B is flowing several hundred kilometers per year (Bindschadler et al., 1987; Whillans et al., 1993). The present inactive state of Ice Stream C may be due to the capture of a major portion of its drainage by another ice stream (Rose, 1979), capture of basal water from Ice Stream C by Ice Stream B, the removal of Ice Stream C's deforming bed, or a combination of these factors (Alley, personal communication).

The bed on which the ice sheet rests is believed to be frozen beneath ice stream divides and thawed beneath ice

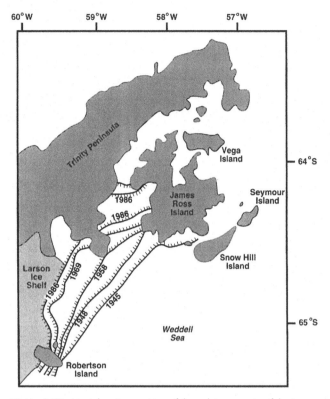

Figure 1.11. Map showing position of the calving margin of the Larsen Ice Shelf over the past several decades. The ice shelf has been retreating at an average rate of almost 1 km/yr since 1945 (modified from Reece, 1950, and Doake, 1982).

Figure 1.12. Aerial photograph of valley glaciers coalescing into a piedmont glacier in the Antarctic Peninsula region. The pristine nature of the ice visible above sea level indicates that terrigenous debris is entering the bays from below sea level.

streams, because frictional heat is higher near the center of ice streams and lowest at ice stream divides (Hughes, 1981). Distinct shear zones, averaging 5 km in width, separate fast-flowing ice streams from adjacent slow-moving ice. These zones are evident at the surface of the ice sheet as bands of crevasses oriented transverse to flow (Shabtaie and Bentley, 1982; Bindschadler et al., 1987) (Fig. 1.6).

The WAIS has a gentle surface profile compared to the EAIS (Figs. 1.3B and 1.7), which indicates that the WAIS is flowing and ablating faster than its East Antarctic counterpart. In turn, this supports glacial models based on basal sliding and "wet-based" conditions for the WAIS (Hughes, 1973, 1977, 1981). Glaciologists remain uncertain as to the exact nature of meltwater at the base of WAIS. Does it occur as a thin (centimeters thick) film of water that spreads under the immense hydrostatic pressure at the base of the ice sheet (Weertman, 1972) or do channelized water systems exist beneath the ice sheet (Hughes, 1981)? Alternatively, meltwater may be incorporated into the sedimentary bed upon which the ice sheet rests (Boulton and Jones, 1979; Alley et al., 1989; Engelhardt et al., 1990). These are important issues, because they bear on the mechanism of ice sheet sliding and ultimately on the stability of the ice sheet (Hughes, 1981).

The relationship between the ice sheet, ice streams, sea level, climate, and the bed of the ice sheet may be determined by examining geomorphic features and subglacial sediments on the continental shelf that are products of the most recent glacial maximum.

Ice Shelves

Nearly 90% of the ice flowing from West Antarctica converges into ice streams (Fig. 1.3A) which are the most

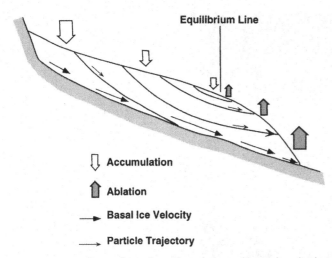

Equilibrium Line

⇩ Accumulation

⇧ Ablation

→ Basal Ice Velocity

→ Particle Trajectory

Figure 1.13. Ice flow within valley glaciers. An equilibrium line divides the zone of net accumulation from the zone of net ablation. Particle trajectories are observed above and below the equilibrium line (from Sugden and John, 1976; reprinted by permission of Barnes and Noble Books, © 1976).

dynamic components of the ice sheet (Hughes, 1973, 1977, 1981; Bindschadler et al., 1987; Denton et al., 1991; Goldstein et al., 1993; Whillans et al., 1993). These ice streams ultimately lose contact with the bed at the grounding line and float as ice shelves and glacier tongues, slowly deforming under their own weight. Modern Antarctic ice shelves are grounded at water depths of up to 1300 m (Keys, 1990).

The modern ice shelves of Antarctica, for the most part, fall into two general categories of glacial drainage: (1) those that are extensions of marine ice sheets (e.g., Ross and Ronne–Filchner ice shelves) and (2) those

nourished almost entirely by direct accretion from precipitation, basal freezing, and glacial ice flowing from predominantly terrestrial sources (e.g., George VI and Larsen ice shelves; Fig. 1.1). The latter group is typically confined to coastal embayments (Fig. 1.1). Larger ice shelves, such as the Ross and Ronne–Filchner ice shelves, are stabilized by basal pinning points (i.e., islands and other topographic highs) that buttress the adjacent marine ice sheet (Hughes, 1973, 1977, 1981).

The large ice streams of West Antarctica converge as they flow into the Weddell and Ross seas, forming vast floating ice shelves (Ronne–Filchner and Ross ice shelves, respectively; Fig. 1.3A). Only one large ice shelf, the Amery Ice Shelf, exists in East Antarctica (Fig. 1.1). Hughes (1977) argued that ice shelves play a key role in stabilizing marine ice sheets by restraining ice stream velocities. This buttressing effect of ice shelves on ice streams is a source of debate among glaciologists.

The Ross Ice Shelf covers 536×10^3 km^2, approximately the size of France, and is slightly larger than the Ronne–Filchner Ice Shelf (532×10^3 km^2; Fig. 1.1). The combined drainage basin of these two ice shelves accounts for roughly 62% of Antarctica's surface area. These ice shelves receive ~53% of the total ice drained from the continent but occupy only 10% of the coastline.

Calving of icebergs, the dominant process of ice shelf ablation, may account for several hundreds of meters of ice lost annually from the ice front (Fig. 1.8). Bottom melting under the Ross and Ronne–Filchner ice shelves is significant only near the ice shelf front, where melt rates reach 10 m/yr (Robin, 1979; Doake, 1985). Conversely, the base of the Amery Ice Shelf is frozen to within 70 km of the ice shelf front (Budd, Corry, and Jacka, 1982). The continual forward advancement of the shelf and the

Figure 1.14. A meltwater stream in the South Shetland Islands. Meltwater streams are rare in Antarctica.

Figure 1.15. Aerial photographs of the (A) Drygalski Ice Tongue and (B) Erebus Glacier Tongue.

removal of ice by iceberg calving keep the margins of the ice shelves in a constant state of flux (Keys, 1990).

Fringing ice shelves, such as the Larsen, Shackleton, West, Fimbul, Riiser-Larsen, and Lazarev ice shelves, are also large features along the Antarctic coast (Fig. 1.1). These fringing ice shelves occupy a significant portion of the coasts of both East and West Antarctica. Because of the lack of confining topography, ice shelf buoyancy and protection offered by sea ice play key roles in maintaining fringing ice shelves.

Confined ice shelves are smaller than fringing ice shelves and are maintained by physiographic constraint within large valleys, bays, and archipelagoes. The coasts of Marie Byrd Land and Palmer Land contain several confined ice shelves, including the George VI, Abbot, and Getz ice shelves (Fig. 1.1).

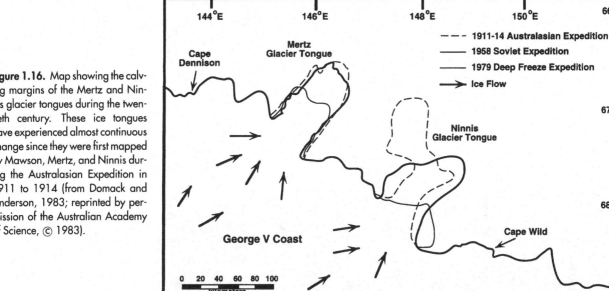

Figure 1.16. Map showing the calving margins of the Mertz and Ninnis glacier tongues during the twentieth century. These ice tongues have experienced almost continuous change since they were first mapped by Mawson, Mertz, and Ninnis during the Australasian Expedition in 1911 to 1914 (from Domack and Anderson, 1983; reprinted by permission of the Australian Academy of Science, © 1983).

Robin and Adie (1964) argued that ice shelves can exist only in subpolar and polar settings, as is the case today. Another criterion for the existence of ice shelves is that their equilibrium line altitude (the level at which accumulation and ablation are in balance) occurs below sea level (Robin, 1988). The warmest (subpolar) atmospheric conditions occur in the Antarctic Peninsula region where ice shelves exist south of the −7°C mean annual isotherm (Fig. 1.9). Along the western side of the Antarctic Peninsula, ice shelves exist in protected regions where valley walls confine flow. On the eastern side of the Antarctic Peninsula, the −7°C mean annual isotherm is situated nearly 350 km farther north than on the western side of the peninsula. This is home to the vast Larsen Ice Shelf (Fig. 1.10).

Virtually all the outlet glaciers flowing eastward from the Antarctic Peninsula flow into the Larsen Ice Shelf. Thus, the ice shelf is sensitive to changes in the glacial regime of these outlet glaciers and the ice caps that nourish them. A perennial canopy of sea ice covering the western Weddell Sea protects the ice shelf from wave erosion. Despite this, the calving wall of the northern Larsen Ice Shelf is retreating at a rate of 1 km/yr; this retreat is attributed to a recent warming trend in the region (Doake,

Figure 1.17. Small "bergy bits" calving from a tidewater glacier on Anvers Island in the Antarctic Peninsula region.

Figure 1.18. A large tabular iceberg in the northwestern Ross Sea. Ice shelves typically are the source of tabular icebergs that are hundreds of square kilometers in area.

Figure 1.19. Mean concentration of icebergs in the southeastern sector of the Southern Ocean. Patterns correspond to mean number of icebergs per 1000 km² (modified from Romanov, 1984).

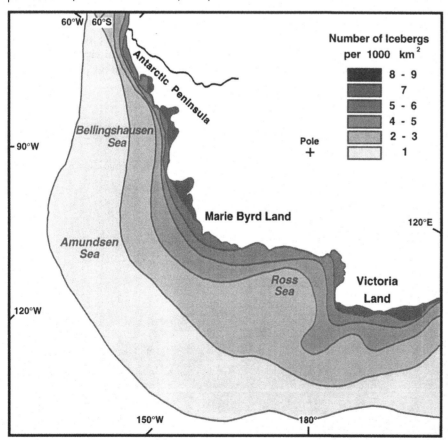

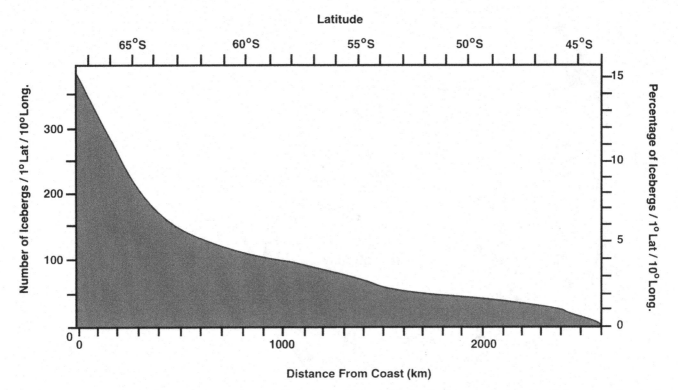

Figure 1.20. Graph of average iceberg population versus degree of latitude and distance from the coast. A large decrease in the iceberg population occurs at ~65° S, with a subsequent gradual decline to a near zero population by 45° S (from Morgan and Budd, 1978; data used to construct the graph are from Shil'nikov, 1969; reprinted by permission of Pergamon Press, Inc., © 1978).

1982) (Fig. 1.11). Whether the stability of outlet glaciers flowing into the Larsen Ice Shelf will be influenced by the retreat of the ice shelf within the next few decades remains to be seen.

The largest ice shelf in the western Antarctic Peninsula region, the George VI Ice Shelf, is an example of an ice shelf confined by valley walls (Fig. 1.9). The George VI Ice Shelf drains ice caps flowing from Palmer Land. This constitutes more than 90% of the ice flowing into Marguerite Bay, with the remainder draining from Alexander Island (Fig. 1.10). Historically, another ice shelf, Wordie Ice Shelf, was a source of ice flowing into Marguerite Bay (Fig. 1.9). However, this ice shelf has disappeared within the past two decades (Doake and Vaughan, 1991).

The base of the George VI Ice Shelf is melting at an average rate of 2 m/yr and is retreating at 1 km/yr (Potter and Paren, 1985). Circulation of relatively warm water masses beneath the ice shelf is believed to be responsible for ice shelf retreat, which indicates the importance of oceanographic circulation in ice shelf stability (Potter and Paren, 1985; Jenkins and Doake, 1991; Jacobs et al., 1992). Hughes (1981) suggested that the thinning of ice shelves could reduce the back-stress exerted by the ice shelf and ultimately lead to the retreat of the grounding line. Locally, portions of the George VI Ice Shelf are thickening, probably in response to increased precipitation rates several hundred years ago. The interplay between thickening and undermelting results in uncertainties concerning the present mass balance of this glacial system (Bishop and Walton, 1981; Potter and Paren, 1985; Jacobs et al., 1996).

Outlet, Valley, and Piedmont Glaciers

The smallest elements of the Antarctic glacial regime are small ice caps, valley glaciers, and outlet glaciers. Valley glaciers are glaciers that flow through mountain valleys, whereas outlet glaciers are constrained to channels by exposed rock as they pass through mountain valleys. At the base of the mountains, valley and outlet glaciers can coalesce to form piedmont glaciers, which are thick continuous sheets of ice. Marine settings with these types of glacial systems include the western Ross Sea along the flank of the TAM and seas adjacent to the Antarctic Peninsula. These settings have valley and outlet glaciers that flow directly into the sea (tidewater glaciers), often forming glacier tongues and piedmont glaciers that extend seaward as fringing ice shelves (Fig. 1.12).

Outlet, valley, and piedmont glaciers are of particular interest to marine geologists due to their relatively rapid response to short-term (decade to century time intervals) climatic changes. The sensitivity of these smaller glacial systems to short-term climatic changes is due to relatively small drainage basins, high accumulation rates, and small thickness compared to ice caps and ice sheets. The bays and fjords into which glaciers drain appear to contain high-resolution sedimentary records of Antarctica's recent climate (Griffith and Anderson, 1989;

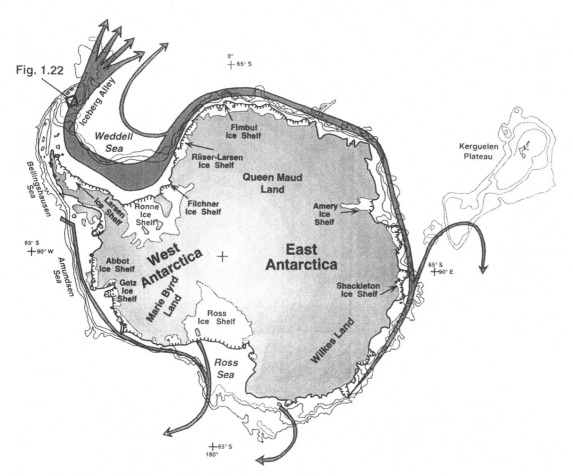

Figure 1.21. Generalized long-term iceberg drift trajectories. Icebergs originating from East Antarctica tend to be concentrated by the East Wind Drift in the Weddell Sea. Bathymetric contour interval is 500 m (modified from Reid and Anderson, 1990).

Domack, 1990; Domack and Williams, 1990; Domack and Ishman, 1993). Proper interpretation of this sedimentary record, however, requires an understanding of those factors that regulate the advance and retreat of Antarctica's outlet, valley, and piedmont glaciers.

Accumulation of valley, outlet, and piedmont glaciers is almost exclusively through direct gain from precipitation. In the TAM, variations in precipitation along the mountain front result in spectacular differences in the glacial setting. Glaciers in the northern part of the mountain belt receive relatively high amounts of precipitation and extend to the sea as tidewater glaciers. In contrast, glaciers of the McMurdo Dry Valleys region (Fig. 1.1) are poorly nourished and most terminate well inland of the coast.

The Antarctic Peninsula experiences annual accumulation rates of 500 to 1000 mm/yr, the highest accumulation rates on the continent (Drewry and Morris, 1993). For glaciers of this region, direct gain by precipitation is the most important form of accumulation (Koerner, 1964; Sadler, 1968; Rundle, 1974; Curl, 1980; Orheim and Govorukha, 1982; Wager and Jamieson, 1983). Other forms of accumulation include rime formation (Koerner,

1964; Mercer, 1967), accumulation of blowing and drifting snow (Sadler, 1968), and direct advection from ice fields and ice caps (Griffith, 1988). Given the strong influence of precipitation patterns on glacier mass balance, it follows that recent observations of retreating ice shelves that valley and outlet glaciers nourish in the Antarctic Peninsula region may reflect changes in precipitation patterns as well as an observed warming trend.

Ablation of Antarctica's valley, outlet, and piedmont glaciers is almost exclusively by calving (Koerner, 1964; Rundle, 1974; Wager and Jamieson, 1983). Minor forms of ablation include direct sublimation, wind advection, and melting. As a result, the equilibrium line of most tidewater glaciers is at the calving line (Fig. 1.13). The only exceptions exist in the South Shetland Islands where relatively warm temperatures promote melting (Fig. 1.14) and in the McMurdo Dry Valleys region where wind ablation and evaporation are important.

Calving is not climate-dependent and, therefore, does not vary systematically along climatic trends. Factors important in the regulation of calving are: (1) the rate of glacial advance; (2) the exposure of glaciers to waves and tidal influence; (3) the relative amounts of protection offered by sea ice to dampen waves; and (4) the geometry of the valleys that confine glaciers. For the most part, sea level and the bathymetry of the bays and fjords control the position of the calving line.

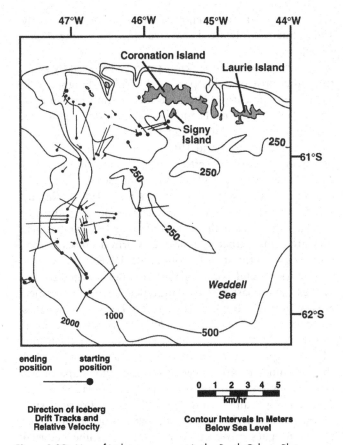

Figure 1.22. Map of iceberg movement in the South Orkney Plateau region recorded using ship's radar over a ten-day period during Operation Deep Freeze 86 (modified from Reid and Anderson, 1990).

Sediment cover is virtually absent in Antarctic fjords; hence, the role of bed deformation in valley glacier movement is considered minimal. The only exceptions may be the valley glaciers of the South Shetland Islands where fjords contain relatively thick sediment accumulations.

Glacier Tongues

Individual glacial outlet systems at the coast may extend seaward as glacier tongues. Though relatively small in areal extent, these systems play a significant role in ice drainage and debris delivery to the ocean (Drewry and Cooper, 1981). All glacier tongues have confined portions in coastal valleys and buoyant regions in open-marine shelf settings. Glacier tongues are among the most dynamic glacial features in Antarctica, undergoing radical changes in size and shape within a few years. Average ice drainage velocities typically exceed 0.5 km/yr and can reach 3.7 km/yr (Lindstrom and Tyler, 1984; MacDonald et al., 1989).

The Drygalski Ice Tongue is 70 km long and 20 km wide, making it the largest glacier tongue in the Ross Sea (Fig. 1.15A). It is the seaward extension of David Glacier that has a drainage basin area of 2.4×10^5 km^2 (Swithinbank, 1988). Satellite imagery of the Drygalski Ice Tongue shows that it has advanced seaward at a velocity ranging from 136 to 912 ± 30 m/yr for the past few decades (Frezzotti, 1992). Another glacier tongue located in the Ross Sea, the Erebus Glacier Tongue (Fig.

Figure 1.23. Side-scan sonograph showing iceberg furrow and gouge marks on the seafloor of the Wilkes Land continental shelf (from Barnes, 1987; reprinted by permission of the Circum-Pacific Council for Energy and Mineral Resources, © 1987).

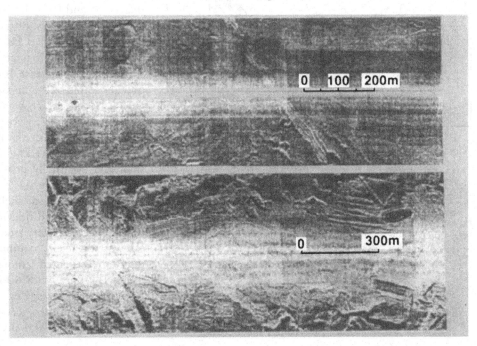

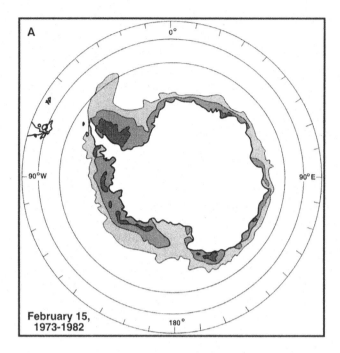

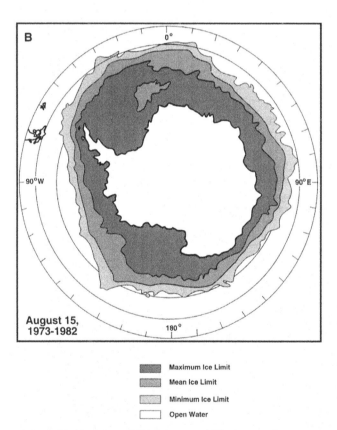

Figure 1.24. Mean austral (A) summer and (B) winter sea-ice concentration around Antarctica. A large seasonal change occurs in sea-ice concentration (from Department of the Navy, 1985).

1.15B), experienced a major calving episode in the early 1940s but has advanced seaward since that time (Keys, 1990).

The Pine Island Bay glacial system is believed by some to be the most unstable part of the WAIS (Hughes, personal communication), because the floating terminus of the ice sheet in Pine Island Bay is unstable. Thwaites Glacier Tongue in Pine Island Bay (Fig. 1.1) doubled in length between 1961 and 1969, then broke off in a series of calving events that left the glacier snout close to the coast. Since 1972, it has been advancing seaward at a rate of up to 3.7 km/yr (Lindstrom and Tyler, 1984). Nearby Pine Island Glacier and its glacier tongue are currently experiencing basal melting rates of 10 to 12 m/yr, because they are being bathed in relatively warm deep water that impinges onto the shelf (Jacobs et al., 1996).

The Mertz and Ninnis glacier tongues, located on the Wilkes Land coast of East Antarctica, drain a combined area of over 4×10^5 km² (Fig. 1.16). The glacier tongues are named after members of Douglas Mawson's coastal research team who lost their lives during the expedition. Maps compiled by Mawson, Mertz, and Ninnis in 1914 show significant differences in the size and shape of the glacier tongues compared to maps from the 1958 Soviet Antarctic Expedition and the 1979 Deep Freeze Expedition (Domack and Anderson, 1983) (Fig. 1.16). Both glacier tongues retreated to near the coastline between 1914 and 1958 and later advanced several tens of kilometers seaward between 1958 and 1979 (Domack and Anderson, 1983).

Icebergs

The dominant form of glacial ablation in Antarctica is calving, which produces icebergs and smaller "bergy bits" (Figs. 1.17 and 1.18). Icebergs calved from ice shelves form large tabular bodies, often hundreds of square kilometers in area (Fig. 1.18). Classification as an iceberg implies that a mass of ice has at least 5 m of ice above the sea surface and is at least 100 m² in area (Keys, 1990). A good example of the massive size that icebergs can attain is Iceberg B-9, which was calved from the Ross Ice Shelf in November 1987. Iceberg B-9 was ~40 by 157 km, or roughly twice the size of the state of Rhode Island. This iceberg traveled ~1 km/day until it collided with the ice shelf and eventually broke into several smaller icebergs (Keys, 1990).

In addition to ice shelf sources, icebergs calve from glacier tongues and tidewater glaciers; the form and size of these icebergs vary tremendously. Glacier tongues and tidewater glaciers, combined with the ice shelves, produce an estimated annual iceberg volume of 0.75 to 3.0×10^3 km³, with 60 to 80% of this ice originating from the ice shelves (Keys, 1990).

The distribution of icebergs around Antarctica has been the subject of several investigations (e.g., Swithinbank,

Figure 1.25. The ice pack surrounding Antarctica is able to move quickly, and often unpredictably, presenting a hazard to research on the continental shelf. Here a ship follows an area in the ice pack along the coast of the Antarctic Peninsula.

McClain, and Little,1977; Morgan and Budd, 1978; Tchernia and Jeannin, 1984; Wadhams,1988;Keys,1990). These studies reveal high concentrations of icebergs in a narrow band surrounding the coast (Fig. 1.19). In general, icebergs drift under the influence of the East Wind Drift, the prevailing surface current around Antarctica (Keys, 1990). Movement of icebergs along the coast is slow and erratic, with the ice masses intermittently grounding. Movement may reach 2 m/sec, but average speeds are usually less (Keys, 1990).

Moving northward away from the coast, there is a rapid decrease in the iceberg population density near 65° S. This is succeeded by a gradual decrease to almost zero near 45° S (Morgan and Budd, 1978) (Fig. 1.20). Superimposed on this latitudinal zonation in iceberg concentration are longitudinal variations caused primarily by irregularities in the shape and bathymetry of the Antarctic coast, especially along the coasts of the Antarctic Peninsula and Victoria Land (Fig. 1.19).

Concurrent with the decrease in iceberg population north of 65° S is a dramatic decrease in average iceberg length by a factor of 2.5 and only a minor decrease in the average height (Morgan and Budd, 1978). This indicates that the constant attack of waves and storms, in addition to warmer surface water, succeeds in breaking most icebergs into smaller pieces. The increase in surface area accelerates attrition and enhances iceberg decay, thereby leading to the rapid decrease in the iceberg population with increasing distance from the coast.

The life span of icebergs is estimated at two to five years; however, an iceberg's estimated life span is only one year if it quickly moves away from the coastal zone. Iceberg longevity may be greater if the iceberg stays near the coast, protected from wave action by pack ice (Keys, 1990).

Studies on the distribution and long-term motion of icebergs in the Southern Ocean were conducted by Swithinbank, McClain, and Little (1977), Morgan and Budd (1978), Tchernia and Jeannin (1984), and Wadhams (1988). Figure 1.21 illustrates the general long-term iceberg drift pattern based on the visual tracking of large icebergs by satellite (1973–1989) and 21 icebergs equipped with radio transponders (Tchernia and Jeannin, 1984). Analysis of the data shows that the major ocean currents appear to control long-term iceberg drift patterns (Tchernia and Jeannin, 1984).

In general, icebergs move counterclockwise, hugging the coast around the continent under the influence of the East Wind Drift (Fig. 1.21). Many larger icebergs eventually move into a large, well-developed clockwise gyre located in the Weddell Sea (Fig. 1.21). Once there, icebergs calved from the large ice shelves that ring the landward margin of the Weddell Sea join older, well-traveled icebergs. Often the icebergs linger in the vicinity of Crary Trough, near the confluence of the Ronne–Filchner Ice Shelf. Icebergs probably become trapped in this area by the eastern edge of the dense ice pack that tends to persist in the western Weddell Sea. Eventually, most of the icebergs penetrate the ice pack and make their way in a narrow zone along the coast of the Antarctic Peninsula. The concentration of icebergs in the Weddell Sea forms an "iceberg alley" along the east coast of the Antarctic Peninsula and into the Scotia Sea (Fig. 1.21).

Icebergs calved in the Bellingshausen and Amundsen seas appear to move predominantly to the southwest under the influence of the East Wind Drift (Fig. 1.21). A

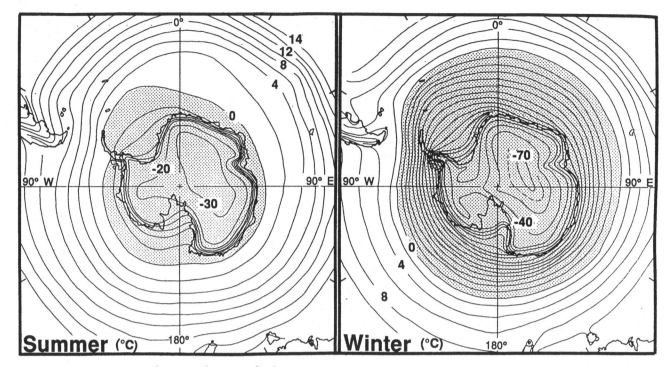

Figure 1.26. Mean summer and winter isotherms (°C) for the Antarctic region. A climatic amelioration of the coastal zone occurs during the summer (modified from Dudeney, 1987).

clockwise gyre located in the Ross Sea and a bathymetric high near 180° (Iselin Bank) deflect icebergs moving along the coast of Marie Byrd Land toward the north (Fig. 1.21). Another gyre located near 150° E along the eastern Wilkes Land margin also deflects icebergs northward.

On a local scale, the movement of icebergs remains unpredictable, especially on a short-term basis. Icebergs of varying size are affected differently by winds and surface currents. Icebergs with different drafts are pushed along by currents at different levels within the surface water column. Keys (1990) notes that many icebergs in the Pacific sector exhibit apparently random behavior in response to eddies, tides, winds, waves, and the Coriolis Force. Low- to medium-velocity winds have a limited effect on the motion of deep-draft icebergs with keels below the thermocline (Crépon, Houssais, and Saint Guily, 1988). Most Antarctic icebergs have drafts deeper than the mixed water layer (~ 40 m).

Observations acquired during Deep Freeze 86 in the South Orkney Plateau area (within iceberg alley) serve to illustrate the unpredictable nature of iceberg motion. During the course of several days, the ship's radar tracked icebergs passing within 30 km of the ship; the result is shown in Figure 1.22 (Reid and Anderson, 1990). These data depict the apparently random drift of icebergs in this area. The drift direction and speed of icebergs did not correlate well with wind direction and speed. The measured speeds of several icebergs exceeded 2 km/hr.

Iceberg drafts can be greater than the thickness of large ice shelves, as determined by studies of modern icebergs and by observing gouges and furrows left by icebergs on the seafloor. Large ice shelves have thickness of ~ 200 to 250 m near their calving margin, and tabular icebergs with drafts up to 330 m have been reported (Orheim, 1980). Studies reveal that icebergs can increase their draft by as much as 50% by tilting or overturning (Keys, 1990). The majority of Antarctic icebergs are not tabular and have been tilted (Keys, 1990).

Side-scan sonar records show abundant iceberg furrows and gouge marks (Fig. 1.23) to depths of about 500 m (Lien, 1981; Barnes, 1987; Barnes and Lien, 1988), but it remains unclear whether these represent modern or relict features. Typically, iceberg gouge features are most common and best developed on bank tops and on the sides of troughs (chapter 3). These gouge features are generally absent in deep troughs. The bathymetry of areas where icebergs frequently ground also provides information on their draft. Inner shelf areas and banks shallower than 300 m tend to constitute the resting grounds for large icebergs.

Sea Ice

The areal extent of sea ice around Antarctica is roughly twice the size of the Antarctic Ice Sheet and experiences a five-fold increase annually (Mullan and Hickman, 1990) (Fig. 1.24). This seasonal variability in the sea-ice coverage is one of the most significant factors regulating the energy balance of the Southern Hemisphere atmosphere and ocean (Mullan and Hickman, 1990; Martinson and Iannuzzi, 1998). This influence is due to the large albedo difference between sea ice and the sea surface, which allows the sea ice to serve as a barrier to energy

Figure 1.27. Rough seas present one of the challenges to working in the Antarctic region.

exchange between the atmosphere and ocean. The position of the sea-ice edge during any given year affects weather patterns and cyclogenesis around Antarctica. The relationship between sea ice and weather patterns is poorly understood (Schwerdtfeger and Kachelhoffer, 1973; Streten and Pike, 1980; Mullan and Hickman, 1990). Finally, the extent of sea ice around Antarctica regulates the penetration of light in surface waters (radiance flux) and therefore primary productivity.

The Navy-NOAA Joint Ice Center (JIC) has prepared weekly sea-ice maps from satellite imagery since 1973 (Fleet Weather Facility Suitland Antarctic Ice Charts). Both spatial and temporal analyses of interannual and seasonal variation of the sea-ice coverage were performed using these data and using the Special Sensor Microwave Imager (SSM/I) data set (Lemke, Trinkl, and Hasselman, 1981; Chiu, 1983; Ropelewski, 1983; Lemke, 1986; Mullan and Hickman, 1990; Reid and Anderson, 1990; Parkinson, 1998). These results are the basis for the following discussion.

The Antarctic sea ice differs in physical and behavioral characteristics from its Arctic counterpart. Arctic sea ice is generally confined to the landlocked Arctic basin. This favors the formation of pressure ridges and multiyear sea ice (Wadhams, 1986). In contrast, the sea-ice cover surrounding Antarctica is everywhere divergent in its general behavior (Wadhams, 1986). Sea-ice development also is affected by winds, which tend to impede its development.

Antarctic sea ice undergoes a vast change in area from roughly 3×10^6 km^2 in the austral spring to 20×10^6 km^2 in the austral autumn (Mullan and Hickman, 1990) (Fig. 1.24). Approximately 85% of the sea ice melts each year; therefore, most of the winter ice is first-year ice. Antarctic sea ice is relatively uniform in thickness, ranging from 1 to 3 m and averaging 1.5 m, because most of the sea-ice cover is first-year ice. Fast ice (ice in a sheet permanently attached to the coast), however, may reach several meters in thickness.

Winds, currents, and storms tend to disperse sea ice and open up new leads (cracks in the sea-ice cover; Fig. 1.25) and polynyas (areas of open water or reduced pack ice located in the midst of concentrated pack ice, possibly maintained by katabatic winds or upwelling currents inhibiting ice formation). This makes ice pack more navigable for ice-strengthened vessels, but these factors also present the greatest hazard to shipping. Sea ice is able to move quickly, often in an unpredictable fashion, under the stress of winds and surface currents. Several ships (e.g., *Deutschland, Endurance, Aurora, Antarctic,* and *Southern Quest*) venturing through leads within the sea ice have become entrapped and crushed when the ice shifted.

Another important characteristic of Antarctic sea-ice cover is its large degree of variability, on both a seasonal and yearly basis (Fig. 1.24). For any particular location, the amount of sea ice present may vary tremendously. Surprisingly, a decrease at one location is typically balanced by an increase at another; therefore, the overall maximum areal ice extent may not change greatly from year to year (Mullan and Hickman, 1990; Parkinson, 1998).

The areal extent of sea-ice cover from austral summer to austral winter increases by a factor of at least 3 times to as much as 30 times. The largest annual fluctuations occur in the Ross and Weddell seas (Fig. 1.24), the sea-ice factories of Antarctica (Lemke et al., 1981; Ropelewski, 1983). The smallest seasonal changes in

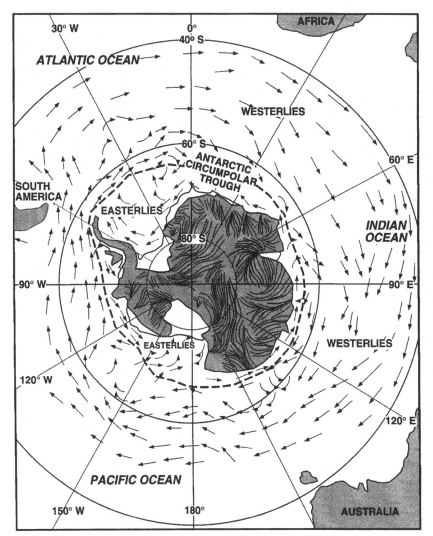

Figure 1.28. Time-averaged surface wind streamlines (lines) of winter flow over Antarctica (modified from Parish and Bromwich, 1987) and surface wind directions (arrows) in the Southern Ocean (modified from Bigg, 1996).

WEATHER AND CLIMATE

Antarctica is the coldest, driest, and windiest continent on Earth (Schwerdtfeger, 1984). The climate is extreme, especially in the interior of the continent, whereas the coastal regions experience more moderate conditions. Temperatures along the coast range from an average austral summer temperature of 0°C to ∼ −20° to −30°C during the austral winter (Fig. 1.26), with a mean annual temperature of ∼ −15°C. Strong winds create rough seas for which the Southern Ocean is notorious (Fig. 1.27). Sea states change suddenly and unpredictably. Low temperatures, violent storms, rough seas, and sea ice combine to form, directly or indirectly, a challenge to the acquisition of marine geologic and geophysical data in the Antarctic region.

As recently as the 1980s, marine investigators in the Antarctic region were as vulnerable to hurricane-strength storms as were Caribbean sailors in the nineteenth century. The German Antarctic North Victoria Land Expedition in the Pennell Coast region was surprised by hurricane "Schepel-Sturm," which reached hurricane strength within an hour and lasted four days, with measured wind speeds exceeding 180 km/hr. The suddenness and severity of the storm caused heavy damage to ship-board and land-based equipment (Kothe et al., 1981). I had a similar experience in 1991 when our ship, R/V *Polar Duke*, was struck by a sudden storm with sustained winds in excess of 100 km/hr.

Modern research vessels are now equipped with weather faxes, which, combined with atmospheric pressure and temperature data, make it possible to predict the onset of cyclonic storms. Successful prediction of storms helps prevent unpleasant days at sea. Because these storms tend to move sea ice, it also is possible to predict when sea-ice conditions may improve in an area to allow collection of marine geophysical data and avoid being trapped in the sea ice.

Knowledge of Antarctica's climate has significantly increased in recent decades with three major stages of data collection. The first stage was the International Geophysical Year (IGY) (July 1957 to December 1958), when an effort was made to increase meteorologic observations from land stations and ships. A total of 39 weather stations operated in Antarctica during the IGY (Taljaard, 1972). The second stage of research began with the first

sea-ice cover occur in the Bellingshausen Sea, in the South Indian Ocean, and in the seas around the Antarctic Peninsula. In general, the movement of sea ice appears to correspond with the prevailing surface currents and winds, indicating that advection of sea ice plays a major role in large-scale sea-ice dynamics around Antarctica (Lemke et al., 1981).

Interannual fluctuations of the maximum Antarctic sea-ice coverage are small (∼ 20%); however, the minimum Antarctic sea-ice cover can vary by more than 100% (Ropelewski, 1983). The minimum sea-ice cover in the Ross Sea is particularly notorious for extreme variability from one austral summer season to the next (Keys, 1990). Seasonal extremes tend to occur at the same time each year, with the annual sea-ice minimum occurring in February and the sea-ice maximum occurring in either August or September (Ropelewski, 1983).

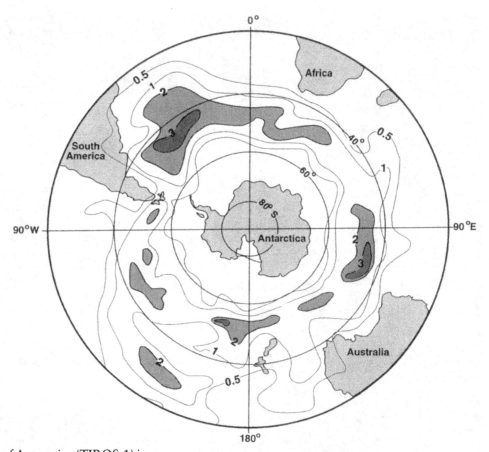

Figure 1.29. Schematic of the Southern Hemisphere depicting the areal frequency of developing cyclones (shaded). The zone of highest storm frequency falls around 50° S (modified from Streten and Troup, 1973).

meteorologic satellite coverage of Antarctica (TIROS-1) in 1960 and was followed by the launching of the NIMBUS satellite in 1968. The third stage began in 1990 and will establish a network of automatic weather stations around Antarctica (Stearns et al., 1993).

Atmospheric circulation around Antarctica follows reasonably consistent patterns. A zone of high pressure exists over the pole throughout the year, forcing air flow from the continental interior. Winds flowing off the continent are strongly influenced by the topography of the ice sheet (Bromwich et al., 1993) (Fig. 1.28). A narrow zone of easterlies exists offshore, driven by continuous cold winds that flow from the continent, and are diverted westward by the Coriolis Force. Between 65° S and 70° S lies the Antarctic Circumpolar Trough, a zone of low pressure. Near the trough, winds are highly variable and cyclonic. From ∼ 30° S to 65° S, the dominant wind pattern consists of a broad band of westerlies.

Many of the climatic variables fluctuate on a semi-annual basis. The edge of the Antarctic Circumpolar Trough moves throughout the year, advancing farther northward in the austral summer (Mullan and Hickman, 1990). The westerly winds of the trough also display a bimodal aspect, being strongest in the austral fall and spring. The immense thermal inertia of the ocean causes air temperature to lag behind the annual solar insolation cycle and causes the austral autumn temperatures over the

ocean to stay warmer longer than those over the continent (Mullan and Hickman, 1990). Sea-ice formation also contributes to the greater warmth of the coastal regions by releasing latent heat and causing a rise in temperature. An increase in storm frequency in the austral autumn tends to move warmer, lower-latitude air into the higher latitudes, thereby contributing to the warmer temperatures experienced along the edge of the continent.

Cyclonic storm systems in the Antarctic usually originate in the middle latitudes, move gradually southward under the influence of the westerlies, and then degrade in the high latitudes. Storm formation and intensification is concentrated at ∼ 50° S (Figs. 1.29 and 1.30). This band of maximum cyclone frequency occurs north of the Polar Front, which marks the center of the Antarctic Circumpolar Current. The position of the pack-ice margin may cause latitudinal shifts of the region with maximum storm frequency (Schwerdtfeger and Kachelhoffer, 1973). The greatest number of storms occurs in the austral winter (Carleton, 1979).

The Antarctic Ice Sheet exerts a strong influence on regional climate. Cooling near the ice sheet's surface creates a temperature inversion on the high interior polar plateau.

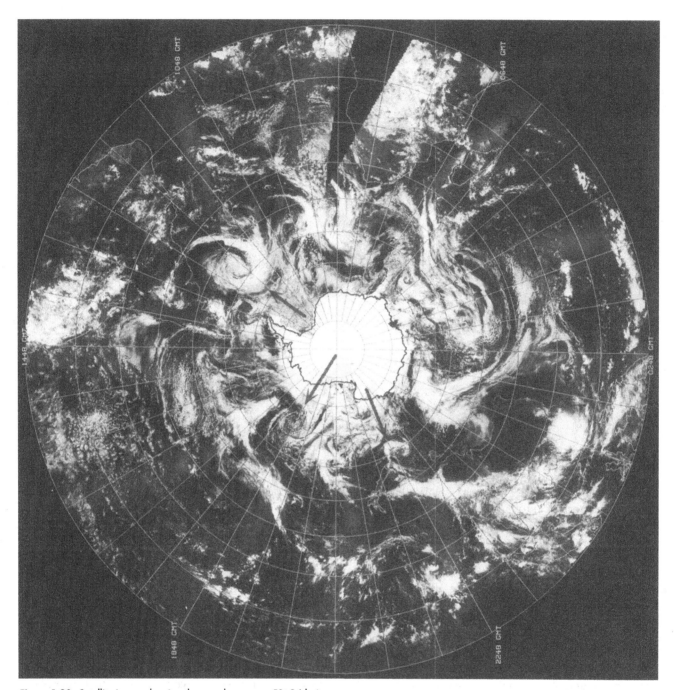

Figure 1.30. Satellite image showing three cyclones near 50° S (designated by arrows) with one located north of the Weddell Sea, another situated north of the Ross Sea, and the third situated south of New Zealand (photograph provided courtesy NOAA; NOAA-2 S.H. SR VIS; January 7, 1974).

This cold air flows in a gravity-driven current, known as an inversion wind, from the continental interior toward the coast (Fig. 1.28). The Coriolis Force gives the winds a net westward flow, resulting in the band of coastal easterlies.

Topographic channeling of the air flow further intensifies strong Antarctic winds, resulting in katabatic winds. Sudden and intense katabatic winds with velocities of 50 to 80 km/hr are often experienced in glacial valleys

(Tauber, 1960). Katabatic winds of the glacial valleys are typically short-lived due to the depletion of the cold air source, whereas the interior of the continent continuously offers ideal conditions for katabatic wind formation (Bromwich et al., 1993). The Cape Denison-Commonwealth Bay region of the Adélie Coast provides an example of the potential power of katabatic flow. Convergent katabatic flow from the EAIS makes this area the windiest spot on Earth. The mean annual wind speed is 80 km/hr, with maximum measured wind velocities exceeding 320 km/hr (Parish, 1981).

Katabatic winds play a key role in sustaining coastal polynyas. One of the best examples of this association is

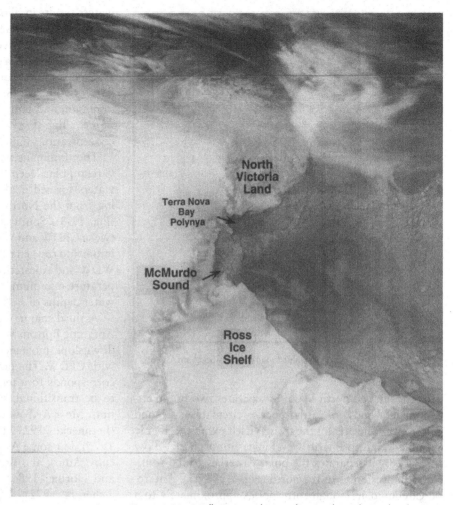

Figure 1.31. Satellite image showing large polynya located in the Terra Nova Bay region (photograph provided courtesy of Dave Bromwich).

the coastal polynya in Terra Nova Bay, western Ross Sea (Bromwich and Kurtz, 1984) (Fig. 1.31). Cold air that converges near the head of Reeves Glacier sustains the katabatic winds of this area (Parish and Bromwich, 1989; Bromwich, Parish, and Zorman, 1990).

Cyclonic storms and katabatic winds can act in concert. As a storm moves inland, the relatively warm, moist air mass associated with the storm acts as a barrier to normal downslope katabatic flow. The passing of the storm removes the barrier, resulting in a violent avalanche of winds down the ice surface.

PHYSICAL OCEANOGRAPHIC SETTING

Oceanography of the Southern Ocean

Oceanographers recognize several circumpolar zones within the Southern Ocean, based mainly on their distinctive vertical stratification of temperature and salinity within the surface and intermediate levels of the water column. Narrow fronts with sharp gradients in water properties separate these zones (Gordon and Goldberg, 1970; Gordon and Molinelli, 1982; Patterson and Whitworth, 1990). These fronts and zones include (from north to south): the Subtropical Front, Subantarctic Zone, Subantarctic Front, Polar Frontal Zone, and Polar Front (Fig. 1.32). Patterson and Whitworth (1990) recognize another zone, the Continental Zone, that includes water masses on the continental shelf. Four oceanic water masses are recognized south of the Polar Front: Antarctic Surface Water, Warm Deep Water, Circumpolar Deep Water, and Antarctic Bottom Water.

Antarctic Surface Water originates near the continent and flows northward until it encounters Subantarctic Surface Water where it begins to sink and mix with intermediate water masses (Fig. 1.33). The region of convergence between these surface water masses is the Antarctic Convergence or Polar Front (Gordon, 1967). North of the Polar Front, surface waters have winter temperatures warmer than 2.0°C. Whereas south of this front, surface waters have winter temperatures below 1.0°C and have lower salinities than surface waters north of the front.

There is a close parallelism between ocean temperatures and surface atmospheric pressure, largely due to the absence of meridional barriers (continents) to atmospheric and oceanic circulation. The Antarctic Circumpolar

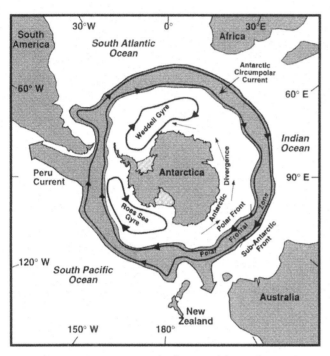

Figure 1.32. Major oceanographic features of the Southern Ocean.

Trough, situated north of 65° S, separates two broad circumpolar bands of atmospheric circulation around Antarctica. The area of the trough itself exhibits a highly variable and primarily cyclonic atmospheric flow (Fig. 1.28). It separates the polar easterlies to the south from the westerlies to the north (Fig. 1.28). The Antarctic Circumpolar Trough corresponds approximately to a region of divergent flow of surface waters, the Antarctic Divergence (Fig. 1.33). North of the Antarctic Divergence, prevailing westerlies drive surface waters toward the east with a northward Ekman component of flow (Fig. 1.32); this is the Antarctic Circumpolar Current (ACC) (Gordon, 1967).

The east-flowing West Wind Drift reaches maximum velocities in the region of the Polar Front and probably extends to the ocean floor with little or no attenuation (Gordon, 1967), as indicated by a nearly circumpolar zone of current ripples, manganese nodules, and scour zones corresponding to the Polar Front (Goodell, 1973; Chapter 6). There is also a strong physiographic influence on deepwater circulation in the Southern Ocean (Sverdrup, Johnson, and Fleming, 1942). For example, the Drake Passage, located between the Antarctic Peninsula and South America, is the site of strong bottom current activity. On a more localized scale, scour zones and ferromanganese nodules occur in gaps between ocean fracture zones (Chapter 6).

South of the Antarctic Divergence, a number of atmospheric low-pressure cells sustain several ocean circulation gyres, except near the continent where polar easterlies drive a westward-flowing current called the East Wind Drift (Deacon, 1937). The westward flow of surface

waters results in two large circulation gyres, the Ross Sea Gyre and Weddell Sea Gyre (Fig. 1.32).

The Antarctic Divergence is an area of upwelling caused by the southward and upward flow of Warm Deep Water (WDW) in response to the northward and downward flow of Antarctic Surface Water (Fig. 1.33). An oxygen minimum and temperature maximum occur at this zone of upwelling (Fig. 1.33). It is also a zone of high nutrient concentrations (Gordon and Molinelli, 1982).

The dominant water mass in the Southern Ocean is Circumpolar Deep Water (CPDW) which is a mixture of waters formed in the Antarctic region and WDW flowing from the North Atlantic, Pacific, and Indian oceans (Fig. 1.33). South of the Polar Front, the boundary between CPDW and WDW is marked by a temperature maximum and oxygen minimum layer. The boundary between WDW and Antarctic Surface Water coincides with a temperature minimum layer (Fig. 1.33). CPDW extends from water depths of ~ 1000 m to between 3200 and 4000 m.

A third major water mass in the Southern Ocean is Antarctic Bottom Water (AABW), which is formed by the downslope movement and mixing of saline shelf water with CPDW. The boundary between CPDW and AABW corresponds to a temperature minimum layer and tends to be transitional, except near areas of AABW formation. Most AABW production occurs in the Weddell Sea (Brennecke, 1921; Deacon, 1937; Carmack and Foster, 1975), but some AABW is produced in the Ross Sea (Jacobs, Amos, and Bruchhausen, 1970; Jacobs, Fairbanks, and Horibe, 1985) and off Wilkes Land (Gordon and Tchernia, 1972).

Oceanography of the Continental Margin

Circulation along the Antarctic continental margin can generally be divided into two types. The first type of circulation occurs along those portions of the continental margin where mixing of dense shelf water with WDW and CPDW results in the formation of Subantarctic Slope Water. This mixing creates AABW further downslope. Mixing of shelf water and WDW is characterized by a downslope displacement of isotherms and a strong thermal gradient near the edge of the continental shelf, indicating "frontal zone mixing" (Carmack and Foster, 1975) (Fig. 1.34A,B). Elsewhere along the margin, shelf water with densities lower than that of WDW flows eastward and northward, overriding CPDW (Fig. 1.34C). A strong thermocline separates relatively warm WDW from cold shelf water and characterizes these regions of minimal water mass mixing (Fig. 1.34C).

WDW is generally warmer (up to 1.0°C) than shelf water and has salinities in excess of 34.65 ppt. Water masses on the continental shelf are cold, typically between −1.5 and −1.9°C. Significant density variations in shelf water masses are primarily the result of salinity differences, because the temperature range is only ~ 0.4°C. Shelf water with a temperature range of −1.5 to −1.9°C must be

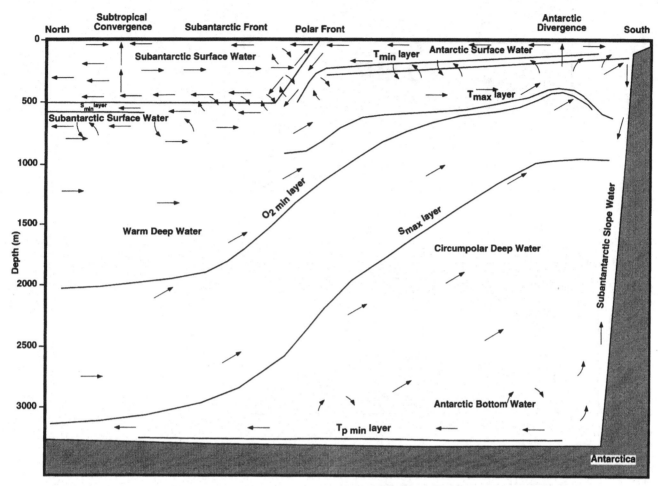

Figure 1.33. Schematic representation of water masses in the Southern Ocean. O_2 min = oxygen minimum; S_{min} = salinity minimum; S_{max} = salinity maximum; T_{min} = temperature minimum; T_{max} = temperature maximum; $T_{p\,min}$ = potential temperature minimum (from Gordon, 1967; reprinted by permission of the American Geographical Society, © 1967).

more saline than 34.50 ppt in order to be more dense than WDW. This saline water mass is referred to as High Salinity Shelf Water (HSSW). Shelf waters with salinities less than 34.50 ppt are called Low Salinity Shelf Water (LSSW).

The primary factors regulating the salinity of shelf waters are the residence time of water masses on the shelf, annual brine production by freezing of sea ice, and dilution from melting sea ice and glacial ice as well as precipitation. Relatively broad continental shelves provide more surface area for seasonal sea-ice formation, and therefore tend to have more saline shelf waters than do narrower shelves. More saline shelf waters tend to accumulate on the western portion of broad shelves and in shelf basins (Jacobs et al., 1970). Gill (1973) argues that repeated episodes of sea-ice formation in a single season, in areas such as the southern Weddell Sea, create the most saline water mass on the Antarctic continental shelf. HSSW also occur in regions of persistent wintertime leads and polynyas where repeated episodes of sea-ice freezing occurs. In contrast, perennial sea-ice cover can result in the presence of LSSW.

Around most of the Antarctic continental margin, WDW flows as a strong boundary current. The upper surface of WDW is often above or at the depth of the continental shelf, which places it in the position to flow onto

the shelf. The salinity of the waters occupying the shelf determines whether WDW flows onto the shelf. HSSW acts as a barrier to onshore flow where WDW encounters this dense HSSW, forcing WDW to override HSSW (Fig. 1.34B). Locally, WDW flows onto the shelf as a warm tongue between HSSW and LSSW (Fig. 1.34A).

Downslope movement of AABW is generally sluggish, with mean speeds of 0.8 to 5.2 cm/sec (Carmack and Foster, 1975). However, where the downslope flow of AABW is focused by canyons and channels, such as offshore of Crary Trough in the southeastern Weddell Sea, flow velocities may be quite strong (Foldvik and Gammelsrød, 1988). In areas where Antarctic Slope Water production takes place, Antarctic Slope Water acts as a barrier that prevents stronger boundary currents from impinging on the continental margin.

Direct measurements of currents on the Antarctic continental margin are scarce, but existing data indicate that the outer shelf and upper slope exhibit much stronger currents

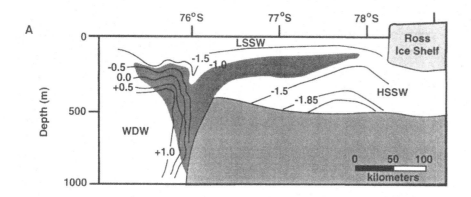

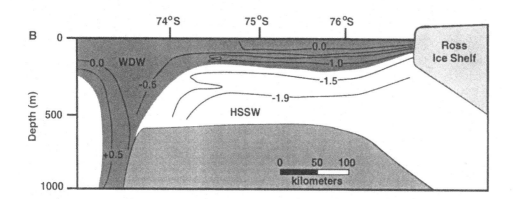

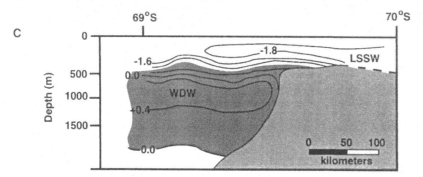

Figure 1.34. Representative potential temperature profiles (°C) from the Antarctic continental margin: (A) profile from the central Ross Sea continental margin showing modified Warm Deep Water (WDW) protruding onto the shelf between High Salinity Shelf Water (HSSW) and Low Salinity Shelf Water (LSSW; modified from Jacobs, Fairbanks, and Horibe, 1985); (B) profile from the western Ross Sea continental margin showing WDW extending across the shelf as a surface water mass (modified from Jacobs, Amos, and Bruchhausen, 1970); and (C) profile from northeastern Weddell Sea continental margin showing WDW impinging onto the outer shelf beneath LSSW (modified from Foster and Carmack, 1976).

than currents on the inner shelf. The strongest currents occur in regions where WDW impinges on the margin. Jacobs, Amos, and Bruchhausen (1970) acquired several short-term (5 to 10 min) current meter readings from ~ 1 m above the seafloor in the Ross Sea. They measured currents in excess of 15 cm/sec at outer shelf stations and currents generally less than 10 cm/sec on the inner shelf.

Pillsbury and Jacobs (1985) collected long-term (several days to one year) current meter data from the Ross Sea. Data from two sites show that near-bottom (4 m above the seafloor) speeds on the outer shelf and slope reached 45 cm/sec, with a mean velocity of 17 cm/sec at two sites (Fig. 1.35). A single, long-term current meter deployment on the inner shelf of the Ross Sea located 42 m above the seafloor recorded current speeds that reached 30 cm/sec, with mean velocities of 8 cm/sec (Pillsbury and Jacobs, 1985) (Fig. 1.35).

Bottom photographs from the upper slope indicate strong bottom currents in both the Ross Sea (Anderson, Brake, and Myers, 1984) and eastern Weddell Sea (Hollister and Elder, 1969). In both these areas, hydrographic profiles indicate that WDW impinges on the upper slope. Bottom photographs from the continental shelf in the

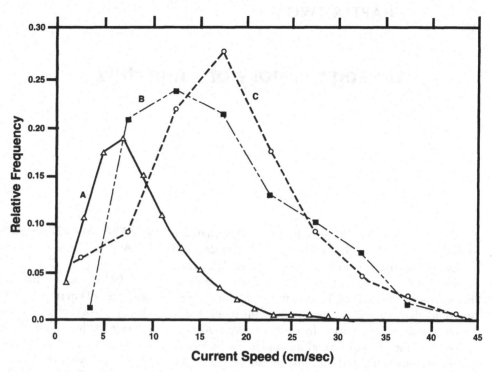

Figure 1.35. Current velocity frequency records from three instruments moored between 4 and 45 m above the seafloor in the Ross Sea. Current meter C was located on the continental slope, meter B on the outer shelf, and meter A on the inner shelf (courtesy of Dale Pillsbury).

Ross Sea and eastern Weddell Sea indicate weak bottom currents.

The predominant surface current in Antarctica is the East Wind Drift, also known as the Antarctic Coastal Current, which flows from east to west (counterclockwise) around the continent (Gordon 1971). Current speeds are variable, but they appear to be strongest near the coast (U.S. Naval Oceanographic Office, 1960) and near the edge of the continental shelf (Carmack and Foster, 1975). Circulation, particularly bottom circulation, is generally weak because of the great depth of the Antarctic shelf. Bottom photographs, scattered current meter measurements, and the fine-grained nature of bottom sediments on the shelf attest to the sluggish nature of bottom currents on the shelf. Bottom photographs and sediment dispersal patterns indicate that wind-driven currents generally influence the seafloor on the tops of banks at water depths above 300 m (Anderson et al., 1983a, 1984). Tidal

currents around Antarctica are relatively weak due to the small tidal range, generally less than 2 m.

The properties of water masses and the circulation beneath ice shelves are poorly understood, but studies indicate a significant exchange of sub-ice shelf meltwater and shelf water masses beneath the Ronne–Filchner (Jenkins and Doake, 1991), George VI (Potter and Paren, 1985), and Ross (Jacobs, Gordon, and Amos, 1979) ice shelves. Current meter records acquired beneath and along the front of the Ross Ice Shelf show dominantly diurnal currents that reach speeds up to 25 cm/sec and have southward and westward flow components (Jacobs et al., 1979; Jacobs and Haines, 1982; Pillsbury and Jacobs, 1985).

CHAPTER TWO

Geologic History of Antarctica

Antarctica is ~14,200,000 km^2 in area, larger than the combined areas of the United States and Mexico, and consists of two major continental blocks. East Antarctica is a large stable continental craton composed mainly of Precambrian metamorphic basement rocks with granitic intrusions that are unconformably overlain by generally flat-lying sedimentary rocks. In contrast, West Antarctica is an archipelago of several microplates with mountainous metamorphic and volcanic terranes. The TAM, one of the major mountain chains of the Earth, mark the boundary zone between East and West Antarctica (Fig. 2.1).

Our geologic knowledge of Antarctica is limited, because an ice sheet averaging 1 to 4 km in thickness covers 98% of the continent. The subglacial topography of the continent is reasonably well-known, due to airborne radio-echo sounding technology. Subglacial topographic maps show that vast portions of the continent lie below sea level. Much of the land would remain below sea level if the ice sheet were to melt and the continent to rebound isostatically (Fig. 1.2). These subglacial basins are part of the Antarctic marine geologic realm.

Given the paucity of outcrops on Antarctica and drill sites on the continental margin, information about the geologic history of the continent draws heavily on published descriptions of strata from contiguous Gondwana continents. This information is used to create a framework for predicting stratigraphic successions that should occur within marine and subglacial basins in Antarctica. This chapter is a summary of the geologic history of Antarctica in the context of its association with other Gondwana continents. It is intended as a reference chapter to provide information about how various portions of the continent and its margin evolved. This chapter also contains information about the breakup history of Gondwana and the development of seaways through and around Antarctica. An examination of the major crustal components of East and West Antarctica initiates this discussion. Chapters 5 and 6 contain detailed discussions of continental margin geology and Antarctica's glacial history, respectively.

GENERAL STRUCTURE AND COMPONENTS OF ANTARCTICA

The TAM separate East and West Antarctica, spanning nearly 3500 km (Fig. 2.1). Support for this division is principally topographic; however, deep crustal investigations provide further information. Seismic refraction and gravity data, as well as geochemical data, reveal an abrupt transition along the eastern front of the TAM (Robinson, 1964; Smithson, 1972; Bentley, 1973; Behrendt et al., 1974; Davey, 1981; Jankowski and Drewry, 1981; Bentley, 1983; Fitzgerald et al., 1986; Kim, McGinnis, and Bowen, 1986; Cooper, Davey, and Cochrane, 1987b; Kalamarides, Berg, and Hank, 1987; Stern and ten Brink, 1989). This abrupt change is considered to represent the boundary between the Precambrian East Antarctic craton and the predominantly Mesozoic-Cenozoic terranes of West Antarctica.

Formation of East Antarctica began in the Archean and ceased, for the most part, by the Cambrian. The West Antarctic continental blocks did not move into place until much later, with most of its developmental history confined to the Mesozoic and Cenozoic. This natural subdivision produces a convenient means for discussing the geology and plate tectonic reconstruction of the Antarctic continent.

Major Components of East Antarctica

The East Antarctic craton forms most of East Antarctica. Its western edge is underlain by a Paleozoic mobile belt that terminates at the present TAM. Unfortunately, ice and time obscure the history of the East Antarctic craton. Most of the detailed descriptions of pre-Paleozoic rocks come from small isolated exposures. In spite of this, information concerning the evolution of the craton emerges from these localized studies. Tingey (1991) provides an excellent review of Antarctica's Precambrian geology.

Formation of the craton began in the Early Archean (Borg et al., 1987) and reached full development by the Middle Proterozoic (Elliot, 1975a; Grikurov, 1982; Black, James, and Harley, 1983). In the Proterozoic, individual

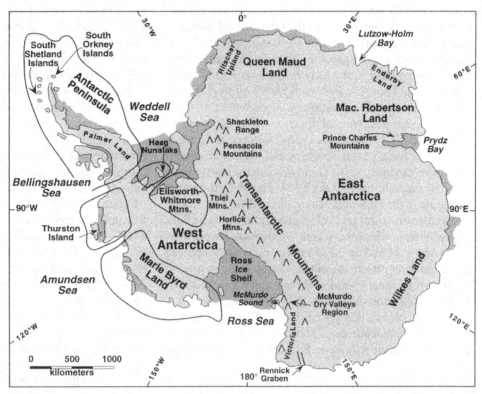

Figure 2.1. Generalized geographic map showing major tectonic features of Antarctica. The four encircled areas represent the four crustal blocks of West Antarctica. A more detailed map that includes most of the localities mentioned in the text appears in the National Geographic Society (1975) map of Antarctica.

continents stabilized and merged into an initial Gondwana-like configuration (Borg and DePaolo, 1994). The Antarctic shield is similar in general lithology and structural style to the shield regions of Australia, South Africa, India, Madagascar, and Sri Lanka (Brown, Campbell, and Crook, 1968; Crawford and Oliver, 1969; Haughton, 1969; Katz, 1973; Hofmann, 1978; Jackson, 1979; Plumb, 1979; Hunter and Pretorius, 1981; Rollinson, Windley, and Ramakrishnan, 1981; Rutland, 1981; Sivaprakash, 1981; Grew and Manton, 1986; Tingey, 1991). There is also evidence that the southwestern United States and East Antarctica were contiguous during the late Precambrian and early Paleozoic. This is known as the Southwest United States–East Antarctic (SWEAT) hypothesis (Dalziel, 1991; Moores, 1991).

Major Components of West Antarctica

Crustal structure, morphology, and geophysical data suggest that West Antarctica is a continental mosaic composed of four discrete or semi-discrete crustal blocks: the Marie Byrd Land, Ellsworth-Whitmore Mountains, Antarctic Peninsula, and Thurston Island blocks (Fig. 2.1). The Haag Nunataks are considered by some to represent a fifth crustal block (Storey et al., 1988; Johnson and Smith, 1992). The exact boundaries and relative behavior of these blocks are not well documented. Based on paleomagnetic analyses, Dalziel and Pankhurst (1987) and Grunow, Kent, and Dalziel (1987b) proposed grouping these blocks, with the exception of Marie Byrd Land, into a single block, Weddellia.

Extensional tectonism was proposed for the zones between some of the West Antarctic microcontinental fragments (Davey, 1981; Jankowski and Drewry, 1981; Dalziel and Elliot, 1982; Jankowski, Drewry, and Behrendt, 1983; Schmidt and Rowley, 1986; Garrett, Herrod, and Mantripp, 1987). Evidence for extension in the Marie Byrd Land sector of West Antarctica includes block faulting and abundant alkaline volcanic rocks (Jankowski and Drewry, 1981; LeMasurier and Rex, 1983). The data also indicate that the West Antarctic blocks may have moved relative to each other and the East Antarctic craton following the breakup of Gondwana (Elliot, 1975a; de Wit, 1977; Jankowski and Drewry, 1981; Dalziel and Elliot, 1982; de Wit et al., 1988; Grunow et al., 1992). At the opposite end of the spectrum, some researchers propose that the blocks can be restored to their original context within Gondwana with little or no displacement between them (Storey et al., 1988).

The Ellsworth-Whitmore Mountains block represents a special problem in Antarctic geology (Schopf, 1969; Elliot, 1975a; Clarkson and Brook, 1977; Dalziel and Elliot, 1982; Craddock, 1983). The Ellsworth and Whitmore mountains are connected by a large rugged highland region with over 2 km of relief. The Whitmore Mountains occur on a high-standing massif that appears to be separated from the Ellsworth Mountains by a narrow trough.

It is not known whether this trough structurally divides the Whitmore Mountains from the Ellsworth Mountains. Radiometric age dates and overall stratigraphic relationships, however, suggest that the Whitmore Mountains are related geologically to the Ellsworth Mountains (Webers et al., 1982). Thus, the Ellsworth and Whitmore mountains are considered to constitute a single block.

The original position and direction of displacement of the Ellsworth-Whitmore Mountains block remain controversial. The rocks of the Ellsworth and Whitmore mountains have obvious ties to Gondwana margin sequences found throughout most of the TAM. However, the structural grain of the Ellsworth Mountains strikes nearly transverse to the structural grain of the TAM and the Cape Fold Belt in Africa, which were contiguous features prior to the Gondwana breakup. In addition, only a single episode of deformation affects the entire stratigraphic sequence, in contrast to three episodes of deformation recorded in the Pensacola Mountains (Fig. 2.1). This leads many researchers to believe that the Ellsworth-Whitmore Mountains block is allochthonous and has experienced significant displacement relative to East Antarctica during the

breakup and dispersal of Gondwana (Schopf, 1969; Dalziel and Elliot, 1982). Grunow, Dalziel, and Kent (1991) argue that strike-slip faulting can account for the 90° counterclockwise rotation of the Ellsworth-Whitmore Mountains block away from East Antarctica.

A series of bedrock highs bounded by steep escarpments separates the Ellsworth-Whitmore Mountains block from the Antarctic Peninsula block to the northwest (Doake, Crabtree, and Dalziel, 1983; Drewry, 1983). Magnetic and gravity models, coupled with apparent horst-and-graben morphology, strongly suggest that northeast to southwest extension has occurred between the Ellsworth-Whitmore Mountains and Antarctic Peninsula blocks (Doake, Crabtree, and Dalziel, 1983; Garrett, Herrod, and Mantripp, 1987). Garrett, Herrod, and Mantripp (1987) suggest that the uplift of the Ellsworth-Whitmore Mountains block is related to this rifting. The time of extension between the Antarctic Peninsula and Ellsworth-Whitmore Mountains blocks remains uncertain, but the abundance of Cenozoic alkali basalts in this region (Garrett and Storey, 1987) in addition to the high relief and rugged topography support a Cenozoic age for much of the movement.

The Marie Byrd Land block is separated from the Ellsworth-Whitmore Mountains block by an extension of the Byrd Subglacial Basin and from the Thurston Island block by a deep, narrow trench occupied by Pine Island Glacier. A similar trench appears to separate the Thurston Island block from the Antarctic Peninsula block. Little is known about the Thurston Island block due to inaccessibility and poor exposure. However, the available data provide evidence that the Thurston Island block is part of a magmatic arc from Gondwana, similar to the Antarctic Peninsula block (Dalziel and Elliot, 1982; Rowley et al., 1983). Although the sub-ice morphology apparently indicates a significant structural break, the Thurston Island block was probably originally located close to the western end of the Antarctic Peninsula block in a location similar to its present position.

OROGENIC EVENTS AFFECTING EAST AND WEST ANTARCTICA

Precambrian Orogenic Events

The cratonic nucleus of East Antarctica has undergone several stages of crustal evolution sinceits apparent formation about 4000 ± 200 Ma during the Napier Orogeny (Sobotovich et al., 1976; Grikurov, 1982; Kamenev, 1982; Ravich, 1982b) (Table 2.1; Fig. 2.2A). Cratonic development around the early nucleus continued with three subsequent Archean craton-building orogenies (Kamenev, 1972; Ellis et al., 1980; Sheraton and Black, 1981; Grikurov, 1982).

Table 2.1. Organic Events in Antarctica

Event	Time	Major Result
Napier Orogeny	4000 ± 200 Ma	formation of cratonic nucleus
Rayner Orogeny	~3500 Ma	thickening of cratonic nuclei
Humboldt Orogeny	~3000 Ma	deformation of miogeoclinal deposits
Insel Orogeny	2650 ± 150 Ma	extensive reworking of old crust
Early Ruker Orogeny	2,000 – 1700 Ma	deformation and greenschist-facies metamorphism
Late Ruker Orogeny or Nimrod Orogeny	1000 ± 150 Ma	widespread plutonism and metamorphism
Separation of Gondwana		
Beardmore Orogeny	633–620 Ma	compressional or convergent (TAM)
Ross Orogeny	early Paleozoic	extension and magmatism followed by intrusion and compression (TAM)
Borchgrevink Event	Devonian-Silurian	metamorphism and igneous activity (TAM)
Peninsula Orogeny	late Paleozoic-early Mesozoic	deformation of late Paleozoic and early Mesozoic rocks in the Antarctic Peninsula region
Late Cretaceous-Cenozoic events	Late Cretaceous-Cenozoic	uplift of TAM

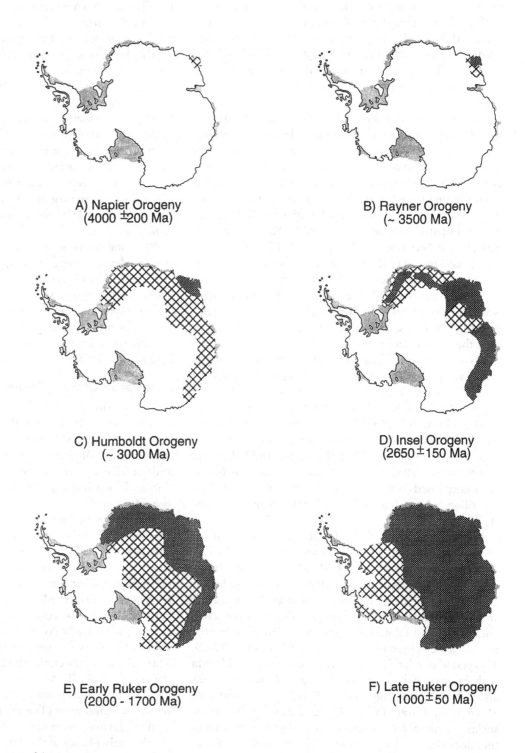

A) Napier Orogeny
(4000 ±200 Ma)

B) Rayner Orogeny
(~ 3500 Ma)

C) Humboldt Orogeny
(~ 3000 Ma)

D) Insel Orogeny
(2650 ±150 Ma)

E) Early Ruker Orogeny
(2000 - 1700 Ma)

F) Late Ruker Orogeny
(1000 ±50 Ma)

Figure 2.2. Paleotectonic development of the Antarctic craton depicting phases of cratonic growth and associated orogenic events (modified from Grikurov, 1982).

Crustal Additions   Pre-Existing Crust

About 3500 Ma, a second event, known as the Rayner Orogeny (Grikurov, 1982; Ravich, 1982b), resulted in the thickening of the cratonic nuclei of East Antarctica (Fig. 2.2B). Following the Rayner Orogeny was a period of quiescence that ended ~ 3000 Ma with the onset of the Humboldt Orogeny (Grikurov, 1982). Rocks representing the Humboldt Orogeny are widespread and are exposed in the Humboldt Mountains of Queen Maud Land, and in Mac. Robertson Land (Fig. 2.2C).

The Insel Orogeny occurred ~ 2650 ± 150 Ma (Grikurov, 1982) and was characterized by high-grade amphibolite metamorphism and the development of large areas of new crust derived from extensive reworking of older crust (Table 2.1). Exposures in Queen Maud Land and in the southern Prince Charles Mountains were used to identify this orogeny (Fedorov and Hofmann, 1982; Grew, 1982; Lopatin and Semenov, 1982; Ravich, 1982b; Ravich and Fedorov, 1982; Tingey, 1982; 1991) (Fig. 2.2D). Grikurov (1982) believes that the Insel Orogeny was the last of the craton-building episodes of East Antarctic history, and that the East Antarctic craton essentially was stabilized by the Late Archean. Elliot (1975a), however, suggests that stabilization of the craton did not occur until the Early Proterozoic.

Several sediment-filled intracratonic miogeoclinal basins developed in the Early Proterozoic, associated with the Early Ruker Orogeny (Grew, 1982; Grikurov, 1982; Tingey, 1982) (Table 2.1; Fig. 2.2E). Exposures exist in the southern Prince Charles Mountains and consist of cross-bedded sandstone, shale, conglomeratic mudstone, and iron formations (Grew, 1982; Grikurov, 1982; Lopatin and Semenov, 1982). These strata were deformed and metamorphosed during the Early Proterozoic (~ 1700 Ma) according to Neethling (1972) and Halpern and Grikurov (1975).

Following the Early Ruker Orogeny, deposition of red beds (possibly coeval with volcanism) occurred in the Ritscher Upland of western Queen Maud Land. In the Shackleton Range, evidence of epicratonic seas and sedimentation typical of stable platforms exists as mature quartz arenites and stromatolitic carbonate beds. Subsidence, sedimentation, and submarine volcanism occurred along the proto-Pacific margin of East Antarctica (Grikurov, 1982). This period of relative quiescence ended ~ 1000 ± 150 Ma with the onset of the Late Ruker Orogeny (Grikurov, 1982), also referred to as the Nimrod Orogeny (Grindley and McDougall, 1969) (Table 2.1; Fig. 2.2F). Uplift was associated with magmatic reworking and partial melting of the older plutonic and metamorphic basement rocks, as well as metamorphism along the proto-Pacific margin of East Antarctica (Tingey, 1991).

The SWEAT hypothesis (Dalziel, 1991; Moores, 1991) holds that the Pacific margins of Laurentia, East Antarctica, and Australia were joined before 750 Ma. Separation of the supercontinents must have been complete by ~ 650 Ma, when deposition of relatively thick sequences of terrigenous and carbonate sediments occurred in subsiding basins along the proto-Pacific margin of East Antarctica. These strata are presently exposed in the TAM (Elliot, 1975b; Grikurov, 1982) and are considered evidence for a rift margin setting at this time.

The Beardmore Orogeny (Grindley and McDougall, 1969), recognized in the central TAM and possibly in the Shackleton Range (Elliot, 1975a), resulted in deformation of older strata into upright to recumbent isoclinal folds or into tight symmetric, asymmetric, and overturned folds (Elliot, 1975a). The style of deformation indicates compressive or convergent tectonic activity (Stump, 1992). Igneous activity associated with the Beardmore Orogeny is primarily recorded by the presence of calc-alkalic pyroclastics that overlie argillite-graywacke sequences in Queen Maud Land, the Horlick Mountains, and the Thiel Mountains (Fig. 2.1).

Radiometric ages indicate a Late Proterozoic age for the Beardmore Orogeny (Tingey, 1991) (Table 2.1). This is supported by observations of Cambrian limestones unconformably overlying rocks deformed during the orogeny (Laird, Mansergh, and Chappell, 1971). Batholithic rocks yield ages of ~ 620 Ma, whereas pyroclastic equivalents are slightly older, with ages of ~ 633 Ma (Gunner, 1976; Faure et al., 1979).

Paleozoic Orogenic Events

The interior portion of East Antarctica has remained, for the most part, tectonically stable since the Precambrian. From the Late Proterozoic until the early Mesozoic, the margin of the East Antarctic shield between the Ross and Weddell seas was subjected to intermittent deformation and intrusive activity. West Antarctica experienced several orogenic events.

Ross Orogeny. Erosion reduced the ancestral TAM (formed by the earlier Beardmore Orogeny), and a broad epicratonic sea advanced across the region during the Early Cambrian. The sea was relatively wide, possibly extending over much of West Antarctica (Elliot, 1975a). Volcaniclastic sequences preserved within the TAM include shallow marine clastics, platform carbonates, and deepwater turbidites (Borg, DePaolo, and Smith, 1990; Stump, 1992, 1995). The Ross Orogeny of the early Paleozoic (~ 500 Ma) (Kleinschmidt et al., 1992; Stump, 1995) influenced the region of the TAM, forming the roots of the present TAM. It was accompanied by continent-wide metamorphism and plutonism (Tingey, 1991).

Bimodal magmatism and extension along the TAM during the early phases of the Ross Orogeny were precursors to more intense tectonism during later phases. During the Late Ross Orogeny, sedimentary rocks of the margin experienced deformation, metamorphism, and intrusion by granitic batholiths (Craddock, 1972; Adams, Gabites, and Grindley, 1982; Borg, 1983; Grindley and Oliver, 1983; Borg, DePaolo, and Smith, 1990; Stump, 1995). The distribution of S-type granitoids within the continent and

I-type granitoids around Antarctica, as well as their geochemical similarity to granitoids of major circum-Pacific batholiths, indicates a westward-dipping subduction zone along the paleo-Pacific border of the East Antarctic craton during the Ross Orogeny (Borg, 1983). Thrusts directed toward the craton are aligned perpendicular to the trend of the TAM (Kleinschmidt et al., 1992). Tessenshon (1994) argues that the evidence for a major compressive plate boundary near the TAM during the Ross Orogeny may be inconsistent with the SWEAT hypothesis.

Striking similarities exist between the deformed sequences of the Ross orogenic belt and the deformed sequences of the Adelaide fold belt in southeastern Australia (Ravich, 1982a). Southeastern Australia was contiguous with Victoria Land before the breakup and dispersal of Gondwana. This suggests that a continuous Ross-Adelaide geosyncline extended more than 8000 km through the modern TAM and southeastern Australia.

West Antarctic sequences with radiometric-age dates indicative of Ross orogenic influence occur in widely scattered outcrops in Marie Byrd Land and the Thurston Island region. Metasedimentary and metavolcanic rocks found in these areas are similar in lithology and style of deformation to the deformed sequences found in the TAM (Craddock, 1972; Wade and Wilbanks, 1972). The age, relationship, and extent of influence of the Ross Orogeny on these deformed rocks in West Antarctica, however, are poorly understood due to limited exposure, structural and stratigraphic complexities, and limited radiometric age data.

Basement rocks composed of marble gneiss, recovered in Deep Sea Drilling Project (DSDP) Site 270 in the eastern Ross Sea, provide an additional link between the Ross Orogeny and West Antarctica (Ford and Barrett, 1975). These rocks are similar to the early Paleozoic Koettlitz Marble that was deformed during the Ross Orogeny and exposed in the TAM of Victoria Land. Goldstrand, Fitzgerald, Redfield, Stump, and Hobbs (1994) point to similarities in early Paleozoic rocks of the Ellsworth Mountains, parts of the Pensacola Mountains, and eastern South Africa as evidence that these regions were situated inland of the zone of deformation that affected the proto-Pacific margin of Gondwana during the Ross Orogeny.

The only evidence for orogenic activity in the interior regions of East Antarctica during the Early Paleozoic exists in the Lützow-Holm Bay area. In that area, high-grade regional metamorphism and associated folding of the Lützow-Holm Complex and the Yamato-Belgica Complex of East Antarctica occurred during the Cambrian (553 ± 6 Ma and 521 ± 9 Ma) (Shiraishi et al., 1994).

Borchgrevink Event. Evidence for a Devonian-Silurian tectonic event in Antarctica, the Borchgrevink Orogeny, is primarily provided by metamorphic and igneous rocks in northern Victoria Land (Grindley and Warren, 1964; Craddock, 1972). Elsewhere in the TAM, shallow marine clastic sediments appear unaffected by the Borchgrevink Orogeny (Elliot, 1975a). Evidence for a mid-Paleozoic tectonic event in West Antarctica consists of a few radiometric dates measured on widely scattered igneous rocks and closely associated metamorphic rocks from Marie Byrd Land, the Thurston Island region, and the Antarctic Peninsula (Halpern, 1968; Adie, 1972; Wade, 1972; Elliot, 1975a).

Research by the German Antarctic North Victoria Land Expedition failed to uncover any evidence of deformation or compressional tectonics, other than intrusive activity, during Devonian-Silurian time (Tessensohn et al., 1981). Therefore, the existence of the Borchgrevink Orogeny was questioned; and in its place, "Borchgrevink Event" was proposed to denote this intrusive activity. Wodzicki and Robert (1986) found evidence of folding in the mid-Paleozoic Bowers Supergroup. The folding, however, is regionally restricted and the deformation is relatively mild without a clear metamorphic overprint. Wodzicki and Robert (1986) concur with the suggestion that Borchgrevink tectonic activity is considered an event rather than an orogeny.

Peninsula Orogeny (Gondwanian and Andean Orogenies). Late Paleozoic to early Mesozoic deformation is recorded primarily in the Antarctic Peninsula (Elliot, 1975a; Barker and Dalziel, 1976; Dalziel, 1981; Dalziel and Elliot, 1982), the Pensacola Mountains (Weddell Orogeny of Ford, 1972), and the Ellsworth Mountains (Ellsworth Orogeny of Craddock 1972, 1982). Elliot (1975a) termed this relatively widespread tectonic event the "Gondwanian Orogeny." This term was first proposed by Du Toit (1937) for deformation in his "Samfrau Geosyncline," which lay along the proto-Pacific margin of the Gondwana supercontinent (Fig. 2.3). The term "Gondwanian Orogeny" traditionally refers to deformation of Late Paleozoic and early Mesozoic sedimentary, metasedimentary, and volcanic rocks underlying a suspected middle Mesozoic unconformity in the Antarctic Peninsula region. This Mesozoic unconformity underlies the Antarctic Peninsula Volcanic Group and was originally believed to separate Gondwanian orogenic deformation from Andean orogenic deformation. Interpretation of initial field work and radiometric age dates indicate that pre-Middle Jurassic basement rocks, which underlie the unconformity in the Antarctic Peninsula, South Shetland Islands, and South Orkney Islands, constitute a convergent margin system (Barker and Dalziel, 1976; Dalziel, 1981; Dalziel and Elliot, 1982) (Fig. 2.1).

A correlative tectonic record exists in South Africa (Cape Orogeny), Australia (New England Orogeny), and South America (Sierra Orogeny). More detailed field analyses, however, revealed that the Antarctic Peninsula metamorphic rocks, previously considered roots of the Gondwanian Orogeny, are deformed plutons of a Mesozoic magmatic arc (Meneilly et al., 1987). The metamorphic grade found in the Antarctic Peninsula region increases

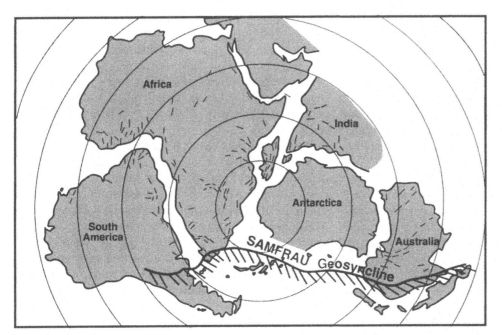

Figure 2.3. Reconstruction of Gondwana during the Paleozoic showing the location of the Samfrau Geosyncline. Short lines indicate the "grain" of Precambrian–early Cambrian rocks (from Du Toit, 1937).

from north to south, progressing from subgreenschist to blueschis (Aitkenhead, 1965; Elliot, 1966; Dalziel, 1982; Gledhill, Rex, and Tanner, 1982). The widespread unconformity originally postulated to separate the Andean and Gondwanian orogenies is not a consistent time plane along the continental margin and does not represent a significant change in tectonic processes (Storey, Thomson, and Meneilly, 1987). In addition, it has become increasingly evident that the Gondwanian fold belt includes deformed regions of widely varying ages (Storey, Thomson, and Meneilly, 1987).

Deposition of thick sequences of graywackes and shales as well as subordinate amounts of conglomerates, arkoses, quartz arenites, cherts, siltstones, and mudstones in basins along the margin of the Antarctic Peninsula mark the initiation of the Peninsula Orogeny in West Antarctica (Adie, 1957; Aitkenhead, 1965; Elliot, 1965, 1966, 1975a; Thomson, Pankhurst, and Clarkson, 1983). Many of these sequences contain interbedded silicic volcanics, basic volcanics, and volcaniclastics, which may include oceanic rocks (Dalziel and Elliot, 1982). The upper Paleozoic sequences are strongly folded and show evidence of polyphase deformation and relatively low-grade regional metamorphism. Early to mid-Cretaceous deformation is best recorded at the southern end of the Antarctic Peninsula (Meneilly et al., 1987), where compression resulted in the folding of upper Jurassic volcaniclastic strata (Barker, Dalziel, and Storey, 1991). Fission track data also indicate significant uplift of the Ellsworth Mountains during the Early Cretaceous (141–117 Ma) (Fitzgerald and Stump, 1991).

Studies of northern Palmer Land revealed that, instead of two separate orogenic cycles, the geology of the Antarctic Peninsula results primarily from the development of a magmatic arc under the influence of several episodes of compressional and extensional tectonics. Meneilly and colleagues (1987) proposed a model for the evolution of this arc that involved peak periods of magmatism and arc compression in the Early Jurassic and Early to mid-Cretaceous. These periods of peak magmatism separate periods of extension in the back arc and arc during the Middle to Late Jurassic and Tertiary. Meneilly and colleagues (1987) concluded that the evolution of the Antarctic Peninsula was influenced primarily by: (1) continuous subduction of Pacific and proto-Pacific crust from the Triassic until the early Tertiary; (2) breakup of Gondwana in the Early Jurassic; (3) extension and back arc basin formation during the Middle to Late Jurassic; (4) arc-continent compression with subsequent back arc closure during the Early to mid-Cretaceous; and (5) ridge crest-trench collision during the Cenozoic, resulting in the cessation of subduction along the Antarctic Peninsula. Initial extension of East and West Antarctica occurred during the Jurassic, accompanied by tholeiitic magnetism in the TAM (Elliot, Kyle, and Pankhurst, 1989). The main phase of extension between East and West Antarctica occurred in the Cretaceous (105–85 Ma) (Lawver and Gahagan, 1994).

Late Cretaceous–Cenozoic Orogenic Events

Geophysical data show that the crust beneath the Ross Sea ranges from 21 to 30 km in thickness, compared to 40 km for adjacent crust underlying the TAM. These data led early researchers to postulate that the West Antarctic crust was thinned by extensional tectonics. Subsequent research, including the acquisition of multichannel seismic reflection profiles, supports this interpretation (Davey,

1981, 1987; Hinz and Block, 1984; Sato et al., 1984; Berg, Hank, and Kalamarides, 1985; McGinnis and Kim, 1985; Kim, McGinnis, and Bowen, 1986; Cooper, Davey, and Behrendt, 1987a; Cooper, Davey, and Cochrane, 1987b; ten Brink et al., 1993). The degree to which uplift of the TAM is associated with rifting in the Ross Sea area remains problematic (Jankowski and Drewry, 1981; Behrendt and Cooper, 1991; Wilson, 1995b); however, a major difference in the timing for the uplift of the TAM and extension within the Ross Sea indicates that the two are not closely linked (Fitzgerald and Baldwin, 1996).

David and Priestley (1914), two geologists on Shackleton's 1907 British Antarctic Expedition, interpreted the TAM to be an enormous structural horst. They were the first to infer that a fault formed the major structural boundary between the subsided Ross Embayment and the TAM. Subsequent research supports a more complex block-faulted structure composed of several blocks with different uplift histories (Gunn and Warren, 1962; Craddock, 1972; Wrenn and Webb, 1982; Smith and Drewry, 1984; Fitzgerald et al., 1986; Gleadow and Fitzgerald, 1987; Fitzgerald, 1994; Wilson, 1995b). Uplift, on the order of 4 to 5 km, has been predominantly vertical, without the folding, thrusting, metamorphism, or andesitic volcanism typically associated with active-margin orogenies.

Apatite fission-track data reveal that uplift of the TAM began in the Early to Middle Eocene (55–50 Ma) and was most active during the Cenozoic, averaging 100 m/Ma (Gleadow and Fitzgerald, 1987; Fitzgerald, 1994). The apatite fission track data also have been used to infer the structure of the TAM. The data reveal an overall structure of a large block dipping gently (1 to 2°) to the west under the polar ice sheet with a step-faulted frontal zone (Fig. 2.4).

Behrendt and Cooper (1991) suggest that the TAM have been rising episodically at rates as high as 1 km/Ma rather than at the mean rate determined by fission-track dates. They further argue that the latest episode of rapid uplift may have occurred since the Pliocene. This is supported by

McKelvey (1991), who argues that the high elevation of Plio-Pleistocene Sirius deposits in the Beardmore area of the TAM (northern Victoria Land) is indicative of uplift rates much higher than those derived from fission-track dates. Fitzgerald's (1994) suggestion that the TAM consist of discrete blocks with different uplift rates provides a logical compromise. He also stresses that fission track data neither confirm nor negate high uplift rates in the Plio-Pleistocene.

The actual mechanics of how the TAM formed remain problematic. Smith and Drewry (1984) suggest that uplift of the TAM occurred as East Antarctica slowly shifted over hot asthenosphere that originally formed beneath West Antarctica. Smith and Drewry's (1984) model shows reasonable agreement between the predicted and observed timing, form, and distribution of uplift in the TAM, but it fails to address the apparently genetic relationship of subsidence in the Ross Embayment, the asymmetric structure of the mountains, and their location along the length of an earlier Paleozoic fold belt.

Fitzgerald, Sandiford, Barrett, and Gleadow (1986) invoke the "passive-rift" model to account for the large, rapid uplift of the TAM. They suggest that rifting was controlled by a fundamental structural asymmetry defined by a shallow crustal penetrative detachment zone dipping to the west beneath the TAM. A profound crustal anisotropy inherited from an earlier collisional event (Ross Orogeny) controls the location and asymmetry of this zone. Invoking a crustal penetrative detachment zone allows the transfer of strain from relatively shallow crustal levels in the Ross Sea region to deeper levels under the mountains, and for subsidence in one region and uplift in the other. This model also accounts for the coincidence of the present location of the TAM with the ancient Paleozoic fold belt created during the Ross Orogeny.

Figure 2.4. Uplift in the McMurdo Dry Valleys region of the TAM relative to a coastal datum. The profile reflects the relative uplift rates and structure of the given segment of the range and does not represent uplift history for the entire TAM (reprinted from Gleadow and Fitzgerald, © 1987, with permission from Elsevier Science Ltd).

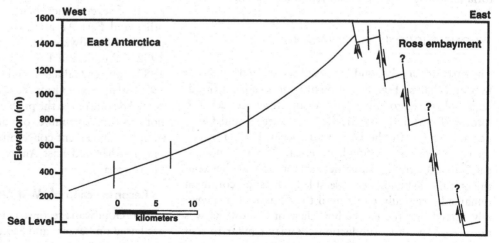

Stern and ten Brink (1989) modeled the lithosphere of East and West Antarctica as two cantilevered elastic beams that were free to move with respect to each other at their adjoining boundary. Flexure along the edge of the East Antarctic boundary resulted in the asymmetric profile of the TAM. This model calls for a more or less continuous depression along the eastern side of the TAM. The Terror Rift, located between McMurdo Sound in the southwest corner of the Ross Sea and 75° S, was initially considered to be this depression; hence, the Stern and ten Brink (1989) model was accepted. The Terror Rift, however, does not extend north of 75° S (Davey and Cooper, 1987), and a later seismic experiment across the Ross Ice Shelf revealed that the Terror Rift does not extend far south of McMurdo Sound (ten Brink et al., 1993). Ten Brink, Bannister, Beaudoin, and Stern (1993) concluded that uplift was caused by thermal buoyancy, as indicated by stretching that preceded the initiation of uplift.

In Victoria Land, the boundary between the TAM and the East Antarctic craton is marked by northwest-southeast-trending structures, the largest of which is Rennick Graben (Fig. 2.1). The deep crustal structure of the region has a northwest-southeast structural fabric, plus a more subtle northeast-southwest set of lineaments that Lodolo and Coren (1994) interpret as evidence for northeastward displacement of Victoria Land relative to the East Antarctic craton. They further suggest that this lateral movement led to the observed northwest-southeast extensional fabric and that this movement occurred during the Mesozoic–early Cenozoic.

Wilson (1995b) points out that brittle fault arrays following the structural boundary between the TAM and the West Antarctic rift system are oriented obliquely to the axis of the mountains and offshore rift basins. He argues that this implies the north to northwest trend of the rift boundary must be inherited from "lithospheric weaknesses" along the ancestral East Antarctic craton. Rifting along the paleosuture has been intermittent; it was most active between 70 and 60 Ma (Behrendt and Cooper, 1991) and ceased following the downfaulting of the Victoria Land Basin. Rifting restarted in the Eocene.

RECONSTRUCTION OF GONDWANA

The early history of Gondwana is obscured in the mists of time and change. Fragmentation of Laurentia from Gondwana is believed to have occurred sometime between 750 (Storey, 1993) and 500 Ma (Moyes, Barton, and Groenewald, 1993). By the Early Ordovician, the major elements of Gondwana (South America, Africa, Madagascar, India, Australia, Antarctica, and several smaller continental blocks) had been assembled. The supercontinent remained essentially intact until the Jurassic. It is generally agreed that the Pacific boundary of Gondwana was the site of an active subduction complex similar to that of the Andean margin today. Antarctica plays a key role in our understanding of the evolution of Gondwana. Its central location within the supercontinent directly ties it to all the other major components.

There have been numerous attempts to reconstruct Gondwana, starting with Du Toit (1937) (Fig. 2.3) and his visionary book, *Our Wandering Continents* (e.g., Dietz and Sproll, 1970; Smith and Hallam, 1970; Barron, Harrison, and Hay, 1978; Norton and Sclater, 1979; Powell, Johnson, and Veevers, 1980; Dalziel and Elliot, 1982; Grindley and Davey, 1982; Schmidt and Rowley, 1986; Barron, 1987; Lawver and Scotese, 1987; de Wit et al., 1988; Lawver, Royer, Sandwell, and Scotese, 1991; Lawver, Gahagan, and Coffin, 1992).

The West Antarctic Problem

One of the central problems that all reconstructions attempt to resolve is how West Antarctica fits into the Gondwanian puzzle, especially with regard to the original orientation of the Antarctic Peninsula and South America (e.g., Lawver, 1984). The volume, *West Antarctica: Problem Child of Gondwanaland,* encapsulates this continuing source of controversy among Antarctic researchers (Dalziel and Elliot, 1982). In the past, various reconstructions for this sector of Gondwana have led either to an unacceptable overlap between the Antarctic Peninsula and South America or to geologically questionable juxtapositions. Attempts at reconstruction primarily reconcile this problem using one of three schemes.

One solution simply ignores the problem of the Antarctic Peninsula overlapping the Falkland Plateau (Dietz and Holden, 1970; Smith and Hallam, 1970; Norton and Sclater, 1979). Another method places the Antarctic Peninsula on the Pacific margin of South America (Barron, Harrison, and Hay, 1978). This orientation, however, conflicts with geologic data that suggest this region of South America was actively engaged in subduction during the Middle Jurassic (Dalziel, 1980). The third solution allows some movement between West Antarctica and East Antarctica. This scheme is based on the increasing body of data that suggests that West Antarctica is a continental mosaic composed of several microplates that moved relative to each other and East Antarctica during the breakup and dispersal of Gondwana (de Wit, 1977; Dalziel and Elliot, 1982; Schmidt and Rowley, 1986; Lawver and Scotese, 1987; Lawver, Gahagan, and Coffin, 1992; Grunow et al., 1992). This recognition has not resulted in a widely accepted resolution to the problem because a consensus does not exist on the number of microplates involved. There is even less agreement concerning their positions relative to one another and East Antarctica during the breakup of Gondwana.

Reconstruction of West Antarctic Blocks

The Gondwana reconstructions presented in this section are based on Lawver and Scotese's (1987) reconstructions

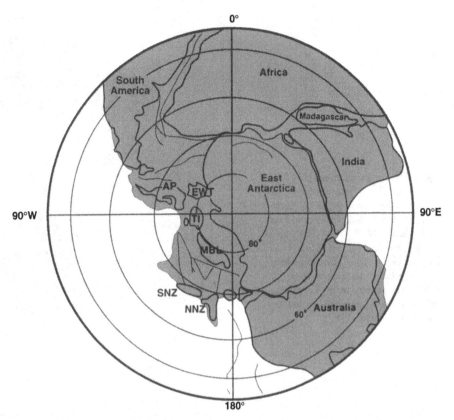

Figure 2.5. Gondwana reconstruction for the Early Jurassic (200 Ma). Blocks are moved relative to a "fixed" East Antarctica. AP = Antarctic Peninsula block; EWT = Ellsworth-Whitmore Mountains block; MBL = Marie Byrd Land block; NNZ = North New Zealand block; SNZ = South New Zealand block; TI = Thurston Island block (reprinted from Lawver et al., 1991, with permission by Cambridge University Press; originally based on Lawver and Scotese, 1987).

and later modifications by Lawver, Gahagan, and Coffin (1992). Lawver and colleagues (1992) illustrate the relative positions of the continents in 10-Ma increments, starting at 200 Ma (Figs. 2.5–2.10). The Lawver and Scotese (1987) model assumes four major blocks for West Antarctica, namely the Marie Byrd Land, Ellsworth-Whitmore Mountains, Antarctic Peninsula, and Thurston Island blocks (Figs. 2.1 and 2.5), with 40 to 50% crustal stretching in the Ross Sea and between the Marie Byrd Land block and East Antarctica. For their reconstruction, Lawver and Scotese (1987) held Africa fixed and employed a Mercator projection. The later reconstructions of Lawver and colleagues (1992) incorporated tectonic lineations deduced from Geosat altimetry data (Fig. 2.11) as well as marine magnetic anomalies and paleomagnetic work of Grunow, Kent, and Dalziel (1987b) and Grunow, Kent, and Dalziel (1991c) (Figs. 2.8–2.10).

Another major problem encountered by researchers in reconstructing the West Antarctic sector of Gondwana is the paleoposition and number of pieces involved in the region joining the Marie Byrd Land block and the Ross Sea with the blocks of New Zealand, Campbell Plateau, and Lord Howe Rise (Fig. 2.8). The controversy surrounding this segment of Gondwana stems from syn- and post-deformation distortion of the crust, structural complexities of the New Zealand blocks, an incomplete record of seafloor magnetic anomalies, and a lack of knowledge with regard to crustal movements during early rifting. An early solution to this problem was to place New Zealand blocks adjacent to the eastern margin of Australia (Smith

and Hallam, 1970). This essentially ignored the large amounts of motion along the Alpine Fault Zone documented by geologic data.

Griffiths (1971) proposed placing the New Zealand blocks adjacent to the Marie Byrd Land block. This placement is consistent with the available paleomagnetic data and geologic evidence (Grindley and Davey, 1982; Cooper et al., 1982). Most subsequent models are similar to the Griffiths (1971) reconstruction but involve slightly different configurations for the positions of the Campbell Plateau and Lord Howe Rise blocks, and greater variation in the number (two to four) and position of crustal fragments that presently constitute New Zealand (Molnar et al., 1975; Grindley and Davey, 1982; Lawver and Scotese, 1987; de Wit et al., 1988; Lawver, Gahagan, and Coffin, 1992) (Fig. 2.8). To address this problem and achieve a tight fit along the Marie Byrd Land and Ross Sea margins, Lawver and Scotese (1987) closed Bounty Trough, shifted Australia slightly clockwise, removed the post-breakup extension in the Ross Sea, and closed Lord Howe Rise and north New Zealand blocks with Australia (Fig. 2.8).

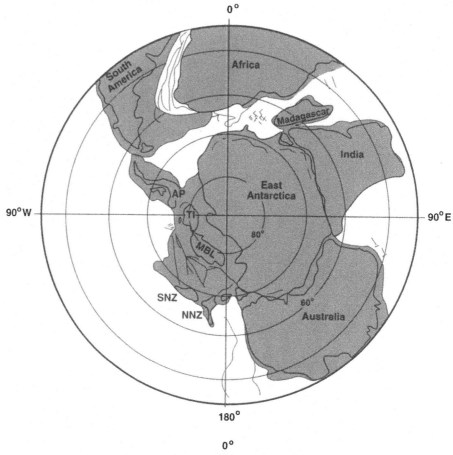

Figure 2.6. Early Cretaceous (130 Ma) paleogeographic reconstruction of southern Gondwana. AP = Antarctic Peninsula block; MBL = Marie Byrd Land block; NNZ = North New Zealand block; SNZ = South New Zealand block; TI = Thurston Island block (reprinted from Lawver et al., © 1991, with permission by Cambridge University Press; originally based on Lawver and Scotese, 1987).

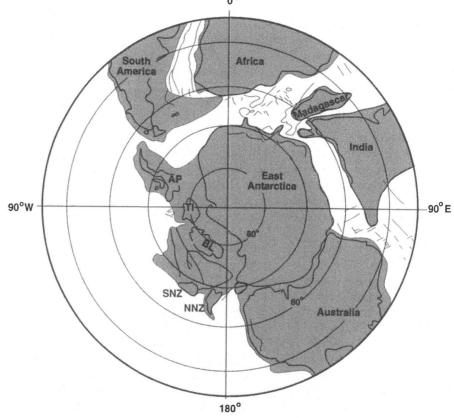

Figure 2.7. Early Cretaceous (118 Ma) paleogeographic reconstruction of southern Gondwana. AP = Antarctic Peninsula block; MBL = Marie Byrd Land block; NNZ = North New Zealand block; SNZ = South New Zealand block; TI = Thurston Island block (reprinted from Lawver et al., © 1991, with permission by Cambridge University Press; originally based on Lawver and Scotese, 1987).

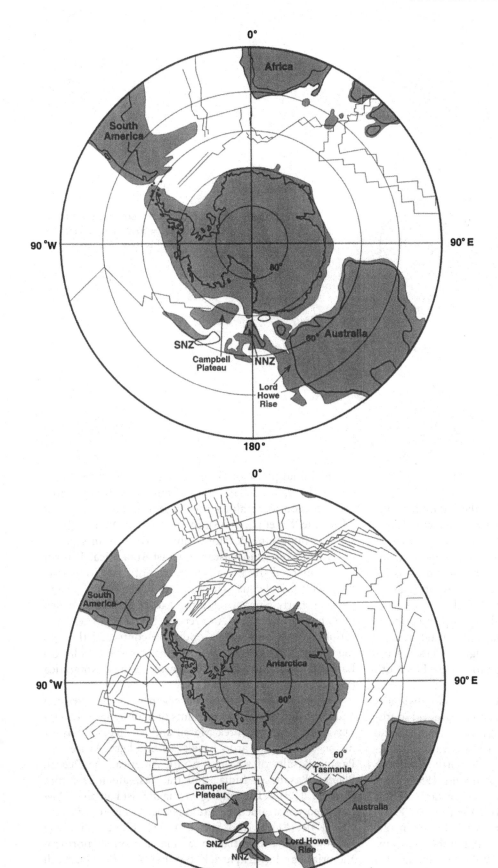

Figure 2.8. Late Cretaceous (80 Ma) paleogeographic reconstruction of the Antarctic region. NNZ = North New Zealand block, SNZ = South New Zealand block (modified from Lawver, Gahagan, and Coffin, 1992).

Figure 2.9. Early Oligocene (30 Ma) paleogeographic reconstruction of the Antarctic region. NNZ = North New Zealand block; SNZ = South New Zealand block (modified from Lawver, Gahagan, and Coffin, 1992).

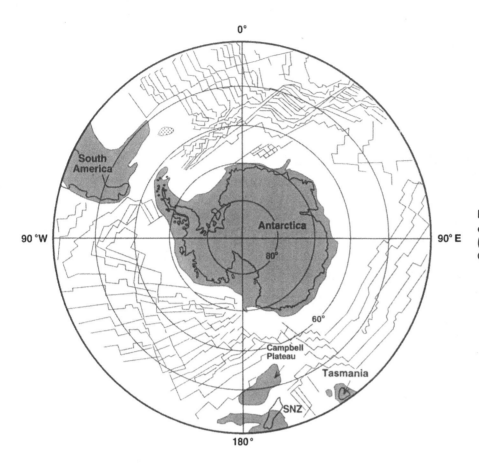

Figure 2.10. Present-day map of the Southern Ocean. SNZ = South New Zealand block (modified from Lawver, Gahagan, and Coffin, 1992).

BREAKUP AND DISPERSAL OF GONDWANA

Antarctica's western margin, including the Antarctic Peninsula and South America, was an active subduction zone rimmed by a magmatic arc during the Mesozoic (Dalziel et al., 1987). The interior "margin" also experienced metamorphic and igneous activity, though the activity was related to a thermal anomaly and not tectonic events (Dalziel et al., 1987). Storey and colleagues (1988) documented evidence for a thermal disturbance in the lithosphere prior to the breakup of Gondwana, that suggests crustal melting and massive underplating of mafic magma. These events resulted in the intrusion of granitic plutons in the Ellsworth-Whitmore Mountains.

Initial stretching between east and west Gondwana is believed to have started in the north and progressed southward (Lawver et al., 1991). Stretching involved right-lateral transtension between east and west Gondwana, including translation and rotation of the Ellsworth-Whitmore Mountains block (Grunow, Dalziel, and Kent, 1991). During this stretching episode, the widespread Ferrar Dolerites were intruded along the present-day TAM (Tingey, 1991; Ford and Himmelberg, 1991; Elliot, 1992). The main axis of stretching is believed to have occurred in the Ross Embayment (Lawver and Gahagan, 1991). Stretching also occurred between the Lord Howe Rise and north New Zealand blocks (Fig. 2.8); otherwise, north New Zealand and Campbell Plateau blocks overlap (Lawver, Gahagan, and Coffin, 1992).

The initial phase of Gondwana separation began during the Permian; a large normal fault system penetrated the supercontinent along the present eastern margin of Africa (de Wit et al., 1988). By the Early to Middle Jurassic, this fault zone developed into a rift system separating Africa from Madagascar and East Antarctica. This rift system marked the initial separation of east Gondwana from west Gondwana. The separation was virtually complete by Early Cretaceous time, except for India, which remained attached to Antarctica (Fig. 2.6). The separation of east and west Gondwana continued through the Early Cretaceous, with rifting between India–Sri Lanka and the Enderby Land margin of East Antarctica (Fig. 2.7).

The breakup of Gondwana began with the intrusion and extrusion of copious amounts of mantle-derived magmas, including the Dufek Massif and Kirkpatrick Basalts in the TAM, the Karoo Basalts of Southern Africa, the Parana Basalts of southern South America, the Deccan Plateau Basalts of India, and the Tasmanian Dolerites (Tingey, 1991). The timing of magmatism in the TAM has been constrained by radiometric dating to 179 ± 7 Ma (Kyle, Elliot, and Sutter, 1981). Elliot and Folan (1986) dated the initiation of magmatism in northern Victoria Land (Kirkpatrick Basalts) as 178 Ma. Both dates correlate well with the onset of Karoo volcanism in South Africa (Kyle, Elliott, and Sutter, 1981) and the emplacement of the Tasmanian dolerites (McDougall, 1961).

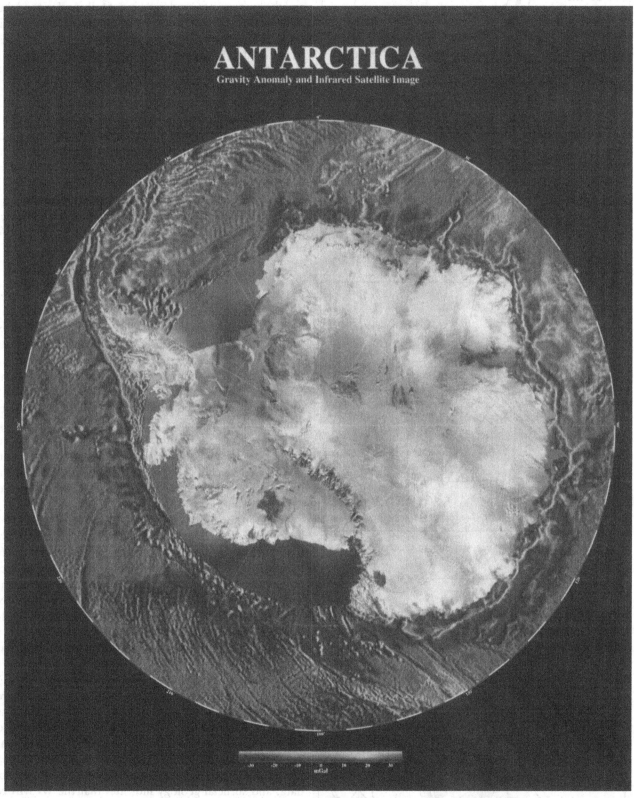

Figure 2.11. Composite satellite image of Geosat gravity anomaly and infrared data (polar stereographic projection south of 60° S). (Copyright 1992 by Scripps Institution of Oceanography and the Institute for Geophysics, University of Texas, Austin; published in Lawver, Dalziel, and Sandwell, 1993).

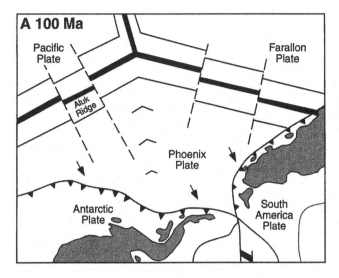

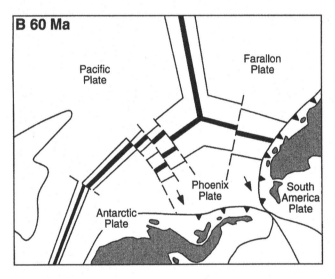

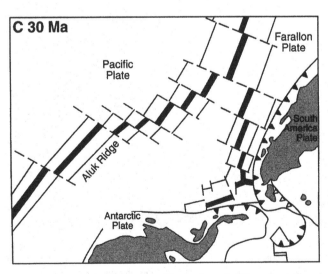

Figure 2.12. Tectonic development of the southeast Pacific region is illustrated at (A) 100 Ma, (B) 60 Ma, and (C) 30 Ma (modified from Barker, Dalziel, and Storey, 1991).

The exact relationship between this igneous activity and the breakup of Gondwana is not fully understood, but some researchers (e.g., Cox, 1978) believe the relationship involves alterations in the mantle due to continuous long-term subduction of oceanic crust beneath the lithospheric cap of Gondwana. The breakup of Gondwana probably relates to the poleward trajectory of the supercontinent during its lifetime. Paleomagnetic and geologic data reveal that Gondwana drifted from tropical latitudes toward the South Pole from the mid-Paleozoic to early Mesozoic.

The tracks of the major components of Gondwana are preserved as magnetic anomalies, which can be dated, as well as oceanic fracture zones and spreading ridges. A good overall review of the spatial evolution of the fragments of Gondwana is presented by Norton and Sclater (1979) with recent updates by Lawver and Scotese (1987) and Lawver, Gahagan, and Coffin (1992).

Southeast Pacific Region

In the southeast Pacific region between the tip of South America and Antarctica, seafloor magnetic anomaly patterns indicate that five separate plates, the Pacific, Phoenix, Antarctic Farallon, and South America plates, existed during the latest Jurassic (Fig. 2.12). An earlier reconstruction by LaBrecque and colleagues (1986) included another plate, the Malvinas Plate, which was supposedly subducted beneath the tip of South America. However, examination of Geosat radar altimetry and seafloor magnetic data from the South Atlantic and Weddell Sea region led to the conclusion that the seafloor in this vicinity was created as a direct consequence of South America and Antarctic plate motion and that the Malvinas Plate never existed (Livermore and Woollett, 1993).

Seafloor spreading along the Pacific-Phoenix spreading center (Aluk Ridge) occurred in the South Pacific. The Phoenix Plate was gradually subducted along the Pacific margin since the Mid-Cretaceous (Fig. 2.12). Subduction of the Aluk Ridge continued through the Late Cretaceous and much of the Cenozoic when the locus of crustal extension shifted to between East and West Antarctica in the Ross Sea region (Lawver and Gahagan, 1994).

Aeromagnetic surveys (Renner, Sturgeon, and Garrett, 1985) define a series of 50 to 100 km wide, long-wavelength positive magnetic anomalies exceeding 500 nanoteslas in amplitude throughout the southeast Pacific region. The largest continuous anomaly, the West Coast Magnetic Anomaly (WCMA), parallels the west coast of the Antarctic Peninsula for a distance of ~ 1300 km (Garrett and Storey, 1987). Correlation of the magnetic anomalies with associated gravity anomalies, seismic refraction data, and surficial geology indicates that the WCMA results from a group of linear batholiths emplaced during Mesozoic-Cenozoic subduction (Garrett, Herrod, and Mantripp, 1987). The WCMA bifurcates at the southern end of Bransfield Strait, suggesting that the Bransfield rift split a subvolcanic batholith.

Weddell Sea Region

Lawver, Gahagan, and Coffin (1992) believe the break-up between South America and Africa resulted in a change in the regional stress field and the formation of a triple junction in the Jurassic. At this time, seafloor spreading was initiated in the Weddell Sea. A prominent escarpment located at the base of the eastern Weddell Sea continental margin, the Explora Escarpment, is believed to coincide with the ancient Gondwana plate boundary separating oceanic crust to the north from crust of unknown origin to the south (Hinz and Krause, 1982; Kristoffersen and Hinz, 1991).

Paleomagnetic data indicate that clockwise rotation of the Antarctic Peninsula relative to East Antarctica between ~175 and ~155 Ma created as much as 1000 km^2 of the Weddell sea floor (Grunow, 1993). By the beginning of the Cretaceous, a northward-propagating rift between South America and Africa, in addition to an east-west-oriented rift system between South America and Antarctica, contributed to the opening of the South Atlantic and the Weddell Basin (Rabinowitz and LaBrecque, 1979; LaBrecque et al., 1986) (Fig. 2.6).

Separation of Antarctica and Australia

Veevers (1986) interpreted the boundary between a large positive magnetic anomaly and the seaward edge of the magnetic quiet zone along the southern margin of Australia to be the continent–ocean boundary. Cande and Mutter (1982) interpreted this anomaly to be the oldest seafloor spreading event. The presence of this anomaly constrained the inception of rifting between Antarctica and Australia to 86 Ma. Veevers's (1986) revised interpretation moves the initiation of spreading between Australia and Antarctica to 95 ± 5 Ma (Veevers, 1986, 1987a, 1990).

Stagg and Wilcox (1992) attempted to constrain the timing of the separation of Antarctica and Australia using seismic reflection data to examine the relationship between continental margin sequences and the oceanic crust. This analysis led to the conclusion that Antarctica and Australia were separated by the Early Cretaceous (125 Ma). Divergence between Antarctica and Australia probably contributed to rifting in the Ross Sea.

Isolation of Antarctica

By the Late Cretaceous, rifting of the New Zealand blocks and eastern Australia had begun, setting the stage for circulation between the Indian and Pacific oceans (Fig. 2.8). Final separation of New Zealand and West Antarctica occurred ~72 Ma (Stock and Molnar, 1987). This ocean passage was not completely cleared until the Early Oligocene, due to the northward drift of Tasmania (Fig. 2.9). These events coincided with the initial separation of South America and the Antarctic Peninsula. By 30 Ma, rifting was under way between all the major elements of Gondwana, resulting in the isolation of Antarctica in the southern polar region for the remainder of the Cenozoic (Fig. 2.9).

Tectonic activity continues today along the Antarctic Peninsula in response to subduction along the Pacific boundary of the continent. Rifting in the Terror Rift of the Ross Embayment has occurred in the late Cenozoic, and perhaps continues today, marking the most recent stage of separation between East and West Antarctica (Cooper et al., 1987a,b; Cooper, 1989).

Increased uniqueness of Antarctic geology, flora, and fauna with respect to their Gondwana counterparts bears witness to the growing isolation of the continent. With the initiation of continental dispersal, the geologic histories of previously adjacent continents diverged. Therefore, the stratigraphic record of these land masses became progressively limited for use in the interpretation of events on Antarctica and along its margins.

ANTARCTIC STRATIGRAPHY

This section provides a detailed discussion of the major stratigraphic units of the Antarctic continent. The outcrop distribution of these units permits construction of a series of paleogeographic maps for late Precambrian through Late Cretaceous time. These maps depict the pre-breakup and early post-breakup paleogeography of Antarctica. The paleogeographic maps can be used to construct idealized stratigraphic columns for any portion of the Antarctic continental margin.

The epochs of the Tertiary are not covered in detail in this chapter. The depositional histories of the basins of Antarctica following the breakup of Gondwana is the subject of Chapter 5, "Continental Margin Evolution."

Precambrian Strata

Plate tectonic reconstruction of the Precambrian lacks the detailed understanding made possible for reconstructions of younger time periods due to the small and/or imprecise data base of paleontologic, paleomagnetic, and geologic criteria upon which paleographic reconstructions are founded. The Gondwana paleocontinent remained largely intact throughout the Proterozoic and early Paleozoic times. Borg and DePaolo (1994) describe many similarities in the early Proterozoic basement rocks of Laurentia, Australia, and Antarctica.

Following the period of full cratonization of the Antarctic shield and prior to deformation during the late Precambrian Beardmore Orogeny, a sequence of platform and basinal sediments was deposited on and adjacent to the East Antarctic craton. Known outcrop locations of the latest Precambrian rocks in Antarctica and on adjoining land masses are shown in Figure 2.13. Figure 2.13 uses stratigraphic information for the late Precambrian in Antarctica and adjoining Gondwana landmasses to

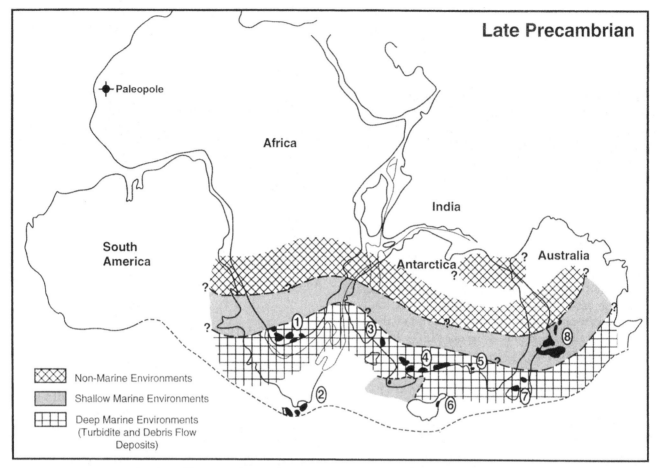

Figure 2.13. Known outcrop locations of late Precambrian sedimentary and metasedimentary rocks (black) and generalized late Precambrian Gondwana paleogeography. Paleoenvironmental designations are inferred. Strata are designated as: 1 = Malmesbury, Gamtoos, Kango, and Kaaimana groups of South Africa; 2 = rocks of unknown age in southern South America; 3 = Turnpike Bluff Fm, 4 = Beardmore Group; 5 = Skelton Group; 6 = Swanson Fm; 7 = Burnei Fm; 8 = Sturtian and Marlnoan series and Pound Subgroup.

depict broad facies relationships. This map should not be considered a paleogeographic "time-slice" because of poor age information on many formations and temporal variability in deposition. Instead, it should be used as a summary of broadly equivalent facies on the ancient supercontinent.

During the late Precambrian, the intact Gondwana supercontinent shifted through a polar position centered in eastern South America and northwestern Africa. Although poorly constrained, older (Sturtian) glacial deposits in southern Australia also suggest a near polar position during late Precambrian. The Pacific margin of the supercontinent consisted of a series of depositional basins that experienced marine conditions during the late Precambrian (Laird, 1991), while marginal- to non-marine conditions existed in the Gondwana highlands of East Antarctica (Fig. 2.13). No glacial deposits of this age are known in Antarctica.

Major deposition during the Precambrian began in a series of deep marine basins extending along the length of the present-day TAM (Elliot, 1975b) (Fig. 2.13). Timing of the initial onset of deposition varied from basin to basin (Stump, 1973). Deep-sea submarine fan sedimentation dominated the infilling processes of these basins.

Cambrian-Ordovician Strata

During Cambrian-Ordovician time, the Gondwana continent underwent a southward shift and Antarctica moved to a lower latitude. Low-lying continental environments existed on the East Antarctic craton, southwestern Australia, southern India, South Africa, and east-central South America, while marine conditions rimmed Gondwana (Fig. 2.14). North-central and west Africa, centered near the pole during the latter half of the Ordovician and into the Silurian, was covered by an extensive ice sheet, the Saharan Ice Sheet, which was similar in scale to the present-day Antarctic ice sheet (Caputo and Crowell, 1985).

Antarctic Strata. In many of the bedrock outcrops of Antarctica, a hiatus representing the Beardmore orogenic event separates the Late Precambrian from Cambrian deposits. A probable exception occurs in the Ellsworth Mountains region, where the basal Heritage Group may represent continuous deposition across the boundary (Laird, 1981).

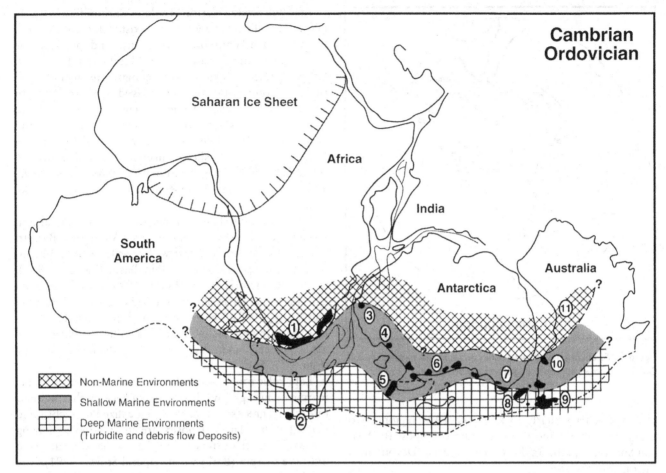

Figure 2.14. Known outcrop locations of Cambrian-Ordovician sedimentary rocks (black) and generalized Cambrian-Ordovician Gondwana paleogeography. Paleoenvironmental designations are derived from Scotese and colleagues (1979). Strata are designated as: 1 = Natal and Table Mountain groups; 2 = deposits of uncertain age; 3 = Urfjell Group; 4 = Blaiklock Glacier Group; 5 = Heritage Group; 6 = Byrd Group; 7 = Skelton and Koettlitz groups; 8 = Dundas Trough; 9 = Kanmantoo Group and Archaeocyatha limestone; 10 = Meatherdaje Beds, Boxing Bay Fm, and Kulpara/Parara limestones; 11 = scattered limestones, evaporites, and redbeds.

Deposition of the Cambrian-Ordovician strata of Antarctica took place as the seas transgressed and eroded the Beardmore orogenic surface (Laird, 1991). Carbonate sedimentation dominated the early phases of deposition while late-stage sedimentation was clastic in nature and accompanied in many cases by felsic volcanism. Deformation and intrusion of these sequences during the early Paleozoic Ross Orogeny are recorded across East and West Antarctica, with the exception of the Ellsworth Mountains and easternmost portions of East Antarctica. Following the Ross Orogeny, the ancestral TAM formed the source for thick continental conglomerates and sandstones that were shed into adjacent basins (Laird, 1991).

Strata of Other Gondwana Continents. Outcrops documenting Cambrian-Ordovician sedimentation on adjoining Gondwana continents exist in New Zealand, Tasmania, Australia, South Africa, and possibly in South America (Fig. 2.14). Shergold, Jago, Cooper, and Laurie (1985) offer a detailed correlation of Cambrian and Early Ordovician strata in Antarctica, Australia, and New Zealand.

Silurian-Devonian Strata

By middle Silurian to Early Devonian time, the Gondwana supercontinent had fully shifted into a south polar position, with the pole itself located in southwestern Africa (Fig. 2.15). A continental environment prevailed across the East Antarctic craton, central Australia, South Africa, and east-central South America (Fig. 2.16). By the Late Devonian, central South America assumed a polar position and was covered by ice masses (Caputo, 1985). The margins of the Gondwana landmass experienced marine conditions (Fig. 2.16).

Antarctic Strata. In Antarctica, deposits of known Silurian-Devonian age occur in outcrops in the Transantarctic, Ellsworth, and Pensacola mountains. Erratics of possible Devonian age have been found in Marie Byrd Land (Fig. 2.16). Documentation for rocks of Silurian age is poor due to a complete lack of fauna (Rowell, Rees, and Braddock, 1987); however, sequences dateable as Devonian often are of sufficient thickness potentially to include Silurian deposits as well. The Beacon Supergroup includes rocks

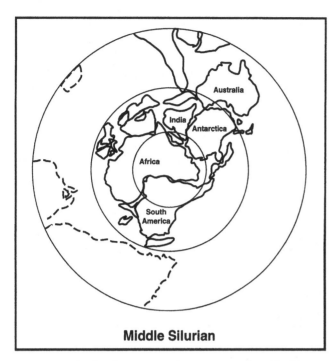

Figure 2.15. South polar stereographic projection of middle Silurian Gondwana paleogeography (modified from Scotese et al., 1979).

of this age, particularly rocks in the TAM region. Two groups, the Taylor and Victoria groups, compose the Beacon Supergroup; the Taylor Group (Silurian-Devonian) is the older of the two.

Strata of Other Gondwana Continents. Only limited exposures of Silurian-Devonian rocks exist on the adjoining Gondwana margins (Fig. 2.16). Silurian-Devonian deposits occur in southeastern Australia, South Africa, and possibly southern South America. In Australia, Silurian-Devonian rocks are found mainly in the southeastern sectors and are dominated by shallow marine limestones and clastics with interspersed volcanics (Sprigg, 1967). Strata in Tasmania also reflect shallow marine conditions, documented by the Gordon River Limestone and Eldon Group. In South Africa, the Bokkeveld and Lower Witteberg groups of the Cape Basin represent Silurian-Devonian strata (Tankard et al., 1982) (Fig. 2.16).

Carboniferous-Permian Strata

During the Carboniferous and Permian, Antarctica occupied a south polar position within the now-combined Gondwana and Laurasia landmasses (Pangea supercontinent; Fig. 2.17). Carboniferous-Permian strata are widespread in coastal Antarctica and on contiguous Gondwana continents, providing the basis for the paleogeographic reconstruction shown in Figure 2.18. These deposits indicate widespread submergence and glaciation of Antarctica, Australia, southeastern India, and southern Africa during the late Carboniferous–Early Permian. Detailed stratigraphic reviews are provided by Elliot (1975b) and Barrett (1981, 1991).

Antarctic Strata. During the late Carboniferous–Early Permian, shallow marine settings existed in Victoria Land, the Ellsworth-Whitmore Mountains, and possibly the Pensacola Mountains and Queen Maud Land (Fig. 2.18). Glacial strata indicate a polar climate and include marine and continental deposits derived from an ice sheet originating from at least one ice center in Victoria Land (Ojakangas and Matsch, 1981; Aitchison, Bradshaw, and Newman, 1988; Miller, 1989; Matsch and Ojakangas, 1991). Retreat of ice sheets during the Early Permian resulted in the deposition of non-marine carbonaceous shales and sandstones within an extensive inland sea or lake (Lindsay, 1970).

By the middle to Late Permian, non-marine alluvial conditions prevailed throughout much of Antarctica (Barrett, 1991). Deposition occurred within at least three unconnected basins, the Nimrod-Ohio Basin (Barrett, 1981), the South Victoria Basin (Elliot, 1975b), and the elongate Takrouna Basin of northern Victoria Land (Collinson and Kemp, 1983). The geometry and isolation of such basins indicate pre-existing highlands, and the existence of actively rising landmasses of the Gondwanian Orogeny (Elliot, 1975a).

Late Paleozoic rocks of the Antarctic Peninsula–Pacific margin include graywacke-shale sequences with conglomerate, chert, greenschist, as well as localized blueschist and mafic basalts. These deposits are interpreted as having formed in trench or trench-slope basins that were accreted onto the margin (Barker, Dalziel, and Storey, 1991). The precise paleogeographic relationship between late Carboniferous turbidite sequences of the Antarctic Peninsula and the correlative marine and non-marine strata previously described is unclear. Elliot (1975b) prefers deposition of the turbidites along the proto-Pacific margin of Gondwana rather than within an inland sea or non-marine water body.

Strata of Other Gondwana Continents. Carboniferous-Permian strata on adjoining Gondwana continents are well represented, particularly in Australia and South Africa (Fig. 2.18). Late Carboniferous deposits in Australia reflect widespread glaciation of that continent, presumably spreading into Australia from an ice sheet, the Gondwana Ice Sheet, centered within Antarctica (Crowell and Frakes, 1971). Strata recording deposition in a trough in eastern Australia indicate glacial–marine and, ultimately, glacial-fluvial conditions (Shoalhaven Group and Bacchus Marsh Tills). Similar deposits (Wynyard, and Maydena formations and Cygnet Coal Measures) record glacial and glacial-marine conditions in Tasmania (Crowell and Frakes, 1971) (Fig. 2.18). Although dominated by non-marine conditions, marine incursions occurred in numerous locations throughout the interior of Australia during this time period (Sprigg, 1967; Crowell and Frakes, 1971). As in Antarctica, post-glacial deposition in southern and eastern Australia was characterized by paralic and non-marine conditions, ultimately resulting in deposition of coal measures.

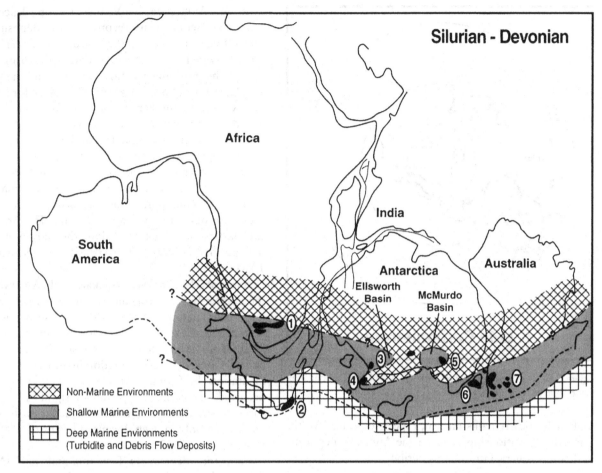

Silurian - Devonian

Non-Marine Environments

Shallow Marine Environments

Deep Marine Environments
(Turbidite and Debris Flow Deposits)

Figure 2.16. Known outcrop locations of Silurian-Devonian sedimentary rocks (black) and generalized Silurian-Devonian Gondwana paleogeography. Strata are designated as: 1 = Lower Witteberg and Bokkeveld groups; 2 = rocks of uncertain age; 3 = Neptune Group and Horlick Fm; 4 = Crashsite Quartzite; 5 = Taylor Group; 6 = Gordon River limestone and Eldon Group; 7 = scattered sandstone, mudstone, and limestone. This paleogeographic reconstruction relies heavily on the works of Sprigg (1967), Scotese and colleagues (1979), and Tankard and colleagues (1982).

Along the coastal region of India, strata of probable time-equivalence to the late Paleozoic deposits of Antarctica are found in the Talchir shale and a probable correlative boulder bed within the Palar Basin (Fig. 2.18). Both units are believed to be glacial and/or glacial-marine in origin (Ahmad, 1981; Sahni, 1982).

The Carboniferous-Permian time period is well represented in the Karoo and Natal basins of southern Africa, where the Dwyka tillites (Fig. 2.18) document widespread glacial and glacial-marine conditions (Hobday, 1982; Tankard et al., 1982; Visser, 1991) (Fig. 2.18). The Dwyka deposits may be correlative to glacial strata in the Ellsworth and Pensacola mountains, which provides further evidence that an ice sheet was partially centered in Antarctica. The post-glacial Permian Ecca mudstones are similar to glacial deposits in Antarctica but are of questionable marine origin (Hart, 1964; McLachlan, 1973). Tankard and colleagues (1982) suggest that these strata were deposited in an extensive landlocked basin receiving periodic marine incursions.

Lower Ecca mudstones, deposited in a linear trough parallel to the present Natal Coast, were succeeded by Middle Ecca regressive shallow deltaic and fluvial/ floodplain deposits. The Middle Ecca deposits are similar to the shoaling-upward Permian sequence recorded in the Victoria Group and equivalent strata of Antarctica. A marine transgression during Upper Ecca time (not correlative to known Antarctic strata) was followed by cessation of downwarping in the Natal Basin. Subsequent Permian alluvial and lacustrine deposits of the lower Beaufort Group, therefore, are restricted to the Karoo Basin. The rising Cape Orogeny (equivalent to the Peninsula Orogeny in Antarctica) strongly influenced deposits of Upper Ecca and lower Beaufort groups (Tankard et al., 1982).

In southern South America, late Paleozoic graywackes, shales, and massive limestones of the Madre de Dios Basin represent potential correlatives to marine sequences (e.g., Trinity Peninsula Group) of the Antarctic Peninsula region (Dalziel and Elliot, 1973). No tillites are known in southern South America, although the Lafonian diamictites (Frakes and Crowell, 1967) of the Falkland Islands may be equivalent to similar strata in the Ellsworth and Pensacola mountains. The relationship of the Falkland Island sequences to deep marine turbidites of South America

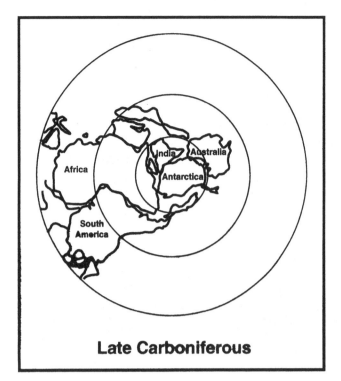

Late Carboniferous

Figure 2.17. South polar stereographic projection of Late Carboniferous Gondwana (modified from Scotese et al., 1979).

and Antarctica remains unclear, especially in light of the unresolved relationship between the Antarctic Peninsula and South America during this period.

Triassic Strata

The Permian–Triassic boundary marks the initiation of the breakup of the Pangea supercontinent. During this time, the intact Gondwana continent shifted northward, moving Antarctica north of the South Pole (Fig. 2.19). Subduction continued along the Pacific margin of Gondwana. Continental conditions prevailed throughout Gondwana, although shallow-marine deposition is documented along the proto-Pacific margin of Antarctica and South America (Fig. 2.20).

Antarctic Strata. In Antarctica, known Triassic strata occur within the TAM trend and along the Antarctic Peninsula (Fig. 2.20). The Trinity Peninsula Group and Legoupil Formation of the peninsula region contain deep and/or shallow marine deposits believed to be partially Triassic in age (Edwards, 1982; Thomson, 1982). The recognition of a Triassic component within these deposits suggests that other unrecognized Triassic rocks may exist in adjacent areas (e.g., South Orkney and South Shetland islands). Although the Trinity Peninsula Group may document deep marine "geosynclinal" deposition (Aitkenhead, 1965; Bell, 1975), invertebrate faunas in the correlative flysch-type sediments of the Legoupil Formation substantiate a shallow marine setting for at least some portions of these units during the Triassic (Edwards, 1982; Thomson, 1982).

In East Antarctica, the Victoria Group (Beacon Supergroup) and equivalents record Triassic deposition. The thickest deposits occur in the Beardmore Glacier area and are subdivided into the Fremouw and Falla formations, although there is some question as to whether the Fremouw and Falla formations represent Triassic or Jurassic strata (Taylor, Taylor, and Farabee, 1988). The strata are similar to older (Permian) alluvial deposits in all respects, except transport direction (Elliot, 1975b; Barrett, 1981).

Pennington and Collinson (1984) contrasted the Permian and Triassic members of the Beacon Supergroup by noting that Triassic units contain volcanic material indicative of contemporaneous volcanism, whereas Permian rocks are devoid of volcanic detritus. The Triassic volcaniclastics that they traced along the margin of the polar plateau share affinities with Triassic volcaniclastics of the TAM.

Strata of Other Gondwana Continents. In Australia, where the Hunter-Bowen orogenic activity essentially ceased by latest Permian time (Sprigg, 1967), Triassic deposition was restricted to the southeastern portion of the continent (Fig. 2.20). Widespread lacustrine deposits, fluvial sandstones, and coal measures record predominantly continental sedimentation. In Tasmania, non-marine sedimentation also occurred, including deposition of significant coal measures (Sprigg, 1967). Triassic marine strata in Australia are known only in deep, localized basins of western Australia (Sprigg, 1967).

Sandstone deposits in Tasmania and the TAM record fluvial sedimentation. While dissimilar on a detailed scale, the strata probably were deposited in the same basin, the Nilsen-Mackay Basin, located between the East Antarctic craton and the Antarctic Peninsula block (Collinson, Kemp, and Eggert, 1987).

Extensive Triassic strata also are known within the South African Karoo Basin (Rust, 1975; Hobday, 1982; Tankard et al., 1982), where continental deposits of the Beaufort and Stormberg groups crop out (Fig. 2.20). The active Cape Fold Belt influenced sedimentation during this time, providing a highland source for alluvial and fluvial sediments in the adjacent foreland basin. Depicting progressive climatic amelioration, depositional environments range from fluvial and lacustrine (Beaufort Group) to fluvial, coal swamp, and eolian/playa (Stormberg Group) (Hobday, 1982). Minor volcanic eruptions recorded within the Stormberg Group as interbedded basalts ultimately gave way to a major episode of volcanism (Early Jurassic), which ended sedimentation in the Karoo Basin (Hobday, 1982).

To date, no rocks of known Triassic age have been located in southern South America. It is possible, in light of the recognition of Triassic faunas within previously described Carboniferous-Permian strata of Antarctica (Edwards, 1982; Thomson, 1982), that Triassic strata may have gone unnoted within possible flysch equivalents in South America.

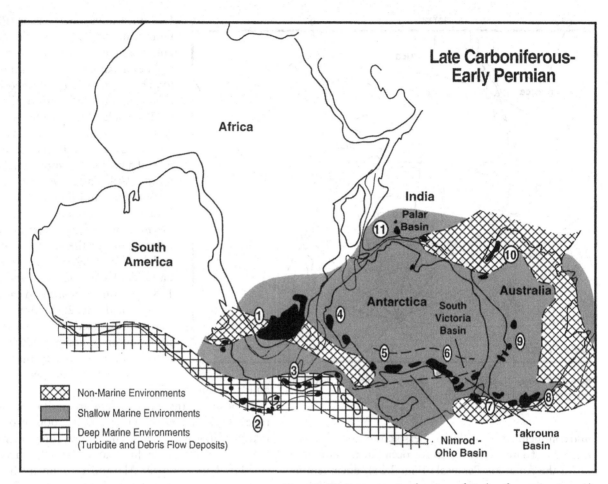

Figure 2.18. Known outcrop locations of Carboniferous-Permian sedimentary rocks (black) and generalized paleogeography of Gondwana for this time interval. The paleogeography of Scotese and colleagues (1979), Hambrey and Harland (1981), and Tankard and colleagues (1982) was used where outcrop exposures are lacking. It should be noted that the precise position of the Antarctic Peninsula with respect to South America is poorly constrained; this increases the difficulty of presenting a detailed paleogeographic reconstruction for that region. Strata are designated as: 1 = Lower Beaufort (?) and Ecca Group of the Karoo and Natal basins; 2 = unnamed strata of the Madre de Dios Basin; 3 = Trinity Peninsula Group and Miers Bluff and Polarstar fms; 4 = Amelang, Theron, and Whichaway fms; 5 = Pecora Fm, Gale Mudstone, Victoria Group, and Whiteout Conglomerate; 6 = Takrouna and Horn Bluff fms; 7 = Wynyard and Maydena fms and Cygnet Coal Measures; 8 = Shoalhaven Group and Bacchus Marsh Tills; 9 = unnamed tills and paralic clays; 10 = Lyones Fm; 11 = Talchir shale and boulder bed.

The general paleogeographic setting during the Triassic in Gondwana was one of continental deposition with marine conditions prevailing along the continental margin (Fig. 2.20). The caution generated with respect to the Carboniferous-Permian paleogeographic maps applies to the paleogeographic interpretations of this time period as well: the exact relationship between the Antarctic Peninsula and South America during the Triassic remains uncertain.

Jurassic–Early Cretaceous Strata

During Jurassic to Early Cretaceous time, global tectonic activity continued in conjunction with the breakup of Pangea (Fig. 2.6). Rifting between east and west Gondwana commenced during the Middle Jurassic, accompanied by an explosive phase of volcanism, and later followed by a quieter phase of flood basalt emplacement (Bradshaw, 1987). Voluminous tholeiitic magmatic activity occurred throughout the TAM, South Africa, and Tasmania, and along the Antarctic Peninsula. In the Antarctic Peninsula region, peaks in magmatic activity occurred in the Early Jurassic and Early Cretaceous (Pankhurst, 1982).

In the TAM, the Dufek Intrusion and the Ferrar Dolerite were emplaced, while the subaerial Kirkpatrick Basalt was erupted locally into a subsiding basin (Bradshaw,

1987). During this period, Antarctica continued moving northward, away from the South Pole. Gondwana experienced continental conditions with the exception of its margin.

Antarctic Strata. To date, there are no recognized Jurassic-Cretaceous sedimentary rocks on the East Antarctic craton (Fig. 2.21). However, cores and dredge samples from the continental shelf have recovered Early Cretaceous coals and siltstones, containing palynomorph assemblages that are virtually identical to those of southeastern Australia (Domack, Fairchild, and Anderson, 1980). Similar deposits were recovered from the Mac. Robertson

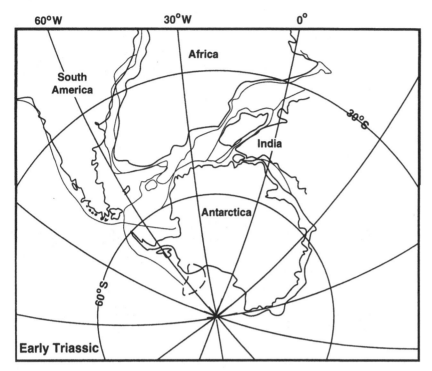

Figure 2.19. Early Triassic Gondwana paleogeography (modified from Lawver, Sclater, and Meinke, 1985).

Land continental shelf (O'Brien, personal communication). Outcrops of late Mesozoic strata occur in the southern part of the Antarctic Peninsula (Fig. 2.21), documenting marine sedimentation along an active margin (Dalziel, de Wit, and Palmer, 1974; Elliot, 1975a, 1983; Farquharson, 1982, 1983; Thomson, 1982, 1983).

Initiated during the Middle to Late Jurassic, the Andean mobile belt controlled sedimentation along the West Antarctic margin throughout the Late Jurassic and Cretaceous. By the Late Jurassic, the Antarctic Peninsula was a narrow magmatic arc bordered on the west by fore-arc basins where fluvial and marine deposits were laid down on an older subduction–accretion complex (Fig. 2.22). To the east (Weddell Sea side) of the magmatic arc, thick sections of volcaniclastic sediments accumulated in back-arc basins. Farther to the east, the landlocked proto–Weddell Sea, accumulated organic-rich mudstones (Fig. 2.23).

The LeMay Group, a probable accretionary complex (Thomson and Tranter, 1986; Tranter, 1987), contains fossils that appear to indicate an Early Jurassic age. A combination of arc-derived clastics and allochthonous oceanic units compose the group. The interbedded basalts and volcaniclastics may document the later-stage filling of a fore-arc basin (Suárez, 1976), prior to the oceanward migration of a volcanic island arc (Edwards, 1982).

In the northern Antarctic Peninsula (Hope Bay), Jurassic to Early Cretaceous conglomeratic strata of the Mount Flora Formation were deposited as alluvial fans. These

fans filled fault block basins (Elliot and Gracanin, 1983). Mount Flora deposits and possible equivalents in the South Orkney Islands (Spence Harbor Conglomerates) most likely formed along an arc extending the length of the Antarctic Peninsula (Fig. 2.22); rocks of the Antarctic Peninsula Volcanic Group represent the arc itself. Palmer Land formed part of a back-arc complex to the east of the arc (Weber and Livschitz, 1985). The alluvial fans and fluvial systems of the Mount Flora Formation could represent the initiation of transport between the volcanic arc of the Antarctic Peninsula and the eastern back-arc basin of Palmer Land (del Valle, Fourcade, and Medina, 1982). Distinct fore-arc and back-arc terrains existed (Fig. 2.22). Activity within the arc progressed from a Middle Jurassic inception in eastern Ellsworth Land to Late Jurassic activity that extended along the length of the peninsula (Elliot and Gracanin, 1983).

Deposits of the Andean fore-arc and/or intra-arc terranes are found in the greater than 5000 m of shallow marine deltaic mudstones, sandstones, conglomerates, and volcanics of the upper Jurassic to lower Cretaceous Fossil Bluff Formation on Alexander Island in the Antarctic Peninsula region (Taylor, Thomson, and Willey, 1979; Thomson, 1982) (Fig. 2.21). These deposits also occur as shallow marine shales, sandstones, and volcaniclastics of the Byers Formation of the South Shetland Islands (Valenzuela and Hervé, 1972; Smellie, 1981). Deep marine flysch-type sediments also occur in the South Shetland Islands, documenting variability in syndepositional upwarp in the area (Thomson, 1982).

The Late Jurassic Latady Formation and its equivalents in the southern Antarctic Peninsula (Fig. 2.21) consist of greater than 800 m of shallow marine fossil-bearing sediments as well as volcaniclastic siltstones, black mudstones, and shales with rare sandstones and very thin (< 2 cm) coal beds (Williams et al., 1972; Rowley and Williams, 1982; Rowley, Kellogg, and Vennum, 1985). These sediments filled the shallow basin (Palmer Land) behind the magmatic arc (Antarctic Peninsula) (Weber and Livschiz, 1985; Weber and Rank, 1987). Extensive back-arc deposits may be concealed beneath the Ronne Ice Shelf and on the continental shelf toward the east (Elliot, 1983).

A continental equivalent of the Latady Formation interfingers with a suite of subaerial calc-alkaline volcanic rocks deposited along the axis of a magmatic arc that existed on the English Coast (southern portion of Antarctic Peninsula) of the Bellingshausen Sea (Rowley, Kellogg, and Vennum, 1985). In the northern Antarctic Peninsula, greater than 550 m of radiolarian-rich black mudstone

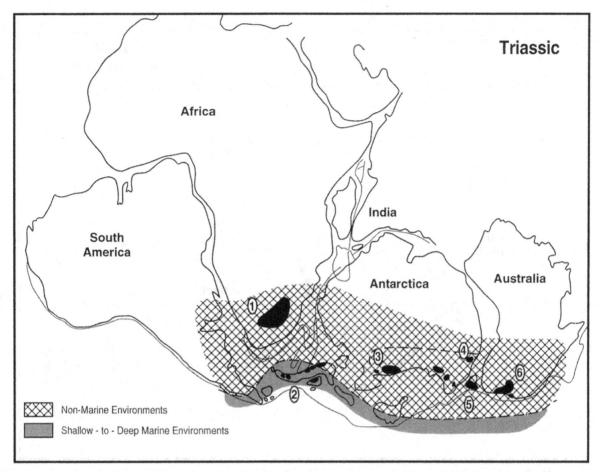

Figure 2.20. Known outcrop locations of Triassic sedimentary rocks (black) and paleogeography of Gondwana during that time. Strata are designated as: 1 = Beaufort and Stormberg groups of the Karoo Basin; 2 = Legoupil Fm and LeMay and Trinity Peninsula groups; 3 and 4 = Victoria Group–Fremouw and Falla Fm; 5 = fluvial and lacustrine strata of Tasmania; 6 = non-marine sandstones and shales of eastern Australia.

and thinly bedded ash layers of the upper Jurassic to lower Cretaceous Nordenskjöld Formation accumulated in back-arc basins extending along the margin of the James Ross Basin (Farquharson, 1982, 1983; Elliot, 1988) (Fig. 2.22). Citing the absence of epiclastic material within the Nordenskjöld Formation, Farquharson (1983) suggests that the magmatic arc was not emergent during deposition of these strata.

In summary, the Antarctic Peninsula was the site of an active volcanic arc during Late Jurassic to Early Cretaceous time (Fig. 2.22). Marine sedimentation occurred along this arc in fore-arc and back-arc settings. Deposition was predominantly within deltaic wedges (in part terrestrial) and deeper marine turbidite fans. A possible non-volcanic offshore source is suggested by transport indicators within fore-arc deposits (Thomson, 1982).

Strata of Other Gondwana Continents. By the Early Cretaceous, Antarctica and Africa were separated (Fig. 2.6), marking the inception of independent stratigraphies for these continents. A physical connection continued to exist in the Late Jurassic to Early Cretaceous between Antarctica, Australia, India, New Zealand, and South America, although continental rifting had begun along the Australian–Antarctica suture (Fig. 2.6). Probable time-equivalent strata on these margins include the non-marine to marine sandstones, shales, and limestones of the Cauvery, Palar,

and Godavaria-Krishna basins of the east Indian continental margin (Sahni, 1982) as well as similar non-marine to marine strata of the Mozambique, Zululand, and Natal basins of southeastern Africa (Dingle and Scrutton, 1974; Hobday, 1982; Tankard et al., 1982) (Fig. 2.22).

Fossil belemnites from the Jurassic Latady Formation of the Antarctic Peninsula show similarities to those of New Zealand, indicating continued marine contact (Mutterlose, 1986). Sharing the same active margin, sedimentation in southern South America and the Antarctic Peninsula region continued to be similar during this time.

The Magallanes Basin area of South America displays a depositional history comparable to that of retro-arc deposits of the Antarctic Peninsula (Dott, Winn, and Smith, 1982; Farquharson, 1982). Continental and marine sandstones (e.g., Springhill Formation; Fig. 2.21) were deposited on the submerged Jurassic volcanic terrane (Serie Tobifera/El Quemado Formation) overlying the Patagonian craton. Within the back-arc basin and the arc itself, flysch sediments (Punta Barossa, Yahgan, Tekenika, and

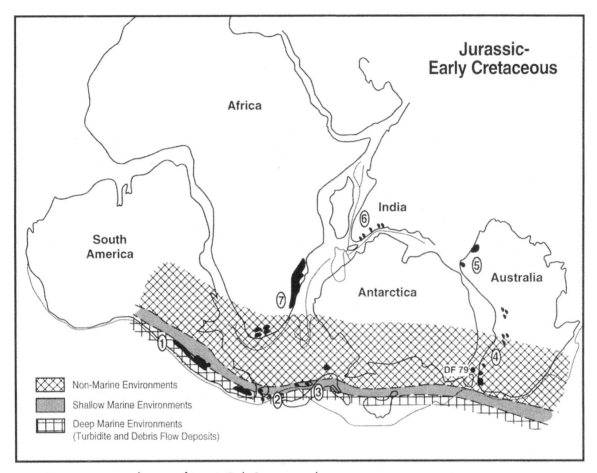

Figure 2.21. Known outcrop locations of Jurassic–Early Cretaceous sedimentary rocks (black) and generalized paleogeography for that time. Strata are designated as: 1 = Magallanes Basin and other Argentine basinal strata including the Punta Barossa, Springhill-Zapata, Comodoro Rivadavia, and Mina del Carmen, Tekenika, and Yahgan fms; 2 = South Georgia strata including the Annenkov Island, Cumberland Bay, and Sandebugten series; 3 = Antarctic Peninsula region strata including the Nordenskjöld, Mount Flora, Latady, Byers, and Fossil Bluff fms; 4 = Otway and Strzelecki groups of southeastern Australia; 5 = unnamed paralic/continental/shallow marine sandstones and shales of western Australia; 6 = non-marine and shallow marine deposits that occur in the Palar, Cauvery, and Godavaria-Krishna basins of southeastern India; 7 = South African strata including the Sena, Maputo, Mzinene, Kirkwood, Enon, Sundays River, and Makatini fms. The DF79 core location is the site where Early Cretaceous siltstones were recovered.

Hardy formations) accumulated to thickness exceeding 5 km, locally reflecting two independent source directions (Dott, Winn, and Smith, 1982). Black laminated shales of the Zapata Formation document euxinic conditions. These deposits are similar in age (Tithonian-Barremian) to those recorded in the Nordenskjöld Formation and related deposits of the Peninsula.

On the south Atlantic island of South Georgia, rocks of Jurassic to Early Cretaceous age consist of thick graywacke, shale, and mudstone sequences of the Sandebugten, Cumberland Bay, and Annenkov Island series (Elliot, 1975a; Tanner and MacDonald, 1982) (Fig. 2.21). The Cumberland Bay Series and the Sandebugten Series

consist of turbidite sandstones and mudstones (Tanner and MacDonald, 1982). The Cumberland Bay Series' thick sequence of volcaniclastic sandstone turbidites is interpreted as having been deposited in an elongate back-arc basin in an aggradational setting (MacDonald, 1986). The Annenkov Island Series consists of radiolarian-rich mudstones and tuffs, which purportedly originated in a back-arc setting relative to a Late Jurassic arc in the southern Andes (Katz, 1973; Dalziel, de Wit, and Palmer 1974).

Correlations by Farquharson (1983) of Late Jurassic to Early Cretaceous black laminated organic shales from DSDP sites and outcrop data from the Antarctic Peninsula, the Magallanes Basin, and other localities in the South Atlantic indicate the extent of an anoxic paleobasin (Fig. 2.23). LaBrecque and colleagues (1986) suggested that restricted circulation and anoxic conditions prevailed for 40 million years after initial rifting when a narrow seaway opened between East and West Antarctica, allowing an influx of oxygen-rich water. Organic shales were drilled in the Weddell Sea during Ocean Drilling Program (ODP) Leg 113 (Barker et al., 1988a).

In Australia, Jurassic and Early Cretaceous strata predominantly occur along the southern coast and continental margin (see summaries by Plumb (1979), Veevers and Evans, (1975), and Williams and Corliss (1982)) (Fig. 2.21). Middle to Early Jurassic rifting, documented by

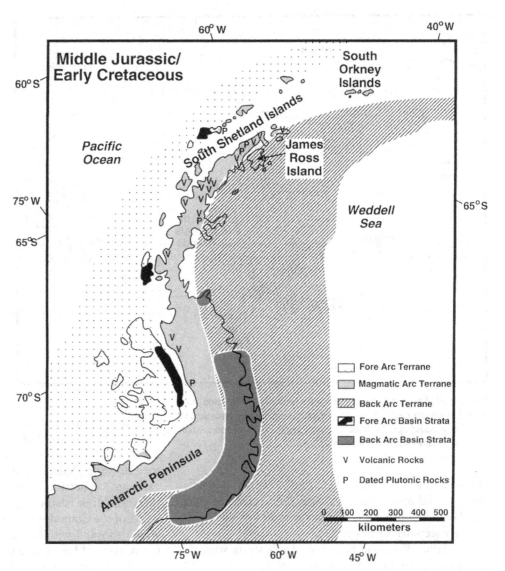

Figure 2.22. Paleogeographic map of the Antarctic Peninsula showing Andean lithotectonic units of Middle Jurassic to Early Cretaceous age (modified from Elliot, 1983).

dolerite extrusions, was followed by the development of rift basins such as the Ellison (Polda) Trough and Otway Rift Valley. Fluvial deposition began as early as the Late Jurassic in the Ellison Trough where continental clastics grade westward into marine carbonates. Fluvial deposition occurred by Early Cretaceous time in the Otway (Otway Group), Bass, Gippsland (Strzelecki Group), Eucla, and Murray basins; the latter two basins contain lower Cretaceous marine strata as well. Veevers and Evans (1975) suggest that deposition of these marine sediments occurred in response to the southward transgression from the Great Artesian Basin, although the marine Aptian strata of the Eucla Basin may indicate incursion along the incipient Antarctica-Australia rift system. Williams and Corliss (1982) date the first evidence of marine incursion along the margin as Cenomanian-Turonian in age.

Late Cretaceous Strata

Dispersal of Gondwana continents continued throughout Late Cretaceous time, fully separating South America, Africa, and India (Fig. 2.8). Antarctica and South Amer-

ica remained joined in the vicinity of the Drake Passage. The precise relationship between Antarctica and Australia is unclear, although separation may have been completed by this time. Subduction was occurring along the Pacific-Antarctic Margin, and uplift was initiated in the TAM. Antarctica became centered at or near the South Pole during this period.

Antarctic Strata. Upper Cretaceous strata are known in the James Ross and Seymour islands vicinity of the Antarctic Peninsula (Fig. 2.24), where they constitute the Marambio (Bibby, 1966; Rinaldi et al., 1978; Rinaldi, 1982) and the Gustav groups (Ineson, Crame, and Thomson, 1986). These strata result from the filling of the James Ross Basin, an euxinic, ensialic back-arc basin (Elliot, 1988).

The Marambio Group (Seymour Island) consists of 1400 m of strata including the Santa Marta, Lopez de Bertodano, Sobral, and Cross Valley formations (Olivero,

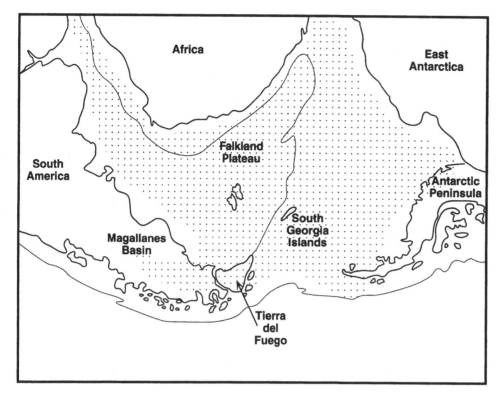

Figure 2.23. Reconstruction of a vast anoxic basin (dotted area) that existed in the present South Atlantic region during the Late Jurassic and Early Cretaceous (reprinted from Farquharson, © 1983, with permission by the Australian Academy of Science).

Scasso, and Rinaldi, 1986; Macellari, 1984). The depositional period of these strata crosses the Cretaceous–Tertiary boundary, providing one of the most complete sedimentary sections for this period.

The James Ross and Seymour islands strata were deposited in a rapidly subsiding back-arc basin (Fig. 2.24) with an areal extent in excess of 10,000 km^2 (del Valle, Fourcade, and Medina, 1982). Sediments record deltaic and estuarine conditions, the steady uplift of the Antarctic Peninsula, and the northwestward migration of the magmatic arc during latest Cretaceous time (Andean Orogeny) (del Valle, Fourcade, and Medina, 1982; Elliot, 1983).

Strata of Other Gondwana Continents. Strata in the Magallanes Basin of South America record upper Cretaceous (Santonian-Campanian) deep marine sedimentation (Cerro Toro and Cerro Matero formations) (Fig. 2.24). These deposits and overlying marine sandstones (Tres Pasos Formation) are interpreted to be part of a large back-arc fan system that prograded southeastward across the Magallanes Basin (Dott, Winn, and Smith, 1982). Shallow marine sediments (upper Tres Pasos and Dorotea formations) of Mastrichtian age document eastward-building deposition in shelf, littoral, and deltaic environments. Dott and others (1982) suggest these deposits correlate with the shallow marine Marambio Group strata of the Antarctic Peninsula region.

Late Cretaceous deposits also exist in other marginal basins of Argentina (Urien and Zambrrano, 1973) (Fig. 2.24). Red beds of probable non-marine origin occur in the Canelones, Salado, and Colorado basins, although an eastward connection to marine conditions may have been present in these basins. In the San Jorge Basin, non-marine strata of the Comodoro Rivadavia and Treból formations were deposited within a landlocked intracratonic basin. Correlation of Late Cretaceous strata between Antarctica and South America provides a general paleogeographic framework for the region, revealing the relationship between submarine fan and deltaic deposits along the Andean Orogeny (Fig. 2.24).

In the marginal basins of southern Australia, the Late Cretaceous was characterized by the progressively eastward development of marine conditions (Veevers and Evans, 1975; Williams and Corliss, 1982). Large sequences of fluvial, deltaic, and marine strata accumulated along the developing margin, separated from overlying Tertiary strata by a pronounced Late Cretaceous erosional unconformity.

Tertiary and Quaternary Strata

During the Tertiary, the continental fragments occupied positions similar to that of the present-day global configuration (Fig. 2.9). Antarctica, centered at the South Pole, separated from South America at the Drake Passage by the Miocene. Antarctica's continental margins were fully developed by the Late Tertiary. Although the precise timing is debated, geologic and thermal isolation resulted with the separation of South America and the Antarctic Peninsula, contributing to the development of a colder climate on the Antarctic continent (Kennett, 1977). Continental conditions prevailed across Antarctica at the beginning of the Tertiary.

Sedimentary strata of Tertiary age are found mainly in the northern Antarctic Peninsula region on James Ross and Seymour islands and as on King George Island in the South Shetland Islands. Birkenmajer (1981a,b, 1984, 1988) provides a review of Tertiary stratigraphy of King George Island. Zinsmeister (1982) and Elliot and Trautman (1982) describe Tertiary strata on James Ross and Seymour islands (Fig. 2.25). Further discussion of the Tertiary strata of Antarctica can be found in Chapter 6, "Antarctica's Glacial History."

Paleocene Strata. The Marambio Group of Seymour Island (northern Antarctic Peninsula) includes the Lopez de Bertodano Formation (Cretaceous to Paleocene) as well as Paleocene Sobral, Wiman, and Cross Valley formations (Macellari, 1984, 1988; Sadler, 1988). The Lopez de Bertodano Formation contains sandy, muddy siltstones deposited in a shallow marine environment (del Valle, Rial, and Rinaldi, 1987; Askin, 1988a,b; Macellari, 1988). The Sobral Formation consists of 225 m of laminated siltstones grading upward into clean cross-bedded, glauconitic sandstones deposited by a prograding delta (Macellari, 1988). Pebbly sandstones and mudstones of the 105-m-thick Cross Valley Formation record deposition within distributary channel and interdistributary environments of a Paleocene deltaic complex (Elliot and Trautman, 1982). Deposition of the Marambio Group and the overlying La Meseta Formation (Eocene) occurred in a progressively, or periodically, tilting basin (Sadler, 1988). Petrographic analyses have shown that these deposits were derived from a magmatic arc terrane, which is consistent with the inferred geologic history of the region (Elliot, Hoffman, and Rieske, 1992). Paleocene strata of King George Island include the Ezcurra Inlet Group (Paleocene). These deposits contain carbonaceous intervals (Birkenmajer, 1987) and a rich pollen-spore assemblage that reflects moist and warm climatic conditions (Stuchlik, 1981).

Eocene Strata. In the vicinity of Seymour Island, Tertiary strata compose the La Meseta Formation of the Seymour Island Group (Andersson, 1906; Elliot and Trautman, 1982; Zinsmeister, 1982). The 450-m-thick La Meseta Formation records shallow water deposition in a deltaic environment (Pezzetti and Krissek, 1986). The formation is subdivided into three units: sandstone/siltstone deposition within the prodelta (Unit I), tide-dominated delta

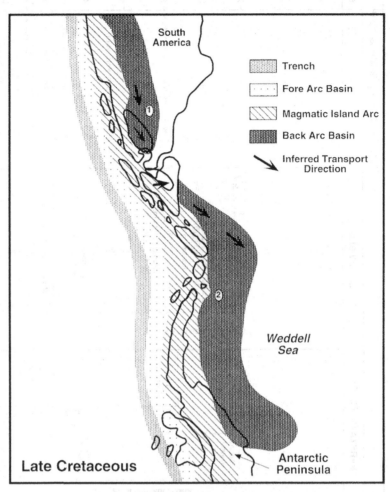

Figure 2.24. Known outcrop locations of Late Cretaceous sedimentary rocks and generalized paleogeography for that time. Strata are designated as: 1 = Magallanes Basin and other Argentine basinal strata including the Cerro Toro, Cerro Matrero, Tres Pasos, Dorotea, Comodoro Rivadavia, and Treból fms; 2 = Gustav and Marambio groups (compiled from Dott, Winn, and Smith, 1982, and Elliot, 1983).

front (Unit II), and protected lagoonal (Unit III) environments (Elliot and Trautman, 1982). Paleontologic work revealed a rich fossil assemblage (Feldmann, 1984; Pezzetti and Krissek, 1986; Doktor et al., 1988). Fossil plant remains include a diverse angiosperm-rich flora and indicate relatively warm conditions, although a shift toward *Nothofagus*-dominated vegetation near the end of the Eocene reflects climatic cooling at this time (Askin, 1992). Deposition of the La Meseta Formation is thought to reflect unroofing of the metamorphic–plutonic complex adjacent to a volcanic arc (Elliot and Trautman, 1982). This formation represents the youngest exposed beds contained in the late Mesozoic to early Cenozoic James Ross basin, and records back-arc deposition along the flank of the Andean cordillera (Elliot, 1988).

Eocene deposits of King George Island include the Fildes Peninsula (Eocene?), Ezcurra Inlet, Point Hennequin, and Dufayel Island groups (Fig. 2.25). These deposits

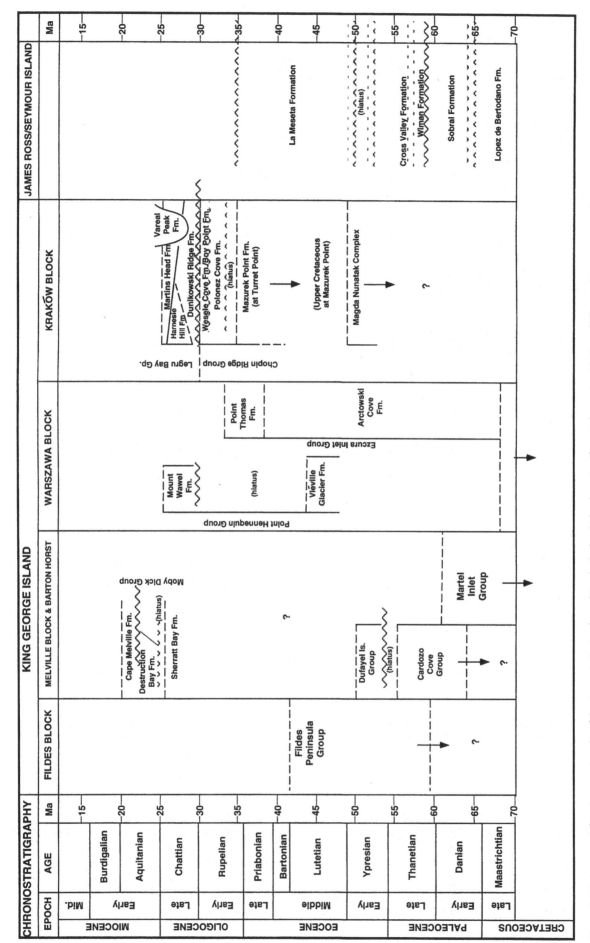

Figure 2.25. Summary of Tertiary stratigraphy of the Antarctic Peninsula and South Shetland Islands region (from Birkenmajer, 1990, and Elliot and Trautman, 1982).

contain carbonaceous intervals (Birkenmajer, 1986, 1987) and a pollen-spore assemblage that reflects warm and moist climatic conditions (Stuchlik, 1981; Czajkowski and Rosler, 1986; Lyra, 1986) in a lowland forest setting (Lyra, 1986). Associated glacial deposits are exposed at Magda Nunatak on King George Island and are believed to represent a widespread Eocene glaciation, the Kraków Glaciation of Birkenmajer, Soliani, and Kawashita (1989). The glacial deposits are overlain by hyaloclastites alternating with flow basalts. The flow basalts have yielded a K-Ar (potassium-argon) age of 49.4 ± 5 Ma (Birkenmajer et al., 1986). An Eocene age for these deposits is supported by coccoliths that occur in the matrix of the hyaloclastite (Birkenmajer, 1991). The apparent absence of glacial deposits in the Eocene strata of Seymour Island, located farther south, may imply that the Eocene glacial deposits of King George Island reflect localized alpine glaciation.

In Admiralty Bay on King George Island, exposures of the Arctowski Cove Formation include freshwater shales, tuffaceous sandstones, conglomerates, and a thin coal seam (Birkenmajer, 1990). A basalt flow overlying this formation yielded a K-Ar age of 37.4 ± 1.1 Ma (Birkenmajer et al., 1986). The *Nothofagus*-pteridophyte spore-pollen assemblage from this formation indicates temperate climatic conditions.

Oligocene Strata. The climate of the Antarctic Peninsula region began to cool by the Oligocene, as indicated by more temperate pollen-spore assemblages within the Polonia Glacier and Point Hennequin groups on King George Island (Birkenmajer, 1987) and probable Oligocene strata of Seymour Island (Elliot and Trautman, 1982). Glacial deposits of the King George Island region belong to the Polonez Cove Formation and include glacial, marine, and volcanic strata deposited in the vicinity of an active magmatic arc (Birkenmajer et al., 1987; Birkenmajer, Soliani, and Kawashita, 1988; Porebski and Gradziński, 1987). The "Polonez Glaciation" was followed by a later (Late Oligocene) glacial episode, the Legru Glaciation (Birkenmajer, Soliani, Kawashita, 1989).

Earlier discussions placed the onset of glaciation in the Antarctic Peninsula region closer to the Oligocene–Miocene boundary (Birkenmajer, 1985, 1987). Gazdzicki (1987) suggested that glaciation was actually initiated in the late Paleocene on the South Shetland Islands. The Pecten Conglomerate, lower member of the Polonez Cove Formation, is a fossiliferous glacial-marine unit. Examination of coccolith assemblages and dating of overlying basalts (Muzurek Formation) yielded an Oligocene age for the Pecten Conglomerate, a unit originally believed to be Pliocene in age (Gazdzicka and Gazdzicki, 1985a,b;

Birkenmajer and Gazdzicki, 1986). The Oligocene glaciers of King George Island existed in a temperate climate and are associated with interglacial deposits that are partially fluvial in origin.

Glacial deposits of the Polonez Cove Formation are deeply incised by fluvial channels belonging to the Wesele Cove Formation, which marks a change to relatively warm, interglacial conditions at this time (Birkenmajer, 1988). The exact age of the Wesele Cove Formation is poorly constrained. These interglacial deposits are overlain by interbedded tills and volcaniclastic debris flows of the Legru Bay Group, indicating another episode of glaciation in the region, the Legru Glaciation (Birkenmajer, 1980). The age of the Legru Bay Group is constrained by K-Ar dates to ~ 30 to 29 Ma (Birkenmajer et al., 1986).

Late Oligocene deposition in the King George Island region is recorded by interbedded basalts and plant-bearing sedimentary deposits of the Point Hennequin Group. Plant fossils in the Mount Wawel Formation consist of a *Nothofagus*-podocarp assemblage indicative of cool temperate climatic conditions (Birkenmajer, 1988). The K-Ar age of the Mount Wawel Formation ranges from 28.3 ± 1.7 to 24.5 ± 0.5 Ma (Birkenmajer et al., 1986).

Oligocene-Miocene Strata. The Moby Dick Group of King George Island consists of fossiliferous marine and glacial-marine deposits (Biernat, Birkenmajer, and Popiel-Barczyk, 1985; Birkenmajer, 1987; Birkenmajer and Luczkowska, 1987). K-Ar dating of dikes indicated an Early Miocene (22 Ma) age for the Moby Dick Group (Gazdzicki and Wrona, 1986; Birkenmajer, Soliani, and Kawashita, 1989). However, examination of planktonic foraminifera from the Cape Melville Formation, part of the Moby Dick Group, suggests a Late Oligocene age for the Moby Dick Group (Abreu, Savini, and Barrocas, 1992).

Glacial deposits in the upper unit of the Cape Melville Formation display a large quantity of ice-rafted debris, often containing Cretaceous fossils. The source for much of the ice-rafted material in the glacial strata was the Antarctic Peninsula region (Birkenmajer and Butkiewicz,1988). These glacial-marine strata cap the shallow marine deposits of the Destruction Bay Member (Cape Melville Formation) (Birkenmajer et al., 1987), which, in turn, overlie a basaltic tuff dated as Early Miocene (Birkenmajer et al., 1988).

Pliocene and Pleistocene Strata Plio-Pleistocene strata are widely scattered in Antarctica and include a wide range of glacial and interglacial deposits. Discussion of Plio-Pleistocene strata and their implications with regard to Antarctica's climatic and glacial history is deferred to Chapter 6, "Antarctica's Glacial History."

CHAPTER THREE

Continental Shelf Geomorphology and Relief-Forming Processes

Those who have studied the Antarctic from ships are aware of the inaccuracy of bathymetric charts, especially for the continental shelf. A general "rule of thumb" for ship captains is to avoid waters less than 100 m in depth. Several ships ignoring this policy have met with disaster, most recently the Argentine tour vessel, *Vaja Parizo*.

Bathymetric data are scarce to absent in many areas, and existing data may date back to the early part of the twentieth century. Even more recent bathymetric charts include data with poor navigational control. Extremely rugged topography, including rock pinnacles that rise to within meters of the sea surface, characterize the inner shelf and inland waters, and these are especially poorly charted. It is not uncommon to discover that an island or the coastline is located 1 to 2 km away from its charted position.

Until the late 1980s, ships working in Antarctic waters relied on satellite navigation systems and older forms of navigation. Accuracy varied from hour to hour and from year to year, depending on the number and locations of satellites in orbit. At their very best, these systems did not provide the accuracy needed to map bathymetric irregularities on the inner continental shelf. This has changed with the advent of the Global Positioning System (GPS), but data acquisition is slow. It will be many years before reliable charts are available, particularly for ice-covered regions.

Several countries have their own bathymetric charts that tend to emphasize areas where research stations are located. Attempts to integrate existing data from different sources have been directed toward compiling regional bathymetric maps. These maps provide the primary data set for examining Antarctica's seafloor relief. Bathymetric data from many expeditions to Antarctica have been compiled in the form of a 1:6,000,000 bathymetric chart, the General Bathymetric Chart of the Oceans (GEBCO maps) (Scott, 1981). Hayes (1991) presents a more recent regional compilation of bathymetric data (1:12,000,000). More detailed regional bathymetric maps have been published by a number of different groups (Table 3.1). The locations of these maps are shown in Figure 3.1. Some of these maps are referred to in the following text to illustrate the physiography and relief-forming processes of the margin.

The more unusual features of the Antarctic continental shelf include its great depth, irregular topography, and landward-sloping profile (Fig. 3.2). The average shelf depth is 500 m (Johnson, Vanney, and Hayes, 1982), approximately eight times the world average (Shepard, 1973). Outer shelf depths average 400 to 500 m, and inner shelf depths are typically to 1200 m. The East Antarctic continental shelf is relatively narrow. The West Antarctic continental shelf is broad on average, and large portions of the shelf are covered by ice shelves or by the WAIS (Fig. 3.1).

This chapter is a brief discussion on isostatic models, which allow for estimation of the magnitude of isostatic depression of the continental shelf. The chapter is also a review of the evidence for glacial erosion of the shelf. It provides a brief overview of geomorphic features on the shelf, and uses several regional examples to illustrate the key features of the shelf and some regional differences in shelf physiography.

WHY IS THE CONTINENTAL SHELF SO DEEP?

Holtedahl (1929) and Shepard (1931) first emphasized the unique morphologic character of high-latitude continental shelves (great depth, rugged topography, and landward-sloping profile), and attributed it to glacial erosion. Others have argued that this unusual topography is due, in large part, to isostatic downwarping of the continental shelf in response to the great weight of the adjacent ice sheet (Zhivago and Lisitsin, 1957). What has been the relative influence of glacial isostasy and erosion in creating the overdeepened and foredeepened topography of the Antarctic continental shelf?

Isostasy

Under the great weight of its ice sheet, the Antarctic continent is depressed in such a manner that the continental surface is bowl-shaped. This isostatic depression extends beyond the coast and influences the continental shelf.

The question of how much isostatic depression is contributing to the great depth and landward gradient of the Antarctic continental shelf can be addressed in two ways. One approach is to calculate the amount of isostatic depression at the ice margin using isostatic equations. Another method involves calculating the degree of isostatic disequilibrium by using gravity data.

Drewry (1983) constructed a map that shows how the Antarctic continent would look if the ice sheet were removed and the continent allowed to adjust isostatically. These reconstructions are based on the assumption that the Earth contains a viscous fluid asthenosphere enclosed by a uniform elastic lithosphere (Brotchie and Silvester, 1969). The calculated magnitude of isostatic depression varies from just under 1 km beneath the central portions of East Antarctica to 0.5 km beneath West Antarctica. Greischar and Bentley (1980) noted that the mean isostatic gravity anomalies in the Ross Sea region are strongly negative near the boundary between West Antarctic inland ice and the Ross Ice Shelf. They interpreted this as being due to a glacio-isostatic imbalance that remained after retreat of the ice sheet from the Ross Sea continental shelf during the Holocene. Their model further indicates that an additional 100 m of uplift will occur before isostatic equilibrium is reached in this region.

Drewry's (1983) rebound map shows ~ 150 m of continental shelf depression in regions where the ice sheet profile is steepest, such as off the Wilkes Land Coast of East Antarctica. This is in close agreement with calculations made by Walcott (1970). Walcott's (1970) results indicate that a continental shelf seaward of a "full-bodied" ice sheet, like that of East Antarctica, will depress about 180 m near the grounding line with a proglacial isostatic depression extending ~ 180 km seaward of this point (Fig. 3.3).

The amount of isostatic depression derived by Drewry (1983) and Walcott (1970) is insufficient to account for the great depth of the Antarctic continental shelf. It is, however, important to recognize that these isostatic models assume that crustal structure and rock properties (mainly density) are uniform across the region under consideration. They do not take into account the possibility of differential subsidence along major structural features or across regions with heterogeneous rock properties and/or different sediment-crustal thickness. The Antarctic continental shelf, like other continental shelves, exhibits major differences in sediment and crustal thickness, basement type, and large tectonic features.

It is therefore reasonable to assume that glacial isostatic depression of the shelf will not conform to the predicted

Table 3.1. Continental shelves with detailed bathymetric maps. See Figure 3.1 for area locations.

Area	Source
Ross Sea	Hayes and Frakes, 1975; Vanney, Falconer, and Johnson, 1981
Sulzberger Bay	Lepley, 1966; Anderson, 1983a
Pennell Coast	Duphorn, 1981; Brake and Anderson, 1983
Antarctic Peninsula shelf	Vanney and Johnson, 1976; Bart and Anderson, 1995
Marguerite Bay	Kennedy and Anderson, 1989
Inland Passage of the Antarctic Peninsula (excluding Bransfield Strait)	Griffith, 1988
Bransfield Strait	Ashcroft, 1974; Jeffers and Anderson, 1990; Grácia, Canals, Farran, Prieto, Sorribas, and GEBRA Team, 1996
Weddell Sea	Johnson, Vanney, Elverhøi, and LaBrecque, 1981; Miller, De Batist, Jokat, Kaul, Steinmetz, Uenzelmann-Neben, and Versteeg, 1990a; Sloan, Lawver, and Anderson, 1995
Wilkes Land shelf	Grinnell, 1971; Vanney and Johnson, 1979; Chase, Seekins, Young, and Eittreim, 1987; Domack and Anderson, 1983
Prydz Bay	O'Brien, Truswell & Burton, 1994
Lützow-Holm Bay	Moriwaki, 1979

configuration. Further, the results of isostatic modeling show that the great depth of the continental shelf cannot be attributed entirely to isostasy, because isostasy, on average, can account for only ~ 25% of the inner shelf depths in most areas. It is also noteworthy that overdeepened and landward-sloping shelf topography characterizes even those portions of the continental shelf where there are no large ice sheets on the adjacent continent, such as the continental shelf of the northern Antarctic Peninsula (Pope and Anderson, 1992) (Fig. 3.2A). Therefore, the great depth of the Antarctic continental shelf must, for the most part, be due to glacial erosion (Anderson, 1991; ten Brink and Cooper, 1992).

Glacial Erosion

Direct evidence of deep glacial erosion on the Antarctic continental shelf is seen in virtually every seismic profile that crosses the shelf (Fig. 3.4). The erosional surfaces range in scale from shelf-wide surfaces covering thousands of square kilometers to smaller troughs that are tens to hundreds of square kilometers in area; the latter are associated with individual outlet and valley glaciers.

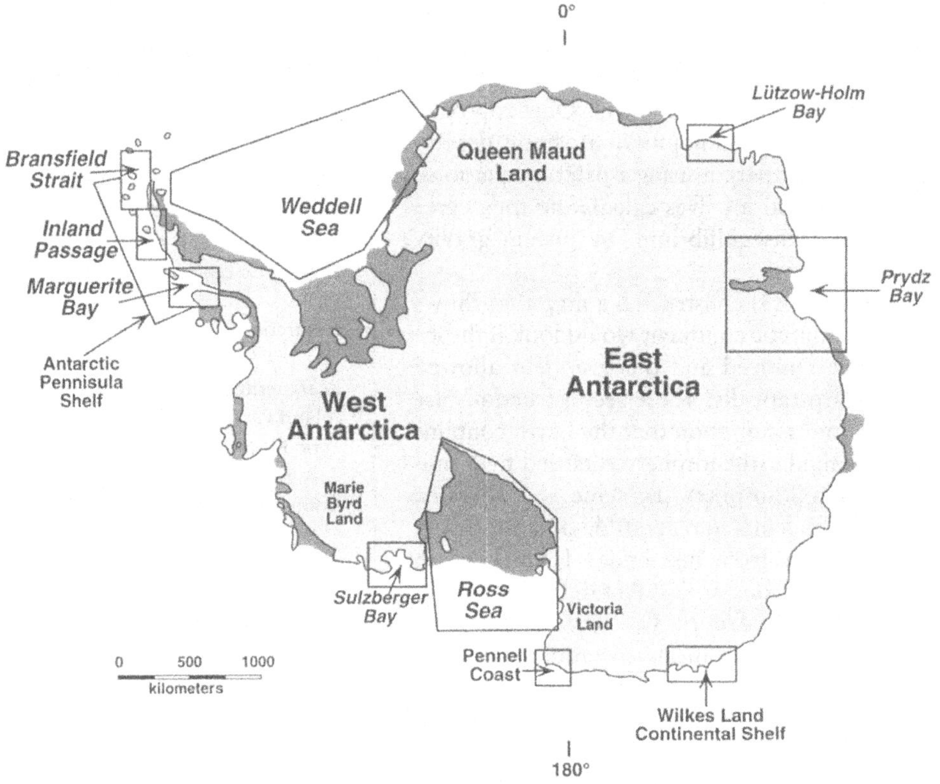

Figure 3.1. Locations of regions of detailed bathymetry maps described in text. Regional bathymetric surveys exist for the entire Antarctic margin. Shaded areas are ice shelves. (See Table 3.1 for more information.)

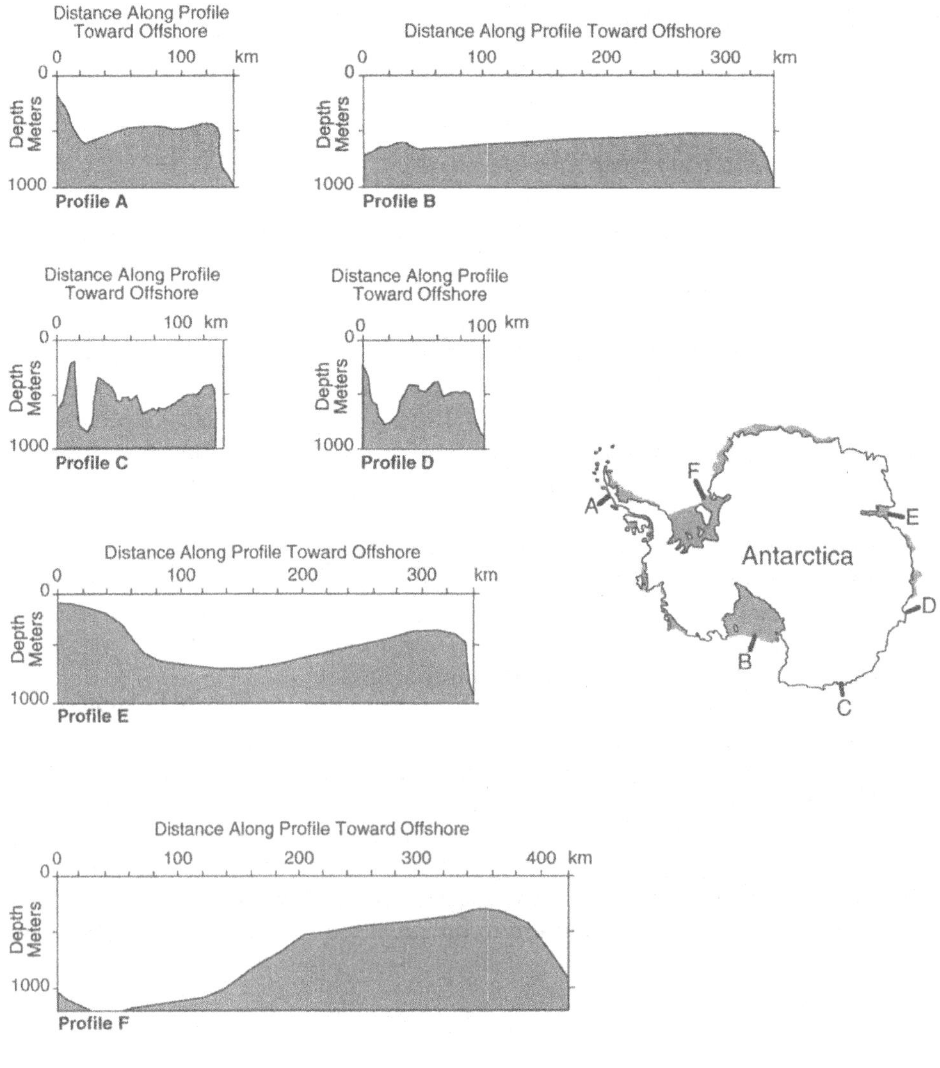

0°

Lützow-Holm Bay

Queen Maud Land

Bransfield Strait

Weddell Sea

Inland Passage

East Antarctica

Prydz Bay

Marguerite Bay

Antarctic Pennisula Shelf

West Antarctica

Marie Byrd Land

Ross Sea

Sulzberger Bay

Victoria Land

0   500   1000
kilometers

Pennell Coast

Wilkes Land Continental Shelf

180°

Distance Along Profile Toward Offshore

0   100   km

Depth Meters

0

1000   **Profile A**

Distance Along Profile Toward Offshore

0   100   200   300   km

Depth Meters

0

1000   **Profile B**

Distance Along Profile Toward Offshore

0   100 km

Depth Meters

0

1000   **Profile C**

Distance Along Profile Toward Offshore

0   100 km

Depth Meters

0

1000   **Profile D**

Figure 3.2. Representative bathymetric profiles from the West (A-B) and East (C-F) Antarctic continental shelves illustrating the great depth, irregular relief, and landward-sloping profile of the shelf.

Distance Along Profile Toward Offshore

0   100   200   300   km

Depth Meters

0

1000   **Profile E**

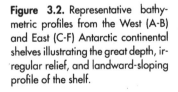

F

A

E

Antarctica

D

B

C

Distance Along Profile Toward Offshore

0   100   200   300   400   km

Depth Meters

0

1000   **Profile F**

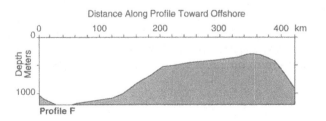

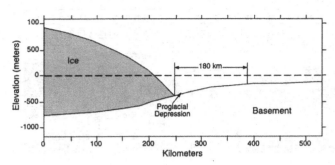

Figure 3.3. Walcott's isostatic model of a continental shelf seaward of a "full-bodied" ice sheet. The proglacial isostatic depression extends ~ 180 km seaward of the grounding line (modified from Walcott, 1970).

Individual glacial unconformities are not always conspicuous on dip-oriented seismic profiles (Fig. 3.4), but strike-oriented seismic profiles leave little doubt that ice sheets have deeply eroded into the shelf. The result is crosscutting unconformities and abrupt lateral truncation of units (Fig. 3.5). Strike-oriented seismic profiles also show that relief on individual glacial unconformities mimics relief on the modern seafloor; individual glacial unconformities typically cut many tens to hundreds of meters into underlying strata, and individual troughs are typically many tens of kilometers in width (Fig. 3.5). Thus, the scale of these unconformities is far greater than fluvial incisions of large rivers such as the Mississippi River (Berryhill, Suter, and Hardin, 1986). In this regard, these glacial unconformities are truly unique seismic stratigraphic surfaces.

Seismic profiles from the central Ross Sea and the northern Antarctic Peninsula continental shelves are used to il-

lustrate the manner in which repeated advance of ice sheets across these continental shelves has resulted in their great depth, rugged bathymetry, and landward-sloping profiles (Fig. 3.4). The offlapping stacking pattern of strata is typical of the Antarctic continental shelf and results from repeated episodes of glacial erosion and deposition. Each time the ice sheet advances across the shelf it erodes older strata on the inner shelf and recycles this material to the outer shelf. More frequent grounding events on the inner shelf have resulted in more pronounced erosion. Amalgamated glacial unconformities characterize the inner shelf, whereas the accumulation of a seaward thickening wedge of sedimentary deposits characterizes the outer shelf (Anderson, 1991; ten Brink and Cooper, 1992; ten Brink, Schneider, and Johnson, 1995). Around most of the continental margin, the inner shelf has been stripped of its sedimentary cover to expose crystalline basement rocks and older sedimentary deposits (Fig. 3.4). The most recent glacial unconformity lies near the seafloor. It is commonly sampled by piston cores and is imaged only by very high-resolution seismic methods. Glacial unconformities provide conclusive evidence that the Antarctic Ice Sheet has, on several occasions, advanced onto the Antarctic continental shelf.

Figure 3.4. Dip-oriented seismic profiles from the (A) Ross Sea (Profile PD90-30) and (B) Antarctic Peninsula (Profile PD88-02) shelves are used to illustrate the offlapping stratal patterns that resulted from repeated ice sheet grounding events on the shelf. Arrows designate larger glacial unconformities. The number of glacial unconformities are underrepresented in dip-oriented sections: glacial unconformities are more conspicuous in strike-oriented profiles (see Fig. 3.5).

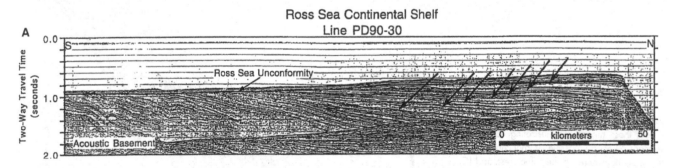

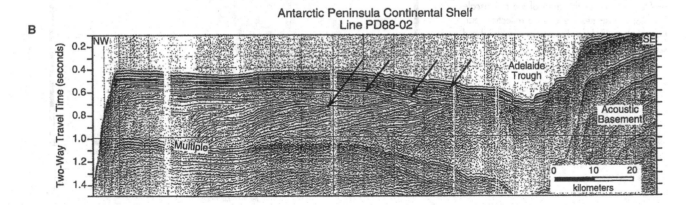

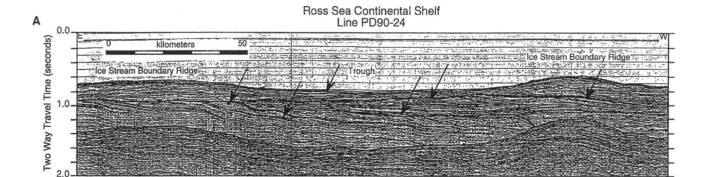

Ross Sea Continental Shelf
Line PD90-24

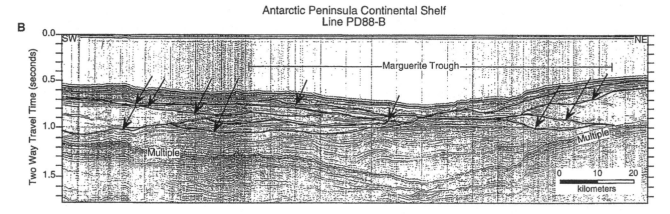

Antarctic Peninsula Continental Shelf
Line PD88-B

Figure 3.5. Strike-oriented seismic profiles from the (A) Ross Sea (PD90-24) and (B) Antarctic Peninsula continental shelves (PD88-B) are used to illustrate the broad-scale cross-cutting relationships caused by glacial erosion on the shelf. The glacial erosion surfaces (designated by arrows) seen on these strike-oriented lines are more conspicuous than on the intersecting dip-oriented lines (see Fig. 3.4). The most recent glacial unconformity is situated just below the seafloor and within the bubble pulse in these records.

Figure 3.6. Multibeam mosaic showing (A) megascale lineations on the Ross Sea continental shelf; (B, opposite) a Chirp subbottom profile across these megaflutes illustrating hummocky seafloor relief and massive internal character of the subbottom beneath these features; and (C, opposite) a portion of a high-resolution seismic profile that crosses the area of the megascale lineations illustrates the chaotic reflection pattern within a till sheet directly beneath the fluted seafloor.

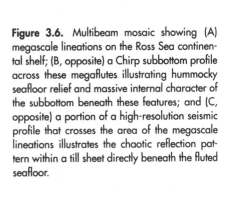

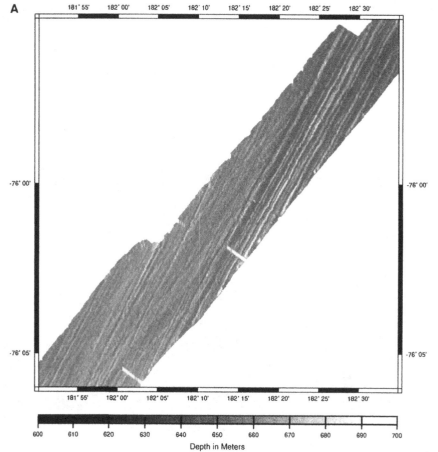

Figure 3.6. (*Continued*)

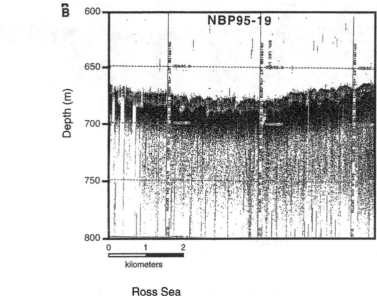

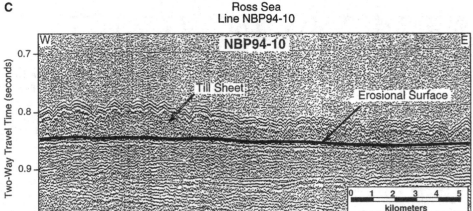

GEOMORPHOLOGY

A highly generalized profile of the Antarctic continental shelf would show rugged relief on the inner shelf, where the ice sheet sculptured bedrock, to a more smooth profile on the outer shelf where the ice sheet advanced across the seaward-thickening wedge of sedimentary strata (Fig. 3.4B). The dominant large-scale geomorphic features on the shelf consist of glacial troughs and ridges and banks between these troughs. The troughs fall into two general groups. Longitudinal troughs trend parallel or slightly oblique to the coast. We will see that these longitudinal troughs represent preferential glacial erosion along geologic boundaries, most commonly the contact between more resistant basement rocks and more easily eroded sedimentary deposits. Figure 3.4B shows a good example in the case of the Adelaide Trough, off the coast of the Antarctic Peninsula. The East Antarctic continent is characterized by extensive longitudinal troughs, which early explorers thought were part of a continuous circum Antarctic trench. We now know that these troughs represent deep glacial erosion within rift basins around this passive margin.

Transverse troughs extend across the shelf and occur where more rapidly flowing converging ice, or ice streams, existed on the shelf during previous glacial maxima. These tend to be more narrow and have V-shaped profiles on the inner shelf where the flow of the ice was guided by geologic features within the basement rocks. As the ice streams flowed across the inner shelf onto the sedimentary wedge of the outer shelf, the flow spread out to produce broader troughs with more U-shaped cross sections (Fig 3.5).

Ridges separate troughs on deeper portions of the continental shelf. The ridge crests are generally rounded and occur at depths of greater than 400 m (Fig. 3.5A). Flat-topped banks commonly exist between troughs on the outer shelf. Troughs on the outer shelf occur at depths between 300 and 100 m. Seismic profiles across banks show that they generally are composed of sedimentary strata that have been planed by glacial erosion. The flanks of these banks provide excellent locations for coring these older strata. The banks are believed to reflect erosion by ice shelves grounded on them during the late stages of ice sheet retreat from the continental shelf (Shipp and Anderson, 1994; Shipp, Anderson, and Domack, in press) (Fig. 3.5B). Modern ice rises, such as Roosevelt Island beneath

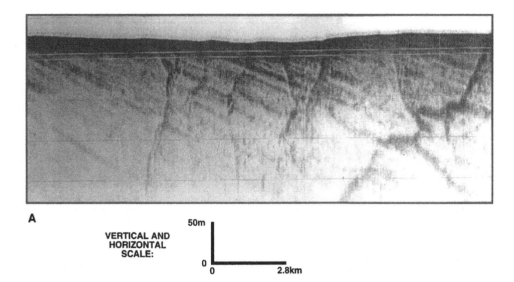

A

VERTICAL AND
HORIZONTAL
SCALE:

50m

0

0   2.8km

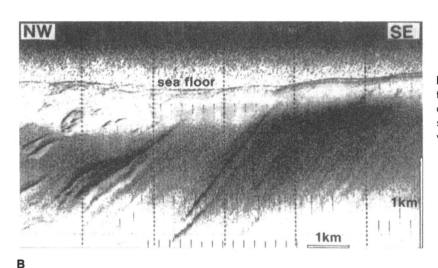

B

Figure 3.7. Side-scan sonar record showing striations that extend along the truogh axis and in the
direction of ice flow. (B) Side-scan sonar record
showing flutes on the Antarctic Peninsula shelf (provided courtesy of Dr. C. Pudsey.)

the Ross Ice Shelf and Berkner Island beneath the Filchner Ice Shelf, are examples of bathymetric highs that are
currently being eroded by these ice shelves.

Subglacial and Grounding Zone Features

The Antarctic seafloor is riddled with geomorphic features formed beneath the ice sheet. Typically, bottom
profiler records from the continental shelf show a rugged
seafloor produced by subglacial processes and/or grounded
icebergs. Smooth seafloors occur in shelf basins where
glacial-marine sediments are accumulating rapidly enough
to mask glacial furrows.

Until fairly recently, few systematic studies of the geomorphic features on the Antarctic seafloor had been conducted, and there was an absence of multibeam (swath
bathymetry) data on the shelf. The first geomorphic analyses relied on comparisons of bottom profiler records with
sparse side-scan sonar records, and on comparisons of
these data with Northern Hemisphere records.

This situation has recently changed. During R/V
Nathaniel B. Palmer's 1995 and 1998 field seasons
(NBP95-01 and NBP98-01 cruises), several thousand
kilometers of multibeam records and several hundred kilometers of deep-tow side-scan sonar data were acquired
in the Ross Sea and on the continental shelf offshore of
Northern Victoria Land. These data are augmented by an
extensive high-resolution seismic data set and over 200
sediment cores acquired during several marine geologic
expeditions during the past 30 years. These combined data
sets, augmented by studies of the Wilkes Land continental shelf (Barnes, 1987), Antarctic Peninsula continental
shelf (Pudsey, Barker, and Larter, 1994), and Prydz Bay
continental shelf (O'Brien, Truswell, and Burton, 1994;
O'Brien and Harris, 1996), provided the primary basis
for the following discussion of geomorphic features on
the continental shelf.

Megascale Glacial Lineations, Striations, and Flutes. Multibeam records show that the Ross Sea continental shelf
is virtually everywhere covered by subglacial scars or

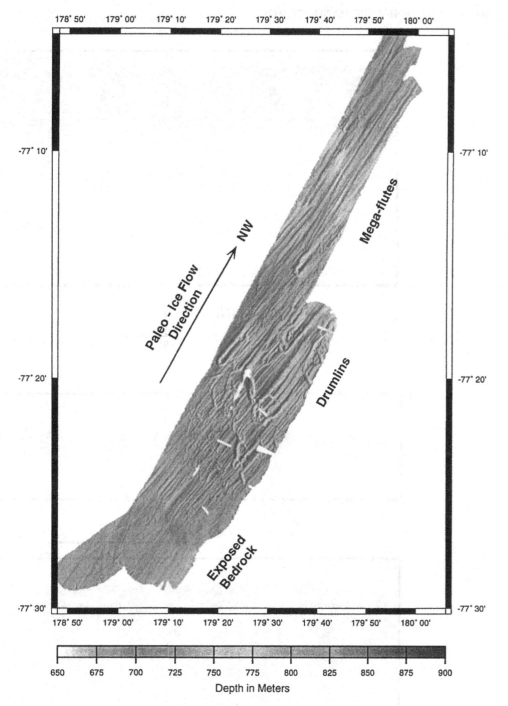

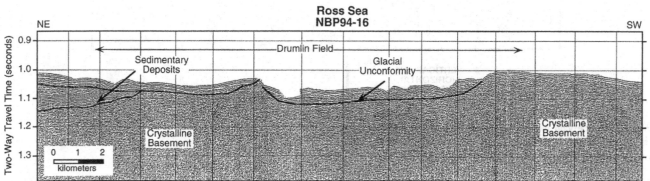

Figure 3.8. (A) Multibeam mosaic showing a drumlin field on the Ross Sea floor, and (B) a segment of a high-resolution seismic profile that extends through the drumlin field in the direction of elongation of the drumlins.

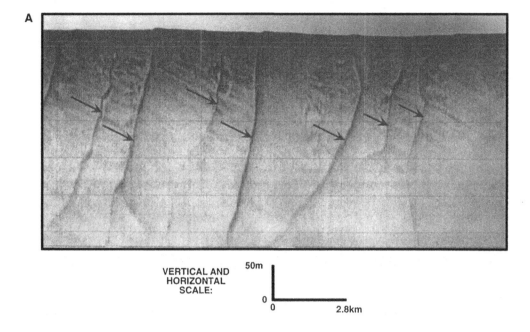

VERTICAL AND
HORIZONTAL
SCALE:

50m

0

0   2.8km

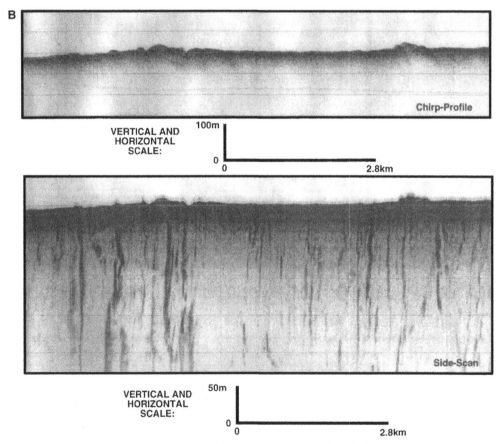

VERTICAL AND
HORIZONTAL
SCALE:

100m

0

0   2.8km

VERTICAL AND
HORIZONTAL
SCALE:

50m

0

0   2.8km

Figure 3.9. Side-scan sonar records illustrating (A) morainal wedges, designated by arrows, with superimposed striations, and (B) morainal ridges imaged by chirp profile and side-scan sonar record.

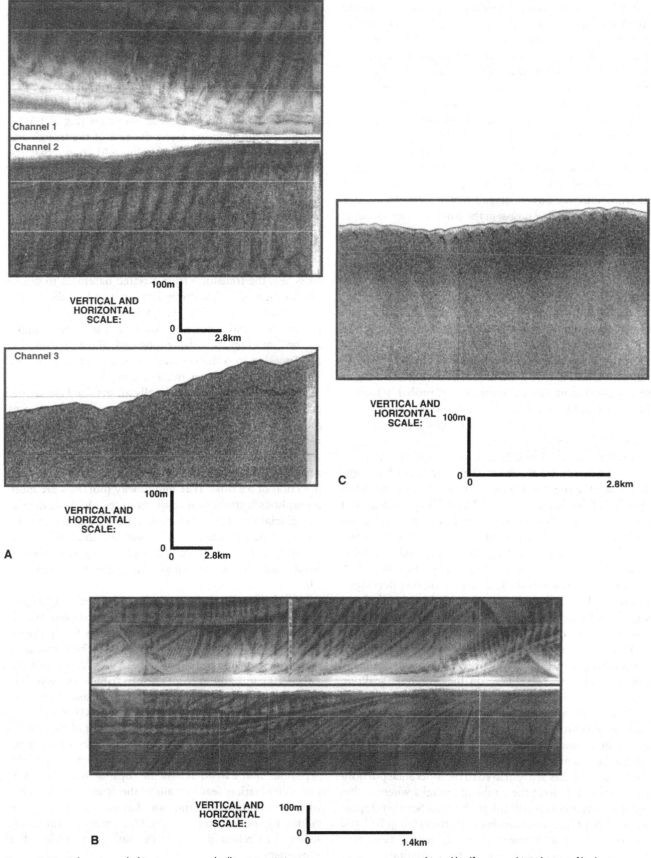

Figure 3.10. Side-scan and chirp sonar records illustrating (A) wave-shaped bedforms interpreted as washboard moraines formed as the ice sheet retreated at a constant rate, (B) large iceberg furrows and striations cutting across wave-shaped bedforms, and (C) chirp profile showing partially buried wasboard moraine.

depositional features, the main exception being banks that
are riddled with iceberg furrows. The most widespread
subglacial features are huge lineations that occur within
and along the flanks of troughs (Fig. 3.6A). These features
are relatively straight, oriented roughly parallel to trough
axes, ~ 100 to 200 m wide, up to 70 km long, and have
relief on the order of tens of meters. They are interpreted as
megascale glacial lineations (Shipp and Anderson, 1997)
based on their similarity to glacial landforms observed in
Landsat images of Canada (Clark, 1993). The 3.5-kHz
subbottom profiler records that cross these features show
a hummocky seafloor with massive internal character (Fig.
3.6B), and high-resolution seismic profiles show a chaotic
internal reflection pattern (Fig. 3.6C).

The megascale lineations on the Ross Sea shelf occur at
the top of a massive glacial unit that occurs within troughs
(Shipp et al., 1994; Shipp, Anderson, and Domack, in
press) (Fig. 3.6C). Piston cores indicate that the upper part
of this unit is composed of "soft" diamicton that is interpreted as a deformation till (Anderson et al., 1992a;
Shipp, Anderson, and Domack, in press). This interpretation is consistent with that of other researchers who have
argued that megaflutes and megascale glacial lineations
are formed by mobile deforming ice (Clark, 1993; Bennett and Glasser, 1996). Clark (1993) even goes so far as
to suggest that megascale lineations are probably formed
beneath rapidly flowing ice streams, which seems to be the
case for the Ross Sea features (Shipp, Anderson, and
Domack, in press).

Another set of strong lineations, or striations, is observed primarily on the inner shelf portions of troughs
and along the edges of banks and troughs in the Ross Sea
(Fig. 3.7A), on the inner shelf off North Victoria Land, and
on the outer shelf of the Antarctic Peninsula continental
shelf. These are erosional features that are on average less
than 30 m wide and tens of kilometers long and have relief
of less than 3 m. In the Ross Sea and on the North Victoria Land continental shelf, striations are best developed
in areas where the ice sheet has eroded into underlying
strata and/or crystalline bedrock, or where the ice sheet
has overridden older moraines.

Flutes are fairly uncommon on the Antarctic Peninsula
continental shelf (Fig. 3.7B) and, to my knowledge, have
not been observed elsewhere on the Antarctic seafloor.
They are common features on the northern Barents Sea
shelf and are interpreted as having been formed on the
surface of a deforming bed (Solheim et al., 1990). It is not
apparent to me why flutes are so rare in the Ross Sea,
where there is abundant evidence for a deforming bed. In
the Ross Sea, flutes are restricted to the inner shelf portions
of troughs and along the flanks of troughs where a thin
sediment layer covers stiff till. In the Ross Sea they appear
to mark the transition zone between the eroding bed and
the deforming bed. Where flutes occur, they are associated
with, and commonly superimposed on, striations.

Drumlins. The NBP95-01 data set revealed an area of
drumlinized topography on the inner shelf of the Ross Sea
(Shipp and Anderson, 1997) (Fig. 3.8A). Individual drumlins are tens of meters high, 100 to 200 m wide, and up to
8 km long. Crescentric scours around the upstream flanks
of the drumlins (Fig. 3.8A) were probably cut by turbulent meltwater flowing around these features (Aario, 1977;
Shaw, 1988). The drumlins extend northward into an area
with megaflutes and megascale lineations (Fig. 3.8A). A
seismic profile acquired along the axis of the trough and
through the center of the drumlin field shows that the
drumlins occur at the boundary between exposed, striated
crystalline basement rocks and a thin layer of sedimentary
deposits (Fig. 3.8B).

There are a number of different models for drumlin
formation, including molding by sediment deformation,
lodgment, and subglacial meltwater, to purely erosive formation (Clark, 1993; Bennett and Glasser, 1996). In the
Ross Sea, the transition from striated basement to drumlins to megascale lineations is interpreted as the transition zone from a "sticky" bed, where the ice sheet is coupled to the seafloor and basal meltwater is confined to the
ice–bed interface, to a deforming bed where meltwater is
incorporated into the bed and the sediments are molded
into megaflutes and megascale lineations.

Moraines. Moraines of virtually all sizes and shapes occur on the Antarctic seafloor. The most common type of
moraines were previously thought to be hummocky moraines. However, as more side-scan sonar records are collected it is becoming apparent that most moraines are
actually elongate features formed transverse to the
direction of ice flow. True hummocky moraines are more
amorphous features that result from remolding of megascale glacial lineations (Fig. 3.6C). The largest hummocky
moraines occur in areas with relatively thick till sheets.
Sediment cores from these features have sampled predominantly soft diamictons that are interpreted as deformation
till.

Another common type of moraine consists of highly
elongate moralal wedges extending transverse to the direction of glacial flow and with steep sides facing the direction of glacial transport (Fig. 3.9A). These moraines
vary widely in size and shape but are most commonly between 100 and 400 m from crest to trough and less than
10 m high, below the resolution multibeam records. In
Victoria Trough in the western Ross Sea, sets of morainal
wedges extend nearly 50 km along the floor of the trough.
The size, shape, and spacing of the moraines vary more
or less randomly along the axis of the trough (Fig. 3.9A).
Striations that extend across the tops of these moraines,
and deformation features along the front sides of some
of the larger moraines, indicate that they have been overridden by the ice sheet (Fig. 3.9A). Larger moraines are
believed to reflect pauses in the rate of grounding line
movement.

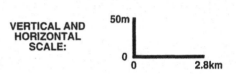

Figure 3.11. Side-scan sonar records illustrating two stages in the evolution of cross-cutting glacial lineations. In the first stage (A) both lineation sets are prominant, and in the second stage (B) the two sets have been blanketed by glacial-marine sediments to produce a V-shaped pattern that is commonly observed in glacial troughs.

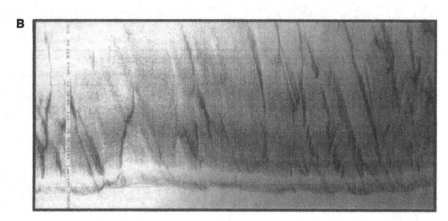

Morainal ridges occur transverse to the direction of ice flow and typically are isolated within troughs. They are on average less than 10 m high and 100 m wide and vary between relatively straight crested and sinuous in form (Fig. 3.9B). Morainal ridges show no signs of having been overridden by the ice sheet.

Side-scan sonar profiles acquired within the axes of glacial troughs in the Ross Sea during NBP 98-01 revealed an extensive area of remarkably symmetric wave-shaped bedforms that extends for at least 50 km along the trough axis (Fig. 3.10A). These bedforms are on average 2 to 3 m high and have wave lengths that average 100 m. They appear like large sand waves on the seafloor; however, piston cores that penetrated into the tops of these bedforms sampled stiff till, indicating a subglacial origin. Chirp profiles did not penetrate these deposits (Fig. 3.10A). In fact, the beds on which these waveforms exist are so stiff that icebergs plowing across them typically erode down to the wavy surface, producing iceberg furrows that look much like a vertebral column (Fig. 3.10B). Subbottom profiler records show extensive areas of western Ross Sea troughs where these wavy surfaces have been draped by glacial-marine sediments (Fig. 3.10C).

The exact origin of these features remains problematic. They are similar to washboard moraines, which are believed to be formed at the retreating grounding line. The

A

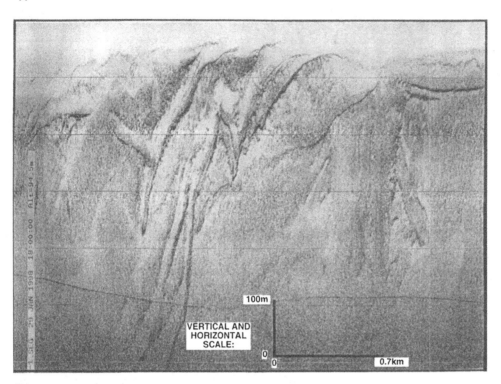

B

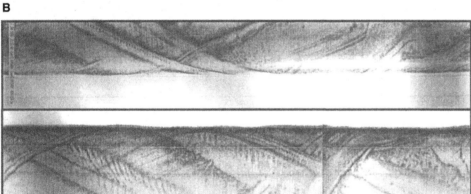

C

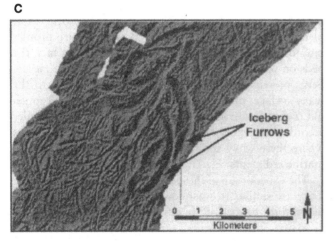

Figure 3.12. Side-scan sonar records illustrating (A) randomly oriented iceberg furrows on the top (above ~300 m) of a bank and (B) cross-cutting lineations with internal ridges and iceberg furrows. These features occur at water depths below ~500 m and were formed by icebergs calved from the retreating wall of the ice sheet. (C) Multibeam record showing large arcuate-shaped iceberg furrows that occur on tops of grounding zone wedges at water depths between 450 and 550 m.

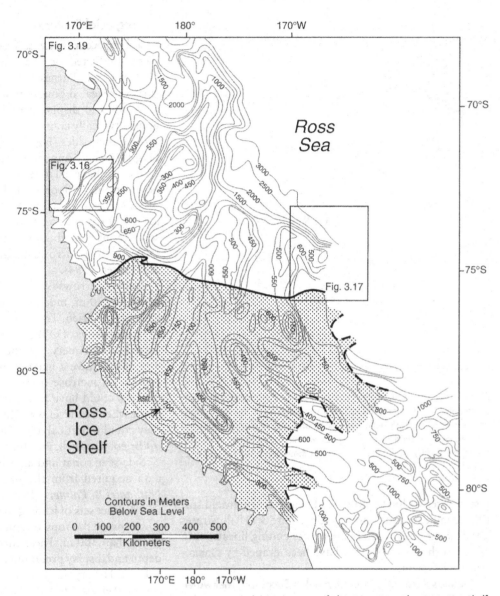

Figure 3.13. Bathymetric map of the Ross Sea. The Ross Ice Shelf grounding line is shown with a dashed line, and the ice shelf location is shaded. (Open shelf bathymetry is modified from Vanney, Falconer, and Johnson, 1981, and sub–ice shelf bathymetry from Bentley, 1991.)

fact that they are composed of till supports a subglacial origin, along with their preservation over such great distances. This, plus the fact that they are draped by glacial-marine sediments, indicates that they were deposited by the retreating ice sheet. The almost constant spacing of individual wave crests is, however, unusual, even for washboard moraines. This implies a fairly constant rate of grounding line retreat. Shipp et al. (in press) argue that they reflect seasonal retreat events.

Cross-cutting Glacial Lineations. ide-scan sonar records from glacial troughs in the western Ross Sea and offshore of North Victoria Land imaged cross-cutting lineations similar to ones illustrated by Clark (1993) in ERS-1 radar images of northern Quebec. These cross-cutting lineations are tens of meters wide, 3 to 10 m high, and 100 to 300 m long (Fig. 3.11A). Figures 3.11A and 3.11B show two stages in the development of cross-cutting glacial lineation. In the first stage (Fig. 3.11A) two sets of lineations are clearly discernible. One set is characterized by transverse bedforms, which are identical to bedforms illustrated by Harris, O'Brien, Quitly, Taylor, Domack, DeSantis, and Raker (1997) that were imaged on the Prydz Bay continental shelf. Harris and colleagues (1997) interpreted these as sediment dunes within the lineations; however,

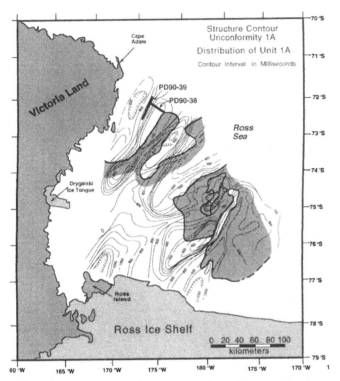

Figure 3.14. Time-structural contour map of Late Pleistocene glacial unconformity 1A cut during the last ice sheet advance to the shelf break. Also shown is the distribution of glacial and glacial-marine sediments (shaded area) resting immediately above this unconformity and illustrating the patchy, regional distribution of these deposits (courtesy of S. Shipp).

piston cores within the Ross Sea lineations penetrated stiff till and indicate a subglacial origin. Figure 3.11B illustrates a later stage of evolution in the cross-cutting lineations in which both sets of features have been draped by glacial-marine sediment.

Features Associated with Channeled Subglacial Meltwater. Notably absent in the NBP95-01 and NBP98-01 multibeam and side-scan sonar records from the Ross Sea floor are tunnel valleys, subglacial braided channels, outwash fans/deltas, and eskers, which would imply channeled subglacial meltwater. In addition, hundreds of piston cores acquired to date on the continental shelf have on only rare occasions recovered graded sands and gravels that might be associated with subglacial meltwater systems. The few exceptions that I know of are cores acquired near the termini of valley and outlet glaciers in bays and fjords (Chapter 4). The virtual absence of meltwater features and deposits on the Antarctic continental shelf is perhaps the most important difference in geomorphic character between the Antarctic continental shelf and Northern Hemisphere glacial terrains. The latter are characterized by a close association of furrows and drumlins with anastomosing tunnel valley systems, eskers, end moraines, and glacial-fluvial deposits (Ashley, Shaw, and Smith, 1985).

Proglacial Features

Iceberg Scour Features. The first side-scan sonar surveys were conducted on the Weddell Sea (Lien, 1981) and Wilkes Land (Barnes, 1987) continental shelves. These surveys revealed gouges that average hundreds of meters in width and display random, crisscrossing tracks. The gouges typically contain grooves with mounds of sediment around their sides (Fig. 1.23) and were interpreted as iceberg furrows. Rounded pot marks, several tens of meters across, are believed to result from occasional impact of icebergs on the seafloor, probably caused by icebergs rolling over (Barnes, 1987) (Fig. 1.23). These surveys showed that iceberg furrows extend to depths of ∼ 500 m. Side-scan sonagraphs from the Antarctic Peninsula continental shelf record iceberg furrows at similar depths (Pudsey, Barker, and Larter, 1994).

Iceberg furrows on the Prydz Bay shelf occur at depths of up to 720 m, much greater than the drafts of modern icebergs (O'Brien, Truswell, and Burton, 1994). Based on the calculations of O'Brien and colleagues (1994), icebergs calved from Amery Ice Shelf with drafts not exceeding 400 m will have a maximum draft of up to 600 m upon rolling, an increase of 50%. Alternatively, these iceberg furrows could have formed during the last glacial maximum when sea level was 120 m lower than at present or they could reflect calving from a much thicker ice margin (O'Brien, Truswell, and Burton, 1994).

Side-scan sonar and multibeam records and multibeam data acquired from the Ross Sea shelf during the R/V *Nathaniel B. Palmer*'s 1995 and 1998 field seasons show two distinct sets of iceberg furrows. One set of iceberg furrows covers the tops of banks, primarily at depths shallower than 300 m. These furrows vary widely in size and depth and display predominantly random drift tracks (Fig. 3.12A). They typically lack any measurable sediment fill.

Another set of larger, straight to arcuate-shaped furrows occurs at depths between 600 m and 400 m (Figs. 3.12B,C). The straighter furrows occur in the deeper portions of troughs, whereas the arcuate furrows are concentrated on the seaward sides of more shallow grounding zone wedges, which mark the former grounding line of the expanded ice sheet (Shipp, Anderson, and Domack, in press). These furrows cut moraines, striations, and megascale glacial lineations and are draped by diatomaceous glacial-marine sediments. They are interpreted as having been produced by large icebergs calved from the grounded calving walls of the expanded ice sheet, and therefore mark the early retreat of the ice sheet (Shipp, Anderson, and Domack, in press).

REGIONAL STUDIES

The physiography and relief-forming processes of the continental shelf vary around the margin due to differences

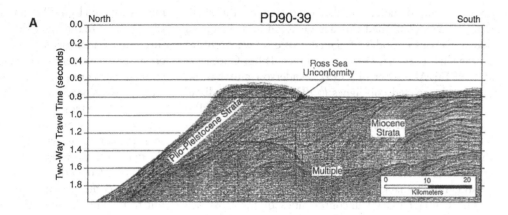

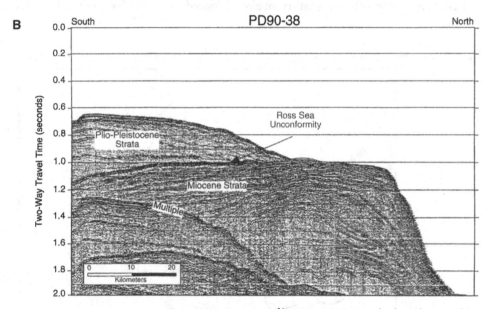

Figure 3.15. Seismic Profile PD90-39 (A) extends along the axis of the westernmost trough in the Ross Sea and crosses a trough mouth fan composed of Plio-Pleistocene strata. Seismic Profile PD90-38 (B), a strike line, extends along the continental shelf edge and crosses the eastern edge of this fan. (See Fig. 3.14 for profile locations.)

in tectonic influences and glacial maritime settings. These effects are best illustrated by contrasting different parts of the continental margin.

Ross Sea

Bathymetry of the Ross Sea, north of the Ross Ice Shelf, has been mapped and described by Lepley (1966), Hayes and Davey (1975), Vanney, Falconer, Johnson (1981), and Shipp et al. (in press). The area beneath the ice shelf has been mapped using an extensive grid of radio-echo sounding data (Bentley and Ostenso, 1961; Heezen, Tharp, and Bentley, 1972; Drewry, 1983; Bentley, 1991).

Seaward of the Ross Ice Shelf, the continental shelf has an average depth of 500 m and displays a typical reverse gradient (Fig. 3.2B). Ridges and troughs oriented northeast-southwest characterize the shelf (Fig. 3.13). These features extend beneath the ice shelf where they become aligned in a northwest-southeast direction. The largest troughs are roughly aligned with major ice streams of the West Antarctic Ice Sheet, which led Hughes (1977)

to conclude that the troughs were carved by the ice streams. Ice streams are characterized by rapid flow and a basal thermal regime conducive to bed erosion (Hughes, 1981). The present grounding line of the Ross Ice Shelf occurs ~700 to 1200 km inland from the shelf break (Fig. 3.13). The grounding line appears to be sinuous, with deep embayments associated with some of the larger troughs.

The first high-resolution seismic data acquired on the Antarctic continental shelf were collected in the Ross Sea as part of *Eltanin* cruises 27 and 32; Houtz and Meijer (1970) and Hayes and Davey (1975) analyzed these data. Hayes and Davey (1975) recognized the Ross Sea Unconformity, an unconformity that appears to extend across the Ross Sea continental shelf (Fig. 3.4A). Houtz and Meijer (1970) estimated that this unconformity results from several hundred meters of glacial erosion on the shelf. Later studies confirmed this deep erosion and suggested that numerous episodes of glacial erosion have occurred on the shelf (Cooper, Davey, and Hinz, 1988; Cooper et al., 1991a; Alonso et al., 1992; Anderson and Bartek, 1992). Alonso and colleagues (1992) used intermediate-resolution seismic records to map the more extensive Plio-Pleistocene glacial unconformities and the units bounded by these surfaces. The NBP95-01 high-resolution data set shows more glacial unconformities and units than are imaged in the intermediate-resolution data set (Shipp, Anderson, and Domack, 1994, in press).

The younger glacial unconformities of the Ross Sea continental shelf converge toward the inner shelf and toward the west to form the Ross Sea Unconformity (Fig. 3.4A). On the outer shelf, the most recent glacial unconfor-

mity is covered by a till sheet that averages 10 to 20 m thick (Fig. 3.6C). On the inner shelf and in the western Ross Sea, this same unconformity cuts into Miocene deposits and crystalline basement rocks (Figs. 3.4A and 3.8B). Miocene strata occur at or near the seafloor throughout most of the western Ross Sea, except on the outer shelf.

Using an extensive high-resolution data set acquired in the western Ross Sea during the 1994, 1995, and 1998 field seasons, Shipp, Anderson, and Domack (in press) mapped the two most recent glacial erosion surfaces and the glacial units bounded by these surfaces. Figure 3.14 shows a time-structure contour map of the youngest unconformity and illustrates the physiography of the shelf at the end of the last ice sheet expansion. Figure 3.14 also shows the distribution of the till sheet resting on this unconformity (Unit 1A; Fig. 3.6C). Unit 1A accounts for only a portion of

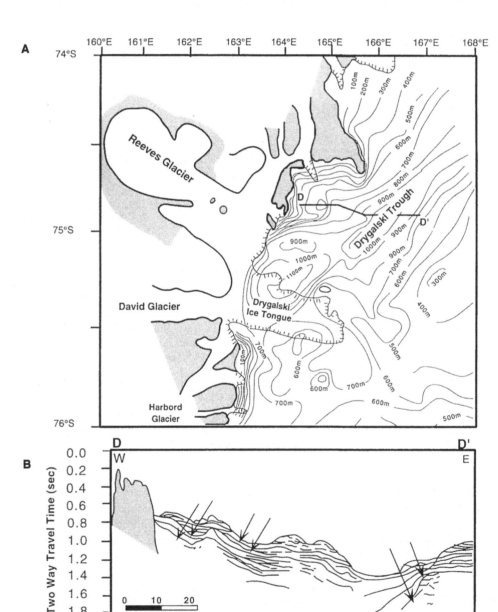

Figure 3.16. (A) Bathymetric map of the Drygalski Trough; (B) interpreted high-resolution seismic profile across the Drygalski Trough showing numerous glacial unconformities (arrows) within the trough (seismic profile from Wong and Christoffel, 1981; copyright by the American Geophysical Union); and (C) glacial paleodrainage map for the Drygalski Trough based on petrographic analysis of tills from the area (modified from Myers, 1982a). The distinguishing rock types for each province are listed.

the total volume of sediment eroded from the shelf in the western Ross Sea. Much of this material was transported seaward of the shelf break and into a large trough mouth fan that is situated on the upper slope, directly offshore of the Victoria Land Trough (Fig. 3.15A, B). Two depositional lobes compose this fan and are interpreted as being associated with converging ice flow across the shelf from East Antarctica (Bart, 1998). Lateral shifts in the locations of these lobes indicate shifts in these ice streams.

The composite subglacial and bathymetric map of the Ross Sea shows several small troughs lying directly offshore of major outlet glaciers that are sourced from the TAM (Fig. 3.13). Immediately seaward of the mountains, the troughs veer to the north, aligning themselves parallel to the prevailing glacial drainage of the region. The Drygalski Trough, the largest of these troughs, is up to 1100 m deep and occurs offshore of David and Reeves glaciers, two of the largest outlet glaciers in Victoria Land (Fig. 3.16A). Single-channel seismic reflection profiles across this trough show several erosional surfaces, implying multiple advances of the East Antarctic Ice Sheet onto the continental shelf (Fig. 3.16B). The northeast trend of the Drygalski Trough indicates that the ice stream flowed to the northeast after reaching the coast. This paleodrainage

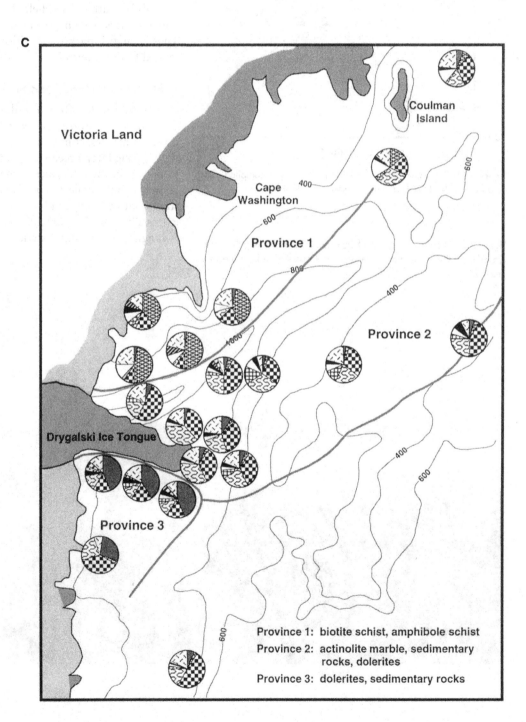

Figure 3.16. (*Continued*)

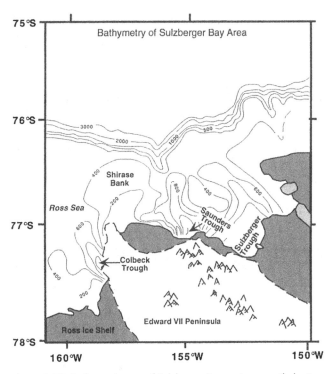

Figure 3.17. Bathymetric map of Sulzberger Bay region compiled using data from Lepley (1966) and from Anderson (1983b).

Figure 3.18. Aerial photograph of the Pennell Coast showing the northern edge of the Transantarctic Mountains. Cape Adare is in the foreground.

reconstruction has been documented by a petrographic analysis of tills in the western Ross Sea (Myers, 1982b; Anderson, Brake, and Meyers, 1984) (Fig. 3.16C).

Sulzberger Bay

Sulzberger Bay lies just east of the Ross Ice Shelf and offshore of Marie Byrd Land (Fig. 3.1). Lepley (1966) first described the bathymetry of the bay. His bathymetric map was modified using data from a Deep Freeze 84 expedition (Anderson, 1983b) (Fig. 3.17). The Sulzberger Trough is the prominent feature on this part of the continental shelf (Fig. 3.17). It is over 900 m deep at the point where it extends beneath the ice shelf. Two other troughs, Colbeck and Saunders troughs, occur on this portion of the shelf. These troughs mark the positions of ice streams that extended from western Marie Byrd Land.

Northern Victoria Land Shelf

Just west of the Ross Sea along the Pennell Coast, the TAM appear to stop abruptly at the sea (Fig. 3.18). Marine geologic surveys, conducted on this part of the continental shelf during Deep Freeze 80 and NBP98-01, acquired data to construct the bathymetric map shown in Figure 3.19. Seismic profiles collected on the eastern portion of the shelf show that the northern limit of the TAM evolves into an extremely rugged, sediment-barren inner shelf (Fig. 3.20). A longitudinal trough extends from Cape Adare to Yule

Bay. The floor of this trough exhibits considerable relief, owing to glacial sculpturing of the highly deformed basement rocks that crop out on the inner shelf. The shelf profile becomes more gentle toward the outer shelf, where a sedimentary wedge exists. Flat-topped banks on the outer shelf are everywhere covered by iceberg furrows. The banks are cut by several U-shaped, transverse troughs. Dennistoun Trough is the largest; it appears to have been the main outlet for ice flowing from the TAM between Cape Adare and the Barnett Glacier (Fig. 3.19).

The Dennistoun Trough was mapped using multibeam swath bathymetry during a recent NBP98-01 cruise. The trough is relatively narrow with steep flanks and has rugged seafloor relief on the inner shelf where highly deformed bedrock is exposed. Where the trough was cut through the sedimentary wedge of the outer shelf, it becomes broader and shallower and has flatter seafloor relief. Several small valleys extend from the adjacent banks into the trough. A side-scan sonar profile collected along the axis of the trough shows striated basement in the landward sector of the trough and remolded moraines in the seaward sector of the trough. There is a virtual absence of geomorphic features that would indicate the presence of a deforming bed within the trough.

The troughs that dissect the continental shelf off the Pennell Coast indicate that ice flowing from Northern Victoria Land previously extended onto the shelf (Brake, 1982; Anderson et al., 1992a). Given the small size of valley glaciers in the region, it seems logical that this expanded ice sheet was nourished by ice flowing across the TAM from East Antarctica.

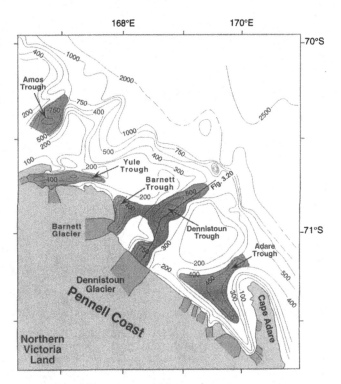

Figure 3.19. Bathymetric map of the continental shelf offshore of the Pennell Coast compiled from bottom profiler records collected during Deep Freeze 80 and multibeam bathymetric data acquired during NB *Palmer* cruise 98-01.

Figure 3.20. Seismic profile across the continental shelf offshore of the Pennell Coast showing deep, landward sloping topography on the inner shelf, associated with the northern limit of the TAM, and more gently seafloor topography of the outer shelf where a seaward-thickening wedge of strata occurs. The profile extends along the axis of Dennistoun Trough (Fig. 3.19).

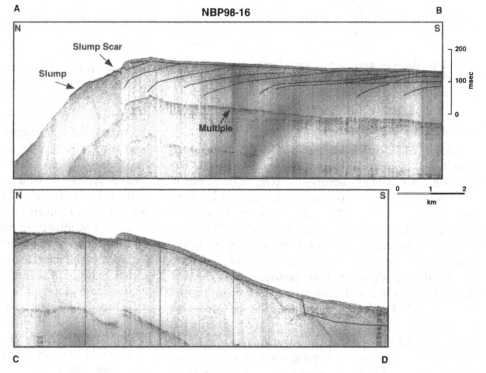

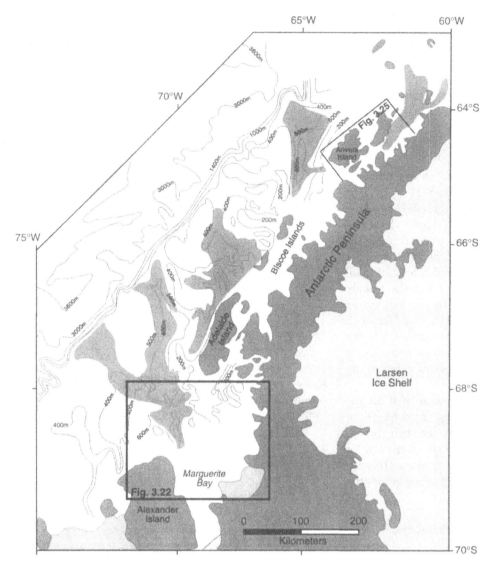

Figure 3.21. Bathymetric map of the Antarctic Peninsula continental shelf north of ~ 69° S. Larger troughs are shaded (compiled from Pope and Anderson, 1992, and Bart and Anderson, 1995).

Antarctic Peninsula Shelf

Because of extensive perennial sea-ice cover, the continental shelf between Sulzberger Bay and Marguerite Bay has remained virtually unsurveyed (Fig. 3.1). North of ~ 69° S, ice conditions are less severe; this area has been surveyed in some detail (Fig. 3.21), including Marguerite Bay and the Inland Passage (Figs. 3.22 and 3.25).

The continental shelf northwest of the Antarctic Peninsula shows broad similarities to other sectors of the Antarctic shelf that lie adjacent to mountainous coasts. The shelf ranges between 400 and 1000 m deep and slopes toward the continent (Fig. 3.2). The width of the shelf decreases northward from 150 km wide offshore of Marguerite Bay to just under 125 km wide offshore of Anvers Island. The outer shelf consists of undulating plains and banks that are dissected by a series of troughs (Fig. 3.21). One of the more prominent troughs, the Marguerite Trough, extends

outward from Marguerite Bay and reaches depths in excess of 1000 m. The trough is relatively narrow and has a V-shaped profile on the inner shelf where it cuts into basement rocks (Fig. 3.23). The Marguerite Trough cuts into a thick wedge of sedimentary deposits on the outer shelf and widens to ~ 70 km (Fig. 3.5B; Profile PD88-B). Another large trough, the Adelaide Trough, trends parallel to the coast of Adelaide Island for nearly 100 km before turning north (Figs. 3.4B and 3.21). Other large troughs include the Biscoe and Anvers troughs.

Seismic profiles from the Antarctic Peninsula continental shelf show the strong influence of geology on the physiography of the shelf. The rugged inner shelf has been completely stripped of sedimentary deposits, revealing rugged bedrock topography. The flatter outer shelf coincides with a seaward-thickening wedge of sedimentary deposits (Fig. 3.4B). The southern portion of the Marguerite Trough is aligned with George VI sound, which is a block-faulted extensional basin (Crabtree, Storey, and Doake, 1985; Maslanyj, 1987). The trough is, on average, 20 km wide and 500 km long (Sugden and Clapperton, 1981). It separates plutonic rocks of the Antarctic Peninsula from sedimentary and metamorphic rocks on Alexander Island (Bell, 1975). The Adelaide Trough cuts into and follows the trend of an inner shelf sedimentary basin (Fig. 3.4B) that is believed to be an old fore-arc basin (Anderson, Pope, and Thomas, 1990). Both troughs change direction on the outer shelf where they incise the sedimentary wedge. A strike-oriented profile (Fig. 3.5B, Profile PD88-B) shows that the troughs have a history of lateral migration and re-incision on the outer shelf (Bart and Anderson, 1995, 1996).

Many large islands occur along the coast of the Antarctic Peninsula. These islands form a chain that is separated from the coast by a deep inland passage extending from Marguerite Bay to the Bransfield Strait (Fig. 3.21). The northeast-southwest trend of the inland passage is a result of the tectonic fabric of the region and preferential glacial erosion along fault zones. Basement rocks and numerous rock pinnacles, located a few meters above and below the sea surface, make up the shallow (< 200 m) platforms that bound the mainland and islands. The inland passage between the mainland and Adelaide Island and the Biscoe Islands is narrow and rugged, which explains why it is poorly charted. These are the most treacherous, yet most scenic, waters in Antarctica (Fig. 3.24).

The inland passage between Anvers and Brabant islands is known as the Gerlache Strait. It makes up the best charted portion of the passage (Fig. 3.25). There is a northeast-southwest trend in the alignment and direction of elongation of the principal islands in this region, collectively known as the Palmer Archipelago. The overall topography and bathymetry of this region show striking linearity and atests to the preferential erosion of tectonic features by glacial ice (Dewart, 1972; Griffith, 1988). Several linear troughs and straits, including the Grandidier Channel, Neumayer Channel, and Gerlache Strait, contribute to this northeast-southwest fabric of the seafloor (Fig. 3.25). A second, roughly east-west orientation of topographic and bathymetric features can be seen in the alignment of the Bismarck Strait, the southwest and northeast coasts of Anvers Island, and within the bays and fjords of the Danco Coast. Both lineation sets correspond to structural features identified onshore.

Hooper (1962) identified a major fault set trending northeast-southwest and another, less prominent set trending more east-west on Anvers and Wiencke islands. He concluded that the northeast-southwest-trending faults were the principal features associated with the uplift of the Graham Land Peninsula in the Anvers Island and Palmer Archipelago areas. Later, Scott (1965) concluded that the northeast-southwest-oriented, linear coastal segments of the region may be fault scarps, perhaps strike-slip faults.

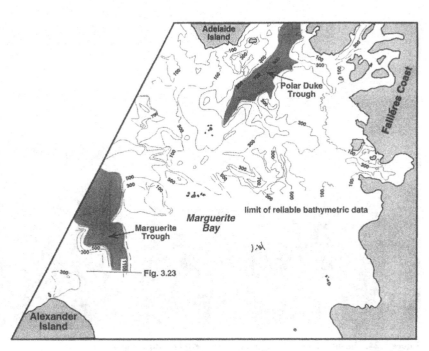

Figure 3.22. Bathymetric map of Marguerite Bay (modified from Kennedy, 1988). Troughs are shaded. (Location of seismic line in Figure 3.23 is also shown.)

Figure 3.23. Deep Freeze 86 high-resolution seismic profile across the Marguerite Trough within Marguerite Bay showing the V-shaped profile of the trough in this area. (See Fig. 3.22 for profile location.) The trough has a more U-shaped profile on the outer shelf where it cuts into sedimentary strata (Fig. 3.5B, Profile PD88-B).

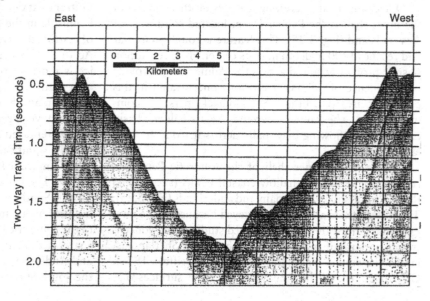

Figure 3.24. The mountainous topography of the Inland Passage region offers some of the most spectacular scenery in Antarctica.

Clearly, the northeast-southwest-trending features of the Danco Coast and Palmer Archipelago regions are part of a more regional structural fabric that extends northward to the Bransfield Basin. The fabric is associated with back-arc extension (González-Ferrán, 1983, 1985; Storey and Garrett, 1985).

The seafloor of the Gerlache Strait displays nearly the same rugged relief as the landmass that surrounds it. Submarine slopes of greater than 20° are common. Water depths range from −270 to −1100 m, and the strait appears to deepen in an almost step-like fashion from southwest to northeast (Fig. 3.25). The Gerlache Strait has an asymmetric cross-sectional profile with a steep, fault bounded western side, similar to, but more narrow than, the Bransfield Basin (Fig. 3.26).

One of the most interesting features on this portion of the inner shelf is the Palmer Deep, located southwest of Anvers Island (Fig. 3.25). This feature is roughly circular and reaches 1200 m depth near its center. High-resolution seismic profiles acquired across the Palmer Deep provide little evidence of the recent erosion concentrated in this particular area (Griffith, 1988; Domack, personal communication). The Palmer Deep may be a small rift basin eroded by ice flowing diagonally across it from the Gerlache Strait.

Unique to the inner shelf of the Antarctic Peninsula region are the broad, shallow coastal banks (strandflats) that occur offshore of the mainland and islands (Fig. 3.25). These strandflats are poorly surveyed because of the navigational hazard they pose. Detailed studies have concentrated on strandflats offshore of the South Shetland Islands (Hansom, 1983) and Anvers Island (Griffith, 1988). Off Anvers Island the strandflats are relict surfaces that

probably formed during relative sea-level stillstands. They occur at two distinct levels, −40 to −80 m, and approximately −130 to −170 m water depth (Griffith, 1988).

Holtedahl (1929) first described the strandflats of the Antarctic Peninsula region. He concluded that they are the products of erosion by fringing glaciers, with sea level controlling the depth of erosion. Strandflats of the Antarctic Peninsula region occur along stretches of the coast where wave energy is high. This is a good generalization for strandflats elsewhere in the world (John and Sugden, 1975). Also, Antarctic strandflats exhibit features, such as stacks (Fig. 3.27) and sea caves, that typically occur on non-glaciated, wave-dominated, rocky coasts. These observations led Griffith (1988) to conclude that the strandflats of the Antarctic Peninsula region are primarily the products of wave erosion. Alternatively, sea ice plays a role in strandflat development. The modern strandflat off Byers Peninsula in the South Shetland Islands is widest in areas of restricted wave fetch, where sea ice persists the longest. According to Hansom (1983), erosion by grounded fast ice is more significant in strandflat development in this area than wave erosion.

The many bays and fjords in the Antarctic Peninsula region have been poorly surveyed. Griffith and Anderson (1989) reported results from reconnaissance work. They found a significant difference between the bays and fjords of the Antarctic Peninsula and those of the South Shetland Islands. These differences are related to the warmer climate on the South Shetland Islands, which results in greater input of meltwater-derived sediment. The bays and fjords of the Antarctic Peninsula region are characterized by their sparse sediment cover and rugged topography (Griffith and Anderson, 1989). In contrast, the bays and fjords of the South Shetland Islands contain greater sediment fill.

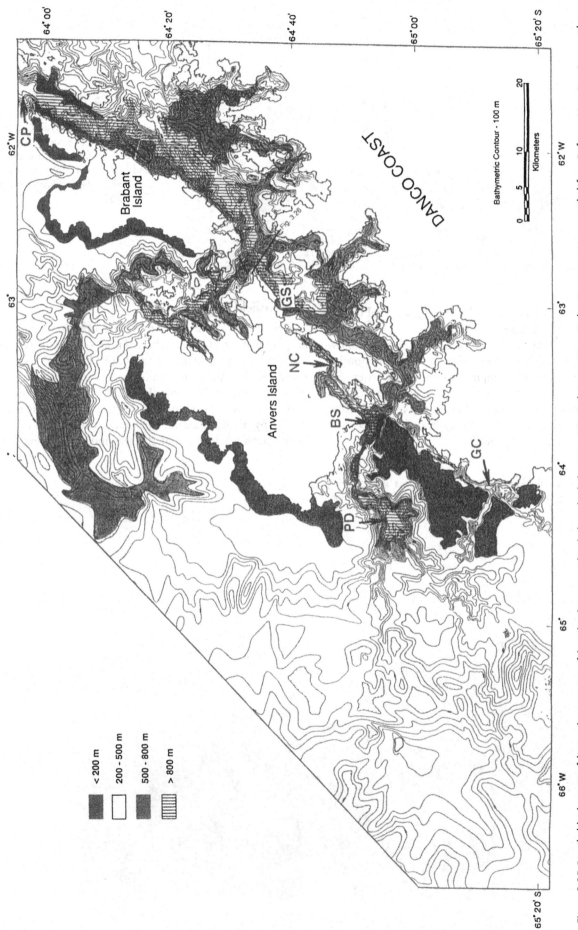

Figure 3.25. Detailed bathymetry of the northern portion of the Inland Passage. The darkest shaded areas are strandflats that occur at −40 to −80 m and −130 to −170 m below mean sea level. Map compiled from Defense Mapping Agency data and data collected during Deep Freeze 82, 83, 85, and 86 (from Griffith, 1988). BS = Bismarck Strait; CP = Croker Passage. PD = Palmer Deep; GC = Grandidier Channel; NC = Neumayer Channel; GS = Gerlache Strait;

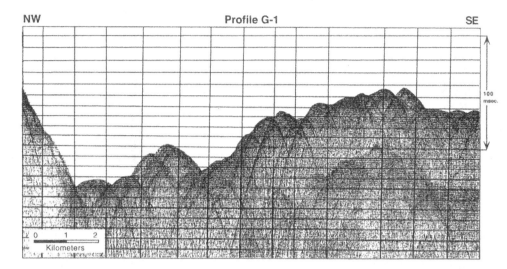

Figure 3.26. High-resolution seismic Profile G-1 collected during Deep Freeze 86 across the Gerlache Strait. The cross section of the strait has a distinctly asymmetric profile and there is a virtual absence of sediment fill. (Profile location shown on Fig. 3.25.)

The sediment masks the underlying rugged glacial topography (Griffith and Anderson, 1989).

Seismic lines from all the bays and fjords surveyed show no significant sediment accumulations at their mouths. This paucity of sediments implies that the bays were eroded early in the glacial history of the region, and the products of erosion have been removed. The eroded material either bypassed the shelf or is too deeply buried to be seen in high-resolution seismic profiles.

Bransfield Strait

The Bransfield Strait is the most prominent physiographic feature of the northern Antarctic Peninsula. It is a deep, elongate extensional basin bordered on the south by the Antarctic Peninsula and on the north by the South Shetland Islands (Fig. 3.28). This region experiences the warmest climate in Antarctica, which is manifested by ablation zones, meltwater streams, and associated outwash features at the terminus of glaciers. The

Figure 3.27. View looking west offshore of King George Island showing sea stacks and rugged coastline that is typical of strandflats in the region.

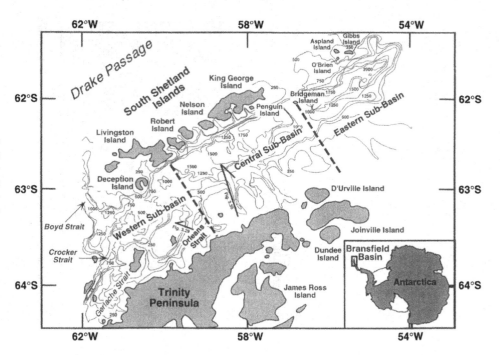

Figure 3.28. Bathymetric map of the Bransfield Strait showing locations of western, central, and eastern subbasins (separated by dashed lines; compiled from Jeffers and Anderson, 1990, and Banfield and Anderson, 1995).

region provides a unique setting for study of the interaction between glacial and tectonic processes. It may offer a modern analog for relatively warm interglacial conditions in Antarctica. The Bransfield Strait was the site of multibeam bathymetric surveys by Spanish investigators during the 1993 to 1994 field season (Grácia et al., 1996), and a later multibeam survey by U.S. scientists (Lawver et al., 1996).

A linear chain of volcanoes and seamounts occurs along the northern edge of the Bransfield Strait. The chain includes Deception Island, Bridgeman Island, and Penguin Island (Fig. 3.28), all of which have experienced recent volcanic activity. These features presumably constitute part of the modern back-arc spreading system (Barker, Dalziel, and Storey, 1991).

The Bransfield Strait is divided into three subbasins with different degrees of basin evolution (Jeffers and Anderson, 1990). The stage of evolution is manifested in their physiographies. The relatively shallow, irregularly shaped western subbasin, situated south and west of Livingston and Deception islands, trends northeast-southwest. It branches northwestward through the Boyd Strait and is bordered to the south by the Gerlache Strait (Fig. 3.28). The abrupt northeastward deepening of the basin from 900 to 1300 m marks the easternmost limit of the western subbasin.

The central subbasin lies south of Robert, Nelson, and King George is lands, and extends northeastward to a bathymetric divide associated with Bridgeman Island (Fig. 3.28). It is bound to the northwest by the steep slope of the South Shetland Islands. On the Antarctic Peninsula side of the subbasin, the shelf break occurs at ~ 250 m.

A second platform gradually deepens from 750 to ~ 900 m. A steeper (9°) slope leads to the basin floor, which extends from 1300 m at the southwest end of the central subbasin to more than 2000 m southeast of King George Island. The continental shelf north of the South Shetland Islands is the shallowest and flattest in Antarctica. The

Figure 3.29. Seismic Profile PD86-4 crosses the Orleans Strait and illustrates the steep, eroded walls of the strait and the hummocky seismic facies that characterize the bottom of the straits in this region. (See Fig. 3.28 for profile location.)

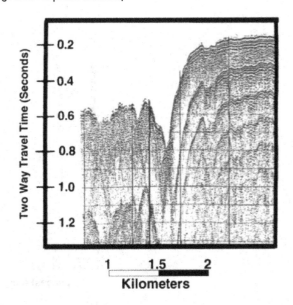

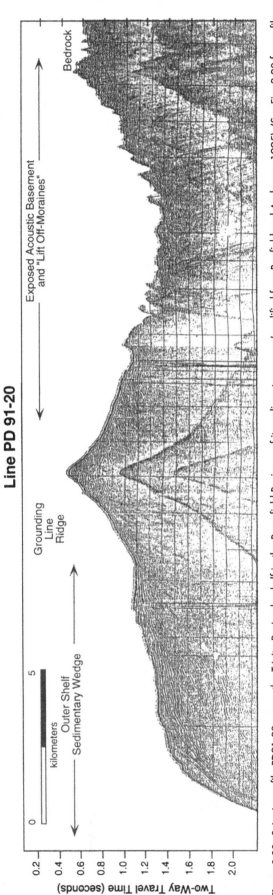

Figure 3.30. Seismic profile PD91-20 across the Trinity Peninsula shelf in the Bransfield Basin. The inner shelf of the Trinity Peninsula is characterized by rugged topography; it has been stripped of its sedimentary cover (modified from Banfield and Anderson, 1995). (See Fig. 3.28 for profile location.)

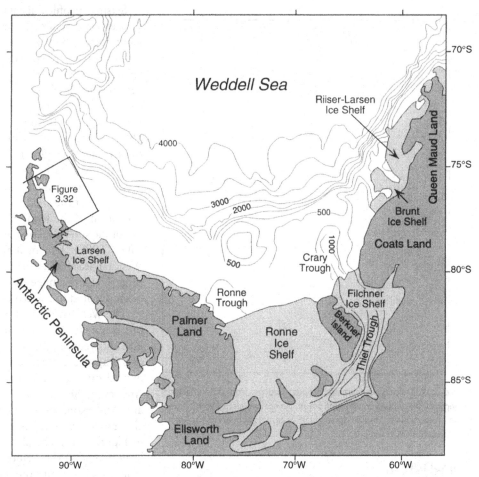

Figure 3.31. Bathymetric and geographic map of the Weddell Sea region (compiled from Johnson et al., 1981, and Behrendt, 1962).

inner shelf, down to about 200 m water depth, is rugged and ice-gouged. The middle and outer shelves are relatively flat, with a seaward gradient.

The eastern subbasin, extending northeast from Bridgeman Island past Elephant and Clarence islands, is narrower than the central subbasin and reaches a depth greater than 2000 m. The morphology of the eastern subbasin is irregular due to the occurrence of many basement highs (Grácia et al., 1996). Several seamounts may represent an extension of the line of submarine volcanoes in the central subbasin. The margins of the eastern basin are free of the large troughs found to the southwest.

The Bransfield Strait connects with the Gerlache Strait via two narrow, fault-bounded passages, Croker Passage and Orléans Strait (Fig. 3.28). Seismic profiles from the Gerlache Strait, Croker Passage, and Orléans Strait show rugged, faulted seafloor topography and a virtual absence of sediment cover on the sides of these straits (Fig. 3.29). During the last glacial maximum, Croker Passage and the Orléans Strait must have been the main outlets for glacial ice flowing northward from the Gerlache Strait into the Bransfield Basin. The Bransfield Strait was the ultimate outlet of glacial ice flowing through the Orléans Strait. Multibeam records reveal a large trough that connects Croker Passage to the Boyd Strait (Canals, personal

communication). Thus, ice draining through these straits was discharged onto the continental shelf.

The shallower shelf areas of the Bransfield Strait region display clear evidence of glacial erosion. Similar to other regions, the inner shelf areas display rugged topography and little sediment cover (Fig. 3.30). Glacial erosion of the outer shelf has been minimal, at least in more recent geologic time (Banfield and Anderson, 1995). Prominent moraine banks occur with troughs between eroded basement and the outer shelf sedimentary wedge (Fig. 3.30).

Within the Bransfield Strait, four troughs aligned perpendicular to the axis of the central subbasin cut into the broad, shallow shelf of the Antarctic Peninsula to a depth of 750 m (Fig. 3.28). The troughs have a V-shaped profile on the inner shelf and U-shaped cross section on the outer shelf. Their orientation is believed to be structurally controlled (Jeffers and Anderson, 1990). Trough mouth fans occur offshore of these troughs (Jeffers and Anderson, 1990; Banfield and Anderson, 1995) (Chapter 4). There are apparently no large glacial troughs extending northward from the South Shetland platform; however, the area is poorly mapped.

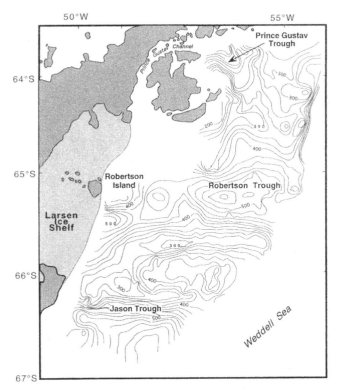

Figure 3.32. Detailed bathymetric map constructed from data collected during marine geologic investigations in the northwestern Weddell Sea (from Sloan, Lawver, and Anderson, 1995; copyright held by the American Geophysical Union).

Weddell Sea

The Weddell Sea is a broad, roughly circular embayment open to the Atlantic Ocean. Its central region reaches depths of 4000 to 5000 m on the abyssal plain (Fig. 3.31). The Weddell Sea is bounded on the east by East Antarctica and its ice sheet, on the south by the West Antarctic Ice Sheet and associated Ronne-Filchner Ice Shelf, and on the west by the Antarctic Peninsula. Mountain glaciers and ice caps flow from the mountains of the Antarctic Peninsula to the coast and into Larsen Ice Shelf.

The eastern Weddell Sea continental shelf is less than 100 km wide and 450 to 600 m deep, and slopes toward the continent (Johnson et al., 1981). Fringing ice shelves, including the Riiser-Larsen and Brunt ice shelves, cover most of the inner portions of the shelf. This is the only area of the Weddell Sea where side-scan sonar data have been gathered (Lien, 1981). These records show iceberg furrows. No linear furrows were detected, although piston cores from the eastern shelf penetrated tills (Anderson et al., 1980b; Anderson et al., 1991a).

The Ronne-Filchner Ice Shelf complex covers the southern portion of the Weddell Sea continental shelf (Fig. 3.31). The ice shelf grounding line occurs some 1100 km south of the ice shelf calving wall. The Crary Trough forms a prominent topographic feature on the southern continental shelf. The term "Crary Trough" refers to the offshore extension of the Thiel Trough (Behrendt, 1962).

Seismic profiles across the Crary Trough show that it was excavated along the contact between folded basement rocks (the roots of the TAM) and seaward-dipping sedimentary deposits (Elverhøi and Maisey, 1983; Miller et al., 1990b). Seismic data from the outer shelf, adjacent to the Crary Trough, show a possible slump scar at the shelf break and a large trough mouth fan, the Crary Fan (Roquelpo-Brouillet, 1982; Kuvaas and Kristoffersen, 1991; Moons et al., 1992). The fan forms the repository for sediments eroded from the Crary Trough.

Although bathymetric data from the southwestern Weddell Sea continental shelf are sparse, the available data indicate that the shelf is not cut by large troughs, with the exception of the Ronne Trough, in the southwestern corner of the shelf (Deep Freeze 68; Haase, 1986; Fechner and Jokat, 1995). This probably results from the fact that the Weddell Sea shelf is virtually bounded by the Thiel Trough (Fig. 3.31); drainage of major ice streams apparently has circumvented the south-central shelf.

The Western Weddell Sea is covered by sea ice throughout the year; few oceanographic surveys have ventured into this region. One exception is the northwestern Weddell Sea, where detailed marine geologic studies were conducted during 1985, 1991, and 1993 (Anderson, 1985; Anderson, Shipp, and Siringan, 1992b; Sloan, Lawver, and Anderson, 1995). The combined data sets from these cruises were used to construct the bathymetric map shown in Figure 3.32. The shelf width averages 200 km. Three large troughs, the Prince Gustav, Robertson, and Jason troughs, accentuate the irregular relief of the shelf. These troughs mark the positions of former ice streams that flowed from the Antarctic Peninsula into the Weddell Sea. Seismic profiles from the area show offlapping strata bounded by unconformities that amalgamate on the inner shelf (Anderson, Shipp, and Siringan, 1992b; Sloan, Lawver, and Anderson, 1995; Chapter 5), similar to other parts of the Antarctic continental shelf.

Wilkes Land

The continental shelf of Wilkes Land averages 100 km in width and displays the typical Antarctic margin profile of a deeper inner shelf shoaling to the outer shelf edge (Grinnell, 1971; Vanney and Johnson, 1979; Chase et al., 1987; Domack et al., 1989b) (Fig. 3.33). The topography is rugged on the inner shelf and it is somewhat smoother on the outer shelf. Four large troughs, the Mertz, Ninnis, Adelie, and Clarie troughs, trend obliquely across the shelf and originate at the mouth of ice streams in the region. These features reach depths of up to 1200 m and exist adjacent to bathymetric highs that rise to within 200 m of the ocean surface. A multichannel seismic profile across the Mertz-Ninnis Trough shows that its southern flank consists of outcropping acoustic basement; the northern flank has seaward-dipping sedimentary deposits (Wannesson et al., 1985). A piston core from the northern flank of the trough penetrated Lower Cretaceous deposits

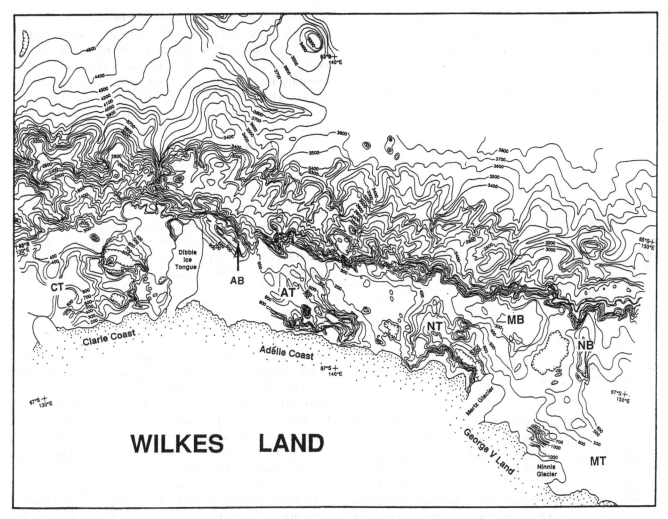

Figure 3.33. Bathymetric map of the continental shelf off Wilkes Land. MT = Mertz Trough; NT = Ninnis Trough, AT = Adelie Trough; CT = Clarie Trough; MB = Mertz Bank; NB = Ninnis Bank; and AB = Adélie Bank (from Chase et al., 1987; reprinted by permission from the Circum-Pacific Council for Energy and Natural Resources, © 1987).

(Domack, Fairchild, and Anderson, 1980). Grab samples from the southern flank of the trough recovered only crystalline rocks (Domack and Anderson, 1983). These observations support the interpretation that this type of trough, with trends parallel or oblique to the coast, was excavated along the contact between crystalline basement rocks and less resistant sedimentary strata (Holtedahl, 1970; Johnson, Vanney, and Hayes, 1982). In this respect, the Mertz and Ninnis troughs are similar to the Crary Trough in the Weddell Sea and Adelaide Trough of the northern Antarctic Peninsula.

Broad, somewhat linear, flat-topped banks occurring at 200 to 400 m depth form prominent features on the middle and outer portions of the continental shelf (Fig. 3.33). Smaller shallow (200 m) knolls and small, roughly circular basins, averaging 700 m in depth, dominate the outer shelf (Vanney and Johnson, 1979; Chase et al., 1987; Domack et al., 1989b). The broad, irregularly shaped banks, such as Mertz Bank, represent portions of the shelf that escaped deep glacial erosion. In contrast, elongate banks, such as Ninnis and Adélie banks, may actually be depositional features (Eittreim, Cooper, and Wannesson, 1995), similar to the ice stream boundary ridges of the eastern Ross Sea shelf (Anderson et al., 1992a).

The Wilkes Land continental shelf was the site of one of the few detailed side-scan sonar surveys to have been conducted in Antarctica (Barnes, 1987). Barnes noted that the seafloor has been furrowed by icebergs to depths up to 500 m (Fig. 1.23). He interpreted the deeper furrows to have been formed during the last glacial maximum by tabular icebergs with drafts of 330 m. Barnes (1987) also mapped lateral moraines that were later shown to represent the grounding line position of the ice sheet during the Late Pleistocene glacial maximum (Domack et al., 1989b). No high-resolution seismic data have been collected on the Wilkes Land continental shelf.

Prydz Bay

Prydz Bay is the seaward extension of a 700-km-long Late Paleozoic(?) and Mesozoic rift basin, the Lambert Graben (Stagg, 1985). The bay occupies a sedimentary basin, the Prydz Bay Basin. The inner shelf portion of

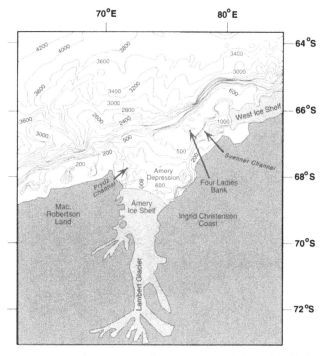

Figure 3.34. Bathymetric map of Prydz Bay and continental shelf offshore Mac. Robertson Land (bathymetry modified from O'Brien, 1994).

the bay is presently covered by the Amery Ice Shelf. The glacial ice flowing into the Amery Ice Shelf represents the largest single drainage outlet for the East Antarctic Ice Sheet, draining an area of 1.09 million km², or about 22% of the East Antarctic Ice Sheet (Allison, 1979). The ice sheet has grounded in Prydz Bay on many occasions since Late Eocene–Early Oligocene time (Barron, Larsen, and Baldauf, 1989a; Hambrey, Ehrmann, and Larsen, 1991). The most recent episode of ice sheet grounding occurred during the Late Pleistocene and ended sometime between 10,700 and 7,300 years ago (Domack, Jull, and Donahue, 1991).

The Prydz Bay shelf is similar to other areas of Antarctica in its great depth and landward-sloping profile (Fig. 3.34). Shelf depths typically range between 400 and 700 m (Stagg, Ramsay, and Whitworth, 1983), with the shelf break occurring between 200 and 600 m. The shelf extends seaward almost 400 km at its widest point. Four Ladies Bank is the largest outer shelf bank in the region, with depths as shallow as 200 m (O'Brien, Truswell, and Burton, 1994).

The deepest portions of the shelf lie closest to Amery Ice Shelf and within the Amery Depression, where depths exceed 800 m (Fig. 3.34). The deeper portion of the Amery Depression eroded into Late Paleozoic and Mesozoic sedimentary rocks (Stagg, 1985; Cooper, Stagg, and Geist, 1991c). The Svenner Channel, with depths reaching 1000 m, parallels the Ingrid Christensen Coast. The Prydz Channel extends from the ice edge, bends to the northwest, and continues to the edge of the shelf (O'Brien, Truswell, and Burton, 1994; O'Brien and Harris, 1996). Depths within the Prydz Channel reach 800 m. In the vicinity of the ice shelf and along the southeastern border where the glaciers of the Ingrid Christensen Coast feed into the bay, the seafloor exhibits typical ruggedness. Here the seafloor is directly underlain by Precambrian crystalline rocks. Where it is underlain by sedimentary deposits, the seafloor is relatively smooth (O'Brien, Truswell, and Burton, 1994).

SUMMARY

The Antarctic continental shelf is characterized by its great depth, irregular topography, and landward-sloping profile. These features are typical of high-latitude continental shelves (Holtedahl, 1929; Shepard, 1931). Glacial erosion has been the main factor contributing to the depth of the continental shelf, whereas isostasy accounts for less than ~ 25% of the depth of the shelf. There is strong geophysical evidence for numerous episodes of ice sheet expansion onto the continental shelf. These expansions caused substantial erosion. Modern glacial troughs are associated with major ice streams and outlet glaciers, and they are the products of multiple glacial advances. A myriad of geomorphic features occur within the troughs. These include megascale glacial lineations, striations, and drumlins.

The flow paths of ice streams and outlet glaciers that extended onto the continental shelf tend to follow geologic boundaries, such as faults and the contacts between crystalline and sedimentary rocks. This geologic influence on ice-flow behavior is strongest on the inner shelf. Once the ice streams reached the thick sedimentary wedge of the outer shelf, they became more seaward-directed and cut broad, U-shaped troughs. The products of glacial erosion on the continental shelf now reside on the continental slope and rise, particularly in trough mouth fans that occur at the mouths of glacial troughs (Chapter 4).

CHAPTER FOUR

Sedimentology

Because of its polar climate, large ice sheets, and highly variable glacial maritime setting, the seafloor surrounding Antarctica offers a unique opportunity for studying glacial-marine sedimentation under extreme climatic conditions. It also provides valuable information about the continent's glacial and climatic record, provided we know how to interpret the sedimentary record properly.

The results of early studies of sediment cores acquired from the Antarctic continental margin are summarized in a number of depositional models that attempt to illustrate sedimentary processes on the continental shelf. Most of these are two-dimensional models.

Carey and Ahmad (1961) published a provocative, widely cited glacial-marine depositional model that inspired most later models, even though their model predates the acquisition of sediment cores on the continental shelf. Anderson's (1972b) model was the first constructed using sediment cores obtained from the Antarctic continental shelf. The model utilizes foraminiferal assemblages and associated core lithologies to define subglacial, sub-ice shelf, and open shelf facies in the eastern Weddell Sea and relates these facies to changing glacial and oceanographic conditions. Later, Anderson and colleagues (1983b) constructed a generalized facies model that emphasizes the different glacial-marine environments on the Antarctic continental shelf. Their model illustrates the general facies architecture associated with advancing and retreating marine ice sheets onto the shelf (based on sediment cores from the Ross Sea), continental shelves that bound mountainous coasts (based on a case study in the Antarctic Peninsula region), and continental shelves where the ice sheet is grounded near the coast (based on an analysis of sediment cores from the Wilkes Land continental shelf).

Elverhøi (1984) studied sediment cores from the vicinity of Riiser-Larsen Ice Shelf in the eastern Weddell Sea and developed a lithofacies model for this region. The model excludes the glacial-marine diamicton facies that Anderson and colleagues (1980b) observed in this area, but includes an outer shelf bioclastic carbonate facies.

Domack (1988) investigated piston cores acquired from the Wilkes Land continental shelf where two large ice streams, the Mertz and Ninnis ice streams, and their as-

sociated glacier tongues are the dominant glacial features. His model concentrates on sedimentary processes and the resulting modern sediments of the shelf. He notes the strong influence of biogenic sedimentation and sediment gravity flow processes. Like the Elverhøi (1984) model, Domack's (1988) model includes an outer shelf residual glacial-marine facies with abundant bioclastic carbonate.

The model of Kellogg and Kellogg (1988) emphasizes the complexity of the transition zone associated with the ice sheet grounding zone. Within this zone, till interfingers with glacial-marine deposits. Kellogg and Kellogg (1988) also infer a zone of little or no deposition beneath the ice shelf.

Each of these models represents a step forward in our understanding of sedimentation on the Antarctic seafloor. Figure 4.1 is an attempt to summarize the salient features of these sedimentation models. All the existing models have the disadvantage of grossly generalizing the environmental settings and inadequate high-resolution seismic and piston core coverage necessary for detailed facies work. As more cores and high-resolution seismic data are added to our data bank, sediment facies models will undoubtedly improve.

This chapter provides an overview of subglacial and glacial-marine sedimentary environments and processes and a description of the main types of sedimentary deposits. A section at the close of this chapter focuses on a regional approach to the Antarctic environment using several case studies.

SUBGLACIAL ENVIRONMENT

Environmental Setting

The subglacial environment is by far the most extensive glacial sedimentary environment in Antarctica today. Sedimentation beneath the ice sheet is, for obvious reasons, poorly understood. However, we can learn much about subglacial sedimentation by examining late Pleistocene subglacial deposits of the continental shelf; ice sheets overrode much of the continental shelf during the last glacial maximum (Kellogg, Truesdale, and Osterman, 1979; Anderson et al., 1980b, 1992a; Elverhøi, 1981; Domack,

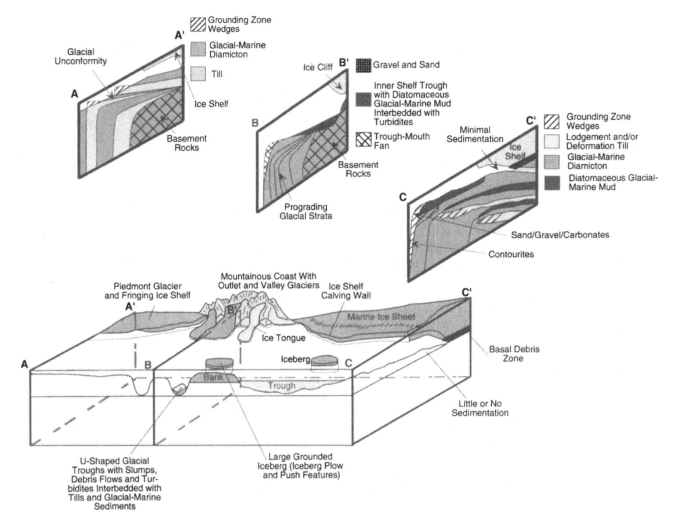

Figure 4.1. Antarctic glacial-marine sedimentation model showing different depositional environments and associated sedimentary facies.

1982; Pope and Anderson, 1992; Pudsey, Barker, and Larter, 1994; Banfield and Anderson, 1995; Chapter 6).

Boulton (1990) argues that there is no reason subglacial processes beneath marine ice sheets should differ from those that occur beneath terrestrial ice sheets. He attributes the greater thickness of marine tills (average 10 to 20 m) compared to terrestrial tills (generally < 10 m) to the fact that ice sheets grounded on the seafloor are resting on soft sedimentary deposits that are more easily eroded and redeposited as till.

In Antarctica, the rate of sediment delivery to the sea by marine ice sheets should be much greater than sediment delivery for terrestrial ice sheets, because marine ice sheets flow much faster than terrestrial ice sheets. They also capture glacial drainage from a large region of the Antarctic. This is manifested by thicker (>1 km thick) accumulation of glacial and glacial-marine deposits on the West Antarctic continental shelf versus the East Antarctic shelf (Chapter 5).

The marine-based WAIS moves hundreds of meters per year. In contrast, the EAIS flows only tens of meters per year (Chapter 1). The more rapid flow of the marine ice sheet is due to converging drainage patterns (Chapter 1) and to the soft, easily deformable bed on which the ice sheet typically rests. This deforming bed reduces basal friction and promotes rapid flow (Alley et al., 1987a, 1987b; Boulton, 1990). Near the grounding line, buoyant forces of the sea may further contribute to acceleration of the ice sheet flow.

In the case of a marine ice sheet, the ice-flow trajectories are directed downward (Fig. 4.2A), a consequence of melting at the base of the ice sheet, faster flow (Fig. 4.2B), and rapid ablation. This downward flow path means that debris transport is essentially confined to the base of the ice sheet. In general, the geophysical and geologic data acquired from the Ross Sea shelf show that the inner shelf has been eroded and the outer shelf shows a change from till grading offshore into glacial-marine sediments, as predicted from the model shown in Figure 4.2A.

According to Hughes (1981), terrestrial ice sheets have a basal thermal regime that is more complex than marine ice sheets; this is an important difference. Basal freezing

A

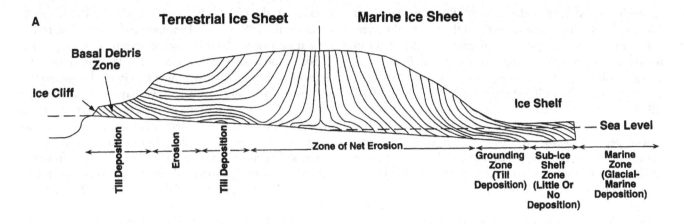

B

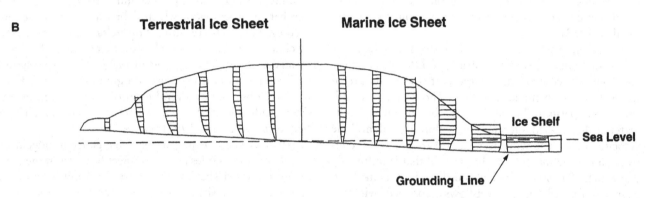

may occur near the margin of a terrestrial ice sheet (Fig. 4.2A), resulting in the creation of a thicker basal debris zone as debris eroded from interior (wet-based) regions of the ice sheet is frozen onto the base of the ice sheet near its margin.

Few areas exist where the base of Antarctic ice sheets can be observed, because they are predominantly grounded at or below sea level. One exception is in the Vestfold Hills of East Antarctica. There, Fitzsimons (1990) examined ice-marginal depositional processes and observed thick basal and englacial debris zones. In this region, meltwater, though limited in quantity, plays a key role in dispersing sediments in the proglacial environment. In fact, Fitzsimons (1990) concluded that the bedforms and deposits of the area closely resemble those of subpolar settings rather than polar settings.

Processes of Subglacial Deposition

Basal debris transported by the ice sheet, or within the bed directly beneath the ice sheet, is deposited as either a lodgement till or deformation till. Debris transported seaward of the grounding zone is deposited from the water column (glacial-marine sedimentation) or near the grounding zone by sediment gravity flow processes. Little sediment transport via meltwater has occurred through and

Figure 4.2. Model depicting (A) glacial flow trajectories of ice particles and different depositional settings and (B) relative flow velocities within the marine (West Antarctica) and terrestrial (East Antarctic Ice Sheet) portions of the Antarctic ice sheet (modified from Hughes, 1981).

seaward of the grounding zone of the Antarctic ice sheets, at least not during the interval of time recorded by sediment cores and high-resolution seismic records.

The first piston cores collected on the Antarctic continental shelf in the eastern Weddell Sea (IWSOE-69 and 70) and on the Ross Sea continental shelf (*Eltanin* cruises 27 and 32) penetrated overcompacted diamictons grading upward and seaward into softer glacial-marine sediment (Anderson, 1972a; Fillon, 1972). Since these cores were collected, hundreds of cores have been collected on the shelf; a significant portion of these have penetrated stiff diamicton. Detailed examination of these deposits led to the conclusion that they are probably lodgement tills (Anderson et al., 1980b; Anderson, Brake, and Myers, 1984; Anderson et al., 1991a).

The term "lodgement" was first used by Flint (1971) to refer to a process whereby slow-pressure melting of flowing ice releases particles from the basal debris zone and allows them to be plastered to the bed under the pressure of the overriding ice. The process involves dewatering under the load of the ice, which is believed to be responsible for the high shear stress of lodgement till (Boulton, Dent, and Morris, 1974). Increased shearing of the bed by the overriding ice and/or incorporation of water into the bed can lead to a transition, both temporal and spatial, from deposition of lodgement till to softer, more water-saturated till, such as deformation till, subglacial melt-out till, and flow till (Boulton, 1970; Boulton and Paul, 1976). Boulton, Dent, and Morris (1974) suggest that a transition in subglacial deposition from deformation- to lodgement-type processes with increasing depth in the bed may also occur as the stress of the overriding ice is attenuated below the threshold necessary for till deformation. Likewise, immobilization may be induced by loss of pore pressure and structural collapse caused by the escape of interstitial water from the bed. Changes in the abundance of meltwater at the ice–bed interface, as well as the pore pressure and water content of the bed, are controlled by the rates of basal melting and escape of meltwater from the system (Muller, 1983).

In reality, the tills sampled from the Antarctic continental shelf must represent a variety of different sediment end members controlled, in large part, by the nature of the bed on which the ice is riding. The degree of lithification, the permeability of the bed, the amount of meltwater at the ice–bed interface and within the bed, and the velocity of the ice flowing over the bed are important controls on the sedimentary qualities of till that is ultimately deposited. The style of subglacial deposition is likely to change spatially if the ice sheet flows across beds with different properties. Temporal changes must occur in subglacial deposition, as either the duration of shear forces increases or the bed thickness and flow velocity vary. The end product would be a stratigraphic progression of deformation till grading downward into lodgement till, or vice versa.

One of the most important factors regulating till deformation is the permeability of the deposit. Till with relatively high permeability is less apt to deform than till with relatively low permeability (Muller, 1983). This is because fluid escape from permeable till results in lower water pressure and correspondingly higher effective pressure. The grain size of the till is particularly important in regulating permeability.

Development of subglacial channels provides an alternate means of discharging water from the subglacial setting, but little evidence of channelized meltwater flow is seen on the Antarctic shelf, in either cored sediment or side-scan sonar records (Chapter 3). The implication is that the meltwater produced by pressure melting at the ice–bed interface is incorporated into the bed. An exception to this may occur in areas where the grounding zone of the ice sheet rested over impermeable bedrock, such as on the inner shelf, or on sand and gravel, such as on banks.

Recent discoveries of a deforming till layer beneath Ice Stream B in the Ross Sea region (Blankenship et al., 1987; Rooney et al., 1987; Alley et al., 1987a,b, 1989; Engelhardt et al., 1990) have demonstrated that sediment deformation offers a highly efficient means of transporting debris at the base of an ice sheet. The layer of deforming till beneath Ice Stream B averages 6 m in thickness and has a smooth upper surface. The lower layer is characterized by flutes that are hundreds of meters wide, up to 13 m deep, and oriented in the direction of flow. These are similar in scale to megascale glacial lineations that occur on the Ross Sea continental shelf (Chaper 3). Cores from the till layer beneath Ice Stream B recovered a water-saturated diamicton with a porosity of 40 ± 1% (Engelhardt et al., 1990).

Alley and colleagues (1987a) estimate average erosion rates on the order of 0.4 mm/yr in the catchment at the base of Ice Stream B. Sediment is transported in a conveyor-belt fashion to the grounding zone and deposited there by a number of processes of which debris flows are probably predominant. The resulting "till delta" is characterized by clinoforms (Fig. 4.3A). They estimate that "till deltas" formed at the grounding zone during the Holocene should be several tens of meters thick. Prograding deposits of this scale are observed in several locations on the continental shelf (Anderson et al., 1992b; Bart and Anderson, 1995; Vanneste and Larter, 1995) (Fig. 4.3B).

Subglacial sedimentation may change from lodgement to deformation modes as the ice sheet flows from the continent and inner shelf, with exposed crystalline bedrock, across the outer shelf where thick sedimentary deposits occur. Glaciologists refer to this as a change from sticky-to deforming-bed conditions (Alley et al., 1994). Not only does the style of subglacial sedimentation change, but the formation of a deformation till can result in a significant increase in sliding velocity of the ice sheet, thereby increasing rates of erosion, transport, and sedimentation (Alley et al., 1989) (Fig. 4.2B).

Marine geologists can contribute to the study of ice sheet dynamics by providing information about the nature of the bed upon which the ice sheet rested during the last glacial maximum when it was grounded on the continental shelf. A significant amount of the relief on the modern shelf is the footprint of the expanded ice sheet from the last glacial maximum. The deposits and geomorphic features at and near the present seafloor bear a record of subglacial processes and conditions at the ice–bed interface during and after ice sheet retreat from the shelf.

Subglacial Seismic Facies

Till Sheets. High-resolution seismic reflection profiles from the Ross Sea, Antarctic Peninsula, and northwestern Weddell Sea continental shelves show a stratigraphy dominated by stacked depositional units that are characterized by sharp, strongly reflective upper and lower erosional boundaries, chaotic to crudely layered internal reflector configurations, and common hyperbolic reflectors

A

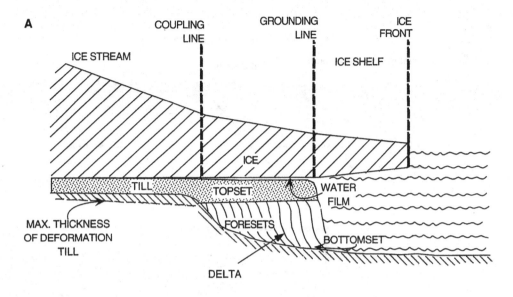

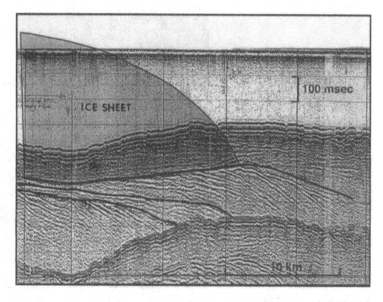

Figure 4.3. (A) Model for "till delta" formation (reprinted from Alley et al., 1987b, with permission from Elsevier Science Ltd), and (B) seismic profile (100 in³ water gun) from the north-western Weddell Sea showing seaward-dipping foresets interpreted as grounding zone deposits. Deposition is focused near the grounding line, and the foresets are composed primarily of slumps, debris flows, and glacial-marine diamicton. As the grounding line advanced seaward it constructed a series of stacked progradational units with foresets that are truncated by landward-sloping glacial unconformities (highlighted).

(Alonso et al., 1992; Anderson et al., 1992a; Bart and Anderson, 1995) (Fig. 4.4A). These units are referred to as till sheets (Shipp et al., 1994; Shipp and Anderson, 1997e). Isopach maps of till sheets in the Ross Sea indicate that individual units average a few tens of meters in thickness and thousands of square kilometers in area (Shipp et al., 1994) (Fig. 3.14). Individual units thicken in a seaward direction and toward the center of troughs and pinch out toward the inner shelf and along the flanks of banks (Shipp, Anderson, and Domack, in press).

Multibeam records of the uppermost till sheet in the Ross Sea show megascale glacial lineations that are oriented roughly parallel to trough axes (Fig. 3.6A). In cross section, these lineations display a mounded microtopography that is typical of the continental shelf seafloor (Fig. 3.6B). Side-scan sonar records show a variety of sub-glacial geomorphic features, and seismic profiles show thrust faults that offset till sheets on the outer shelf (Fig. 4.4B). These thrusts are indicative of a frozen region near the outer portion of the ice sheet that permitted upward movement of ice-flow lines and lifting of strata. These combined geomorphic and structural features lend support to the deforming bed mechanism for deposition of till sheets.

Ice Stream Boundary Ridges. Seismic profiles oriented east-west across the eastern Ross Sea continental shelf show ridge and trough topography. The trough axes display clear evidence of erosion, which is believed to be concentrated along the flow axes of paleo-ice streams. The trough boundaries are marked by a series of broad (average 50 km), high (average 150 msec) ridges that display an internal reflection character ranging from chaotic

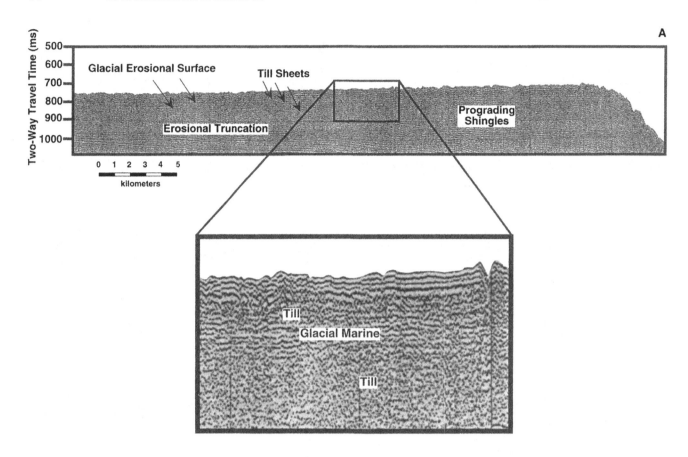

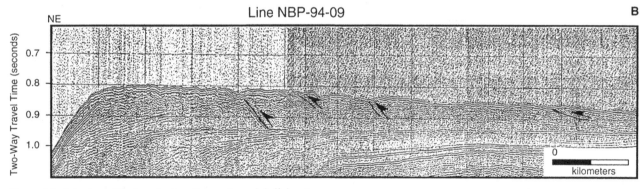

Figure 4.4. Seismic profiles from the Ross Sea continental shelf showing (A) stacked till sheets and (B) thrust faults offsetting these till sheets.

to stratified (Fig. 4.5A). Megascale glacial lineations characterize the seafloor. Discrete stacked packages that laterally accrete toward the trough axis compose the ridges. These accreting units are interpreted to be derived from subglacial sedimentation along the ice stream boundary where flow velocities decelerate causing deposition of debris that is eroded from the trough axis; therefore, they are called ice stream boundary ridges (Anderson et al., 1992a). Other paleo-ice stream boundaries are marked by erosional scarps (Fig. 4.5B), which perhaps implies that two adjacent ice streams were flowing at very different rates (Shipp and Anderson, 1997d). Ice stream boundary ridges have not been observed elsewhere on the Antarctic

continental shelf, at least not in the northwestern Weddell Sea and Antarctic Peninsula region, and no known corollary exists in the Northern Hemisphere. Perhaps they are unique to the eastern Ross Sea continental shelf because of ice stream maturity and/or the highly variable flow rates of ice streams in that area.

Sediments of the Subglacial Environment

Massive diamicton is one of the most widespread late Pleistocene lithofacies on the continental shelf. It has been sampled via piston cores from till sheets and ice stream boundary ridges.

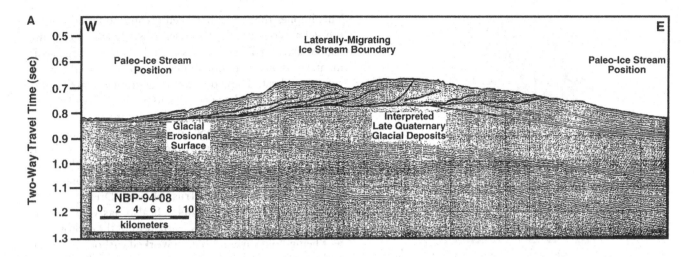

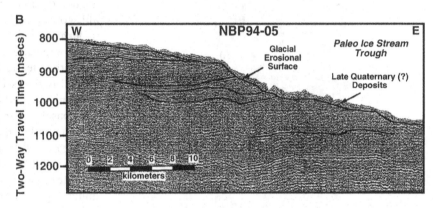

Figure 4.5. Interpreted seismic profiles from the Ross Sea continental shelf showing (A) an ice stream boundary ridge and (B) an erosional scarp associated with an ice stream boundary. Ice stream boundary ridges are believed to form at the boundary of two actively flowing ice streams, whereas erosional scarps are believed to form at the boundary between an actively flowing ice stream and a dormant ice stream.

Barrett (1975) examined diamictons recovered in DSDP sites 270 through 273 from the Ross Sea continental shelf. He noted the similarity of these diamictons to tills from Ohio. Detailed analyses of diamictons collected in piston cores from the continental shelves of the Ross Sea (Kellogg, Truesdale, and Osterman, 1979; Anderson et al., 1980b, 1984, 1992a), Weddell Sea (Anderson, 1972b; Anderson et al., 1980b; Elverhøi, 1981; Haase, 1986), and Wilkes Land (Anderson et al., 1980b; Domack, 1982) led to the consensus that these diamictons include tills. The tills are similar to some glacial-marine sediments and debris flows; therefore, distinguishing between these sediment types may be very difficult (Chriss and Frakes, 1972; Kurtz and Anderson, 1979; Anderson et al., 1980b, 1984, 1992a). Tills should possess the following characteristics: (1) lack of sorting; (2) absence of marine fossils or presence of only reworked marine fossils; (3) general lack of stratification; (4) random pebble fabric perpendicular to the bedding plane, but preferred grain fabric along bedding surfaces; (5) rounded pebble shape relative to glacial-marine sediment; (6) textural and mineralogic homogeneity within individual units, which is generally indicated by

fairly constant magnetic susceptibility values; (7) little or no organic carbon; and (8) petrographic similarity within a given petrographic province.

Sedimentary Structures. X-radiographs from dozens of piston cores that penetrated tills typically show no stratification or, in rare cases, crude alignment of grains and pebbles (Fig. 4.6). Upper contacts with younger glacial-marine sediments tend to be rather sharp; however, we are not able to observe the basal contacts of tills because cores taken to date have not sampled them.

Grain Size. Grain size distribution curves for tills from the eastern Weddell Sea, Ross Sea, and Wilkes Land continental shelves are illustrated in Figure 4.7. Weddell Sea tills show a total lack of sorting. Ross Sea tills contain a more fine-grained matrix than Weddell Sea tills, whereas tills from the Wilkes Land shelf contain greater sand concentrations. These variations in grain size distributions are believed to reflect differences in source and transport distance. Weddell Sea and Wilkes Land shelf tills were collected within a few tens of kilometers of the continent. In contrast, Ross Sea tills were collected several hundred kilometers from possible source areas of crystalline rocks.

Figure 4.6. X-radiograph of a till showing lack of stratification and random pebble fabric relative to the bedding plane.

This implies that there has been greater comminution of the debris constituting Ross Sea tills. Furthermore, the source area for Ross Sea tills includes several large sedimentary basins that contributed silt and clay to the overriding ice sheet. These differences in grain size are important in controlling the porosity and permeability of the till and, therefore, its tendency for deformation.

Fossil and Organic Carbon Content. Till may contain rare microfossils including reworked foraminifera (Milam and Anderson, 1981), radiolarians, diatoms (Kellogg, Truesdale, and Osterman, 1979), as well as pollen and spores (Truswell, 1983; Truswell and Drewry, 1984; Truswell and Anderson, 1984). It also has very low organic carbon concentrations, with most of the carbon content being "old carbon" (i.e., reworked carbon). Organic carbon concentrations within a given core should vary slightly, whereas glacial-marine sediments should have greater organic carbon concentrations and display variable concentrations within a unit (Fig. 4.8).

Textural and Mineralogic Homogeneity. Strong textural and compositional homogeneity within individual units is another feature of till, especially when compared to glacial-marine sediment (Fig. 4.9A,B). This homogeneity results from the continuous mixing process associated with basal transport, both within the ice and within deforming till layers. Compositional homogeneity occurs in all size ranges, from clay particle to pebble size (Anderson et al., 1980b). Magnetic susceptibility measurements reflect the homogeneity of till when compared to the variable mineralogic composition of iceberg-rafted glacial-marine units (Fig. 4.9C).

Particle Fabric. Pebble fabric studies are difficult to perform on piston cores due to the small diameter of the cores and changes in pebble alignment caused by the coring process. Such studies are limited to measurements of elongate pebbles relative to the bedding plane. Even this type of measurement must be done with caution, because pebbles may be reoriented during the coring process. Still, pebble fabric is such an important criterion for analysis of ancient glacial deposits that it warrants some examination by those who deal with modern and late Pleistocene deposits.

Kurtz and Anderson (1979) examined pebble fabric relative to the bedding plane of tills and debris flows from the Ross and Weddell seas. Domack (1982) conducted similar two-dimensional pebble orientation analyses on tills from the Wilkes Land continental shelf. Domack (1982) also attempted to measure the three-dimensional orientation of the long axes of pebbles in a single piston core by using pairs of X-radiographs taken perpendicular to one another. His results showed a near vertical orientation of pebbles in some tills. Results of a recent pebble fabric analysis of Ross Sea tills have shown that deformation till may be characterized by more vertical alignment of pebbles relative to lodgement tills (Licht, personal communication).

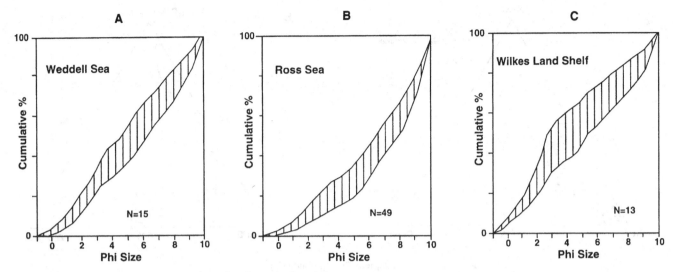

Figure 4.7. Representative grain size envelopes for till samples from (A) Weddell, (B) Ross, and (C) Wilkes Land continental shelves. N = number of samples analyzed (from Anderson and Molnia, 1989; reprinted by permission from the American Geophysical Union, © 1989).

Figure 4.8. Downcore concentrations in organic carbon in till and glacial-marine sediment (provided courtesy of Eugene Domack).

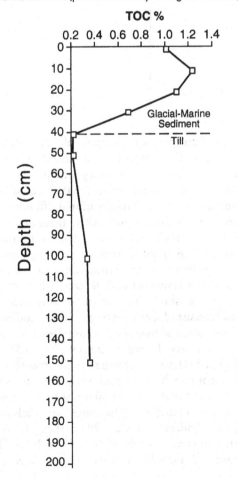

The most successful attempt at measuring pebble fabric in sediment cores from the Antarctic continental shelf was that of Hambrey, Ehrmann, and Larsen (1991), who conducted fabric analyses on core from ODP Site 742 in Prydz Bay, East Antarctica. The analyses included grain orientations on horizontally cut faces of core. The results show variations in grain fabric, from nearly random to strongly preferred orientations (Fig. 4.10). Their data show fairly strong preferred orientations in diamictons, which are interpreted as lodgement and water-lain tills with weak to random alignments in glacial-marine sediments (Fig. 4.10).

Silt and sand grain fabric analyses of Antarctic sediments are not reported in the literature, but this is probably a good criterion for distinguishing till from glacial-marine sediment. More important, microfabric analysis of till may provide valuable information about the interaction between the ice sheet and its bed. Muller (1983) argues that it is at this scale that details of the lodgement and dewatering process should be recorded.

Particle Shape and Surface Features. The shapes of pebbles from diamicton collected on the continental shelf and slope were characterized by Barrett (1975), Anderson and colleagues (1980b), Domack (1982), and Kuhn and colleagues (1993). Pebbles from tills yield medial roundness values using Krumbein's (1941) procedure. When the roundness-sphericity values for pebbles from tills are plotted, they fall within Boulton's (1978) basal transport field (Fig. 4.11A). Of the pebbles examined by Anderson and colleagues (1980b) and Domack (1982), ~ 12% were found to be highly striated and ~ 80% were faceted.

The shapes and surface textures of sand grains from the Weddell Sea and Ross Sea tills were examined by Mazzullo and Anderson (1987) using an automated Fourier method for shape analyses and scanning electron microscopy for surface texture characterization. Their results show greater roundness for grains taken from tills compared to grains

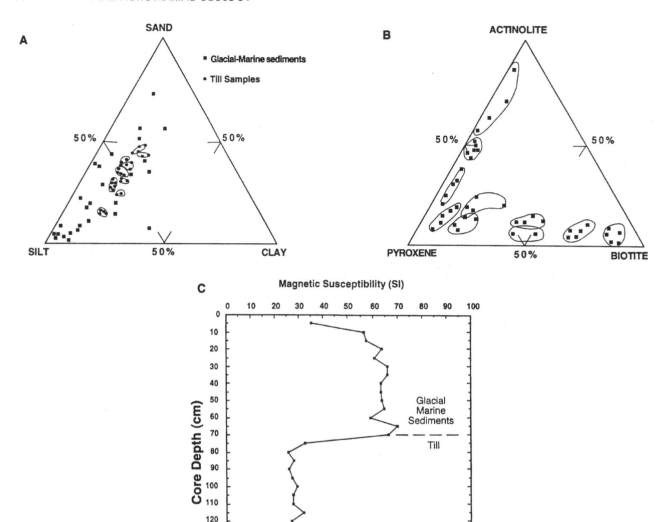

Figure 4.9. Ternary plots illustrating textural and mineralogic homogeneity in Ross Sea tills and glacial-marine sediments: (A) sand/silt/clay ratios and (B) heavy mineral ratios for samples taken at 50-cm intervals in piston cores (Myers, 1982a). Till samples (dots) from individual units are circled in (A). A wide scatter in textural properties of surface glacial-marine sediments (squares) versus tills is observed. These petrographic differences are manifested in (C) magnetic susceptibility data, which show less variability in tills relative to glacial-marine sediments.

from glacial-marine units. This was attributed to greater abrasion and edge attrition during glacial transport, which is presumed to be greater for subglacial transport than for englacial and supraglacial transport (Boulton, 1978). Scanning electron microscopy shows that typical glacial surface textures, such as surficial fracture textures and breakage blocks, are commonly found on grains from both basal tills and glacial-marine sediments (Fig. 4.12 A–C).

Petrologic Provenance. Since the turn of the twentieth century, glacial geologists have utilized petrographic data (mineral and clast content) to identify till and to determine the source area of these deposits, thus providing a basis for construction of glacial paleodrainage maps. The concept of petrologic provinces necessitates that glacial sediments exhibit spatial as well as compositional affinities due to the fact that ice sheets transport debris along discrete flow lines. Hence, the rock and mineral composition of a till reflects the geology along that flow line. The underlying assumption in this type of petrographic analysis is that marine ice sheets transport and deposit sediment at their base in a fashion similar to that of terrestrial ice sheets. Tills should consist of fewer varieties of rocks and minerals than iceberg-rafted sediments, because the source area of ice-rafted debris may be quite extensive (Fig. 4.9).

Domack (1982) demonstrated that tills from the Wilkes Land continental shelf can be positively correlated with exposures of continental rocks. Similar results were obtained from studies of diamicton on the continental shelves of the Weddell Sea (Anderson et al., 1991a) (Fig. 4.13), Ross Sea (Anderson et al., 1980b, 1984, 1992a; Jahns, 1994), and Antarctic Peninsula, as well as the adjacent Mar-

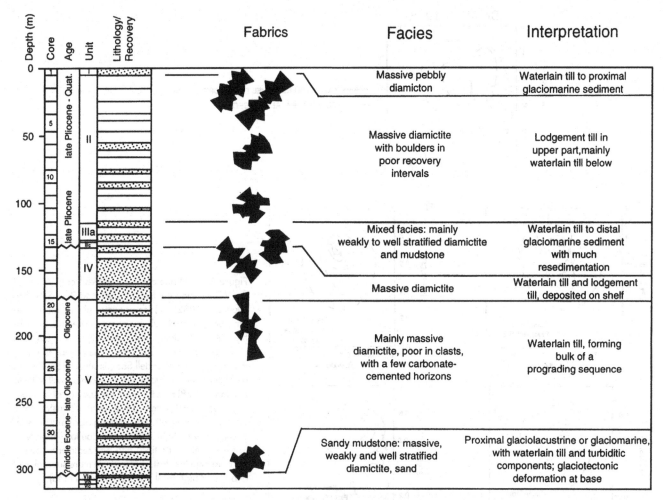

Figure 4.10. Stratigraphy, sedimentology, and interpretation of the cored interval from ODP Site 742 in Prydz Bay illustrating results from a pebble fabric analysis by Hambrey, Ehrmann, and Larsen (1991). The lodgement and water-lain tills (possible deformation till) both have stronger pebble fabrics compared to glacial-marine deposits.

guerite Bay (Kennedy and Anderson, 1989). These studies demonstrate that discrete till provinces can be mapped over distances of tens and even hundreds of kilometers across the continental shelves. The provenance studies provide the most compelling sedimentologic evidence for the former presence of ice sheets in these areas and provide a means of reconstructing the paleodrainage of former ice sheets (Fig. 4.13).

Shear Strength. Until recently it was believed that one of the diagnostic features of tills on the Antarctic continental shelf was their high cohesive strength, which was attributed to overcompaction by ice (Anderson et al., 1980b). We now realize that tills may remain quite soft, either if pore waters are not allowed to escape during sedimentation or if the till is subject to post-depositional deformation (Engelhardt et al., 1990). Both types of till may be common on the Antarctic continental shelf.

Muller (1983) points out that the full load of the ice sheet is expressed as a combination of lithostatic and hydrostatic pressure, and sediment measures the total load of the ice only if pore water escapes freely from the system. We therefore should avoid relating sediment consolidation to ice thickness. However, I believe that shear strength can be a useful criterion for distinguishing different types of

diamicton (glacial-marine sediments, lodgement till, and deformation till) when used in conjunction with the other previously described criteria.

To date, most of the diamictons sampled from the Ross Sea, eastern Weddell Sea, Wilkes Land, and Antarctic Peninsula continental shelves have high shear strengths relative to glacial-marine sediments. Indeed, bent core barrels often provide the first evidence that a piston core has penetrated a till (Fig. 4.14). Anderson and colleagues (1980b) and Kurtz, Anderson, Balshaw, and Cole (1979) measured the shear strengths of tills from the Weddell Sea, Ross Sea, and Wilkes Land continental shelves. They obtained values of greater than 2.0 kg/cm for tills, whereas glacial-marine sediments typically have strengths of less than 1.5 kg/cm.

Since Kurtz and colleagues (1979) completed their study, additional shear strength data have been acquired on dozens of cores from the Ross Sea and Antarctic Peninsula continental shelves. As more data are gathered, some trends in shear strength are emerging, and these patterns

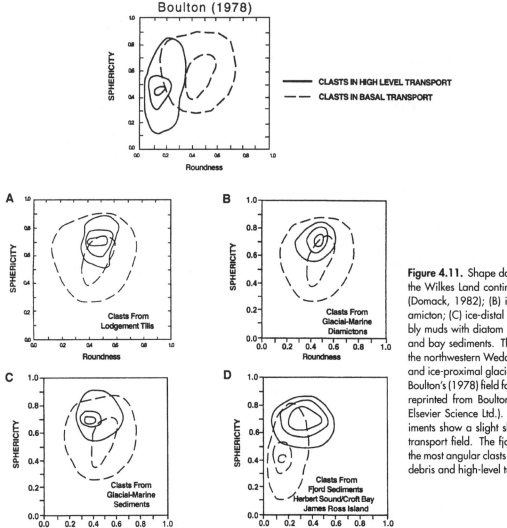

Figure 4.11. Shape data for pebbles from: (A) tills of the Wilkes Land continental shelf and Ross Sea shelf (Domack, 1982); (B) ice-proximal glacial-marine diamicton; (C) ice-distal glacial-marine sediments (pebbly muds with diatom tests); and (D) clasts from fjord and bay sediments. The latter three groups are from the northwestern Weddell Sea (Smith, 1985). The tills and ice-proximal glacial-marine sediments all fall into Boulton's (1978) field for basal transport (dashed area; reprinted from Boulton, 1978 with permission from Elsevier Science Ltd.). The distal glacial-marine sediments show a slight shift toward Boulton's high-level transport field. The fjord and bay sediments contain the most angular clasts and indicate more superglacial debris and high-level transport (dashed area).

are used in the following classification scheme for diamictons of the Antarctic continental shelf: glacial-marine sediments typically display a fairly linear increase in shear strength with increasing burial depth, except where lithologic changes result in abrupt differences. Tills that display high shear strength (> 3.0 kg/cm) and have fairly uniform shear strength with depth in a given unit are interpreted as lodgement tills (Fig. 4.15A). Tills with shear strengths ranging between 1.5 and 5.0 kg/cm and variable shear strengths with depth within a given unit are interpreted as deformation tills (Fig. 4.15B).

Ross Sea diamicton shear strength data have been examined in detail to see if there is a pattern to the areal distribution of lodgement till and deformation till relative to, for example, the axis of glacial troughs. In general, stiff lodgement tills occur most commonly on the inner shelf and along the flanks of troughs and are associated with striations and flutes (Chapter 3). Soft deformation till is more common in the outer shelf sectors of troughs, and is characterized by hummocky seafloor and megascale glacial lineations.

Summary

At least two till types are recognized on the Antarctic shelf: stiff or "overcompacted" till that is assumed to be deposited by lodgement processes (lodgement till), and soft till that is believed to be deposited from a deforming till layer (deformation till). Distinguishing lodgement from deformation till is important because these deposits yield crucial information about the behavior of past ice sheets. Lodgement and deformation tills are probably end members to a more transitional range of processes and deposits, but this two-part classification will have to suffice until more detailed work leads to criteria for distinguishing other intermediate till types.

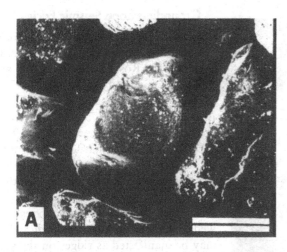

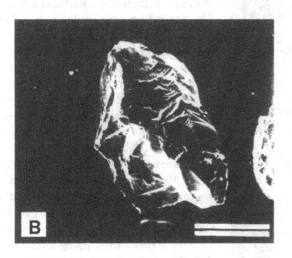

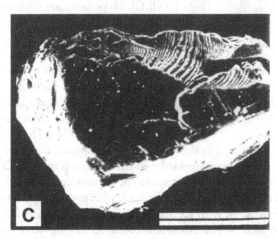

Figure 4.12. Photomicrographs of sand grains from Ross Sea till. Grains (A) with rounded to subrounded shapes are not uncommon. Photomicrographs (B) and (C) illustrate striations and facets on angular sand grains. The angular grains are most abundant in tills and glacial-marine sediments of the Antarctic continental shelf (Mazzullo and Anderson, 1987).

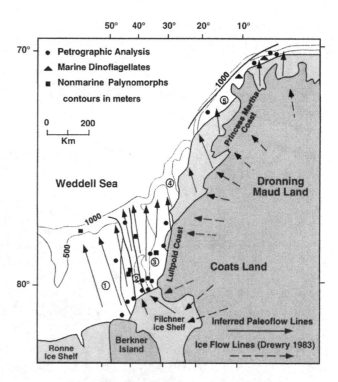

	Province 1	Province 2	Province 3	Province 4	Province 5
Cores	G 15, 2-22-2, 2-29-1	G 1, 3-1-1	G 6, G 2, 3-7-1	G 17, 2-20-1, 3-1-2, 3-3-1	16, 28, 29, 30, 3-18-1, 3-17-1
Location	West of Crary Trough	Central Crary Trough	Eastern Edge of Crary Trough and Basement High	East of Crary Trough	Princess Martha Coast
Pebble Content	(Qtz Arenite,> Qtz Wacke),^ & (Basalt/ Diabase/ Andesite)	Graywacke, Meta-Qtz, Slate/ Phyllite, Shale, >>Arkose, & >>High Grade Met.	Diorite, >Granite, >>Meta-Qtz, & >> Qtz-Wacke	Basalt/ Diabase/ Andesite	Basalt/ Diabase/ Andesite
Sand Content	Qtz/ Feldspar, Graywacke,>> Sed. & Met. Rockfrags.	Qtz/ Feldspar, Graywacke,>> Sed. & Met. Rockfrags.	Qtz/ Feldspar, Graywacke,>> Sed. & Met. Rockfrags.	Qtz/ Feldspar, Graywacke,>> Sed. & Met. Rockfrags. Volcanic, >Basic, >Intermed.	Volcanic, >Basic,> Intermediate
Palynomorph Content	Terrestrial	Terrestrial	Terrestrial	Terrestrial	Marine

Figure 4.13. Mineralogic provinces for Weddell Sea tills based on factor analysis of petrographic data for samples taken at 50-cm intervals in cores. Recycled dinoflagellates and palynomorphs also indicate different source areas for tills of the region (modified from Anderson et al., 1991a).

GROUNDING ZONE ENVIRONMENT

Environmental Setting

For purposes of this discussion, the "grounding zone" refers to the transition zone where ice may be locally and intermittently coupled to the bed and where the bed may be frozen, freezing, melting, or melted (Hughes, 1981; Menzies, 1995). Doake and colleagues (1987) have shown that the grounding zone of the Filchner Ice Shelf is a broad

Figure 4.14. Bent core barrels are often the first indication that a till has been sampled.

Grounding Zone Seismic Facies

Grounding Zone Wedges. Results from a seismic facies analysis of late Pleistocene strata on the Ross Sea (Anderson et al., 1992a; Shipp and Anderson, 1997c), Weddell Sea (Elverhøi and Maisey, 1983; Anderson et al., 1992b), and Antarctic Peninsula (Bart and Anderson, 1995, 1996; Vanneste and Larter, 1995) continental shelves have led to the recognition of geomorphic features and deposits that are believed to record grounding zones of the last glacial maximum. Extensive, wedge-shaped depositional bodies that may be manifested as ridges on the seafloor are prominent among these features. The grounding zone wedge in the Ross Sea has been mapped for tens of kilometers along-strike (Shipp and Anderson, 1997c) (Fig. 4.16A). Grounding zone wedges also occur on the Antarctic Peninsula shelf (Fig. 4.16B), on the northwestern Weddell Sea shelf (Fig. 4.16C), and possibly on the eastern Weddell Sea shelf (Elverhøi and Maisey, 1983).

Grounding zone wedges range from 50 to 100 m thick and reach 80 km in width. Seismic records show massive, layered, and chaotic reflection patterns (Fig. 4.16A–C) and crude to distinct, low-angle (<3°) seaward-dipping foresets that downlap onto glacial erosion surfaces (Fig. 4.16B,C). Hyperbola are common. In strike sections, the wedges pinch and swell, whereas they typically terminate by downlap on the underlying erosional surface in dip sections. Multibeam records across the wedges in the Ross Sea show abundant megascale lineations that extend to their seaward edge (Shipp and Anderson, 1997a) (Fig. 3.7A). The multibeam records also show large arcuate-shaped iceberg furrows on the landward sides of the wedges (Fig. 3.12C).

A mechanism for formation of grounding zone wedges has been proposed by Alley and colleagues (1989) based on their work with Ice Stream B in the Ross Sea. According to their till delta model, erosion rates along ice streams are high and require relatively rapid deposition at or near the grounding line. As the grounding zone advances, the sediment wedge progrades, depositing top set beds composed of till, as well as foreset and bottomset beds composed of sediment gravity flow deposits (Fig. 4.3). The size and shape of the deltas are predominantly controlled by the availability of accommodation space on the shelf and the duration of the grounding event at any given location on the shelf. This mechanism is capable of developing prograding wedges that are tens of meters thick and extend

(∼ 100 km wide) area where the ice shelf is locally grounded. Likewise, Blankenship and colleagues (1987) demonstrated that there is an extensive area landward of the Ross ice Shelf grounding zone where glacial ice is locally coupled to the bed.

Grounding zones are a highly dynamic setting. Significant changes in grounding zones probably occur at rates of 50 to 200 years (Alley et al., 1987a,b). From this we would expect sedimentation within this environment to include a wide spectrum of processes ranging from subglacial to glacial-marine and sediment gravity flow processes. In addition, meltwater may contribute to sedimentation, although there is little evidence that this is the case.

across several kilometers of the shelf in 1000 years (Alley et al., 1989). The apparent lack of interfingering relationships between the wedges and glacial-marine strata, given the resolution of our data, distinguishes grounding zone wedges from till tongues (King and Fader, 1986).

Grounding Line Ridges. High-resolution seismic profiles and side-scan sonar records from the Bransfield Basin show glacially striated bedrock at the seafloor and relatively thick sedimentary strata in troughs and associated trough mouth fans (Banfield and Anderson, 1995) (Fig. 4.17). Conspicuous within the troughs are linear mounds of sediment, interpreted as grounding line ridges (Figs. 3.30, 4.18). Detailed mapping of these ridges revealed that they extend across the width of the troughs. The ridges occur at approximately the same mid-shelf location and water depth in different troughs (Fig. 4.17). The number, size, shape, and seismic facies do, however, vary from trough to trough. Stratification ranges from unstratified to clinoforms that dip either away from or toward the paleoglacial center and, in some cases, display reversals in dip directions and a back-stepping character (Fig. 4.18). To my knowledge, the only other documentation of these features was by Wong and Christoffell (1981), who mapped a large ridge extending along the coastal margin of the Drygalski Trough in the Ross Sea.

Grounding line ridges are presumed to be formed by a combination of subglacial deposition and sediment mass movement near the grounding line of ice streams and outlet glaciers. Unlike grounding zone wedges, they show evidence for ice pushing and grounding line retreat and intertongue with glacial-marine deposits (Fig. 4.18). They do not show significant progradation. They are virtually identical to morainal ridges that occur in Northern Hemisphere regions (Banfield, 1994; Banfield and Anderson, 1995) and in the inland passage of Chile (DaSilva, Anderson, and Stavers, 1997). Their overall geometry implies that grounding line ridges are deposited at a relatively stationary, but perhaps short-lived, grounding line position, and that the grounding zone is narrow. This is perhaps related to their low-latitude position in Antarctica and may explain why they are so similar to morainal ridges of the Northern Hemisphere and Chilean Inland Passage.

Sediment of the Grounding Zone Environment

Glacial-marine deposits of the Antarctic seafloor encompass a wide range of sediment types, including diamictons that are virtually identical to basal till and diatomaceous ooze with only minor amounts of ice-rafted material. The relative role played by glacial versus marine processes controls the composition and texture of these sediments.

Little is known about sedimentary facies within the grounding zone environment, but it is assumed that prominent lithofacies include subglacial till, sediment gravity flows, and glacial-marine diamicton. Anderson, Kurtz,

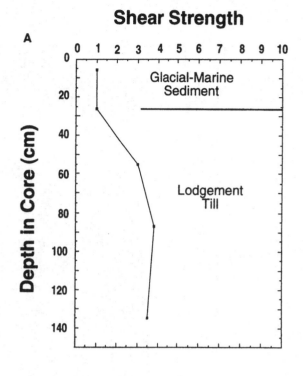

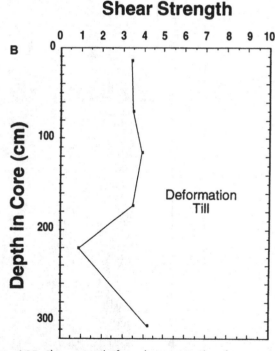

Figure 4.15. Shear strength (from shear vein) values for: (A) glacial-marine sediments resting on lodgement till (relatively constant shear strength with depth) and (B) deformation till showing variable shear strength with depth in the core.

Domack, and Balshaw (1980b) and Anderson, Kennedy, Smith, and Domack (1991c) have described diamicton in the Ross and Weddell seas that they believe was deposited within the grounding zone environment. They call this "transitional glacial-marine sediment" because it was deposited in the highly dynamic zone that is transitional

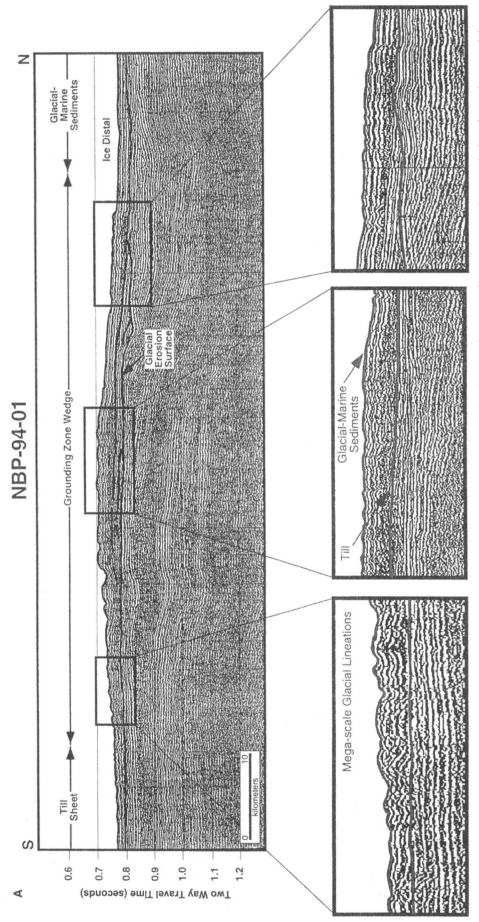

Figure 4.16. Seismic profiles showing grounding zone wedges from the (A) Ross Sea (NBP94-01), (B) Antarctic Peninsula shelf (PD90-56), and (C) northwestern Weddell Sea shelf (PD91-22). The Ross Sea grounding zone wedge (A) has a chaotic reflection pattern that grades seaward into acoustically layered glacial-marine deposits. Profile PD90-56 (B) shows stacked wedges with seaward-dipping foresets bounded by erosional surfaces. Profile PD91-22 (C) shows a single grounding zone wedge with a chaotic internal reflection character.

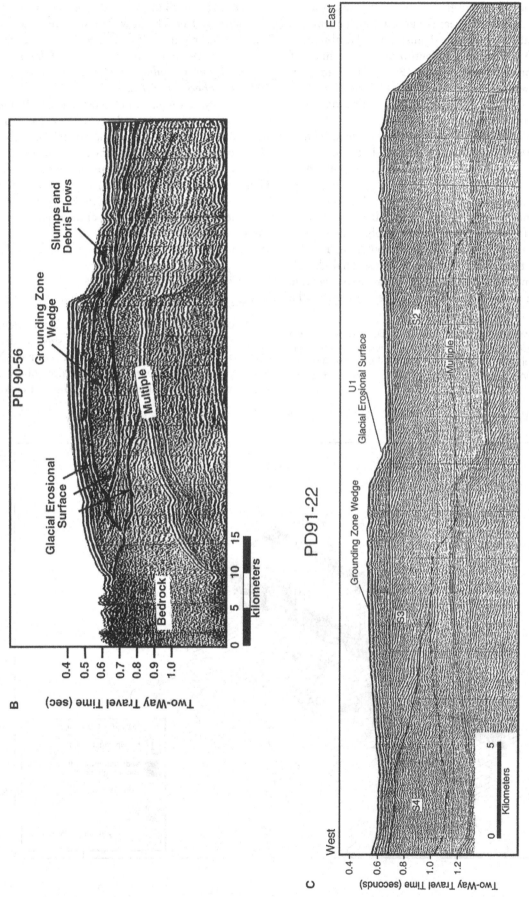

Figure 4.16. (*Continued*)

between the ice shelf and the ice sheet. It is also the region where basal melting near the grounding line removes most, if not all, basal debris. Transitional glacial-marine sediments are probably equivalent to the water-lain till of Hambrey, Ehrmann, and Larsen (1991). The diamicton is, in many ways, similar to till in that it tends to: (1) be either massive or display very crude stratification (Fig. 4.19); (2) be unsorted to very poorly sorted (Fig. 4.20A); (3) be texturally and mineralogically homogeneous within individual units; and (4) have pebble shapes that plot within Boulton's (1978) subglacial transport field (Fig. 4.11B).

Glacial-marine diamicton differs from lodgement till in that it is not overcompacted, and it has a more random pebble fabric that results from pebbles having settled through the water column (Anderson et al., 1980b). The main distinguishing criterion of transitional glacial-marine sediments is the presence of a sparse benthic foraminiferal assemblage believed to be endemic to the sub-ice shelf environment (Anderson, 1975a; Milam and Anderson, 1981). This low-diversity fauna consists of a few robust, highly calcified species that occur in low abundance. Dominant species include *Globocassidulina crassa*, *Globocassidulina subglobosa*, *Cibicides refulgens*, and *Eherenbergina hystrix*.

Hambrey, Ehrmann, and Larsen's (1991) detailed analyses of diamicton in ODP Leg 119 Site 742 cores from Prydz Bay include water-lain till, which is probably equivalent to transitional glacial-marine sediment and shows fairly strong to weak alignment of pebbles (Fig. 4.10). Hambrey and colleagues (1991) suggest that variations in grain fabric may reflect changes in the style of sedimentation due to alternate grounding and decoupling of the ice sheet.

ICE SHELF ENVIRONMENT

Environmental Setting

According to Drewry (1986), the fate of debris that reaches the grounding line of an ice sheet is dependent upon the vertical distribution of that debris in the ice sheet and ice shelf, the bottom melt rate at any point on the ice shelf, and the upflow line sedimentation-melt history. The

Figure 4.17. Distribution of seismic facies in the Bransfield Basin (from Banfield and Anderson, 1995; reprinted with permission from the American Geophysical Union, © 1995). Also shown is the location of seismic profile PD91-20A (gray line). (See Fig. 4.18 for examples of seismic facies.)

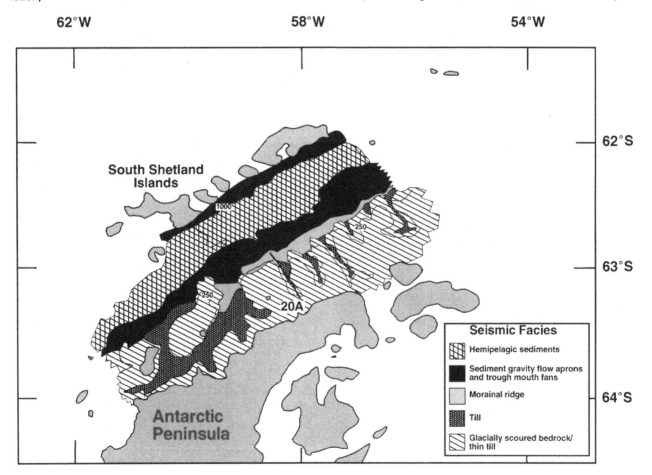

vertical distribution of debris in the ice shelf is controlled by the internal flow regime of the ice sheet and ice shelf.

Existing models for the dynamics and basal thermal regime of ice shelves are predominantly theoretical, and there is a need to gather base-line data to test these models. There is a general consensus among glaciologists and physical oceanographers that the thermal regime at the base of the Ross and Ronne-Filchner ice shelves is one where melting prevails seaward of the grounding line (Fig. 4.2), regardless of whether the ice shelf is advancing or retreating (Gill, 1973; Robin, 1979; Thomas, 1979; Hughes, 1981; Drewry and Cooper, 1981). Calculated basal melt rates are on the order of a few meters per year with an upper limit of 30 m/yr (Gill, 1973; Robin, 1979). Flow rates near the grounding line are estimated to be tens to hundreds of meters per year (Drewry and Cooper, 1981). This implies that a basal debris zone 10 to 20 m thick will be melted within a few kilometers of the grounding line (Drewry and Cooper, 1981). The sediment deposited within this grounding line proximal zone is probably diamicton (transitional glacial-marine sediment) and it may become quite thick if the grounding zone remains stable for long periods of time. The melt-out of basal debris near the grounding zone is supported by the absence of basal debris layers in ice cores from the Ross (Zotikov, Zagorodnov, and Raidovsky, 1980; Drewry and Cooper, 1981) and Lazarev ice shelves (Kolobov and Savatyugin, 1982).

Substantial basal freezing occurs seaward of the Amery Ice Shelf grounding line (Budd, Corry, and Jacka, 1982). Thus, it is more likely to transport basal debris to its calving margin than the Ross and Ronne-Filchner ice shelves. This difference is probably due to differences in the sub–ice shelf oceanography, particularly in the temperature and residence time of water masses that flow beneath these ice shelves (Gill, 1973; Robin, 1979).

Drewry and Cooper (1981) suggest that the thermal regime of small ice shelves is more conducive to basal freezing and the entrainment of basal debris layers compared to large ice shelves. This concept is supported by the occurrence of debris layers on some of Antarctica's small ice shelves (Debenham, 1919; Kellogg and Kellogg, 1987), and by observations of sediment-laden icebergs believed to be derived from small ice shelves and ice tongues (Anderson, Domack, and Kurtz, 1980a) (Fig. 4.21A,B).

The distribution of sediment within an ice shelf is also controlled by the proximity of the grounding line to valley and outlet glaciers, which may supply englacial and supraglacial debris. As this distance decreases, there is greater vertical distribution of sediment in the ice shelf, and debris is transported farther onto the continental shelf. Ice shelves connected to marine ice sheets probably lack englacial and supraglacial debris because the sources of such debris may be widely scattered. For example, the grounding line of the Ross Ice Shelf is separated from potential sources of englacial and supraglacial debris by hundreds of kilometers, excluding sources from Ross and Roosevelt islands.

The McMurdo, Larsen, and George VI ice shelves are directly nourished by outlet and valley glaciers. These ice shelves are more likely to entrain and transport debris at all levels within the ice. Pedly, Paren, and Potter (1988) describe basal debris layers over portions of the George VI Ice Shelf. Supraglacial debris layers occur

Figure 4.18. Seismic profile PD91-20A from the Bransfield Basin shows back-stepping grounding line ridges (see Fig. 4.17 for profile location) (from Banfield and Anderson, 1995; reprinted with permission from the American Geophysical Union, © 1995).

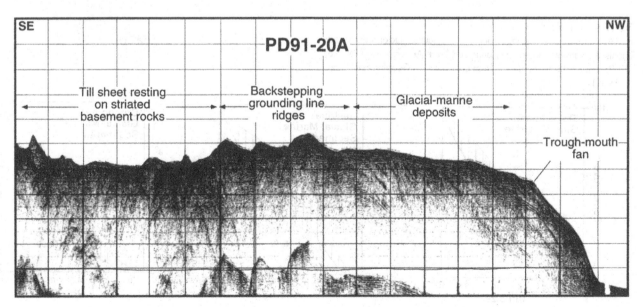

on the McMurdo Ice Shelf (Kellogg, Kellogg, and Stuiver, 1990) and on small fringing ice shelves of the Antarctic Peninsula region (Fig. 4.22).

Sediment of the Ice Shelf Environment

Sub-ice shelf deposits are too thin to be imaged on seismic records, hence our discussion of these deposits is limited to the deposits themselves. Sampling beneath Antarctic ice shelves has been limited to a few observations beneath the Ross (Webb et al., 1979) and Lazarev ice shelves (Kolobov and Savatyugin, 1982).

In 1977 and 1978, several gravity cores were acquired beneath the Ross Ice Shelf as part of the Ross Ice Shelf Project (RISP) (Webb et al., 1979). The cores penetrated a thin (< 20 cm thick), "soft" diamicton containing sedimentary clasts consisting predominantly of diatomaceous ooze and claystone (Webb et al., 1979). The surface sediments overlie an overcompacted diamicton of similar composition, except that it contains siliceous and calcareous microfossils. The lower overcompacted diamicton unit was interpreted as till by Anderson and colleagues (1980b). The only other sediment core acquired beneath a modern ice shelf was collected beneath the Lazarev Ice Shelf (Kolobov and Savatyugin, 1982). This sample was a 28-cm-long core that penetrated greenish-gray, unstratified silty clay (90% by weight) with a slight admixture of sand grains (< 10%) and sparse diatom fragments.

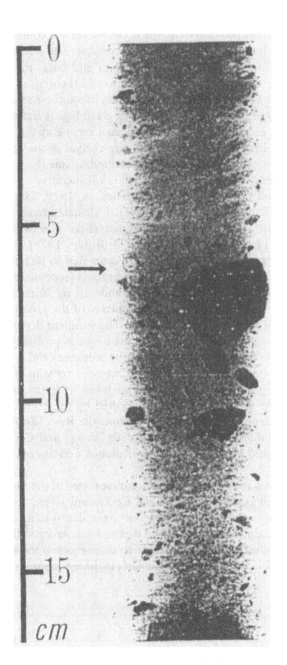

Figure 4.19. (right). X-radiograph of transitional glacial-marine deposit. The deposit is massive below the arrow and shows crude stratification above the arrow.

Figure 4.20. (below). Representative cumulative grain size envelopes for (A) transitional, (B) residual, and (C) compound glacial-marine sediments. N = number of samples analyzed (from Anderson and Molnia, 1989; reprinted with permission from the American Geophysical Union, © 1989).

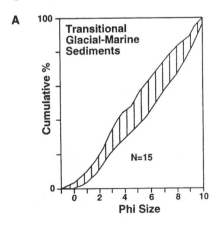

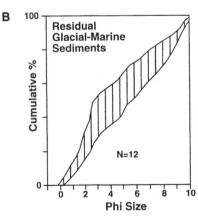

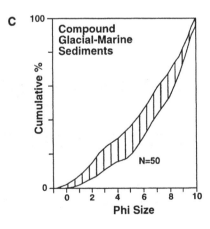

Figure 4.20. (A) Basal debris layers in an overturned iceberg calved from the Mertz Ice Tongue. The minimum thickness of the debris layer was 3 m. (B) Ice sample taken from the same iceberg. The average debris content of ice samples taken from the iceberg was 7%, and pebbles consisted of a single rock type (black shale) (from Anderson, Domack, and Kurtz, 1980a).

Figure 4.22. Surface debris layer on a small fringing ice shelf in the James Ross Island area, Antarctic Peninsula, and sediment-laden icebergs calving from this ice shelf. This part of the ice shelf has vanished since this photograph was taken in 1985.

Piston cores from Marguerite, Pine Island, and Sulzberger bays penetrated a general stratigraphy consisting of, from bottom to top, diamicton, silt with minor amounts of ice-rafted debris and microfossils, and glacial-marine sediments or traction current deposits with abundant microfossils (Anderson et al., 1991c). Detailed analysis of Marguerite Bay sediments led to the conclusion that the diamicton included both till and transitional glacial-marine sediment and that the silt unit is a sub–ice shelf deposit (Kennedy and Anderson, 1989). The presence of an ice shelf facies in these areas is attributed to the fact that these three bays are confined ice shelf settings. The retreat of the grounding lines in these bays since the last glacial maximum has apparently been slow enough to allow accumulation of relatively thick ice shelf deposits (Anderson et al., 1991c). Sub–ice shelf sediments also exist on the Ross Sea continental shelf, but these deposits are thin and have previously gone unnoticed. They include a basal unit with abundant pea-size sedimentary clasts grading upward into a silt unit with minor amounts of ice-rafted debris (Domack et al., in press).

Ice Shelf Lithofacies

Since 1980, detailed marine geologic surveys have been conducted in three very different ice shelf settings. These include: (1) large polar ice shelves with unconfined flow; (2) small polar ice shelves with confined flow; and (3) fringing ice shelves with unconfined flow. High-resolution seismic profiles and piston cores collected during these surveys provide the basis for contrasting facies relationships (Anderson et al., 1991c). Figure 4.23 shows facies models for the different ice shelf settings.

Large Polar Ice Shelves – Unconfined Flow (Ross Ice Shelf). Unconfined ice shelves are pinned on islands and banks on the continental shelf. They serve to buttress the ice sheet, which is grounded at a greater depth. The West Antarctic Ice Sheet and its associated Ross and Ronne-Filchner ice shelves fit this case, even today (Fig. 4.23A). The ice shelf grounding zones of the Ross and Ronne-Filchner ice shelves occur at depths generally greater than 700 m. During the last glacial maximum, these ice shelves extended farther north and onto the outer continental shelf in Ross (Anderson et al., 1980b; Anderson, Brake, and Myers, 1984) and Weddell seas (Anderson et al., 1980b, 1991c; Elverhøi, 1981).

Evidence of a grounded ice sheet on virtually the entire Ross Sea continental shelf during the late Pleistocene exists in the form of till, megascale glacial lineations, and

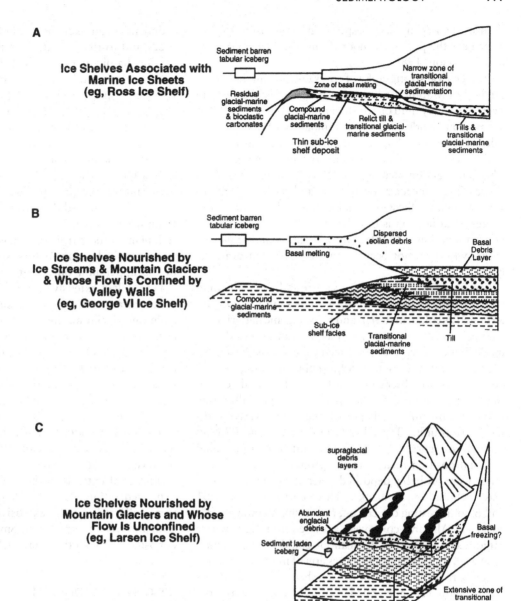

A

Ice Shelves Associated with Marine Ice Sheets (eg, Ross Ice Shelf)

B

Ice Shelves Nourished by Ice Streams & Mountain Glaciers & Whose Flow is Confined by Valley Walls (eg, George VI Ice Shelf)

C

Ice Shelves Nourished by Mountain Glaciers and Whose Flow Is Unconfined (eg, Larsen Ice Shelf)

Figure 4.23. Sedimentation model illustrating facies associated with three different ice shelf settings: (A) ice shelves associated with marine ice sheets, (B) ice shelves nourished by ice streams and mountain glaciers with confined flow, and (C) ice shelves nourished by mountain glaciers with unconfined flow (modified from Anderson and Molnia, 1989).

striations on the seafloor, widespread glacial erosion surfaces, and grounding zone features (Anderson et al., 1992a; Shipp and Anderson, 1997a,b,c,d). The general stratigraphy of piston cores from most of the Ross Sea continental shelf consists of till overlain by transitional glacial-marine sediments. Detailed analysis of cores has shown that the transitional glacial-marine unit may include a muddy unit with relatively minor amounts of ice-rafted material. Domack (personal communication) believes that this may be a sub–ice shelf deposit. Diatomaceous glacial-marine sediment caps the sequence, indicating seasonally open-marine conditions. Meltwater deposits are virtually lacking. Contacts between units are relatively sharp. This stratigraphic succession indicates retreat of the ice (Fig. 4.23A).

Small Polar Ice Shelves – Confined Flow (George VI Ice Shelf). The George VI Ice Shelf receives ice draining from a

number of different sources including Antarctica's northernmost ice cap, the Palmer Land Ice Cap. It differs from the Larsen and Ross ice shelves in that it is confined to a broad glacial valley, George VI Sound, and there are numerous shallow banks offshore of the ice shelf that may have served as pinning points during glacial maxima (Fig. 3.22). On the basis of theoretical modeling and geomorphologic observations, Stuiver and colleagues (1981) placed Marguerite Bay under 2000 m of grounded ice during the last glacial maximum. Even more compelling evidence for an ice sheet grounded in the bay exists in the form of lodgement till at depths of up to 726 m on the floor of the bay and a widespread glacial erosion surface at or near the present seafloor (Kennedy, 1988; Kennedy and Anderson, 1989).

The general stratigraphy of cores taken from the southern half of Marguerite Bay consists of diatomaceous mud overlying sandy mud on top of diamicton (Fig. 4.23B). Diatomaceous muds consist predominantly of diatom frustules, sponge spicules, and silt-sized quartz and feldspar grains, with subordinate amounts of sand-sized material. The sand fraction includes a well-sorted, current-derived mode. The concentration of this sand mode appears to vary systematically with water depth and distance from shore. Stratification is usually either lacking or exists as faint laminations. Pebbles are rare and include volcanic rocks, which do not occur within the ice shelf drainage basin.

Two types of ice shelf deposits are found: (1) a silty mud with a small, relatively well-sorted fine sand fraction (interpreted as eolian material) and rare biogenic material (B2 facies); and (2) a sandy-silty mud with an unsorted sand mode that is interpreted to be ice-rafted debris (B1 facies) (Kennedy and Anderson, 1989). Pebbles are rare, and exotic pebbles are absent in both facies. The mud rests in sharp contact upon diamicton (transitional glacial-marine sediment and till) and is gradational with overlying diatomaceous units. Typically, B1 mud grades into B2 mud upward in a core. This suggests a shift from a proximal to distal position relative to the grounding zone. These muds were probably deposited when the grounding zone was anchored on the banks. Three cores penetrated B1 and B2 mud resting directly on, and in sharp contact with, basal till. All three cores were acquired in troughs where rapid decoupling from the seafloor could have occurred. However, the stratigraphic succession of most piston cores suggests that the retreat of the ice shelf was gradual. As the grounding zone retreated, so did the zone of basal debris. Large portions of the bay floor lay beneath an ice shelf, which was virtually barren of debris. This was the environment in which terrigenous mud (B1 and B2 facies) was deposited over transitional glacial-marine sediments (Fig. 4.23B). Meltwater flowing from beneath the ice shelf grounding zone supplied the great quantity of terrigenous silt that composes this sediment. The concentration of sand-size, ice-rafted debris in the sediment gradually decreases in an offshore direction. This is attributed to the fact that the debris represents supraglacial (eolian) material that has descended through the ice sheet and ice shelf, eventually to be deposited seaward of the grounding zone. The glacial maritime setting during the deposition of diatomaceous mud, the most widespread modern sediment type, presumably resembled the present setting in Marguerite Bay.

Large Fringing Ice Shelves – Unconfined Flow (Larsen Ice Shelf). The Larsen Ice Shelf is directly nourished by valley glaciers and outlet glaciers flowing from the Palmer Land Ice Cap (Fig. 1.10). This situation is conducive to sediment entrainment at higher levels in the ice shelf (Fig. 4.22) and to the formation of thick, widespread glacial-marine diamicton on the shelf (Fig. 4.23C).

The continental shelf offshore of the northern Larsen Ice Shelf was the site of a detailed marine geologic survey during Deep Freeze 85 and a later marine geophysical survey in 1990 (Anderson et al., 1991c). Many of the sediment cores collected during this survey were acquired on the inner shelf, where the ice shelf was located only a few decades ago (Fig. 1.11). Detailed sedimentologic and petrographic analyses of diamictons acquired in piston cores showed them to be glacial-marine sediments (Smith, 1985). A thin layer of fossiliferous glacial-marine sediments rests sharply on glacial-marine diamictons that are virtually lacking in fossils. Diatoms are abundant in the upper unit only north of Seymour Island where sea ice is less extensive during summer months. To the south, forams are common in surface sediments. The fossiliferous glacial-marine sediments contain a well-sorted, current-derived sand and coarse silt mode that does not occur in the underlying glacial-marine diamictons. Smith (1985) illustrated that the size and concentration of this mode varies in a systematic fashion with water depth and distance from land. So, unlike the glacial-marine diamictons it overlies, the fossiliferous glacial-marine sediment bears textural evidence that marine currents were instrumental in its deposition. None of the cores penetrated tills; however, seismic records show a prominent glacial unconformity that extends to the shelf break and a grounding zone wedge (Fig. 4.16C). The glacial-marine diamictons were collected over an extensive area of the continental shelf and were interpreted by Smith (1985) as sub-ice shelf deposits, although they may include deformation tills.

ICE CLIFF ENVIRONMENT

Ice cliffs occur along the coast where glacial flow is slow enough for waves to erode the ice front as quickly as it advances and/or calving is sufficient to remove the heavily crevassed ice from the glacier soon after it reaches the shoreline. Ice cliffs are typical of areas with divergent glacial drainage, which includes much of the East Antarctic coastline and stretches of the Antarctic Peninsula (Fig. 4.24). Ice cliffs are predominantly grounded at sea level, but along the Wilkes Land Coast they are grounded at depths between 175 and 425 m (Robin, 1979). Where an ice cliff is grounded near sea level, waves erode basal debris from the ice and redistribute the sediment (Drewry and Cooper, 1981). Where the base of the ice cliff is submerged, sediment-laden bergy bits are produced by wave erosion (Anderson, Domack, and Kurtz, 1980a). Because of their small size, these bergy bits will likely melt and deposit their debris near the coast. A coarse, gravel and bioclastic gravel facies is associated with ice cliff proximal settings in the western Ross Sea (Anderson, Brake,

Figure 4.24. Aerial photograph showing an ice cliff in Andvord Bay and a gravel ridge situated just offshore of the ice cliff.

and Meyers, 1984), offshore of the Wilkes Land Coast (Anderson et al., 1983b), and in the Antarctic Peninsula region (Anderson et al., 1983a).

ICE-FREE COASTS

Ice-free coasts are a rarity in Antarctica; the processes and sedimentary facies associated with these environments essentially are unknown (Fig. 4.25). They are most widespread along the northern part of the Antarctic Peninsula and the South Shetland Islands. To my knowledge, there have been no detailed process-oriented sedimentologic studies conducted on Antarctica's ice-free coasts and associated offshore environments. These settings are characterized by raised beaches that provide excellent settings for sedimentologic research.

PROGLACIAL MARINE ENVIRONMENT

The Continental Shelf

From a sedimentologic standpoint, the most important differences between the Antarctic continental shelf and other continental shelves of the world are that the Antarctic continental shelf: (1) is covered by sea ice most of the year, and in many areas sea ice covers the shelf throughout the year; (2) is deep, averaging 500 m, and in some areas reaching depths of 1400 m; (3) displays considerable topographic relief as a result of glacial erosion – in some areas gradients on the shelf (up to 15°) are greater than those on the adjacent continental slope; (4) has virtually no fluvial discharge into the sea; (5) has few wave-dominated coastal zones – in most regions the coast is covered by glacial and pack ice; and (6) has a terrigenous sediment supply to the sea, which is essentially restricted to glacial processes.

Because of its unique features, sedimentation on the Antarctic continental shelf is different from other shelves, with the possible exception of high-latitude regions of the Arctic. The unusual nature and distribution of surficial sediments reflects these differences.

For the vast portion of the Antarctic continental margin, the extreme shelf depth greatly reduces any impact of wave- and wind-driven currents on the seafloor. Furthermore, tidal currents are considered sluggish on the Antarctic shelf. In the absence of these agents, marine sedimentation is primarily controlled by geostrophic currents, impinging deep-sea currents, and sediment gravity flow processes.

Glacial-Marine Seismic Facies

High-resolution seismic profiles from the continental shelf show that younger (Pleistocene) strata are dominated by massive and chaotic seismic units that are bounded by glacial unconformities. These are interpreted as subglacial (till sheets) and grounding zone facies (Figs. 4.4 and 4.16). On the outer shelf, these seismic facies grade seaward into acoustically layered seismic facies that are interpreted as

Figure 4.25. Gravelly beaches are a common, yet poorly understood, sedimentary environment in the Antarctic Peninsula and along the Victoria Land coast in the western Ross Sea. Elsewhere in Antarctica, beaches are rare.

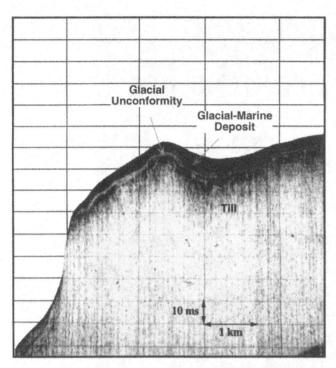

Figure 4.26. High-resolution (minisparker) seismic profile across the shelf break of the western Ross Sea showing acoustically layered glacial-marine deposits resting sharply on till.

glacial-marine deposits (Figs. 4.16A and 4.18). Figure 4.26 shows acoustically layered glacial-marine deposits resting directly on subglacial deposits and illustrates the dramatic difference between these seismic facies.

Pliocene and older strata on the continental shelf include a greater proportion of glacial-marine seismic facies, which is attributed to less frequent ice sheet grounding events and associated erosion of interglacial (glacialmarine) deposits. In general, seismic facies associated with glacial-marine deposits display a relatively continuous, layered internal character and the same kinds of draping, onlapping, and offlapping geometries that occur in other types of marine strata (Alonso et al., 1992; Anderson and Bartek, 1992; Bart and Anderson, 1995).

Sediment of the Proglacial Shelf Setting

The main surface sediment types found on the continental shelf include: (1) relict tills; (2) glacial-marine sediments (> 10% ice-rafted debris); (3) siliceous mud and ooze, which consists of 10 to 30% and more than 30% biogenic silica, respectively, and only small quantities of ice-rafted debris (< 10%); (4) bioclastic carbonates; (5) terrigenous silt and clay with little or no biogenic silica (< 10%) or ice-rafted debris (< 10%); (6) moderately sorted to well-sorted sand deposited by marine currents; and (7) sediment gravity flow deposits.

Detailed sedimentologic investigations have been conducted on the Ross Sea continental shelf (Chriss and Frakes, 1972; Anderson, Brake, and Myers, 1984; Dunbar et al., 1985), including McMurdo Sound (Glasby et

al., 1975; Barrett, Pyne, and Ward, 1983; Bartek and Anderson, 1991); in Sulzberger Bay (Anderson, Brake, and Myers, 1984); on the Pennell Coast continental shelf (Brake, 1982; Anderson, Brake, and Myers, 1984); in the Weddell Sea (Anderson, 1972a; Anderson et al., 1980b; Haase, 1986); in the Bransfield Strait region (Anderson et al., 1983a; Singer, 1987; Banfield, 1994; Yoon et al., 1994); and on the Wilkes Land continental shelf (Milam and Anderson, 1981; Domack, 1982; Anderson et al., 1983b; Dunbar et al., 1985). These studies provide the basis for the following discussion of shelf sediments.

Glacial-Marine Sediment. Chriss and Frakes (1972) conducted a factor analysis of sediment samples collected from the tops of piston cores from the Ross Sea. They identified a number of different glacial-marine sediment types and concluded that the observed sediment reflects differences in the relative influence of glacial and marine sedimentation. Later, Anderson and colleagues (1980b) examined modern and ancient deposits collected in piston cores from the Weddell Sea, the Ross Sea, and the Wilkes Land continental shelves. In addition to transitional glacial-marine deposits, Anderson and colleagues (1980b) divided the remaining glacial-marine sediments of these regions into two broad categories, compound glacial-marine sediments and residual glacial-marine sediments.

Compound glacial-marine sediment includes mud and diatomaceous mud with subordinate amounts (usually < 10%) of ice-rafted material. These deposits result from combined ice-rafting and settling from suspension of fine-grained sediments, hence the name "compound glacial-marine sediment" (Anderson et al., 1980b). They are widespread on the deep (usually > 300 m) continental shelf. In fact, they are the most widespread modern glacial-marine sediments on the Antarctic proglacial shelf.

Compound glacial-marine sediments differ from basal till and transitional glacial-marine sediments in their better sorting (Fig. 4.20C), presence of stratification and bioturbation, lack of mineralogic and textural homogeneity within units (Fig. 4.9), more diverse suite of ice-rafted minerals and pebble lithologies, variable magnetic susceptibility (Fig. 4.9), greater and more diverse fossil content (diatoms may constitute a significant portion of the sediment), relatively high organic carbon content (Fig. 4.8), and more random pebble fabric (Anderson et al., 1980b; Anderson, Brake, and Myers, 1984; Anderson et al., 1991c; Hambrey, Ehrmann, and Larsen, 1991) (Fig. 4.10). Pebble shape varies within compound glacial-marine sediments and may not significantly vary from tills and transitional glacial-marine sediments (Kuhn et al., 1993) (Fig. 4.11C). Pebble shapes are more angular in compound glacial-marine sediments acquired along mountainous coasts, such as along the Victoria Land coast in the Ross Sea and off the Antarctic Peninsula (Domack, Fairchild, and Anderson, 1980) (Fig. 4.11D). The large biogenic component of compound glacial-marine sediment

Figure 4.27. Bioclastic carbonate sample collected in nearshore waters offshore of Wilkes Land (courtesy of E. Domack).

makes it unique, compared to Northern Hemisphere glacial-marine sediment (Anderson and Ashley, 1991). This is attributed to a lack of dilution by terrigenous material from meltwater sources in Antarctica (Domack, 1988; Anderson and Ashley, 1991).

The other major group of glacial-marine sediments, residual glacial-marine sediments, consist almost entirely of coarse ice-rafted debris and bioclastic material, including fragments of bryozoans, mollusks, echinoderms, corals, and foraminifera. This sediment results from ice rafting under the influence of marine currents that are strong enough to winnow fine-grained materials, hence their coarse skewness and poor sorting (Fig. 4.20B). Alternatively, residual glacial-marine sediments are deposited in areas where bottom mixing by organisms is sufficiently great to allow fine-grained material to be eroded by bottom currents (Singer and Anderson, 1984). Winnowing by bottom currents occurs mainly on shallow banks and at the shelf edge where strong impinging deep-sea currents occur (Anderson et al., 1983a; Anderson, Brake, and Myers, 1984; Anderson et al., 1991c). Residual glacial-marine sediment is the most difficult glacial-marine sediment type to core; therefore, it is underrepresented in core collections from the continental shelf.

Siliceous Mud and Ooze. Siliceous ooze and mud occur in shelf basins in every area surveyed to date, except the Weddell Sea. The reason for its absence in the Weddell Sea is not fully understood, but must, in part, be due to severe sea-ice conditions in the region that limit the production of biogenic silica. Elsewhere, concentrations of biogenic silica in surface sediment frequently reach 30 to 40% (Dunbar, Leventer, and Stockton, 1989). The high silica concentration indicates that the Antarctic ooze is a true ooze; in fact, it is the only siliceous ooze known to occur on a continental shelf. The Antarctic ooze is easily recognized by its olive green color. Diatoms are the dominant source of siliceous material; sponge spicules are the next most abundant component. Laminations are common. High-resolution subbottom profiles show that the siliceous mud and ooze compose a draping unit whose thickness varies due to irregularities in the underlying glacially scoured surface (Shipp et al., in press).

Anderson, Brake, and Myers (1984), Dunbar and colleagues (1985), and DeMaster, Nelson, Harden, and Nittrouer (1991) examined the factors that influence the distribution of siliceous sediment on the Antarctic continental shelf. The primary factors are sea-ice cover (the dominant limiting factor on primary productivity in Antarctic shelf waters) and reworking and dispersal by marine currents. Sediment trap experiments indicate that only a fraction of the biogenic silica produced in surface waters reaches the seafloor with dissolution of this material occurring predominantly in the upper 50 m of the water column (Dunbar, Leventer, and Stockton, 1989).

DeMaster and colleagues (1991) and Harden, DeMaster, and Nittrouer (1992) have measured accumulation rates for Holocene diatomaceous sediments on the Ross Sea continental shelf and in the Antarctic Peninsula region using ^{210}Pb and ^{14}C. Rates range from 0.2 to 5.0 mm/yr. From their analysis, they concluded that 25 to 50% of

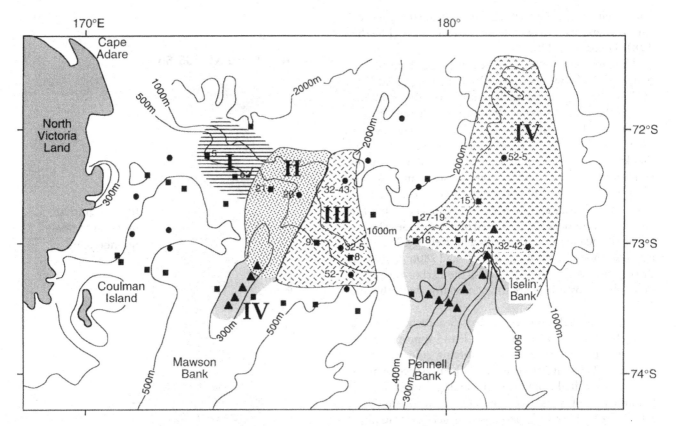

Figure 4.28. Distribution of carbonate facies in the northwestern Ross Sea. Squares indicate core locations for DF 87 piston cores; circles represent locations of *Eltanin* and DF 80 cores; triangles mark NB *Palmer* 98-01 grab sample locations. The facies include: I = barnacle-foraminifer facies; II = muddy bryozoan-barnacle-pelecypod-foraminifer facies; III = bryozoan-barnacle-foraminifer facies; and IV = foraminiferal sand facies (modified from Taviani et al., 1993).

the gross silica production in the surface waters is preserved in the sea bed. Indeed, DeMaster (1981) believes that the shelf basins of Antarctica may be accumulating enough biogenic silica to be of importance in the global silica budget.

Carbonates. Lisitzin (1962) first described carbonate sediments on the Antarctic continental shelf of the Davis Sea and near Lars Christensen Coast. This sediment consists dominantly of bryozoans, with lesser amounts of foraminifera and calcareous worm tubes. Bioclastic carbonates also occur on banks in the northwestern Ross Sea (Taviani, Reid, and Anderson, 1993), on the Wilkes Land continental shelf (Domack, 1988), and on the Weddell Sea outer continental shelf (Elverhøi, 1984). These deposits occur on the shallower portions of the shelf and on the shelf edge, where marine currents winnow fine-grained terrigenous debris. The bioclastic carbonates consist of the remains of bryozoans, barnacles, ostracods, forams, pelecypods, gastropods, and corals (Fig. 4.27).

Piston cores containing carbonate deposits were recovered from the outer continental shelf of the northwestern Ross Sea during *Eltanin* cruises 27, 32, and 52, and later during Deep Freeze 87 and *Nathaniel B. Palmer* cruises 95-01 and 98-01. Carbonate sediments occur along the outer continental shelf and upper slope, and on the seaward faces of the Mawson and Iselin banks (Fig. 4.28). These deposits primarily consist of varying amounts of fragmented bryozoans, corals, pelecypods, gastropods, barnacles, and foraminifers with echinoids and ostracods

constituting secondary components. Calcium carbonate content may reach up to 80% of the total sediment. The remaining fraction consists of siliciclastic particles (ranging in size from pebbles to silt) and biogenic silica (predominantly sponge spicules and diatoms). Ice-rafted pebbles are common. The carbonate sediment shows no evidence of significant dissolution, and even thin aragonite skeletal remains appear well preserved.

A series of four distinct carbonate facies, trending east-west along the outer shelf margin of the northwestern Ross Sea, was identified on the basis of faunal composition including barnacle-foraminifer, muddy bryozoan-barnacle-pelecypod-foraminifer, bryozoan-barnacle-foraminifer, and foraminiferal sand facies (Taviani, Reid, and Anderson, 1993) (Fig. 4.28). The latter facies consists of a mixture of foraminifera, terrigenous sand, and bioclastic carbonate debris. The relative percentage of carbonate versus terrigenous sand varies widely across a bank.

Radiocarbon dates on Ross Sea carbonates indicate that carbonate sedimentation was active during the late Pleistocene (Taviani, Reid, and Anderson, 1993). Radiocarbon dates on carbonate sediments from the Wilkes

Land shelf and the Weddell Sea shelf indicate that these calcareous deposits also are late Pleistocene in age (Elverhøi, 1981; Domack, 1988).

The carbonates of the Ross Sea shelf and slope occur at water depths between 265 and 2000 m. Lingle and Clark (1979) estimate that the shelf was 75 to 100 m shallower during the glacial maxima than at present; thus, the environment of deposition of Ross Sea carbonates was an outer shelf setting with water depths greater than 165 m. Planktonic foraminifers indicate seasonally open-marine conditions. Deeper water carbonates (slope and rise samples) were deposited by sediment gravity flow processes (Anderson, Kurtz, and Weaver, 1979b). The only other late Pleistocene deposits recovered on the Ross Sea shelf include till and glacial-marine sediment (Kellogg, Truesdale, and Osterman, 1979; Anderson et al., 1980b; Anderson, Brake, and Myers, 1984). Thus, the carbonates of the outer shelf most likely are time-correlative with this material.

Compositional analyses of Ross Sea carbonates (Taviani, Reid, and Anderson, 1993) lend support to previously recognized criteria for identifying cold water carbonates. These criteria include: (1) the presence of an associated ice-rafted component (including dropstones); (2) a dominance of calcite relative to other carbonate minerals (the remaining fraction consists solely of aragonite); (3) allochems that are entirely skeletal; and (4) heavy oxygen isotopic compositions (in the range of +3.0 to +5.1% Pee Dee Bellumnite (PDB)). The heavy values are consistent with precipitation of calcium carbonate at subfreezing temperatures in ambient waters with isotopic compositions close to standard mean ocean water (SMOW).

Traction Current Deposits. Piston cores and grab samples acquired from shallow banks (less than about 400 m) on virtually every portion of the Antarctic continental shelf sampled to date have recovered moderately sorted to well-sorted sand with rare ice-rafted clasts (Fig. 4.29A). These sand bodies range in extent from a few hundred kilometers to over 10,000 km^2 in the case of the Iselin Bank sand sheet in the western Ross Sea. These sands and muddy sands typically contain abundant carbonate debris and foraminifera, and they grade laterally into carbonates or residual glacial-marine sediments. They extend from bank tops to over 600 m water depth.

Individual grains range from angular to spherical. Grain size generally decreases from the seaward edges of banks toward the landward portions of the banks and from east to west across the banks. On the North Victoria Land shelf, sand bodies that blanket the banks in this region are composed of volcanic sand that is derived from Cape Adare, which occurs at the eastern edge of the shelf. These sands have been transported toward the west a distance of over 60 km from this source.

Side-scan sonar records from bank tops show irregular sand waves. Attempts to core these sand bodies have met with little success, so their thickness is not known. Side-

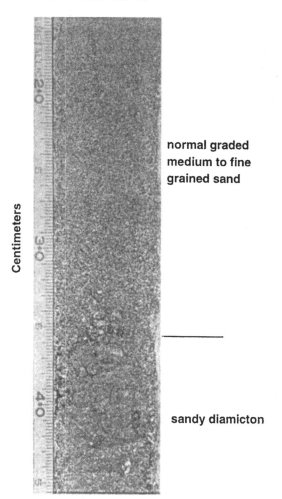

A   Core NBP95-55

normal graded medium to fine grained sand

sandy diamicton

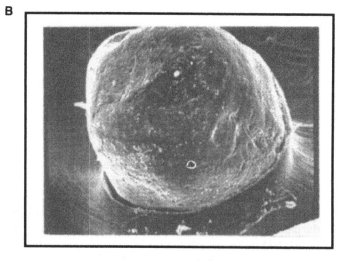

B

Figure 4.29. (A) Photograph of core NBP95-55 from the seaward edge of Mawson Bank in the Ross Sea showing sand body that blankets the bank. Photomicrograph of (B) an eolian quartz grain from the Weddell Sea continental shelf showing spherical shape.

scan sonar records from the Mawson Bank in the Ross Sea indicate that the sands are thick enough to infill iceberg furrows that are several meters deep.

Traction current deposits also occur on the outer shelf and upper slope in those areas where strong boundary currents impinge on the continental shelf, such as in the western Ross Sea (Anderson, Brake, and Myers, 1984) and along the Wilkes Land continental shelf (Dunbar et al., 1985).

Eolian Sediments. Barrett, Pyne, and Ward (1983) and later, Bartek and Anderson (1991) described sands and muddy sands from the McMurdo Sound area that they interpret as having an eolian transport component derived from nearby ice-free valleys of Victoria Land. The dominant mode of transport involves debris being blown onto the surface of sea ice and then being deposited on the seafloor during the summer when the sea ice melts. This process probably was more widespread during previous interglacials when larger portions of the coast were ice-free.

One of the most intriguing discoveries of eolian material on the Antarctic continental shelf was made in the late 1950s during Operation High Jump. Samples acquired along the front of the Filchner Ice Shelf were found to contain eolian quartz sand, which was extremely well sorted with high sphericity and frosted surface texture (Rex, Margolis, and Murray, 1970) (Fig. 4.29B). Scanning electron microscopic analyses of this sand by Rex and colleagues (1970) showed no evidence of glacial surface features. They argued that these eolian sands had been deposited in place and lowered to their present depth (300 to 500 m) by isostatic flexure. Later, Anderson (1971) acquired piston cores from the area and found the sand deposits to be at least 2.5 m thick. The most likely source of these sands is Berkner Island, which is located to the south beneath the Filchner Ice Shelf.

Sediment Gravity Flow Deposits. The irregular glacial topography of the Antarctic continental shelf is of extreme importance in regulating sediment dispersal and gravity related transport. Because local gradients on the glacially scoured continental shelf typically range from 5 to 15°, sediment gravity flows can be initiated at shallow depths as well as on the continental slope and rise. In addition to the high relief on the Antarctic seafloor, other factors contribute to sediment mass movement, not the least of which is the plowing effect of grounded ice sheets and icebergs.

The Antarctic continental margin is largely aseismic (with the exception of the Bransfield Basin and McMurdo Sound); therefore, seismic triggering of sediment gravity flows is not considered an important mechanism. Glacial isostasy, however, is undoubtedly an important triggering mechanism for sediment mass movement (Anderson, Kurtz, and Weaver, 1979). During a period of retreat of the ice sheet from the continental shelf, the shelf is uplifted by as much as 300 m, depending on the thickness of the grounded ice sheet, its marginal profile, and location of its grounding line (Fig. 3.3). Rates of isostatic uplift are probably on the order of 5 to 10 cm/yr, based on rates of isostatic response to deglaciation in Canada and Greenland (Andrews, 1968; ten Brink, 1974).

Seismic records acquired along the flanks of glacial troughs typically show slump features, and piston cores from these troughs commonly penetrate slump and debris flow deposits (Kurtz and Anderson, 1979; Anderson et al., 1983b; Anderson, Brake, and Myers, 1984; Wright, Anderson, and Fisco, 1983). Debris flows are typically diamictons with sedimentary clasts, including diatomaceous mud clasts from overlying units and displaced fossils from bank tops. Otherwise, the debris flow deposits are similar to glacial and glacial-marine diamictons (Kurtz and Anderson, 1979).

Graded and non-graded sands containing sedimentary clasts and displaced fossils are interpreted as turbidites and grain flow deposits. These deposits occur in troughs adjacent to banks with sandy surface sediments, such as along the western flank of the Crary Trough in the Weddell Sea, on the flanks of the Mertz Bank on the Wilkes Land continental shelf, and on the flanks of the Pennell and Mawson banks in the Ross Sea. The inner continental shelf of Antarctica is also characterized by steep gradients, and turbidites and grain flow deposits are common in these areas.

Bartek and Anderson (1991) have described a range of sediment gravity flow facies from piston cores collected along the flanks of Ross Island in McMurdo Sound. These are entirely volcaniclastic deposits, and the most common facies include disorganized gravels that grade upward into graded gravels and sands. The McMurdo Sound sediment gravity flows involve grain flow and high-concentration turbidity current transport. Facies relationships are poorly defined, and there is no consistent pattern in the style of sediment gravity flow deposits with increasing distance from shore. Bartek and Anderson (1991) argue that a lack of flow channelization has resulted in the close association of proximal and distal facies near the base of the slope.

In summary, sediment gravity flow processes play a key role in redistributing glacial deposits on the continental shelf. The close association of glacial-marine deposits and sediment gravity flow deposits on the Antarctic continental shelf, as well as on the continental slope and rise, is important. The association of sediment gravity flow deposits and diamictites or, for that matter, the association of diamictites and carbonates in ancient strata has been erroneously used as a criterion to argue against a glacial origin (Schermerhorn, 1974).

Iceberg Turbates. Side-scan sonar records from the Antarctic continental shelf show abundant iceberg furrows concentrated mainly at depths shallower than 400 m (Lien, 1981; Barnes, 1987; Chapter 3). Mounds of plowed sediment occur along the flanks of these furrows (Fig. 3.12A). Sediment cores from the shelf occasionally show evidence of disturbance in the form of convolute bedding and sedimentary clasts that are possibly due to iceberg plowing, especially if the cores are collected from the relatively flat bank tops. The term "iceberg turbate" (Vorren et al., 1983) is used to describe such deposits. These deposits probably include diamictons that are similar in character to debris flows and deformation till.

BAY AND FJORD ENVIRONMENTS

Environmental Setting

A substantial body of literature has been compiled on the subject of fjords, but this literature deals almost exclusively with northern-latitude fjords, particularly those of the Canadian Arctic, Greenland, and Scandinavia (Powell, 1983a; Syvitski and Blakeney, 1983; Powell and Molnia, 1989; Syvitski, 1989). Bays and fjords constitute an important component of the glacial-marine environment of the Antarctic Peninsula region and the Victoria Land coast, but they have only recently been studied in detail (Griffith and Anderson, 1989; Park et al., 1989; Domack, 1990; Domack and Williams, 1990; Domack and Ishman, 1993; Domack and McClennen, 1996). This work is an important step in extracting the decadal-scale and century-scale climatic record from the expanded Holocene sedimentary sections that exist in bays, fjords, and shelf basins.

During the past several years, investigations of fjord sedimentation have been conducted in three areas along the northern portion of the Antarctic Peninsula, the South Shetland Islands, Palmer Archipelago, and Danco Coast. Data were gathered from over a dozen embayments in the northern Antarctic Peninsula and its offshore archipelago (Smith, 1985; Singer, 1987; Griffith, 1988; Domack, 1990). These surveys include aerial mapping of the glacial-marine setting, high-resolution seismic-reflection profiling, grab sampling, and piston coring. Representative bays from each of the three major areas were chosen to illustrate regional trends in sedimentation (Griffith and Anderson, 1989). Complementary studies by Domack (1990), Domack and Williams (1990), Domack and Ishman (1993), and Domack and McClennen (1996) concentrated on sedimentation processes occurring within these bays and fjords.

The Antarctic Peninsula is a region of strong climatic gradients, in both temperature and precipitation (Chapter 1). These climatic differences are reflected in the glacial-marine settings of the region and provide an ideal modern setting in which to contrast fjord sedimentation in different climatic and glacial settings (Griffith and Anderson, 1989; Domack and Ishman, 1993; Domack and McClennen, 1996).

Because calving is the dominant ablationary mechanism in Antarctica, the calving line (represented by the omnipresent ice cliff) also represents the glacier's equilibrium line (Fig. 1.13). Thus, debris entrained at high levels in the glacier will be transported downward in the glacier as it flows down valley. This, in part, accounts for the apparent absence of sediment in Antarctic valley and out-

let glaciers (Fig. 4.30A), although basal debris layers are common in glaciers that terminate on land (Fig. 4.30B).

During several of my cruises to the bays and fjords of the Antarctic Peninsula region, many sediment-laden icebergs and bergy bits were observed. These included englacial debris layers (Fig. 4.31A) and supraglacial debris layers (Fig. 4.31B). The small size of these sediment-laden icebergs and bergy bits, as well as their locations within the fjords, indicates a local origin. On one occasion, I was fortunate to observe a calving event that resulted in debris-laden ice bobbing to the surface from beneath an otherwise clean outlet glacier. These observations suggest that there may be large amounts of glacially derived debris at the base of the tidewater glaciers. This material may be flushed from beneath the glacier by meltwater or dumped at the glacial terminus. If this is true, then much of the sediment entering the marine environment may do so from below sea level where it cannot be observed (Rundle, 1974; Griffith and Anderson, 1989).

Domack and Williams (1990) have gathered in situ temperature, salinity, and turbidity measurements from three fjords of the Danco Coast and have measured tongues of turbid water at several levels in the water column near the terminus of tidewater glaciers (Fig. 4.32). Their sedimentation model calls for the generation of these tongues at the base of the glacier. These low-density turbid water tongues then move upward to levels where they flow along surfaces of roughly equal density (Fig. 4.32). Thus, sediment dispersal occurs at several layers in the water column, as well as at the seafloor via sediment gravity flow processes.

Sediments and Facies of Bays and Fjords

Domack and McClennen (1996) describe five factors that govern sedimentation in fjords of the Antarctic Peninsula region including: (1) proximity to glacial ice front – glacier-proximal areas receive a greater proportion of ice-rafted debris, which masks biogenic material and organic carbon; (2) sea ice – sediment in areas with more persistent sea ice is relatively depleted in biogenic silica and organic carbon; (3) bathymetry – areas shallower than ~400 m tend to be strongly iceberg turbated and are floored by gravel lags and palimpsest or relict sediments, and below ~400 m, fine-grained siliceous muds with ice-rafted debris accumulate; (4) climate – climate controls the amount of melting that takes place across the glacier surface and the thermal regime within the glacier, which, in turn, controls meltwater discharge. Fjords in warmer regions such as the South Shetland Islands have a dominant fine-grained terrigenous component that is derived from meltwater discharge; and (5) glacial drainage basin size – fjords with larger glacial systems tend to have greater terrigenous sediment components than do fjords with smaller glacial drainage systems.

The following subsections provide case studies of several fjords with different climate and glacial settings.

A

B

Figure 4.30. Photograph of (A) a clean calving wall of a valley glacier in the Antarctic Peninsula region (with a helicopter in the foreground, for scale); (B) a basal debris layer in an Antarctic Peninsula valley glacier.

Figure 4.31. (A) Englacial debris layers in an iceberg drifting off the Antarctic Peninsula coast; (B) supraglacial debris on a small iceberg in the Inland Passage.

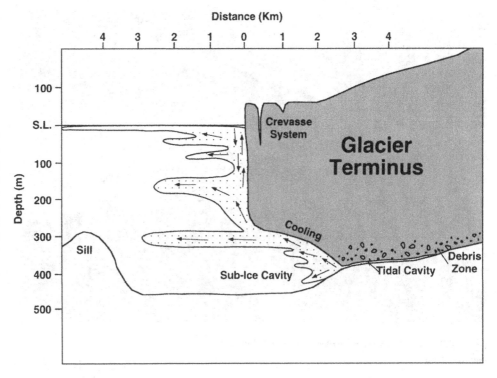

Figure 4.32. Model of cold water tongue generation beneath a sub-glacial marine cavity (from Domack and Williams, 1990; reprinted with permission from the American Geophysical Union, © 1990). Dotted pattern indicates the distribution of suspended terrigenous particulate matter.

Warm Subpolar Bays and Fjords (Maxwell Bay). The bays and fjords of the South Shetland Islands experience the warmest climate of Antarctica (Fig. 1.9). Mean summer temperatures on King George Island average 2°C, and the glaciologic setting is characterized by elevated equilibrium lines (~ 150 m above sea level; Domack and Ishman, 1993). Strong estuarine circulation and periodically strong bottom currents characterize the bay (Domack and Ishman, 1993). The coast consists mainly of rocky beaches rimmed by ice cliffs that descend from small ice caps on King George Island and Nelson Island (Fig. 4.33). Large tidewater glaciers are predominantly confined to embayments, and many of these glaciers terminate on land. Meltwater streams flowing from these land-locked glaciers form small fan-deltas (Fig. 4.34).

Maxwell Bay, the largest bay in the South Shetland Islands region, has been studied in some detail and is used to illustrate the general style of sedimentation in the subpolar

Figure 4.33. Aerial view of Maxwell Bay showing both land-locked and tidewater glaciers with debris on glacier surfaces.

Figure 4.34. Aerial view of meltwater streams and fan-delta on King George Island.

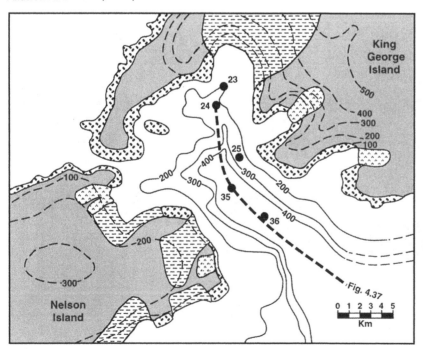

Figure 4.35. Glacial setting of Maxwell Bay. Dashed contours are ice elevations in meters. Dots designate core locations; bold dashed line indicates location of the seismic profile shown in Figure 4.37. (Patterns used to define the glacial types are provided in Fig. 4.36) (modified from Anderson and Molnia, 1989).

bays and fjords of this region. The glacial-marine setting of Maxwell Bay is shown in Figure 4.35. The various glacier types that occur in Maxwell Bay, as well as other bays and fjords of the Antarctic Peninsula region, are outlined in Figure 4.36.

Surface waters in Maxwell Bay are sediment-laden during summer months, when plumes are observed near the termini of tidewater glaciers. Bathymetric data show that the bottom of Maxwell Bay is relatively flat near the center of the bay (Fig. 4.35), but the flanks of the bay are characterized by very rugged topography (Park et al., 1989).

A seismic profile acquired along the axis of the bay (Fig. 4.37) shows rugged basement topography in the upper reaches of the bay, whereas the deeper central and outer portions of the bay are being filled by acoustically layered sediments. These deposits attain a thickness of greater than 180 msec ($\sim$ 120 m) near the head of the bay.

Piston cores from Maxwell Bay penetrated up to 6 m of poorly sorted, massive to crudely laminated, terrigenous mud. Thin (2 to 5 mm), fining-upward, silty or sandy layers occur randomly throughout the cores, but their bases do not appear to be erosional. X-radiographs show no evidence of winnowed surfaces, cross-bedding, or bioturbation (Griffith, 1988). These cores are essentially devoid of siliceous biogenic material, and ice-rafted debris is rare. The sediment resembles that acquired in southern Chilean fjords

(DaSilva, 1995) and is interpreted as having been deposited by settling from suspension of meltwater-derived silts (Griffith and Anderson, 1989). The general character of sediments, seismic facies, and relatively thick accumulations observed in Maxwell Bay are also observed in Admiralty Bay and other bays of the South Shetland Islands (Griffith and Anderson, 1989, unpublished data).

Wet Polar Bays and Fjords (Lapeyrère Bay, Palmer Archipelago). The bays and fjords of the Palmer Archipelago experience colder conditions and higher precipitation than the South Shetland Islands (Fig. 1.9). Lapeyrère Bay has been studied in the greatest detail (Fig. 4.38).

The glacial setting of Lapeyrère Bay is dominated by a single, large outlet glacier, Iliad Glacier (Fig. 4.38). Two deep basins (> 600 m water depth), partially separated by an intervening high, characterize the bay floor (Fig. 4.38). A short, steep drop-off leads from near the head of the bay down into the inner basin, which has a sloping, uneven bottom and is flanked by steep sidewalls. An outer basin, which also has steep sidewalls, is separated from a trough that extends seaward from Dallmann Bay by a partial sill across the confluence.

The seismic line acquired in Lapeyrère Bay extends to within 300 m of the Iliad Glacier terminus. The profile records the rough, relatively shallow bottom (217 m) near the terminus and illustrates the moderately strong, but uneven reflectors overlying acoustic basement (Fig. 4.39). A core (Core 128) from atop the bay head platform penetrated a gravelly sand overlain by pebbly terrigenous mud. The bulk composition of Core 128 is similar to proximal deposits in bays fed by tidewater glaciers (Powell, 1983; Gilbert, 1982; Elverhøi, 1984) and results from both a meltwater contribution and material that is rafted directly from the glacier front. This is also a zone of intense iceberg turbation.

The entire inner basin of Lapeyrère Bay appears to be floored by sediment gravity flow deposits, primarily turbidites (Griffith and Anderson, 1989). The chaotic and hummocky surface at the base of the steep slope leading into the basin is indicative of amalgamated slumps (Piper and Iuliucci, 1978). Core 129 sampled muddy sand with inverse grading interpreted as a grain flow deposit. Other cores from the basin floor penetrated graded sand and silt units, interpreted as turbidites. In general, grain size (of the dominant mode) decreases, sorting improves, and acoustic reflectors become stronger away from the glacier terminus.

The seismic line reveals acoustically layered sediments in the inner basin of Lapeyrère Bay. Core 130, from the deepest part of the inner basin sampled this acoustically laminated package (Fig. 4.39). It recovered a well-sorted sand overlain by a laminated mud containing only rare diatoms. This sequence is interpreted as amalgamated turbidites and glacial-marine sediments. Thus, there is a progression from ice-proximal deposits and amalgamated slumps with coarse mass flow deposits near the glacier

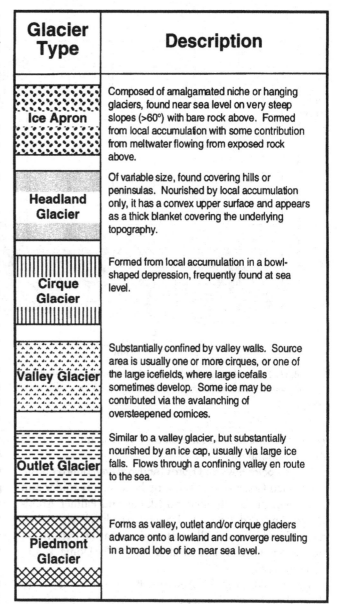

Glacier Type	Description
Ice Apron	Composed of amalgamated niche or hanging glaciers, found near sea level on very steep slopes (>60°) with bare rock above. Formed from local accumulation with some contribution from meltwater flowing from exposed rock above.
Headland Glacier	Of variable size, found covering hills or peninsulas. Nourished by local accumulation only, it has a convex upper surface and appears as a thick blanket covering the underlying topography.
Cirque Glacier	Formed from local accumulation in a bowl-shaped depression, frequently found at sea level.
Valley Glacier	Substantially confined by valley walls. Source area is usually one or more cirques, or one of the large icefields, where large icefalls sometimes develop. Some ice may be contributed via the avalanching of oversteepened cornices.
Outlet Glacier	Similar to a valley glacier, but substantially nourished by an ice cap, usually via large ice falls. Flows through a confining valley en route to the sea.
Piedmont Glacier	Forms as valley, outlet and/or cirque glaciers advance onto a lowland and converge resulting in a broad lobe of ice near sea level.

Figure 4.36. Classification of glacial systems in the Antarctic Peninsula region (modified from Griffith and Anderson, 1989).

terminus to better-sorted, finer-grained turbidites in more distal locations within the inner basin.

The high mud content and faint lamination of Core 131, taken from the mid-bay high (Fig. 4.39), reflect sedimentation from suspension, probably from episodic meltwater discharge. The relatively abundant ice-rafted debris and greater amount of diatomaceous material in this core results from a lower relative rate of terrigenous sedimentation on top of the high, out of the path of density-driven underflows.

The mouth of Lapeyrère Bay is floored by acoustically layered sediment (Fig. 4.39). Core 132, collected at the mouth of the bay in a water depth of 560 m, recovered over 5 m of sandy diatomaceous mud. The relative rate of fine-grained terrigenous sedimentation is so low in this

SE   N

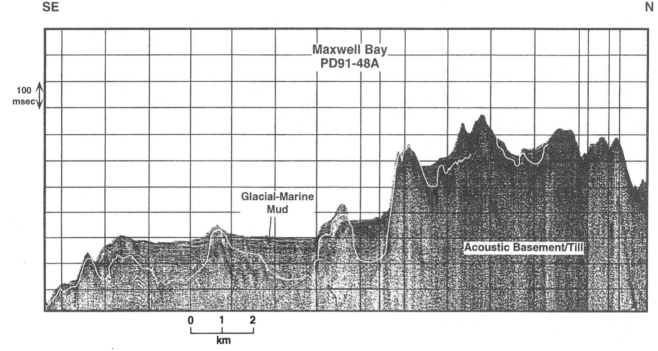

Figure 4.37. Seismic (minisparker) Profile PD91-48A from Maxwell Bay illustrates seismic facies within the bay. (See Fig. 4.35 for profile location.)

region that the deposits are diatomaceous, and ice-rafted debris is abundant.

Dry Subpolar Bays and Fjords (Andvord Bay, Danco Coast). The Danco Coast of the Antarctic Peninsula is colder and dryer than the South Shetland Islands and Palmer Archipelago. Andvord Bay of the Danco Coast has been studied in detail and is used to illustrate sedimentation patterns in the region (Griffith and Anderson, 1989; Domack and Ishman, 1993).

The glacial setting of Andvord Bay is somewhat complicated because tidewater glaciers descend along the entire perimeter of the bay (Fig. 4.40). It is dominated at its head by several large outlet glaciers, but the sidewalls are home to a greater variety of glacial types. Four large valley glaciers drain into the eastern side of the bay, including one that merges indistinctly into the easternmost outlet glacier at the southern end. Between these valley glaciers are cirque glaciers, ice aprons, and small headland glaciers (Fig. 4.40). The western side of the bay is even more complicated, with several smaller valley glaciers interspersed between cirques, headland glaciers, ice aprons, and a piedmont glacier.

Well-defined troughs are associated with the large outlet glaciers at the head of the bay (Fig. 4.40). The trough in Lester Cove reaches a depth in excess of 500 m. Many submarine promontories project into the center of the bay.

The seismic line from Andvord Bay (Fig. 4.41) shows exposed acoustic basement in the upper reaches of the bay. A chaotic reflection pattern near the seafloor may indicate the presence of a thin till layer. The outer portion of the bay has a draping, acoustically laminated unit that thins toward the bay's mouth.

All the piston cores from the inner portion of Andvord Bay (Cores 141–147) (Fig. 4.40) penetrated sandy diatomaceous mud with scattered pebbles. The sand fraction is unsorted. Crude laminations are visible in X-radiographs of cores, but these laminae do not correspond to sandy intervals. Cores from the outer part of the bay (Cores 136, 137, and 149) (Fig. 4.40) penetrated muddy diatomaceous ooze with scattered pebbles. The concentration of ice-rafted sand and gravel decreases toward the mouth of the bay. Domack and McClennen (1996) obtained radiocarbon dates from Kasten cores in Andvord Bay that yielded average sediment accumulation rates for outer bay diatomaceous oozes of 1.5 to 1.8 mm/yr. The generally thin sediment cover in the upper portion of the bay and the significant biogenic and organic carbon content of sediments in the lower half of the bay (Domack and Ishman, 1993) indicate a low rate of terrigenous influx from the glaciers at the head of the bay. This is supported by the paucity of sediment gravity flow deposits in cores.

In direct contrast to Lapeyrère Bay, dense sediment-laden underflows are probably rare or nonexistent in Andvord Bay. If we consider only the glacial setting, we might find this to be surprising, because there are no visible differences between the glaciers in these two bays. In Andvord Bay, the uniform draping nature of surficial deposits and the similarity of sediment composition (based on piston cores) in the outer portions of the bay indicate that fine-grained sediments are being efficiently dispersed throughout the bay in high-level plumes. The poor sorting

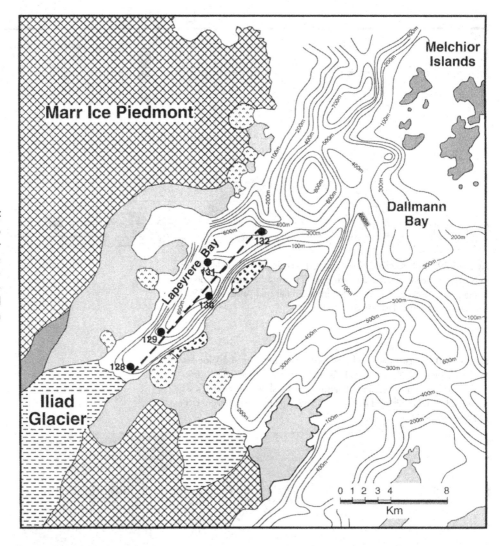

Figure 4.38. Glacial setting of Lapeyrère Bay. (See Fig. 4.36 for the classification key of glacier types.) Dots designate core locations; dashed line indicates location of seismic profile shown in Figure 4.39 (reprinted from Griffith and Anderson, 1989, with permission from Elsevier Science Ltd.)

and overall fine-grain size of the sediment reflects settling from suspension in a quiescent environment. No evidence of current activity exists.

Griffith and Anderson (1989) suggest that differences in the style of sedimentation in Andvord Bay and Lapeyrère Bay may be due to climatic difference in the two regions. The Palmer Archipelago (Lapeyrère Bay area) receives considerably more precipitation, including rain, than the Danco Coast (Andvord Bay area). It is possible that glaciers of the Palmer Archipelago experience greater and more episodic meltwater discharge than do glaciers of the Danco Coast, and that this results in more frequent turbidity currents in bays like Lapeyrère Bay.

Figure 4.42 summarizes the facies associations described above. Notably absent in the bays and fjords of the Antarctic Peninsula region, excluding perhaps the South Shetland Islands, are medial moraines and fan deltas that are associated with more temperate fjords and bays (Da Silva, Anderson, and Stravers, 1997). Only one example of a tunnel valley has been described in the literature, and that is in Brialmont Cove (Domack, 1990).

CONTINENTAL SLOPE AND RISE ENVIRONMENT

Bathymetry

Detailed mapping of the Antarctic continental slope and rise has been conducted in only a few areas, namely, off Wilkes Land (Chase et al., 1987) (Fig. 3.33) and in the eastern Weddell Sea (Miller et al., 1990b). Regional bathymetric maps (GEBCO, 1981; Hayes, 1991) show that the continental slope and rise are heavily dissected by submarine canyons and channels (Fig. 3.33). The upper slope tends to be quite steep with gradients of up to 25°. Considerable hummocky relief results from sediment mass movement (slumps and debris flows). Most canyon heads occur below the upper slope, but there are notable exceptions, such as the Wegener Canyon in the Weddell Sea. The Wegener Canyon has been mapped in greatest detail, including a multibeam survey (Miller et al., 1990b). Seismic records from the Wegener Canyon and several other canyons show evidence that these canyons have recently been eroded (Fig. 4.43), and turbidites at or near the seafloor downslope of these canyons corroborate this

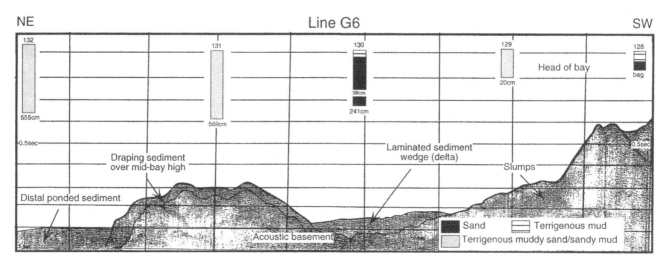

Figure 4.39. High-resolution (minisparker) seismic profile and lithologic logs of piston cores from Lapeyrère Bay. (Figure 4.38 shows locations of the profile and cores.) (Reprinted from Griffith and Anderson, 1989 with permission from Elsevier Science Ltd.)

Figure 4.40. Glacial setting of Andvord Bay. (See Fig. 4.36 for key to classification of glacial types.) Dots designate core locations; thick dashed line indicates location of seismic profile shown in Figure 4.41 (reprinted from Griffith and Anderson, 1989, with permission from Elsevier Science Ltd.).

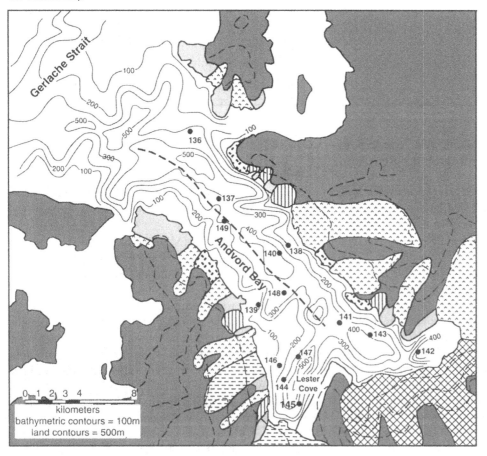

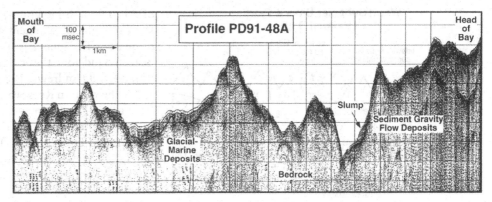

Figure 4.41. High-resolution (minisparker) seismic profile from Andvord Bay. (Figure 4.40 shows the location of this profile.)

recent activity. Indeed, one of the most interesting problems yet to be resolved is how these canyons remain active given the great depth of the continental shelf. Clearly, fluvial incision during eustatic lowstands can be ruled out as a means of canyon incision and reactivation.

A multibeam survey of the central Ross Sea upper slope revealed an extensive network of gullies that begin abruptly at the shelf break, converge downslope, and then bifurcate into small distributary channels that feed upper slope fans (Fig. 4.44). The size and scale of these gullies is similar to features that exist on the upper slope in the eastern and northwestern Weddell Sea, on the Antarctic Peninsula slope, and on the slope offshore of Wilkes Land (S. Cande, personal communication). In the Ross Sea, the gullies occur immediately downslope of a prominent grounding zone wedge. These gullies and the small upper slope fans they feed are most likely products of subglacial underflows that were initiated at the grounding line of the ice sheet.

The lower slope is characterized by more gentle gradients, except where canyons exist. Multibeam records from the Ross Sea illustrate the generally irregular relief of the lower slope, which results from complex networks of channels that radiate seaward (Fig. 4.44B).

The Antarctic continental rise is a relatively broad feature, relative to other rises of the world. It is characterized by generally smooth seafloor topography, except where channel–levee complexes and sediment drifts occur. Seismic records from the Antarctic continental rise reveal a dramatic increase in continental rise growth and development, primarily by sediment mass movement that is believed to coincide with deep erosion of the continental shelf by expending ice sheet (Chapter 5).

Sediment Gravity Flow Deposits

Repeated ice sheet grounding on the continental shelf has resulted in deep erosion of the continental shelf (Chapter 3). Much of the debris excavated from the shelf now resides on the continental slope and rise, as well as in deep-sea fans.

In the absence of rivers and a wave-dominated coastal zone, sediments are delivered unsorted to the Antarctic continental shelf. There is no direct sourcing of submarine canyons by rivers during eustatic lowstands. Surprisingly, well-sorted turbidites are common on the deep seafloor; indeed, they occur all around Antarctica (Wright, Anderson, and Fisco, 1983).

One means of producing well-sorted turbidites from poorly sorted glacial and glacial-marine sediments involves sediment gravity flow transition. Slumps are sheared gradually during downslope transport until they begin to flow internally as debris flows (Hampton, 1972). Eventually, sufficient water is added to the flow to create turbulence, and a turbidity current develops. This process of sediment gravity flow transition is one of the most important mechanisms for generating sorted sediments from glacial and glacial-marine deposits on the Antarctic seafloor (Wright and Anderson, 1982).

High-resolution seismic records across the continental margin typically display gradations in seismic facies that indicate downslope sediment gravity flow transitions. The eastern Weddell Sea was the site of a detailed study of sediment gravity flow processes on the Antarctic continental margin and abyssal floor (Wright and Anderson, 1982; Anderson, Wright, and Andrews, 1986). This study included acquisition of high-resolution seismic (small airgun), 3.5-kHz subbottom profiler, and 12-kHz bottom profiler records along with piston cores from the observed seismic facies. Figures 4.45, 4.46, and 4.47 summarize the downslope transition in seismic facies and associated lithofacies.

Seismic records from the upper slope display many of the typical features of slides and slumps, including folded strata, tensional depressions, and slump scars (Fig. 4.45). These slump deposits consist of virtually unmodified till that has a high shear strength (Kurtz and Anderson, 1979) (Cores 1578-16 and 1578-30; Fig. 4.47).

Warm Subpolar Bays & Fjords
(Maxwell Bay)

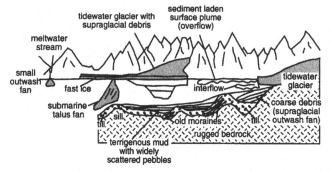

Wet-Polar Bays & Fjords
(Lapeyrère Bay)

Dry-Polar Bays & Fjords
(Andvord Bay)

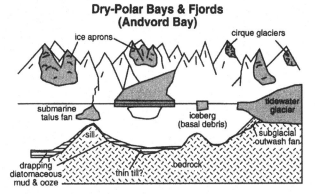

Figure 4.42. Facies models for Antarctic bays and fjords (from Anderson and Molnia, 1989; reprinted with permission from the American Geophysical Union, © 1989).

Farther downslope, massive, hummocky deposits, interpreted as debris flows, reside (Figs. 4.45 and 4.46). This seismic facies is characterized by acoustically transparent to chaotic reflection patterns. Piston cores sampled sandy diamictons interbedded with thin graded sands (Cores 1578-17 and 1578-31; Fig. 4.47). On the lower slope, the seismic facies include alternating acoustically laminated and transparent deposits (Fig. 4.46). Piston cores from this seismic facies penetrated thin (generally < 40 cm) graded, sandy turbidites interbedded with hemipelagic sediment (Cores 1578-19 and 1578-32; Fig. 4.47).

One of the most striking discoveries from this work was the degree to which sediment gravity flow transition from till to sandy turbidite results in an increased mineralogic maturity for the final deposit (Wright and Anderson, 1982). Original slumped sediment consists

of 25 to 47% lithic grains, whereas turbidites generated from this glacial sediment are composed predominantly of quartz. The quartz is believed to be derived from the Beacon Sandstone that exists beneath the EAIS in Queen Maud Land (Andrews, 1984).

On the western Ross Sea slope, turbidites grade upward from gravel-size fragments of bryozoans, corals, barnacle plates, and mollusks into foraminiferal sand (Wright, Anderson, and Fisco, 1983). This sediment is derived from outer shelf banks that are covered by bioclastic gravels and sands (Taviani, Reid, and Anderson, 1993) (Fig. 4.28).

Oceanographic Influences on Sedimentation

The dominant oceanographic influence on slope sedimentation is that exerted by circumpolar boundary currents (Anderson, Kurtz, and Weaver, 1979). Where these currents impinge on the upper slope, coarse bottom sediments exist, and bottom photographs show signs of strong bottom current activity (Hollister and Elder, 1969; Anderson, Brake, and Myers, 1984). Where dense water flows off the shelf and mixes with and/or displaces WDW, more sluggish bottom currents are indicated by fine-grained sediments on the upper slope (Anderson, Kurtz, and Weaver, 1979). The lower slope and continental rise are characterized by weak bottom currents and fine-grained sediments, excluding sediment gravity flow deposits. The influence of impinging deep-sea currents on continental margin sediments is discussed in greater detail later in this chapter.

Sediments of the Continental Slope and Rise

There have been few detailed investigations of glacial-marine and hemipelagic sediments on the Antarctic continental slope and rise. Anderson, Kurtz, and Weaver (1979) examined sediments in piston cores from the continental slope and rise of the Weddell and Ross seas. More recent studies of slope sediments in the eastern Weddell Sea were conducted by Ehrmann, Melles, Kuhn, and Grobe (1992) and Grobe, Huybrechts, and Fütterer (1993).

Piston cores from the Ross and Weddell seas penetrated interbedded glacial-marine sediments and sediment gravity flow deposits. The cores displayed considerable downcore variability (Anderson, Kurtz, and Weaver, 1979; Ehrmann et al., 1992; Grobe, Huybrechts, and Fütterer, 1993). Glacial-marine sediments are massive and poorly sorted and contain large (5 to 50%) amounts of ice-rafted sand and scattered pebbles. These glacial-marine sediments are interbedded with laminated, crudely sorted to moderately sorted silt and clayey silt with very little ice-rafted debris and massive mud with little or no ice-rafted debris. The contacts between these deposits are generally sharp (Fig. 4.48). The laminated units typically display a bimodal grain size distribution, and individual laminations vary from 0.5 to 1.0 cm thick (Anderson, Kurtz, and Weaver, 1979). Ice-rafted debris, when present, tends to be concentrated within discrete thin interbeds. Thin

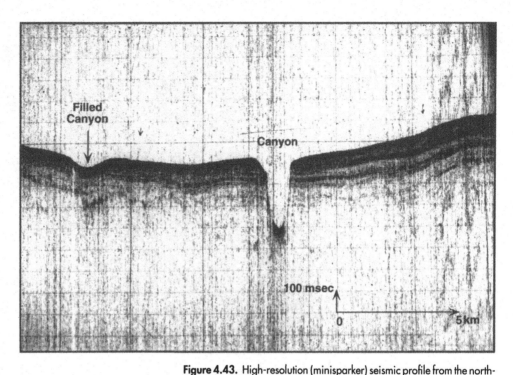

Figure 4.43. High-resolution (minisparker) seismic profile from the northwestern Ross Sea upper continental slope showing canyon cutting through glacial-marine deposits.

beds with millimeter-scale sedimentary clasts, primarily hemipelagic sediment, are believed to be derived from high-concentration turbidity currents. Microfossils are common to abundant in the massive units and are rare to absent in the laminated units. In general, the grain size of individual silt laminae decreases, and their sorting increases downslope. The differences in these sediment types are assumed to reflect differences in rates of sedimentation. A virtual absence of ice-rafted debris and microfossils in the laminated silts implies relatively fast accumulation, which led Anderson and colleagues (1979b) to interpret them as contourites and fine-grained turbidites. Thin interbeds of coarse silt and very fine sand support this interpretation.

ABYSSAL ENVIRONMENT

A number of factors contribute to the distribution of surface sediment on the Antarctic deep seafloor. Among these are the offshore flux of fine-grained terrigenous sediment from the continent (by ocean currents and turbidity currents), surface productivity of biogenic organisms, dissolution of biogenic sediment at depth, reworking by marine currents, and, to a lesser degree, ice rafting and authigenic sediment formation (Lisitzin, 1962, 1972; Nayudu, 1971; Goodell, 1973; Anderson, 1990). The distributions of sedimentary facies can be related to major oceanographic fronts, bottom current circulation patterns, water depth, seafloor topography, and distance from the continent.

Sediment of the Abyssal Environment

Several workers have mapped and described the distribution pattern of abyssal sediment on the Antarctic

continental rise and abyssal floor (Lisitzin, 1962, 1972; Goodell, 1973; Nayudu, 1971; Piper et al., 1985; Anderson, 1990; McCoy, 1991). The sediment distribution map shown in Figure 4.49 is from McCoy (1991). It is highly generalized, but is suitable for illustrating the broad patterns in abyssal sedimentation around Antarctica. Also shown in Figure 4.49 are the locations of abyssal areas (by figure number) where detailed sedimentologic studies (focusing on surface sediments) have been conducted. These areas are discussed later, in the case studies section of this chapter.

The Antarctic continent is producing an abundance of terrigenous silt and clay, which occupies a broad belt around the continent (Fig. 4.49). Hemipelagic sediments are typically poorly sorted and polymodal, and consist predominantly of detrital quartz. These sediments include laminated and massive, frequently bioturbated, units. Clay minerals are typically well-crystallized chlorite and illite (Goodell, 1973). Ice-rafted debris, volcanic ash, and ferromanganese micronodules are minor components.

The mineralogies of seafloor surface sediment in the area of the Antarctic Peninsula show that the peninsula is a major source of terrigenous material for the southeast Pacific Ocean (Edwards, 1968). The subpolar glacial-maritime setting of that region is conducive to glacial erosion and transport of sediment.

The belt of terrigenous sediment surrounding the continent is broader around West Antarctica than East Antarctica (Fig. 4.49). This is explained, in part, by the fact that the glacial systems that drain into the West Antarctic

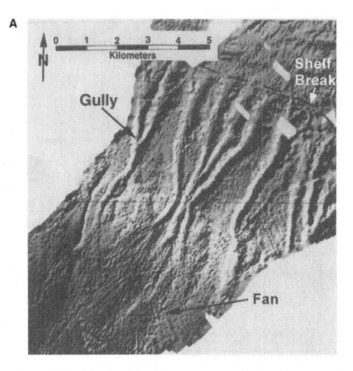

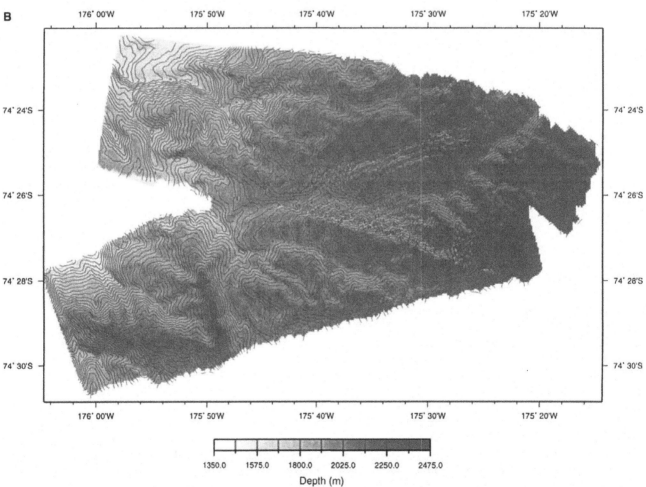

Figure 4.44. (A) Multibeam mosaic from the Ross Sea upper continental slope showing gullies and upper slope fans; (B) multibeam record of the lower slope in the eastern Ross Sea showing irregular relief caused by diverging canyons.

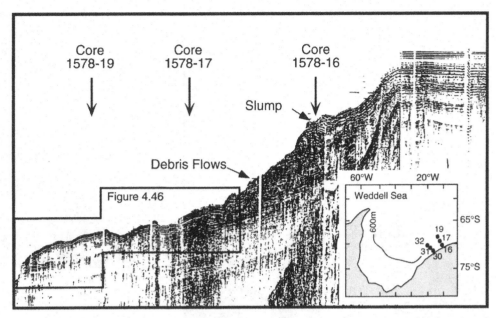

Figure 4.45. Representative seismic facies associated with sediment mass movement on the Antarctic continental slope. This particular example is from the eastern Weddell Sea (inset map). Insert shows the location of the 3.5-kHz subbottom profile shown in Figure 4.46. Also shown are the locations of piston cores collected along two transects and illustrated in Figure 4.47.

margin have a larger drainage basin and, therefore, a larger sediment source area than glaciers that drain into any given portion of the East Antarctic margin. The exception to this is the Prydz Bay region of East Antarctica, which receives sediments from the vast Amery drainage system. Furthermore, the glacial setting of West Antarctica is characterized by faster flow rates, and is therefore more conducive to sediment erosion and transport than the terrestrial glacial regime of East Antarctica. Ultimately, these differences are manifested in the distribution of fine-grained terrigenous sediment around the continent.

The seaward extent of the terrigenous sediment belt around Antarctica is also a function of the efficiency with which this sediment is being transported offshore by marine processes, a subject about which we know relatively little. It also is clear that turbidity currents have played a significant role in delivering terrigenous sediment to those portions of the deep seafloor located near Antarctica (Nayudu, 1971; Payne, Conolly, and Abbott, 1972; Tucholke et al., 1976; Wright, Anderson, and Fisco, 1983; Anderson, Wright, and Andrews, 1986; Porebski, Meischner, and Görlich, 1991).

In general, terrigenous sediments become finer in an offshore direction away from the Antarctic continent. This simply reflects increasing distance from the sediment source. The ice-rafted component decreases sharply away from the Antarctic continental shelf (Anderson, Kurtz, and Weaver, 1979), and ice-rafted debris is virtually lacking in bottom sediment north of the Antarctic Convergence (Goodell, 1973). This paucity of ice-rafted sediment is due to the accelerated decay rate of icebergs as they are subjected to the relatively warmer (> 0°C) and rougher surface waters off the continental shelf (Figs. 1.19 and 1.20). Icebergs also tend to drift parallel to the continent once they encounter the circumpolar currents seaward of the shelf break.

North of the terrigenous silt and clay belt that surrounds Antarctica, there is a broad belt of siliceous ooze (Fig. 4.49), which consists primarily of diatom frustules resulting from high surface productivity. Its presence is associated with the region between the Antarctic Divergence and Antarctic Polar Front, where cyclonic motion and upwelling of intermediate water masses dominate surface circulation (Goodell, 1973) (Fig. 1.32). The belt also marks the northern limit of terrigenous sediment transport by turbidity currents. The boundary between siliceous ooze and hemipelagic sediment also corresponds roughly to a major divergence in abyssal circulation (Goodell, 1973). South of this boundary, abyssal currents flow in a westward direction along the Antarctic continental margin. To the north of this boundary, abyssal flow is more eastward (Chapter 1). Much of the terrigenous sediment that is transported northward from the Antarctic continental margin is apparently entrained by these westward-flowing currents, thereby restricting their northward distribution (Anderson, 1990). North of this boundary, siliceous sediment is transported to the east during its final descent to the seafloor.

DeMaster and colleagues (1991) measured rates of siliceous sediment accumulation beneath the Antarctic Polar Front. They found rates as high as 20 to 180 cm/kyr, and estimated that ~25 to 50% of the gross silica production in surface waters is preserved on the seafloor.

The northern boundary of the siliceous ooze belt marks the northern limit of bottom waters that are undersaturated with calcium carbonate. The concentration of

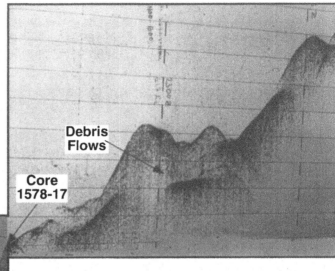

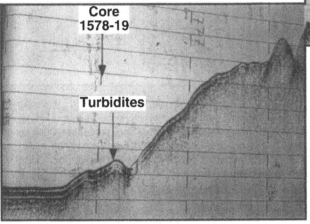

Figure 4.46. 3.5-kHz subbottom profiler record from the eastern Weddell Sea continental slope showing acoustic character of debris flow deposits (hummocky, transparent bodies) grading downslope into acoustically layered and transparent units that piston cores showed to be interbedded hemipelagic sediments and turbidites.

siliceous material decreases and calcareous material increases toward the Antarctic Convergence. This zone of mixed siliceous and calcareous ooze occupies a 200- to 300-km-wide belt on the northern margins of the Antarctic abyssal plains (Goodell et al., 1973) (Fig. 4.49). North of this belt, the sediment is dominated by either pelagic clay or calcareous ooze, where calcareous biogenic sediment production is sufficient to mask siliceous sediments (Lisitsyn, 1972). The distribution of calcareous sediment on the subantarctic deep seafloor is strongly depth-dependent. In the southern Indian Ocean, the calcium carbonate compensation depth (CCD) occurs at a depth of ~ 4500 m, in the South Pacific at ~ 5100 m, and in the South Atlantic at ~ 5400 m (Lisitsyn, 1972).

Brown pelagic clay is accumulating in portions of the abyssal floor that fall below the CCD. Unlike terrigenous sediments that occur adjacent to the continent, this pelagic clay consists almost exclusively of clay minerals, zeolites, and detrital grains (Skornyakova and Petelin, 1967; Nayudu, 1971). Montmorillonite is the dominant clay mineral. The dominant zeolite is phillipsite, which may constitute 50% of the sediment by volume. Pyroxene, pla-

gioclase, and opaques dominate the mineral grain component (Nayudu, 1971).

CASE STUDIES

Proper interpretation of the marine sedimentary record in Antarctic regions requires an understanding of sedimentary processes and sedimentary response to environmental change. Toward this goal, detailed sedimentologic studies have been conducted in a number of different sectors of the Antarctic region. These case studies provide a range of end-member settings where sedimentary facies can be related to the physical and biologic setting. The following sections summarize the results of these studies by region.

Ross Sea

Because the Ross Sea is the most thoroughly studied region of the Antarctic continental shelf, initial discussion of oceanographic influences on sedimentation will begin there. The oceanographic controls on sedimentation in the Ross Sea are essentially representative of other portions of the continental shelf that have been studied to date.

The Ross Sea continental shelf is characterized by northeast-southwest-oriented banks and troughs (Fig. 3.13). Water depths on the shelf range from ~ 300 m to just over 700 m, and the shelf has a landward slope

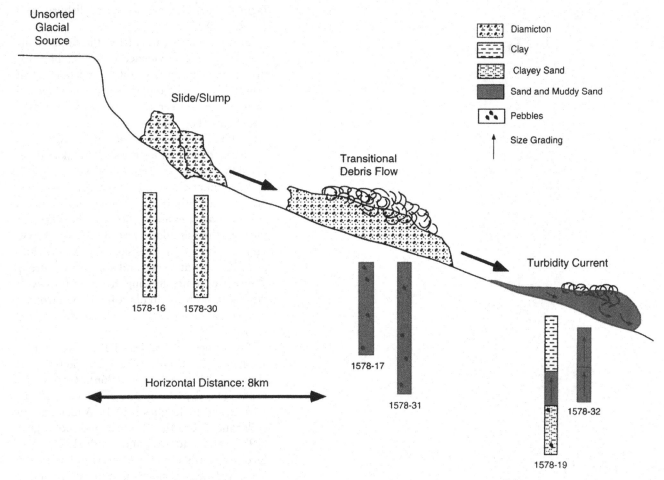

Figure 4.47. Downslope sediment gravity flow transitional processes based on seismic data and sediment cores from the northeastern Weddell Sea continental slope (modified from Wright and Anderson, 1982).

(Chapter 3). The Ross Sea continental slope is divided into two distinct physiographic portions, with the Pennell Bank forming the dividing line. East of the Pennell Bank, the slope is relatively gentle (average gradient of 2°). West of the Pennell Bank, the slope is steeper and more irregular in outline. The upper slope averages 5°, and the lower slope averages 1.5°. This portion of the slope is dissected by several submarine canyons (Fig. 4.43).

Oceanography. Three main water masses dominate on the Ross Sea continental shelf: LSSW, HSSW, and a modified form of Circumpolar Deep Water termed "Warm Core Water" (WMCO) (Jacobs, Fairbanks, and Horibe, 1985). A fourth water mass consists of surface water that occupies the upper layer of the water column. The temperature and salinity of this surface water changes seasonally, mainly in response to sea ice freezing and thawing. Jacobs, Fairbanks, and Horibe (1985) subdivide the shelf water column into more water masses, but division into the three main water masses listed above is suitable for relating oceanographic processes to sedimentation.

WMCO is identified on the continental shelf because of its relatively warm temperature (−0.05 to −1.15°C). LSSW and HSSW are colder (−1.50 to −1.92°C) than WMCO. A 34.50-ppt salinity boundary subdivides LSSW and HSSW. WMCO will override HSSW as it flows onto

the shelf to form a "warm tongue" (Fig. 4.50A). In the western Ross Sea, WMCO that flows onto the shelf is impeded by HSSW in this region, although some WMCO does flow above the HSSW and mixes with surface water (Fig. 4.50B). Thus, circulation on the shelf is largely regulated by the salinity of the shelf water column, which, in turn, is controlled by sea-ice production (Chapter 1).

One of the more conspicuous oceanographic features of the Ross Sea is the increase in salinity of the shelf water column from east to west across the shelf (Fig. 4.50C). This is due to greater rates of brine production in the western Ross Sea and to the general east to west transport of shelf waters (Klepikov and Grigor'yev, 1966; Jacobs, Fairbanks, and Horibe, 1985).

Few direct current measurements have been made near the seafloor of the Ross Sea. From a number of brief (5 to 10 min) measurements 1 m above the seafloor, Jacobs, Amos, and Bruchhausen (1970) reported currents in excess of 15 cm/sec on the outer shelf and generally less than 10 cm/sec on the inner shelf. More recent measurements, some continuing an entire year, also show stronger near-bottom flow near the continental shelf break (Jacobs,

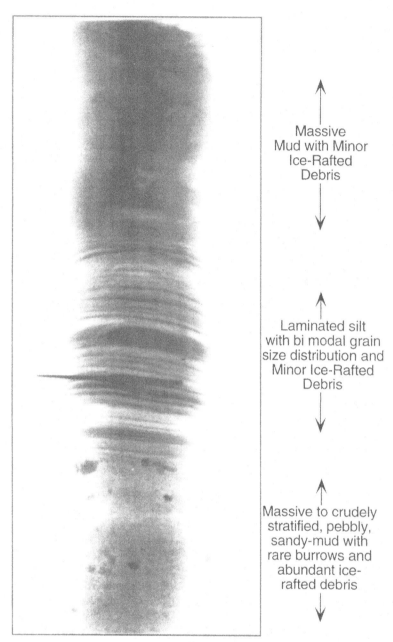

Massive
Mud with Minor
Ice-Rafted
Debris

Laminated silt
with bi modal grain
size distribution and
Minor Ice-Rafted
Debris

Massive to crudely
stratified, pebbly,
sandy-mud with
rare burrows and
abundant ice-
rafted debris

Figure 4.48. X-radiograph showing abrupt changes in sediment character in slope deposits from the Weddell Sea continental slope.

Fairbanks, and Horibe, 1985) (Fig. 1.35). Current speeds at an inner shelf site, 42 m above the seafloor, had a mean velocity of ~ 8 cm/sec and maximum velocities less than 15 cm/sec during most of the measured interval. In contrast, current meters on the upper slope and shelf break (52.3 and 52.4) (Fig. 1.35) recorded much higher peak and average velocities, with mean velocities ranging between 17 and 18 cm/sec. Current meter records along and under the Ross Ice Shelf show dominantly diurnal currents that have a generally southward and westward flow component and reach up to 25 cm/sec (Jacobs, Gordon, and Amos, 1979;

Jacobs and Haines, 1982; Pillsbury and Jacobs, 1985).

Bottom photographs from the floor of the Ross Sea record evidence of weak, bi-directional currents on the middle and inner shelf (Fig. 4.51A). The outer shelf is characterized by gravel lags covered by erect and encrusting organisms (Fig. 4.51B) and by rippled sands (Fig. 4.51C). Directional indicators reflect flow from east to west along the shelf break and upper slope (Anderson, Brake, and Myers, 1984).

In summary, direct (bottom current meter) and indirect (bottom photographs) current measurements from the floor of the Ross Sea indicate that circulation of HSSW and LSSW on the inner and middle shelf is typically sluggish. On the outer shelf and upper slope, strong contour currents flowing from east to west dominate bottom circulation. These stronger currents are associated with the WMCO water mass.

Sedimentology. Studies of Ross Sea surface sediment have been conducted by Stetson and Upson (1937), Kennett (1966), Chriss and Frakes (1972), Glasby and colleagues (1975), Barrett and colleagues (1983), Anderson and colleagues (1984), Dunbar and colleagues (1985), and Anderson and Smith (1989). Surface sediments contain mixtures of unsorted ice-rafted debris, siliceous biogenic material, calcareous shell material, and terrigenous silt and clay. The concentration of these various components across the shelf reflects the relative influence of glacial, oceanographic, and biologic processes (Chriss and Frakes, 1972).

The sea bed of the Ross Sea shelf break-upper slope is covered by sandy gravel (residual glacial-marine sediments), sand, and muddy sand (Fig. 4.52). Similar coarse-grained deposits also occur on banks. The thickness of these deposits is uncertain because of the difficulty of coring them. Compositionally, they consist of ice-rafted, coarse lithic sand and gravel, and calcareous bioclastic material. The relative amounts of these components vary. The bioclastic component constitutes between less than 5% and 90% of the sediment, and generally increases from east to west along the outer shelf (Taviani, Reid, and Anderson, 1993) (Fig. 4.28).

Thin, fairly extensive sand bodies occur on the shelf near Edward VII Peninsula and on the banks of the western Ross Sea (Fig. 4.52). The sand bodies occur mainly in water depths up to 400 m, except on the shelf break upper slope where they occur at depths up to 600 m. They are interbedded with, and grade laterally into, residual glacial-marine sediments. Bottom photographs from the

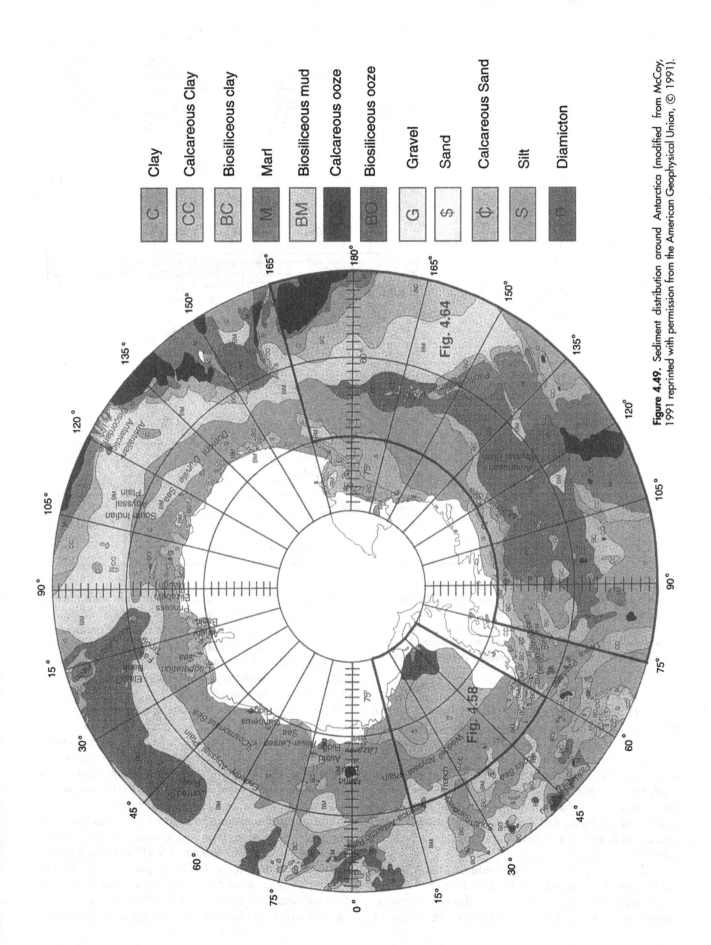

Figure 4.49. Sediment distribution around Antarctica (modified from McCoy, 1991 reprinted with permission from the American Geophysical Union, © 1991).

C		Clay
CC		Calcareous Clay
BC		Biosiliceous clay
M		Marl
BM		Biosiliceous mud
BO		Calcareous ooze
BO		Biosiliceous ooze
G		Gravel
$		Sand
¢		Calcareous Sand
S		Silt
D		Diamicton

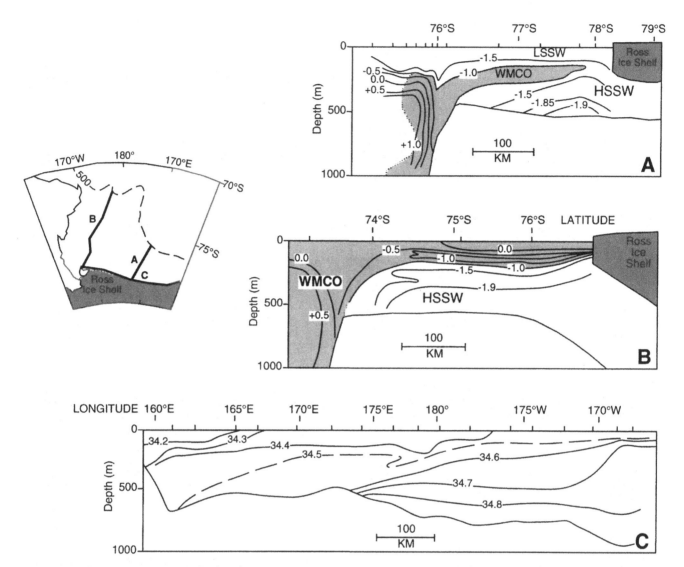

Figure 4.50. (A and B) Temperature profiles from the Ross Sea continental shelf showing different levels of intruding Warm Core Water (WMCO; shaded area). (C) Salinity profile along the front of the Ross Ice Shelf. The 34.5-ppt isohaline surface is highlighted (dashed), because this is the level above which WMCO will flow onto the continental shelf (profiles from Pillsbury and Jacobs, 1985, reprinted with permission from the American Geophysical Union, © 1985).

bank tops do not show ripples or other bedforms. Rather, they show dense benthic organisms, primarily bryozoans (Fig. 4.51C). The sands are interpreted as traction current deposits that are being derived from the poorly sorted residual glacial-marine sediments on the banks and outer shelf, and possibly from pre-existing sand bodies formed beneath ice shelves that were grounded on the banks. The absence of bedforms and dense coverage of benthic organisms indicates that bedload transport on the banks is currently inactive.

Grain size data for the coarse sediment that blankets the shelf break-slope and shelf banks indicate bottom current velocities in excess of 15 cm/sec and current velocities 100 cm above the bed that exceed 25 cm/sec (Anderson and Smith, 1989). This is consistent with peak current meter measurements of velocities in excess of 25 cm/sec measured 4 m above the seafloor on the shelf break-slope (Fig. 1.35). These sands provide a dramatic illustration of the competence of marine currents at sorting even poorly sorted deposits with the aid of biologic mixing of the bed (Singer and Anderson, 1984).

Piston cores from the lower slope penetrated poorly sorted sandy, pebbly mud, and laminated mud, which shows no evidence of winnowing. Marine currents, therefore, appear to be effectively winnowing sediment on the upper slope, but not on the lower slope in this region. This indicates that the boundary currents are strongest where they impinge on the upper slope-shelf break.

Fine-grained sediments composed of terrigenous silt, very fine sand, diatom frustules, sponge spicules, and unsorted ice-rafted debris (generally < 10%) blanket the inner Ross Sea shelf (Fig. 4.52). In general, the concentrations of biogenic silica and organic carbon increase to the south and west across the shelf (Dunbar et al., 1985). This increase is attributed to reworking and redistribution

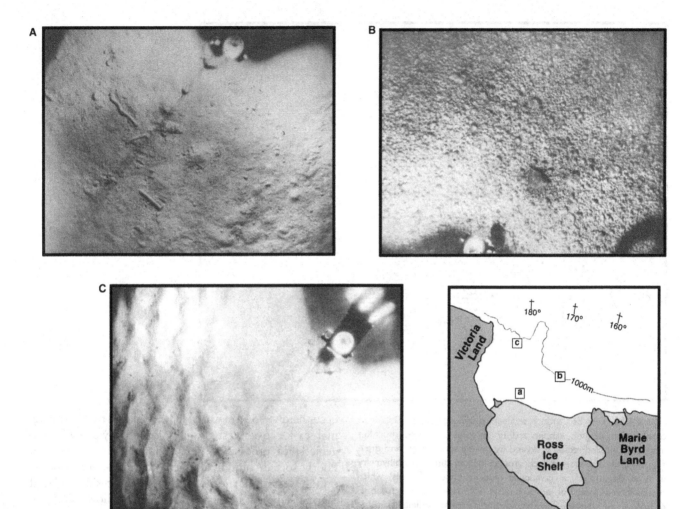

Figure 4.51. Bottom photographs from the Ross Sea showing (A) muddy seafloor on the inner shelf where bottom currents are weak, (B) gravelly lag deposits, and (C) ripples on the outer shelf where strong boundary currents exist.

of these components by marine currents, which generally flow to the south and west, and also to less severe sea-ice cover in the western Ross Sea (Dunbar et al., 1985; Kellogg and Kellogg, 1988). Near-bottom current velocities on the inner shelf are generally weak and, 95% of the time, fall below the suspension threshold velocity for very fine sand (25 cm/sec). Fine-grained sediments swept from the outer shelf and banks by the impinging WMCO accumulate as a draping unit on the deeper portions of the inner shelf where sluggish bottom currents exist. The thickness of these diatomeceous muds and oozes varies by several meters locally, a result of the fact that these sediments are draping very irregular seafloor that has been scoured by glacial ice and icebergs.

The other controlling agent on sediment dispersal in the Ross Sea is sea-ice cover, which regulates surface productivity and, therefore, biogenic sediment production. During the austral summer, sea-ice cover is more severe in the eastern Ross Sea than the western Ross Sea. The sea-ice distribution results in higher biogenic sediment production in the western Ross Sea (Dunbar et al., 1985). The general westward flow of surface waters in the Ross Sea also contributes to higher diatomaceous concentrations in

compound glacial-marine sediments of the western Ross Sea (Anderson, Brake, and Myers, 1984).

In summary, the broad-scale sediment distribution patterns in the Ross Sea show strong influences of oceanographic processes and sea-ice production. Impinging boundary currents winnow fine-grained sediment from the outer shelf and upper slope, and on bank tops, producing gravelly lag deposits and rippled sands. Some of the fine-grained material is swept onto the shelf by the impinging current, where it accumulates on the more quiescent seafloor. The diatom component of surface sediments is strongly regulated by seasonal sea-ice development, whereby accumulation is greatest in the western Ross Sea where open seas are more prevalent during the austral summer.

North Victoria Land Continental Shelf

Located offshore of the Pennell Coast is one of the shallowest portions of the East Antarctic continental shelf. The majority of the shelf lies within a water depth of

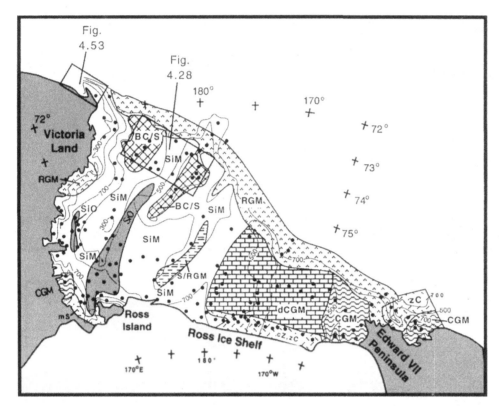

Figure 4.52. Ross Sea surface sediment distribution map. Dots indicate locations of grab samples and trigger cores used to construct this map. The stippled region is floored by muddy sand, sand, and gravelly sand, and therefore reflects stronger bottom currents. Bathymetry is in meters. CGM = compound glacial-marine sediments, which predominantly consist of terrigenous silt with minor ice-rafted debris; dCGM = diatomaceous compound glacial-marine sediments; cZ = clayey silt; zC = silty clay; mS = muddy sand; RGM = residual glacial-marine sediments, which predominantly consist of ice-rafted sand and gravel; SiM = siliceous mud; SiO = siliceous ooze; S = sand; and BC = bioclastic carbonate (modified from Anderson, Brake, and Myers, 1984).

300 m, with the exception of troughs. During the 1980 Deep Freeze cruise to the area, piston cores and grab samples were collected from the continental shelf and upper slope in the region (Anderson and Kurtz, 1980) (Fig. 4.53). One of the objectives of this survey was to examine sedimentation on a portion of the shelf that is bounded by mountains where valley and outlet glaciers flow into the sea. Later, in 1998, the area was revisited as part of the *Nathaniel B. Palmer* cruise 98-01. The objective of this cruise was to examine the glacial history of the shelf. During this cruise, high-resolution seismic data, multibeam data, deep tow side-scan sonar/chirper data, and sediment cores and grab samples were collected along with a single hydrographic profile across the shelf and slope. The combined data sets from these two cruises provide the basis for the following discussion.

Sedimentology. Sand, muddy sand, and residual glacial-marine sediments blanket the outer shelf, and finer-grained sediments occupy the inner shelf (Fig. 4.53). Siliceous muds

occur below ~ 500 m water depth in glacial troughs. Perennial sea-ice cover west of ~ 167°E probably prohibits biogenic silica production and accumulation on the shelf. Poorly sorted compound glacial-marine sediment occurs on the western portion of the shelf, and residual glacial-marine sediment accumulates on the shallow (< 250 m) portions of the shelf.

Volcaniclastic material derived from Cape Adare forms the dominant component of sands that blanket the bank east of Cape Adare and the outer shelf. The volcanic material was transported up to 60 km to the west by a contour current whose speed near the seafloor must occasionally exceed 25 cm/sec, based on size distribution of the sand fraction (Anderson, Brake, and Myers, 1984). The hydrographic profile acquired during our 1998 cruise shows modified WDW extending well onto the eastern shelf.

In summary, the broad-scale surface sediment distribution patterns on the continental shelf off the Pennell Coast are similar to those in the Ross Sea. The outer shelf, upper slope, and banks are blanketed by residual glacial-marine sediments and traction current deposits, which indicate that these areas are influenced by impinging boundary currents. The inner shelf is covered by fine-grained sediments, including siliceous mud in troughs. The diatomaceous component of bottom sediments decreases from east to west due to more severe sea-ice cover in that direction.

Wilkes Land Continental Shelf and Slope

During Deep Freeze 79, piston cores, grab samples, and physical oceanographic profiles were acquired along

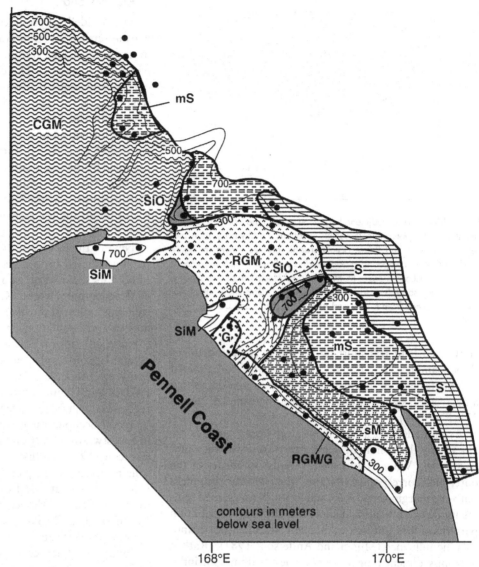

Figure 4.53. Surface sediments of the continental shelf offshore of the Pennell Coast. Dots indicate sample locations. Bathymetry is in meters. CGM = compound glacial-marine sediments; RGM = residual glacial-marine sediments; SiM = siliceous mud; sM = sandy mud; SiO = siliceous ooze; S = sand (volcaniclastic sand derived from Cape Adare); mS = muddy sand; G = gravel, and Ba = basement outcrop (modified from Anderson and Molnia, 1989).

several transects extending from the George V Coast, East Antarctica to the continental slope (Anderson et al., 1979) (Figs. 4.54 and 4.55). One objective of this project was to investigate sedimentation on a portion of the continental shelf that is bounded by the East Antarctic Ice Sheet. The glacial-maritime setting along most of the coast is characterized by ice cliffs, with the exception of the Mertz and Ninnis glacier tongues. Together, these glacier tongues represent significant ice drainage from the Wilkes Land region.

The seafloor physiography is dominated by a large trough parallel to the coast, two transverse troughs, and outer shelf banks. The upper slope is steep and dissected by numerous submarine canyons (Fig. 3.33).

The physical oceanographic setting of the Wilkes Land continental shelf is similar to that of the Ross Sea. Modified WMCO extends onto the shelf as a diluted "warm tongue" at intermediate depths (Fig. 4.55).

Sedimentology. Surface sediments and foraminiferal distribution patterns within these sediments were studied by

Milam and Anderson (1981) and Dunbar and colleagues (1985). The upper slope–shelf break of the Wilkes Land shelf is covered by sands and gravelly sands (residual glacial-marine sediments) with decreasing grain size in an onshore direction (Fig. 4.54). Siliceous muds and oozes blanket the inner shelf troughs. Two grab samples collected from the steep inner shelf above 150 m water depth consist of gravel and gravel-size bioclastic material.

The relationship between the water mass structure and grain size patterns across the Wilkes Land shelf is illustrated using samples acquired along Profile B (Fig. 4.56). Moderately sorted to well-sorted sands with dominant modes in the 1.25-phi to 2.75-phi size range indicate

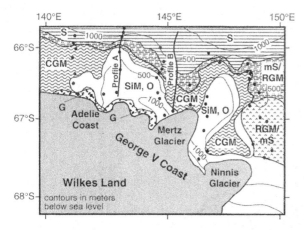

Figure 4.54. Surface sediment distribution map for the Wilkes Land continental shelf. Dots indicate sample locations. Bathymetry is in meters. Profiles A and B are oceanographic profiles shown in Figure 4.55. CGM = compound glacial-marine sediments; SiM = siliceous mud; SiO = siliceous ooze; RGM = residual glacial-marine sediments; mS = muddy sand; S = sand; and G = gravel. Dots show sample locations (modified from Milam and Anderson, 1981, and Dunbar et al., 1985).

active bed load transport on the outer shelf and upper slope (samples 22, 24, and 25) (Fig. 4.56). Fine-grained sediments blanket the inner shelf, and grain size data show a decrease in size in an onshelf direction (samples 26 and 27) (Fig. 4.56).

As in the Ross Sea, impinging deep ocean currents are apparently the dominant sedimentary agents on the outer part of the Wilkes Land shelf. Sandy sediments of the outer shelf correspond to the area where WMCO impinges onto the margin (Figs. 4.55 and 4.56). Fine-grained sediments are accumulating in inner shelf basins. Distribution patterns of foraminifera on the shelf indicate little oceanographic influence (Milam and Anderson, 1981). Rather, bottom sediment type appears to be the dominant influence on foraminiferal ecology.

Figure 4.55. Representative hydrographic profiles from the Wilkes Land continental shelf. Warm Core Water (WMCO) is shown by a stippled pattern (data provided courtesy of Stan Jacobs).

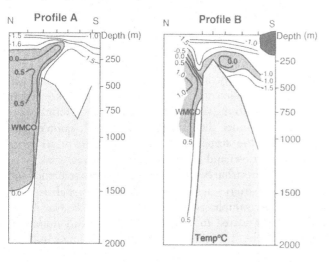

Weddell Sea

The Weddell Sea provides one of the most varied glacial-marine settings in Antarctica. It is bounded on the east by the EAIS, which terminates at the sea as ice cliffs and fringing ice shelves. The continental shelf in the area is narrow and slopes landward, and the continental slope is steep (Fig. 3.31). The southern Weddell Sea shelf is partly covered by the Ronne-Filchner Ice Shelf. The southwestern shelf is wide and is separated from East Antarctica by the deep Crary Trough (Fig. 4.57). The western shelf is bounded by the mountainous Antarctic Peninsula. Glacial drainage from the peninsula into the Weddell Sea is primarily captured by the Larsen Ice Shelf. The western shelf is also broad and is dissected by several transverse troughs (Fig. 3.32).

Next to the Ross Sea, the Weddell Sea is the most thoroughly sampled and studied region in Antarctica, in terms of both its sedimentology and its physical oceanography.

Oceanography. Figure 4.57 provides a highly generalized illustration of the distribution of major bottom water masses and circulation patterns in the Weddell Sea, and is based on physical oceanographic data acquired during several oceanographic expeditions over a period of several years (Anderson, 1975b; Carmack and Foster, 1975; Foldvik and Gammelsrød, 1988). Four water masses are recognized (Chapter 1): LSSW, HSSW, WDW, and AABW.

LSSW occupies the eastern continental shelf and flows in a westwardly direction as a pronounced coastal current (Deacon, 1937; Hollister and Elder, 1969) at velocities ranging from 10 to 30 cm/sec (Klepikov, 1963). West of ~ 30°W, this relatively fresh-water mass flows along the outer shelf as a V-shaped water mass (Gill, 1973). The freshness of the LSSW is predominantly due to seasonal melting of sea ice and to meltwater contributions from the Brunt and Filchner ice shelves (Anderson, 1975b; Doake, 1985). HSSW forms on the broad continental shelf of the southern Weddell Sea and flows in a northwestwardly direction across the continental shelf and onto the slope (Fig. 4.57). WDW is the dominant deep water mass in the eastern Weddell Sea and flows along the eastern continental slope as a boundary current.

West of ~ 40°W, HSSW is sufficiently dense to mix with WDW to form AABW (Seabrooke, Hufford, and Elder, 1971; Carmack and Foster, 1975). AABW also is produced along the narrow (~ 100 km wide) stretch of the continental shelf seaward of the Crary Trough, where dense water flowing from the vicinity of the trough, beneath LSSW, flows onto the slope (Foldvik and Gammelsrød, 1988). Within this region, AABW flows down slope within a canyon at velocities in excess of 100 cm/sec. Elsewhere, downslope flow of AABW is sluggish (Carmack and Foster, 1975; Weber, Bonani, and Fütterer, 1994). Newly formed AABW flows across the continental rise and Weddell Abyssal Plain in a northeastward direction at an average velocity of 2.8 cm/sec (Carmack and Foster, 1975).

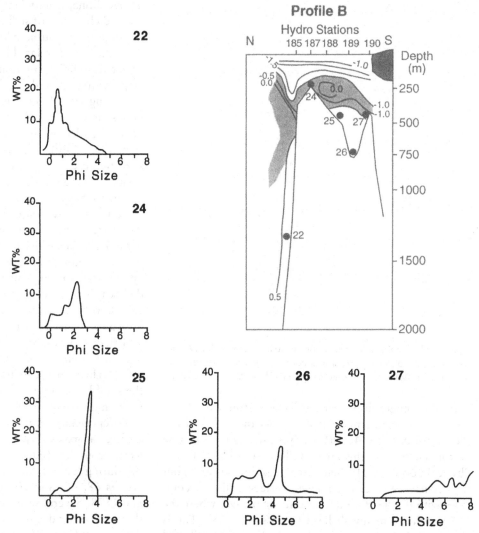

Figure 4.56. Grain size frequency curves for surface sediment samples collected along hydrographic Profile B (Fig. 4.54), illustrating the association of sands with impinging tongue of Warm Core Water (WMCO, shaded) and decreasing size of the dominant transport mode in an onshore direction.

Sedimentology of the Continental Shelf and Slope. The distribution pattern of modern sediment on the outer shelf and slope of the eastern Weddell Sea correlates well with the circulation patterns described above. The continental shelf off Princess Martha Coast is blanketed by compound glacial-marine sediments (Fig. 4.58) resting on till (Anderson et al., 1980b, 1990; Elverhøi, 1981). In this area, LSSW occupies the continental shelf, and the WDW boundary current does not impinge onto the shelf (Fig. 1.34C). Surface sediment on the upper continental slope, where WDW impinges on the seafloor, consists of sandy gravel (residual glacial-marine sediment) including bioclastic carbonate (Anderson, Kurtz, and Weaver, 1979; Elverhøi and Roaldset, 1983; Grobe, Huybrechts, and Fütterer, 1993). Bottom photographs indicate moderate to strong bottom current activity (Hollister and Elder, 1969). Sediment gravity flow deposits, including poorly sorted debris flow deposits and graded sands, cover most of the upper slope (Wright and Anderson, 1982; Grobe, Huybrechts, and Fütterer, 1993) (Fig. 4.47). Interbedded hemipelagic sediment, contourites, and sediment gravity flow deposits cover the middle and lower slope

(Anderson, Kurtz, and Weaver, 1979; Grobe, Huybrechts, and Fütterer, 1993). The hemipelagic sediment is massive to crudely stratified and contains relatively high concentrations of ice-rafted debris and biogenic material (Fig. 4.48). In contrast, contourites are laminated and contain little or no biogenic material and ice-rafted sand or gravel (Anderson, Kurtz, and Weaver, 1979; Elverhøi and Roaldset, 1983; Grobe, Huybrechts, and Fütterer, 1993). Fine-grained sediments, in conjunction with bottom photographs from the area, indicate sluggish bottom current activity along the lower slope and continental rise.

Farther south, along the eastern Weddell Sea continental margin, off the Coats Land coast, surface sediments on the shallow portions of the continental shelf (< 350 m) consist of interbedded calcareous gravel, sand, and residual glacial-marine sediments (Angino and Andrews, 1968; Anderson, 1972b; Anderson et al., 1983a) (Fig. 4.58), and

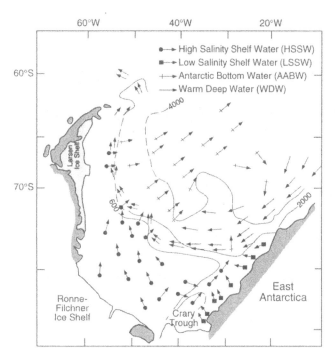

Figure 4.57. Distribution of major bottom water masses in the Weddell Sea, based on a study of physical oceanographic data collected over several years. Bathymetry is in meters (modified from Anderson, 1975b).

bottom photographs reveal a gravelly bottom with abundant encrusting organisms (Fig. 4.59A) and deflected sessile organisms (Fig. 4.59B), implying strong to moderate bottom currents in this area (Hollister and Elder, 1969). Physical oceanographic data are sparse for this area, but the occurrence of coarse-grained surface sediment coincides with that portion of the continental shelf where the LSSW is the dominant shelf water mass (Fig. 4.57) This is also the shallowest area where samples have been collected in the eastern Weddell Sea.

Compound glacial-marine sediments, with little or no biogenic material, blanket the southwestern continental shelf of the Weddell Sea (Anderson, 1972b; Fütterer and Melles, 1990). Bottom photographs indicate sluggish bottom currents in this area (Hollister and Elder, 1969) (Fig. 4.59C). The only exception is the shelf offshore of Berkner Island, where residual glacial-marine sediment and sand, including eolian sand, occurs (Fig. 4.58). Foraminiferal (*Globoquadrina pachyderma*) mud covers the upper continental slope of the southern Weddell Sea. Locally, a gravelly lag deposit occurs (Fütterer and Melles, 1990). The concentration of forams varies with depth in cores, and foraminiferal mud overlies compound glacial-marine sediment and debris flows (Anderson, Kurtz, and Weaver 1979; Kurtz and Anderson, 1979). Piston cores from the lower slope penetrated alternating units of laminated silt, with little or no biogenic material and ice-rafted debris, and massive, bioturbated silt with greater concentrations of ice-rafted and biogenic material. These deposits are interpreted as contourites and hemipelagic material, respectively (Anderson, Kurtz, and Weaver, 1979;

Weber, Bonani, and Fütterer, 1994). The general fine-grained character of sediments on the outer shelf and upper slope in the southern Weddell Sea implies weak bottom currents. This is related to oceanographic mixing and associated AABW formation in this area, where flow on the slope is sluggish and in a downslope direction.

During 1982 Deep Freeze operations, piston cores and grab samples were collected from the northwestern corner of the Weddell Sea to examine sediment distribution patterns (Fig. 4.60). This region is unique to the Antarctic shelf because Seymour Island is virtually ice-free, and its eastern coast is exposed to wave attack during the austral summer when the sea ice melts. The sediment distribution patterns also are unique relative to other portions of the Antarctic shelf. North of Seymour Island, fine-grained, compound glacial-marine sediment blankets the shelf (Fig. 4.60). Offshore, fining occurs east and south of the island and indicates marine-current transport is toward the Weddell Sea. This is the general direction of prevailing winds for the region. Grain size studies indicate that currents are causing bed load transport to a depth of ∼200 m, and that very fine sand and silt are transported several tens of kilometers offshore before being deposited (Fig. 4.60). The coarsest sediment (sand and gravel) occurs at the shelf break–upper slope and is apparently associated with an impinging boundary current in this region.

Sedimentology of the Continental Rise and Abyssal Plain Surface Sediments. Piston cores and gravity cores were collected from the Weddell Sea abyssal floor and continental rise during *Eltanin* cruises 7 and 12, *Glacier* (Deep Freeze) cruises 68, 69, and 70, and *Islas Orcadas* cruises 1277 and 1578. Several scientists conducted detailed studies of these and other deep-sea cores from the Weddell Sea (Anderson, 1972a,b, 1975a,b; Wright, 1980; DeFelice and Wise, 1981; Wright and Anderson, 1982; Fisco, 1982; Ledbetter and Ciesielski, 1982, 1986; Anderson, Wright, and Andrews, 1986; Pudsey, 1992; Pudsey, Barker, and Larter, 1994).

The sediment collected from the continental rise and abyssal plain typically varies from terrigenous mud, with rare microfossils in the southern region, to diatomaceous mud and ooze in the northern portion of the abyssal plain (Fig. 4.58). Hemipelagic sediment predominantly consists of massive, terrigenous silt and clayey silt in which the concentrations of ice-rafted debris and biogenic material vary spatially and temporally. The calcareous component consists almost entirely of the planktonic foraminifer *Neogloboquadrina pachyderma*. The benthic foraminifer assemblage is dominated by two arenaceous species, *Cyclammina pusilla* and *Cribrostomoides subglobosus* (Anderson, 1975a), but these are restricted to surface sediments. The CCD is multibathic, ranging from shelf to abyssal depths (Anderson, 1975b). Radiolarians are rare, and diatoms are restricted to the northern portion of the sea (Kaharoeddin et al., 1979, 1980; Pudsey, Barker, and Hamilton, 1988).

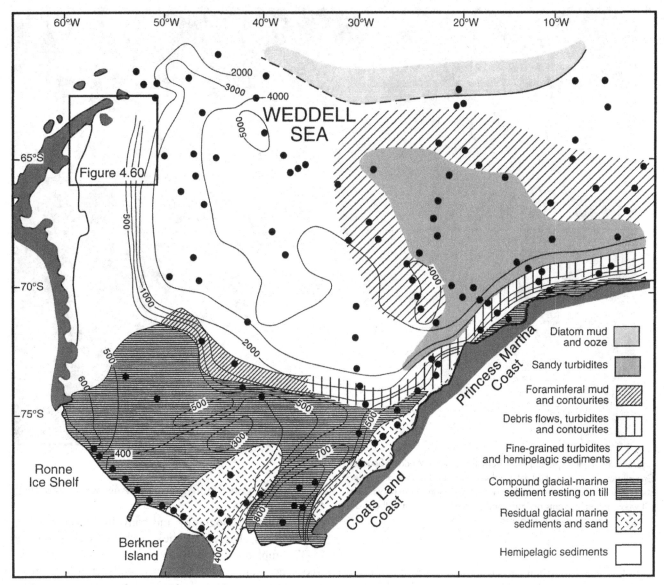

Figure 4.58. Near-surface sediment distribution patterns in the Weddell Sea based on piston and gravity cores. Dots indicate core locations (*Eltanin, Islas Orcades, Glacier,* and *Polarstern* cores). Sediment distribution patterns on the southwestern shelf, along the front of the Ronne Ice Shelf, are based on the work of Fütterer and Melles (1990).

The distribution of calcareous versus arenaceous foraminifera in the surface sediments of the Weddell Sea shows a strong correlation to the distribution of major bottom water masses (Anderson, 1975b). This is attributed, in part, to the dissolution of calcareous tests by more corrosive WDW and HSSW. Calcareous foraminifera are predominantly restricted to the upper slope and shelf in the eastern Weddell Sea, and presently occur at abyssal depths only in the western Weddell Sea.

Rugged topography, due to the presence of numerous seamounts, characterizes the northern part of the Weddell Abyssal Plain. Piston cores collected near these features typically contain interbedded hemipelagic sediment, fine-grained turbidites, siliceous ooze, and debris flows (Fisco, 1982). Convolute bedding and sedimentary clasts concentrated in layers distinguish debris flows. Sedimentary clasts usually include a significantly larger volcaniclastic component than pelagic or hemipelagic sediments of the area (Kaharoeddin et al., 1979, 1980). Sedimen-

tary clasts also consist of biogenic ooze (usually siliceous), mud, and, less frequently, silt and sand. Clast compositions typically differ from the sediment types that bound the debris flow deposits. Most clasts are soft, but partially lithified sedimentary clasts do occur. These deposits must reflect a sediment transport mechanism intermediate between high-concentration turbidites and nearly turbulent debris flows.

Sedimentology of the Weddell Fan. During the Shackleton Expedition of 1915, the resident geologist described several dredge samples from the northeastern Weddell Abyssal Plain. He discovered that several of the grabs had collected well-sorted sand (Wordie, 1921). Unfortunately, the

Figure 4.59. Bottom photographs from the eastern Weddell Sea. (A) Gravelly lag deposits and (B) deflected sessile organisms indicate moderate to strong bottom currents at depths shallower than ~ 350 m, whereas (C) weak bottom currents are indicated at depths > 350 m in the Crary Trough and on the southwestern continental shelf.

samples were lost when ice crushed the *Endurance*, but a single piston core, collected in the region during the 1970 IWSOE expedition, penetrated 165 cm of well-sorted quartz sand (Anderson, 1972a). This intriguing find prompted a more detailed survey of the region during the 1977 and 1978 *Islas Orcadas* expeditions (Wright and Anderson, 1982; Anderson, Wright, and Andrews, 1986). The primary result of the later investigations was the discovery of the Weddell Fan.

Single-channel seismic records and piston cores collected during *Islas Orcadas* cruises 1277 and 1578 (Fig. 4.61) provided evidence that turbidity currents are an important sedimentary agent on the eastern abyssal plain of the Weddell Sea. These turbidity currents source the Weddell Fan. The upper fan is characterized by numerous, deep, V-shaped canyons (Fig. 4.62, Profile 1578A), the largest one being Wegener Canyon (Miller et al., 1990b). These canyons are confined to the middle-lower slope and converge at a large submarine channel (Deutschland Channel) at the base of the slope (Fig. 4.62, Profile 1578-B). Sediment cores from the Deutschland Channel penetrated

disorganized gravel and graded gravels with sand (Fig. 4.62, Core 15-38).

Hummocky topography and convolute seismic reflections (slumps) characterize the slope base (Fig. 4.46). Core 15-37 sampled the hummocky unit (Fig. 4.62) and recovered hemipelagic sediment alternating with massive mud units containing rip-up clasts. The latter are interpreted as debris flow deposits. A large channel-levee system occurs on the western side of this channel (Fig. 4.62, Profile 1578-C). Cores from the levees penetrated interbedded laminated silt and massive bioturbated silt with sedimentary clasts and thin graded sand beds (Fig. 4.62, Core 15-39).

Acoustically laminated strata occur on the abyssal floor (Fig. 4.63, Profile 1578-D). Individual reflectors can be traced for relatively long distances across the abyssal floor. These acoustically laminated deposits lap onto, and are ponded against, abyssal hills in the northern part of the basin, leaving sediment-free deeps behind the highs (Fig. 4.63).

Piston cores from the mid-fan recovered relatively thick, graded sand units (Fig. 4.63; Cores 15-42, 43, and 44). Individual sand units range from a few centimeters to several meters thick, are normally graded, and display mineralogic grading from lithic fragments to quartz sand. From piston cores in the region, Anderson and colleagues (1986) estimated that the extent of the sandy mid-fan portion of

Figure 4.59. (*Continued*)

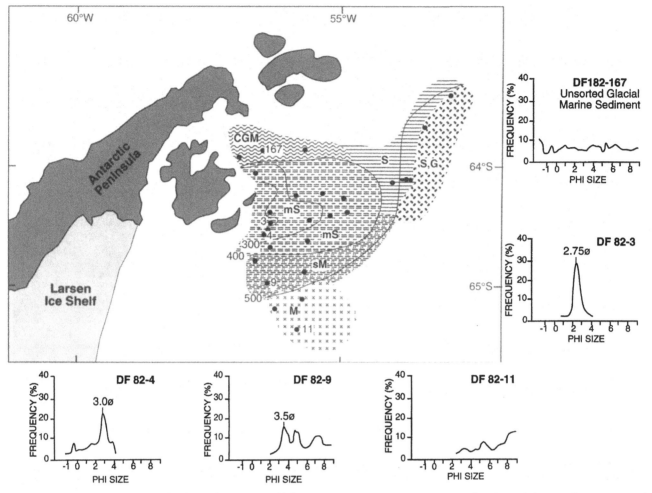

Figure 4.60. Sediment distribution for the northwestern Weddell Sea. Also shown are grain size curves illustrating fining away from the islands. Dots indicate sample locations. Bathymetry is in meters. CGM = compound glacial-marine sediments; M = mud; Sm = sandy mud; S = sand; mS = muddy sand.

the Weddell Fan encompasses an area of more than 290,000 km². A drill site from the outer portion of the fan (ODP Leg 113, Site 694) penetrated nearly 300 m of Mid- to Late Miocene through Pleistocene strata composed largely of turbidites (Kennett and Barker, 1990). Thus, the Weddell Fan is a long-lived feature dating back to at least Mid-Miocene time.

The great extent of sandy turbidites in the northeastern Weddell Sea implies that fine-grained turbidites are even more widespread within the associated suprafan environment. Fisco (1982) demonstrated this through examination of fine-grained sediments of the region. She described moderately sorted, graded, coarse silt (4.0 to 6.5 phi) turbidites and finer-grained (finer than 6.5 phi) laminated turbidites. Laminations are 1 to 5 mm thick. Individual graded units typically range from 5 to 30 cm in thickness and exhibit sharp basal contacts. Bioturbation, microfossils, and ice-rafted debris are generally absent. The graded sand and associated fine-grained turbidites occupy an ex-

tensive (0.75 million km²) area of the northeastern Weddell Sea (Fig. 4.61). The Weddell Fan is now recognized as part of a larger fan complex that includes the Crary Fan to the south; the combined areas of these fans make this fan complex one of the largest in the world (Anderson, Wright, and Andrews, 1986).

The Crary Fan was first recognized in the subsurface of the southeastern Weddell Sea using seismic data acquired during the 1977 and 1979 Norwegian Antarctic Research Expeditions (NARE). Early work on this fan focused on its seismic stratigraphy (Kuvaas and Kristoffersen, 1991; Moons et al., 1992). Additional seismic data were collected as part of a joint German-Belgian survey, and the combined data sets were used to map the lateral extent and internal character of the fan (De Batist et al., 1994) (Fig. 4.61). The Crary Fan extends across the Weddell Abyssal Plain for at least 600 km, obtains a maximum thickness of 1500 m, and displays classic channel-levee complexes where individual levees are hundreds of meters thick (De Batist et al., 1994).

Kuhn and Weber (1993), Melles and Kuhn (1993), and Weber and colleagues (1994) mapped the modern channel-levee systems of the Crary Fan and characterized seismic facies using high-resolution (parasound) subbottom profiles. The largest canyon, Deutschland Canyon, extends to

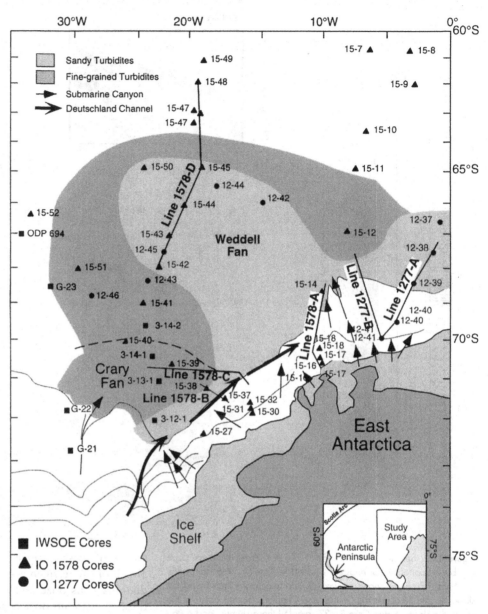

Figure 4.61. Map of the eastern Weddell Sea showing locations of seismic lines and sediment cores collected as part of a 1978 and 1979 study of the Weddell Fan. Also shown is the area of the Weddell Fan (Anderson, Wright, and Andrews, 1986) and Crary Fan (De Batist et al., 1994). Large arrows show Deutschland Canyon, and smaller arrows depict other canyons. Lines indicate location of seismic profiles shown in Figures 4.62 and 4.63.

the north along the base of the slope off Queen Maud Land (Fig. 4.61). A number of submarine canyons converge with the Deutschland Canyon along the East Antarctic continental margin.

The mechanism for supplying sediment to the Weddell and Crary fans remains problematic. How are turbidites delivered to the abyssal plain from a continental shelf that: (1) is deeper than 400 m; (2) has a landward-sloping profile; and (3) is floored by poorly sorted glacial and glacial-marine material? Seismic data and piston cores indicate that the Weddell Fan was active during the recent past, probably during the last glacial maximum. Anderson and colleagues (1986) suggest, on the basis of similarities in mineralogies and rock types from turbidites and those from glacial sediment on the continental shelf, that the turbidites are directly sourced from subglacial meltwater streams when the EAIS grounds at the edge of the continental shelf.

Southeast Pacific

Considerable data (piston cores, bottom photos, high-resolution seismic data, and nephelometer profiles) have been acquired in the southeast Pacific sector of the Southern Ocean and in the southwest Weddell Sea sector. Therefore, this region is used to characterize sedimentation processes on the Subantarctic abyssal floor. The following discussion is based on the surface sediment distribution map shown in Figure 4.64.

A wide belt of terrigenous sediment, extending north from the Antarctic margin, characterizes the southeast

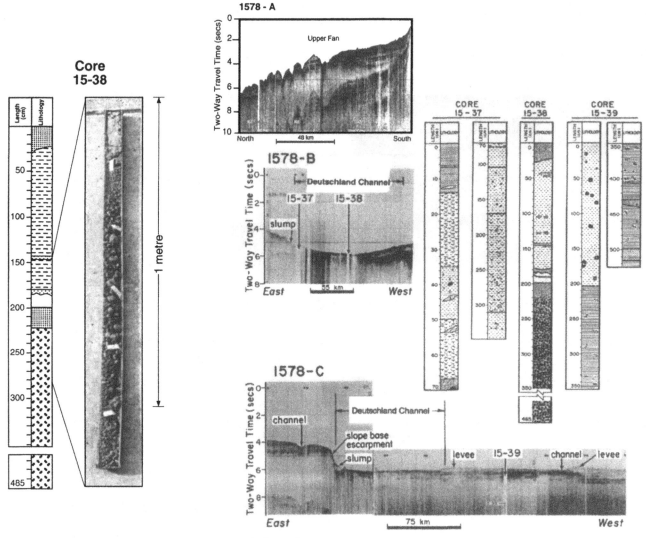

Figure 4.62. Seismic (airgun) Profiles 1578 A–C cross the lower slope and basin floor of the northeastern Weddell Sea (see Fig. 4.61 for profile locations). Profile A shows several submarine canyons that dissect the slope. Profile B crosses the Deutschland Channel. Slumps occur at the toe of the slope. Piston Core 15-37 sampled these slumps, and piston Core 15-38 sampled the disorganized gravels that fill the Deutschland Channel at this location. Seismic profile C crosses the channel–levee complex, and Core 15-39 illustrates the lithologic character of the levee deposits, which include interbedded massive muds with small rip-up clasts and laminated silts.

Pacific sector of the subantarctic (Fig. 4.64). Detailed analyses of DSDP Leg 35 cores (Tucholke et al., 1976; Piper and Brisco, 1975) and piston cores (Baegi, 1985; Wright, Anderson, and Fisco, 1983) have shown that these terrigenous sediments include hemipelagic sediments, fine-grained turbidites and associated overbank deposits, and contourites. Piper and Brisco (1975) provide a number of criteria that can be used to distinguish these different modes of origin.

Mineralogic analysis of terrigenous sediments of the continental slope and rise of the southeast Pacific sector indicate that the Antarctic Peninsula is the principal source of these terrigenous sediments (Edwards, 1968). North of the terrigenous sediment belt is a broad belt of diatomaceous ooze and mixed diatomaceous-calcareous ooze. Calcareous ooze occurs above ~ 4000 m, the depth of the CCD in this region, and pelagic clay occurs below this depth in the northern sector of the Southeast Pacific.

Pelagic clays are enriched in minerals and elements of volcanic origin (Nayudu, 1971; Zemmels, 1978), which are probably being derived from oceanic seamounts and ridges.

Submarine Fans and Sediment Drifts. The continental slopes and rises of the Amundsen and Bellingshausen seas are extensive and dissected by numerous submarine canyons (Dangeard, Vanney, and Johnson, 1977). The canyons extend well out onto the abyssal floor and form major conduits through which terrigenous sediment is supplied to the deep seafloor (Dangeard, Vanney, and Johnson, 1977; Tucholke and Houtz, 1976). The role of turbidity currents in delivering terrigenous sediment to the deep seafloor is expressed by the presence of fans that occur on the Bellingshausen continental rise and abyssal plain (Tucholke and Houtz, 1976; Tucholke et al., 1976;

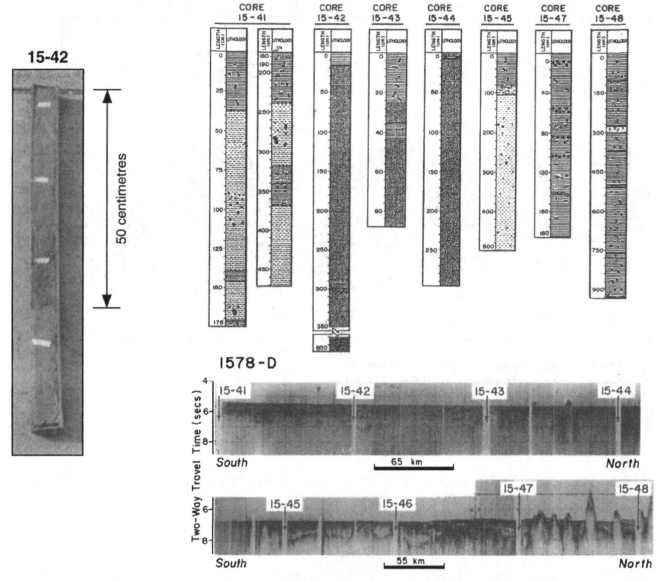

15-42

Figure 4.63. Seismic Profile 1578-D across the northern portion of the Weddell Abyssal Plain showing alternating acoustically transparent and laminated units that, in the northern part of the abyssal plain, are ponded against seamounts. Also shown are lithologic logs for piston cores collected along this transect and a photograph of graded sands in Core 15-42. (See Fig. 4.61 for profile location.)

Dangeard, Vanney, and Johnson, 1977; Tucholke, 1977; Wright, Anderson, and Fisco, 1983; Kagami, Kuramochi, and Shima, 1991). Two ancient submarine fans, the Charcot and Palmer fans, have been mapped using seismic reflection data (Dangeard, Vanney, and Johnson 1977; Tucholke and Houtz, 1976). The fans contain typical levee–channel complexes, as well as large sediment drifts and smaller-scale migrating sediment dunes (Tucholke, 1977; McGinnis and Hayes, 1994; Rebesco et al., 1994; Chapter 5). DSDP Leg 35 drill sites on the Bellingshausen continental rise sampled the fan deposits of this region. They penetrated thick turbidite sequences with well-sorted quartz sand (Tucholke et al., 1976). Anderson, Bartek, and Thomas (1991b) suggest that these well-sorted, quartz-rich turbidites were deposited during the initial stages of glaciation on the Antarctic Peninsula continental shelf.

Sedimentologic and petrographic analyses of piston cores from the region were conducted by Baegi (1985) and Wright and colleagues (1983). These studies led to the identification of four elongate fans at or near the seafloor (Fig. 4.65). Disorganized gravel and graded gravel to sand units occupy the channels. These sediments include glacially striated grains and show little evidence of prolonged weathering. Individual fans have distinct rock and mineral compositions. The data suggest direct input of debris to canyon heads by glaciers during a previous glacial maximum when ice was grounded at or near the shelf break (Wright, Anderson, and Fisco, 1983; Baegi, 1985; Pope and Anderson, 1992). Piston cores collected from interchannel areas contain finely laminated, very fine sand, silt, and clay interpreted as overbank (levee) deposits (Wright, Anderson, and Fisco, 1983).

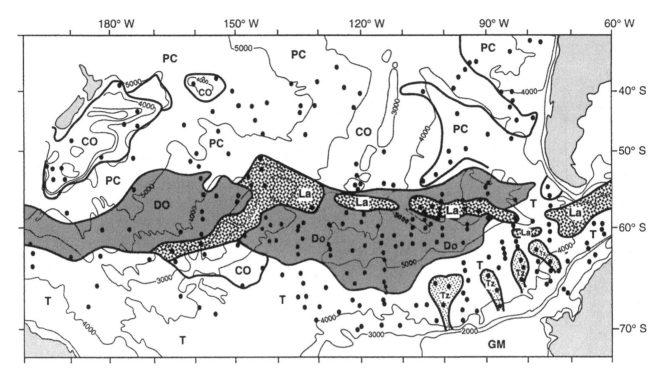

Figure 4.64. Surface sediment distribution map for the southeastern Pacific Ocean. Dots show sample locations. Bathymetry is in meters. T = terrigenous silts and clays; Tz = sandy turbidites; DO = diatomaceous ooze and mud; CO = calcareous ooze and mud; PC = pelagic clay; La = gravelly or sandy lag deposits (modified from Anderson, 1990c).

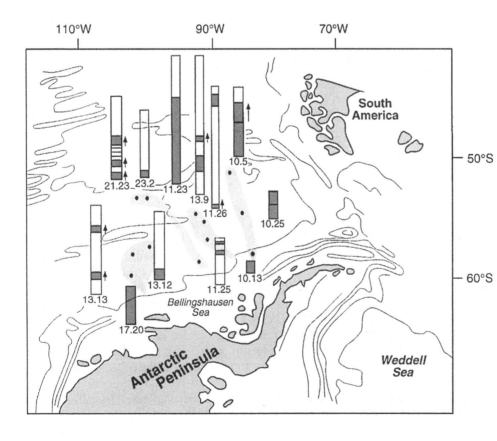

Figure 4.65. Distribution of modern elongate, gravelly, and sandy deep-sea fans in the southeast Pacific Ocean based on the occurrence of graded gravel and sand in piston cores. Shaded portions of the lithologic logs indicate either sand or gravel. Arrows designate normally graded units (from Baegi, 1985).

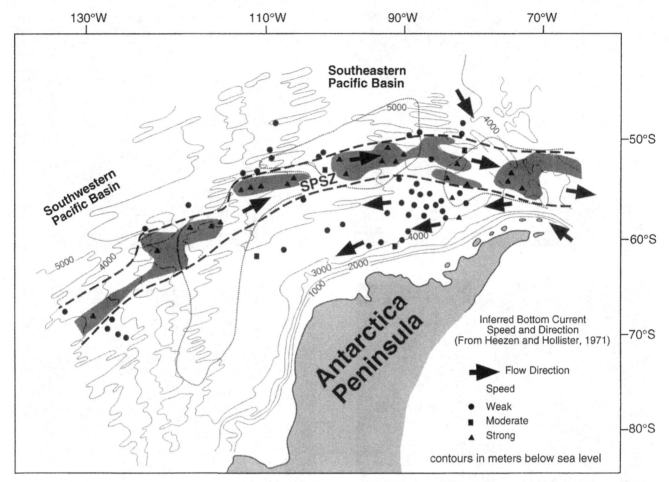

Figure 4.66. Map showing the distribution of scour zones and manganese nodules, as well as inferred bottom current information (from bottom photographs) for the southeastern Pacific (from Anderson, 1990). Areas where lag deposits and rippled sands occur are shaded. The dotted lines show the limits of manganese nodules (Piper et al., 1985). The dashed lines show the limits of the South Pacific Scour Zone (SPSZ; from Heezen and Hollister, 1971). Also shown are inferred bottom current directions and speed (from Heezen and Hollister, 1971). Bathymetry is in meters.

Bottom Current Influence on Abyssal Sedimentation. Bottom currents have had a profound influence on sedimentation in the southeast Pacific region. Bottom photographs from the region show three east-west-oriented belts of bottom current activity. Photographs taken on the abyssal floor nearest the Antarctic continental margin display evidence for weak bottom currents (Heezan and Hollister, 1971) (Fig. 4.66). This area corresponds approximately to the belt of fine-grained terrigenous sediment north of the margin (Fig. 4.64). North of this low-energy depositional belt, photographs display evidence of strong to moderate bottom currents in the form of deflected sessile organisms, scour around pebbles and nodules, current lineations, and ripples (Fig. 4.67A). A few photographs taken in the vicinity of oceanic fracture zones and the mid-ocean ridge show exposed basement rocks (Fig. 4.67B), which correspond to an extensive field of manganese nodules (Fig. 4.67C). Magnetostratigraphic analysis of piston cores from the South Pacific suggests that Brunhes age sediment is thinner, and locally absent, in this sector of the abyssal floor (Goodell et al., 1968). This is referred to as the Southeast Pacific Scour Zone (SPSZ).

The SPSZ is associated with eastward transport of bottom water through gaps in the Pacific-Antarctic Ridge, along the deepest part of the abyssal plain, and through the Drake Passage (Heezen and Hollister, 1971) (Fig. 4.66). North of this scour zone, a belt of weak bottom current energy exists, and fine-grained siliceous ooze and pelagic mud blankets the seafloor (Fig. 4.67D). A prominent bottom nepheloid layer occurs throughout the southeastern Pacific basin (Anderson, 1990).

Ferromanganese Deposits. The distribution and densities of manganese nodules on the seafloor of the southwestern Pacific have been mapped by a number of investigators (Goodell, 1973; Goodell, Meylan, and Grant, 1971; Glasby, 1976; Piper et al., 1985). Goodell and colleagues (1971) mapped a continuous belt of ferromanganese concretions south of the Antarctic Convergence. They noted that this belt is associated with a variety of sediment types

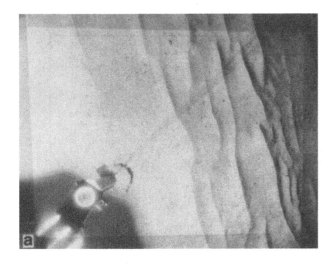

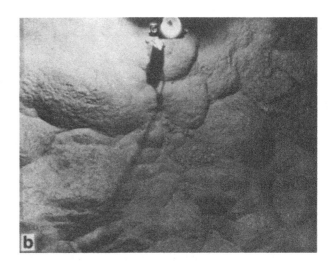

Figure 4.67. (A) Ripples, (B) exposed basement rocks, (C) manganese nodules, and (D) muddy bottom with tracks and trails of benthic organisms from the southeast Pacific basin.

and corresponded to an area of intense bottom scour, based on bottom photographs (Fig. 4.67C). Large areas of concretions lie predominantly below 4000 m, below the CCD (Glasby, 1976), although smaller concentrations are found at a wide depth range (Goodell, Meylan, and Grant, 1971).

Goodell (1973) and Goodell and colleagues (1976) emphasized the importance of low sediment accumulation rates in nodule formation. Once nodules cover the seafloor, the nodules provide surface roughness, which causes fluid turbulence as currents flow over the bed and retards the settling of fine-grained sediment. Thus, the presence of manganese nodules provides an environment conducive to their continued growth.

Several studies have concentrated on geochemistry of southwestern Pacific sediments and ferromanganese deposits (Goodell, Meylan, and Grant, 1971; Margolis and Burns, 1976; Glasby, 1976). These studies have led to a better understanding of factors that regulate the formation and distribution of metalliferous deposits on the southeastern Pacific seafloor. Glasby (1976) has reviewed the geochemistry of southeastern Pacific ferromanganese deposits.

CHAPTER FIVE

Continental Margin Evolution

Knowledge of the geology of the Antarctic continental margin draws on extrapolations of marine and aerial geophysical surveys, the few drill sites in the region, the geology of neighboring Gondwana continents, and information gained from land-based geologic studies. The amount of information dramatically varies for different regions because of the differences in the number, quality, and types of surveys conducted. These combined conditions make the Antarctic continental shelf a formidable challenge to geologists and geophysicists attempting to unravel its geologic history.

This chapter presents results of marine geologic and geophysical surveys conducted to date on the Antarctic continental margin. The objective is to describe the tectonic and stratigraphic evolution of those parts of the margin that have been studied (Fig. 5.1). The continental margins of East Antarctica are all passive margins resulting from the separation of Africa, India, and Australia from Antarctica. The West Antarctic continental margin has had a more complex evolution and remains the more poorly understood of the two margins. Discussion will first focus on the West Antarctic continental margin, starting with the Ross Sea and followed by the Pacific-Antarctic Margin and western Weddell Sea. Discussions of the East Antarctic continental margins include the Queen Maud Land, Wilkes Land, and Prydz Bay margins. Each region is discussed separately, and each section includes a description of the tectonic history and stratigraphy.

ROSS SEA

The Ross Sea forms a large embayment on the Antarctic coast at the boundary between East and West Antarctica. It is bounded on the west by the TAM of Victoria Land, on the south by the Ross Ice Shelf, on the east by Marie Byrd Land, and on the north by the Pacific Ocean (Fig. 5.1). The Ross Embayment and associated basins formed in response to extensional stresses generated mostly during the breakup of Gondwana. More multichannel seismic (MCS) and high-resolution seismic reflection data have been collected in the Ross Sea than in any other portion of the Antarctic continental shelf. It is also the only portion of

the West Antarctic continental shelf that has been successfully drilled.

Geologic History and Tectonic Setting

High heat flow as well as the presence of grabens and thinned continental crust (4 to 27 km, as compared to 40 km in the TAM and intermediate thickness in Marie Byrd Land) have led several researchers to characterize the Ross Embayment as a broad rift (1000 km wide) (Blackman, Von Herzen, and Lawver, 1987; Davey, 1987; Davey and Cooper, 1987; Cooper, Davey, and Hinz, 1988). Behrendt and Cooper (1991) estimate that approximately 350 km of extension has occurred in the Ross Sea. The formation of this rift has been related to the breakup of East Gondwana, although later extension did occur (Davey, 1981; Grindley and Davey, 1982; Cooper, 1989; Cooper et al., 1987a,b; Cooper, Davey, and Hinz, 1991; Lawver and Scotese, 1987; Lawver et al., 1991; Davey and Brancolini, 1995; Chapter 2).

Prior to the breakup of Gondwana, the present Ross Sea and Marie Byrd Land region were connected to the New Zealand microcontinent, which included the present Campbell Plateau, Chatham Rise, and Lord Howe Rise (Barron and Harrison, 1979; Grindley and Davey, 1982; Davey and Brancolini, 1995) (Fig. 5.2). During the latest Paleozoic and throughout the Triassic, the area of the present Ross Sea formed a broad continental platform over which an extensive blanket of alluvial deposits, the Beacon Supergroup, was shed (Barrett, 1981). Initial rifting began in the Middle Jurassic (175 Ma), accompanied by extrusion and intrusion of tholeiitic mafic rocks (Ferrar Dolerite and Kirkpatrick Basalt) over widespread areas (Davey 1987). The first major rifting event occurred during the Early Cretaceous (Davey and Brancolini, 1995). Approximately 40 to 50% crustal extension during the early rifting period resulted in the development of rift grabens trending parallel (north-northeast to south-southwest) to the axis of spreading (Lawver and Scotese, 1987). The New Zealand microcontinent remained fixed to Gondwana until approximately 80 Ma (Christoffel and Falconer, 1972; Molnar et al., 1975; Davey, 1981; Lawver et al., 1991). During the breakup that spanned most of the Cretaceous, deposition of nonmarine and shallow

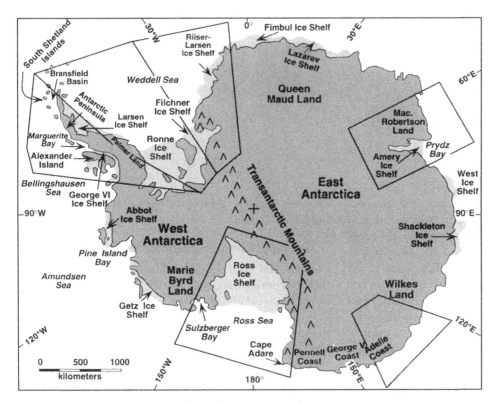

Figure 5.1. Geographic map and locations of study areas discussed in Chapter 5.

Figure 5.2. Model of plate and landmass positions in the Ross Sea region prior to the breakup of Gondwana (modified from Lawver and Scotese, 1987).

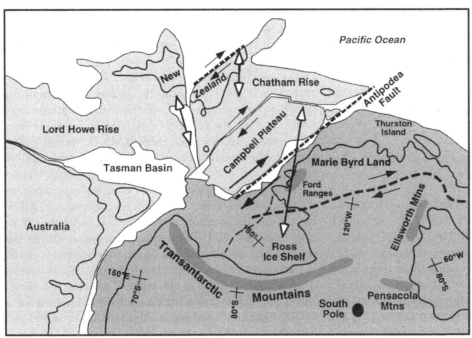

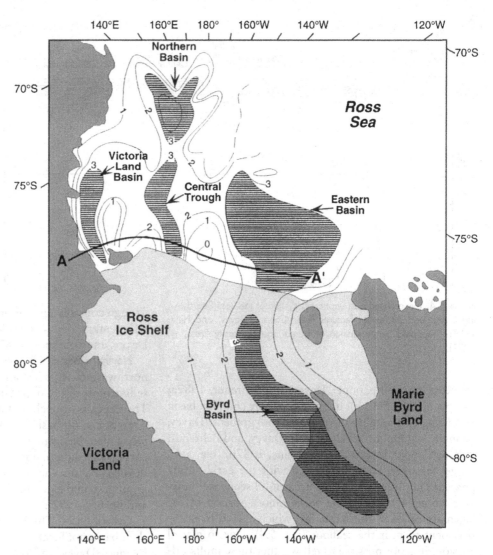

Figure 5.3. Locations of sedimentary basins (horizontal bars) in the Ross Sea. Total sediment thickness isopach contour interval is in kilometers. Location of cross section A-A' in Figure 5.4 is also shown (modified from Davey, 1987).

marine clastic sediments occurred in rift basins, probably in association with layered volcanic rocks. This episode of extension and basin filling ended with rifting between Marie Byrd Land and the Campbell Plateau (Fig. 5.2). Final separation of Tasmania and the South Tasman Rise from northern Victoria Land occurred by 40 Ma (Lawver et al., 1991).

Cooper, Davey, and Hinz (1988) suggest that a second, Cenozoic (post-Eocene(?)) stage of rifting strongly influenced the evolution of the Ross Sea continental shelf and its sedimentary basins. This later rifting event was associated with a major episode of plate reorganization during the Eocene (Molnar et al., 1975; Stock and Molnar, 1987; Cande and Mutter, 1982), as well as with uplift of the TAM (Fitzgerald, 1992, 1994), volcanism in Victoria Land and Marie Byrd Land, right–lateral strike–slip motion (Salvini et al., 1997), and gentle regional subsidence of the Ross Sea depocenters (Cooper, Davey, and Hinz, 1991b). Late Cenozoic extension has been confined to the Terror Rift in southern Victoria Land (Cooper et al., 1991a). Alkaline volcanism has occurred throughout the western Ross Sea–Victoria Land region during most of the Cenozoic.

Five sedimentary basins in the Ross Sea region have been identified through seismic refraction, magnetic, and gravity surveys (Davey, 1981; Hinz and Block, 1984; Bentley, 1991; Cooper et al., 1991b; Cooper, Barker, and Brancolini, 1995) (Figs. 5.3 and 5.4). These basins have an elongate morphology, with their long axes oriented roughly north-northeast to south-southwest. All five basins are associated with rift grabens (Cooper, Davey, and Hinz, 1988). Byrd Basin lies entirely beneath the Ross Ice Shelf, so its geology remains obscure. The other basins lie on the open shelf; however, they may extend beneath the Ross Ice Shelf (Bentley, 1991). The parallelism between these basins and the structural fabric of the TAM suggests that they initially formed in the Jurassic (Wilson, 1991).

Victoria Land Basin. Victoria Land Basin, located in the western Ross Sea adjacent to the TAM, extends from the mid-shelf to perhaps several hundred kilometers beneath

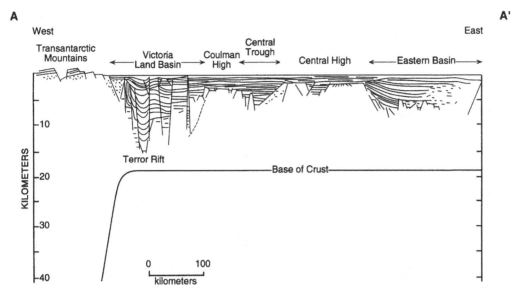

Figure 5.4. Generalized east-west cross section of the Ross Sea showing basins and rift-related features (from Cooper, Davey, and Hinz, 1991b; reprinted by permission of Cambridge University Press, © 1991).

the Ross Ice Shelf (Cooper, Davey, and Behrendt, 1987a; Cooper et al., 1991a,b). It is a structurally complex basin consisting of a number of subbasins and grabens (Davey, Bennett, and Houtz, 1982; Cooper, Davey, and Behrendt, 1987a; Cooper, Davey, and Cochrane, 1987b) (Fig. 5.4). Large basement faults associated with the early rifting phase do not offset the younger basin-fill sequence; however, younger strata of the westernmost portion of the basin are offset by normal faults, which extend from the basement to near the seafloor (Fig. 5.4). Due to glacial erosion, it is not possible to tell whether these faults offset Plio-Pleistocene deposits (Anderson and Bartek, 1992). The faults formed in association with the uplift of the TAM during the Late Cenozoic (Cooper, Davey, and Behrendt, 1987a; Cooper, Davey, and Cochrane, 1987b; Cooper et al., 1991b). Sedimentary deposits within Victoria Land Basin reach a thickness of 14 km (Cooper, Davey, and Hinz, 1988).

Central Trough. The Central Trough is a nearly symmetric, 80-km-wide graben. It coincides approximately with a prominent north-south-oriented positive gravity anomaly, which Hayes and Davey (1975) argued marks an area of crustal thinning and/or high-density intrusions into the crust underneath the graben. Recently acquired seismic data (Cooper, Barker, and Brancolini, 1995) suggest that the crust is 17 to 18 km thick within the center of the trough and 21 to 24 km thick beneath the flanking ridges. The eastern margin of the Central Trough possibly coincides with a major transform separating the north-south-oriented rift from an east-west-oriented spreading center in the eastern Ross Sea (Davey, 1981). The proposed transform parallels the Campbell Fracture Zone to the north, along which 300 to 330 km of dextral offset has occurred (Davey and Christoffel, 1978). The Central

Trough contains up to 6 or 7 km of virtually undeformed sedimentary deposits (Hinz and Block, 1984; Cooper, Davey, and Behrendt, 1987a).

Northern Basin. The Northern Basin is located in the northwestern Ross Sea and contains in excess of 3 km of sedimentary fill (Cooper, Barker, and Brancolini, 1995). The deposits are believed to include Mesozoic, primarily Cretaceous, rift basin fill, possibly Paleogene deposits. An incomplete Neogene (Pre-Middle Miocene) sequence occurs in the southern portion of the basin, but Late Miocene through Pleistocene strata compose a prominent trough mouth fan in the northern portion of the basin (Anderson and Bartek, 1992).

Eastern Basin. The Eastern Basin constitutes the only shelf basin not located entirely within an early rift graben (Cooper, Davey, and Hinz, 1988). The total sedimentary section is up to 7 km thick (Davey, Bennett, and Houtz, 1982; Cooper, Davey, and Hinz, 1988). Subsidence beyond the graben margins has been attributed to crustal downwarping in response to the load of Late Paleogene and Neogene glacial sediments deposited in the region (Davey, Bennett, and Houtz, 1982).

Byrd Basin. The presence of the Byrd Basin beneath the Ross Ice Shelf (Fig. 5.3) is inferred from seismic refraction data and aeromagnetic data, which suggest the presence of ~2 km of sedimentary deposits (Bentley, 1991). Truswell (1983) and Truswell and Drewry (1984) suggest that abundant recycled palynomorphs of Late Cretaceous through Tertiary age found in Quaternary glacial and glacial-marine sediments on the Ross Sea continental shelf originated in the Byrd Basin.

Lithostratigraphy

In 1973, participants on DSDP Leg 28 drilled four sites on the Ross Sea continental shelf to gain insight into the glacial history of Antarctica (Hayes and Frakes, 1975) (Fig. 5.5). DSDP Sites 270–272 collectively penetrated a total of 1121 m into an offlapping sequence of strata

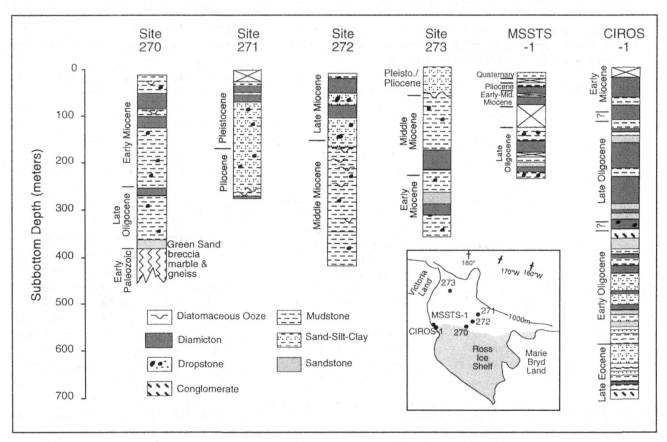

Figure 5.5. Generalized lithologic descriptions of cores from DSDP Sites 270-273, MSSTS-1, and CIROS-1 (after Hayes and Frakes, 1975; Balshaw, 1981; Barrett and McKelvey, 1986; Robinson et al., 1987).

located in the Eastern Basin of the Ross Sea. Site 273 is located in the western Ross Sea and sampled primarily Early and Middle Miocene strata. Core recovery at all four sites was poor. Detailed descriptions and analyses of the DSDP Leg 28 cores are presented by Hayes and Frakes (1975).

Cores in the Victoria Land Basin were acquired during the McMurdo Sound Sediment and Tectonic Studies-1 (MSSTS-1) project (Barrett, 1986a) and the Cenozoic Investigations of the Ross Sea-1 (CIROS-1) project (Barrett et al., 1989) (Fig. 5.5). The objectives of these drilling projects included dating the onset of glaciation in the western Ross Sea and gaining a better understanding of the uplift of the TAM. Another objective of these projects was to fill the void in the stratigraphic record between Early Jurassic Ferrar magmatism to early Miocene McMurdo Volcanics (Barrett, 1986a,b).

MSSTS-1 drilled to 227 m below the seafloor with 46% core recovery (Barrett, 1986a); CIROS-1 drilled to 702 m below the seafloor with 98% recovery (Robinson et al., 1987). Ages were assigned to the interval above 200 m in CIROS-1 by correlation with the MSSTS-1 lithostratigraphy, whereas ages for the interval below 200 m were determined by diatom (Robinson et al., 1987) and dinoflagellate biostratigraphy (Hannah, 1994). Barrett (1986b, 1989) provides detailed descriptions of the MSSTS-1 and CIROS-1 cores.

The succession of sedimentary strata sampled at DSDP drill sites on the eastern Ross Sea continental shelf (Sites

270–272) rests on crystalline basement rocks, which at Site 270 consist of foliated marble and calc-silicate gneiss of early Paleozoic age (Hayes and Frakes, 1975) (Fig. 5.5). Nearly 30 m of Oligocene sedimentary breccia overlies these basement rocks, and this breccia is overlain by 1.8 m of Oligocene glauconitic sandstone that bears no evidence of glaciation at that time. Late Oligocene to Early Miocene silty claystone with scattered ice-rafted granules and pebbles rest above the glauconitic sandstone unit and provide the first signs of glaciation in the region. A thick sequence of interbedded diamictite, mudstone, and diatomaceous glacial-marine mudstones compose the Early Miocene and younger section (Fig. 5.5). Detailed analysis of the Miocene deposits in all three sites by Balshaw (1981) led to the interpretation that the diamictites are either tills or ice-proximal glacial-marine deposits. This interpretation was later supported by Hambrey (1993).

DSDP Site 273 in the western Ross Sea penetrated early and Middle Miocene glacial-marine deposits that are unconformably overlain by Plio-Pleistocene glacial-marine deposits (Fig. 5.5).

The stratigraphic succession recovered from MSSTS-1 consists primarily of interbedded diamictites and dropstone-bearing mudstones (Fig. 5.5). Barrett (1986b) believes that the MSSTS-1 core documents numerous glacial

Table 5.1. Seismic stratigraphy of Victoria Land Basin based on an analysis of MCS data and intermediate-resolution data (IRS). See Chapter 2 for descriptions of stratigraphic units in the region

MCS	IRS	Age	Interpretation
V 1	A-J	Late Miocene to Recent	Glacial-marine sediment and till (DSDP Site 273, Fig. 5.5)
V 2	K-O	Early-Mid Miocene	Glacial-marine sediment and till (DSDP Site 273 and CIROS-1, Fig. 5.5)
V 3	P-R	Late Oligocene Early Miocene	Glacial-fluvial sandstone and mudstone (CIROS-1 and MSSTS-1, Fig. 5.5)
V 4	S-T	Mid Eocene Early Oligocene	Shallow marine and deep-water glacial-marine sediments
V 5		Late Cretaceous to early Paleogene	Marine? sedimentary
V 6		Jurassic to late Cretaceous	Volcanic rocks and rift-basin fill (non-marine and shallow marine)
V 7		early Paleozoic to Jurassic	Variable-Includes most of the rocks found in the TAM (Chapter 2)

Seismic stratigraphy of MCS data is from Cooper et al. (1987a) and the seismic stratigraphy of the intermediate-resolution data is from Bartek et al. (1996).

events during the past 30 Ma when grounded ice advanced into McMurdo Sound, leaving lodgement till as a vestige of its presence in the sound. In fact, the entire succession is believed to have been deposited by glacial and glacial-marine sedimentary processes.

CIROS-1 consists of interbedded diamictite, mudstone, sandstone, and conglomerate. The strata range in age from Early Miocene to Middle (?) Late Eocene (Hannah, 1994) (Fig. 5.5). Robinson and colleagues (1987) indicate that the cored sequence is readily divided into: (1) an upper interval (0 to 366 m) composed primarily of mudstone with dropstones and diamictite and (2) a lower sequence (366 to 702 m) consisting of interbedded sandstone and mudstone with minor amounts of diamictite and conglomerate. Barrett and colleagues (1987) found that the Early Oligocene section of this core contains faceted, striated, subangular to subrounded dropstones. It was thought that these dropstones indicated glacial activity and ice calving at sea level in this region as early as the Middle Eocene (Hannah, 1994), although the age of the oldest glacial-marine sediments may be Late Eocene (Barrett, personal communication).

Seismic Stratigraphy

Detailed accounts of the seismic stratigraphy of the sedimentary basins of the Ross Embayment have been presented by Hinz and Block (1984), Sato and colleagues

(1984), Cooper and colleagues (1987a, 1990a), Bartek and colleagues (1991), Anderson and Bartek (1992), Alonso and colleagues (1992), De Santis and colleagues (1995), Brancolini, Cooper, and Coren (1995), and Bartek and colleagues (1996). The following subsections summarize these works. Figure 5.6 provides the locations of seismic lines referred to in the text.

The Antarctic Offshore Acoustic Stratigraphy (ANTOSTRAT) program succeeded in bringing together several different seismic data sets that have led to both the development of a regional seismic stratigraphic scenario for the Ross Sea and the compilation of structure maps depicting the thickness and distribution of major seismic units (Alonso et al., 1992; Brancolini, Cooper, and Coren, 1995; De Santis et al., 1995). Regional correlation of seismic units is difficult, especially for younger (Plio-Pleistocene) sequences, because of the discontinuous distribution of sequences on the shelf. For this reason, the seismic sequences of the Victoria Land Basin in the western Ross Sea are treated separately (Table 5.1) from those of the eastern Ross Sea (Table 5.2). The most complete seismic stratigraphic record of the Ross Sea continental shelf evolution occurs in the Eastern Basin, so this discussion will focus on this area.

Victoria Land Basin. MCS records reveal seven major seismic units in the western Ross Sea (Cooper, Davey, and Behrendt, 1987a; Cooper et al., 1991a) (Fig. 5.7A, units V1-V7 in Line 407). Higher-resolution seismic data acquired from the McMurdo Sound region show as many as 20 unconformity-bounded seismic sequences. Information from the MSSTS-1 and CIROS-1 sites was used to help constrain the ages of these units (Henrys et al., 1994; Barrett et al., 1995; Bartek et al., 1996) (Table 5.1; Fig. 5.7B). Unit V7, the acoustic basement in this region, was faulted during the early rift phase of the breakup of Gondwana. Unit V6 includes the rift basin fill deposits and associated volcanic rocks and is believed to be Jurassic through Late Cretaceous in age. The paleogeographic reconstruction for this time interval indicates that sedimentary strata of Unit V6 consist of non-marine and shallow marine deposits (Fig. 2.21). Onlap of units V5-V1 onto the acoustic basement, uniform spacing and lateral continuity of the reflectors, and a lack of deformation of these units by basement faults (Fig. 5.7C) suggest that deposition of units V5-V1 occurred under marine conditions along the axis of a previously formed rift basin (Cooper, Davey, and Behrendt, 1987a).

Table 5.2. Seismic sequences of the eastern Ross Sea. See descriptions of DSDP Sites 270-272 in Figure 5.5 for more detailed lithologic information on the region

MCS	IRS	Unconformity	Age	Interpretation
RSS-8			Plio-Pleistocene	Subglacial and glacial-marine
	Unit 1			
		RSU1		
RSS-7			Plio-Pleistocene	Subglacial and glacial-marine
	Units 7-2			
		RSU2		
RSS-6			Late Miocene-Early Pliocene	Subglacial and glacial-marine
	Unit 8			
		RSU3		
RSS-5	Unit 9		Mid-Late Miocene	Glacial-marine
		RSU4		
RSS-4			Early Miocene	Subglacial(?) and glacial-marine
	Unit 10			
		RSU4A		
RSS-3			Early Miocene	Marine and glacial-marine
	Unit 11			
		RSU5		
RSS-2			Late Oligocene-Early Miocene	Subglacial(?) and glacial-marine
	(Unit 12)			
		RSU6		
RSS-1			Cretaceous(?)-Oligocene(?)	Marine
	Unit 13			
Basement				

Seismic sequences modified from Anderson and Bartek (1992), De Santis et al. (1995), and Brancolini et al. (1995).

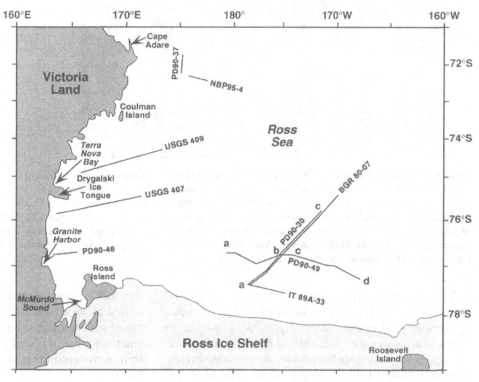

Figure 5.6. Location of Ross Sea seismic profiles referred to in text.

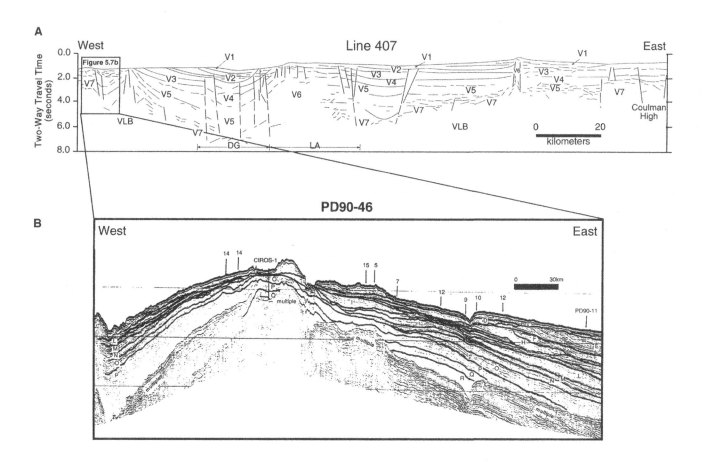

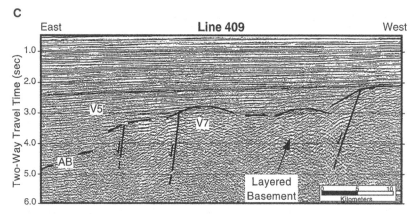

Figure 5.7. Representative seismic profiles from the Victoria Land Basin: (A) Interpreted USGS Line 407 showing seismic sequences, illustrating faulting of Late Cenozoic strata during the late-stage rifting in the Victoria Land Basin (from Cooper, Davey, and Behrendt, 1987a; line reproduced with the permission of the Circum-Pacific Council for Energy and Mineral Resources). DG = Discovery Graben; LA = Lee Arch; VLB = Victoria Land Basin; numbered V labels refer to the stratigraphic units of Cooper and colleagues (1987a; Table 5.1). (B) Interpreted seismic Profile PD90-46 showing sequences of the Victoria Land Basin in the McMurdo Sound region (reprinted from Bartek et al., 1996, with permission from Elsevier Science Ltd.; Table 5.1). (C) Portion of multichannel seismic Line 409 showing faults in basement. These faults do not offset overlaying strata (Cooper, Davey, and Behrendt, 1987a). (See Fig. 5.6 for profile locations.)

The higher-resolution data set shows a progressive change in the character of seismic units and bounding surfaces upward in the section (Fig. 5.7B). Lower in the section, sequences are characterized by fairly continuous reflections, and sequence boundaries are relatively flat, except on the inner shelf, where possible troughs occur (i.e., sequence boundaries P/O and O/N). Higher in the section, sequence boundaries include landward-sloping surfaces

that shift basinward, and individual sequences are characterized by more discontinuous reflections, often showing broad-scale cross-bedding. This change in seismic facies and the nature of bounding surfaces indicate a basinward shift in the grounding line of the ice sheet (Bartek et al., 1996).

Central Trough. Seismic records from the Central Trough indicate that the stratigraphy and structure of the trough are similar to those of the Victoria Land and Eastern basins. The rift basin-fill sequence is nearly horizontal (Fig. 5.4) and probably consists of Cretaceous to Early Paleogene continental and shallow marine deposits. A prominent unconformity (RSU6), interpreted to be Middle Oligocene in age, separates this sequence from the overlying glacial-marine sequence (post–Late Oligocene; Cooper, Davey, and Behrendt, 1987a). Deformation of the strata within the basin is minimal, suggesting that only one phase of rifting occurred in the Central Trough (Cooper, Davey, and Behrendt, 1987a).

Northern Basin. The Northern Basin is characterized by a thick succession of Middle Miocene and older strata resting beneath a prominent Mid-Miocene unconformity, which is the amalgamated RSU3 and RSU2 unconformities. These older strata have experienced significant deformation in the northern portion of the basin.

Bart (1998) conducted a detailed seismic stratigraphic analysis of the Northern Basin using a grid of intermediate-resolution seismic data collected during 1994 and 1995 cruises to the region. His work focused on the Plio-Pleistocene section, which includes a prominent trough mouth fan, the Victoria Trough Mouth Fan. He recognized and mapped 10 seismic units separated by glacial unconformities, reflecting at least 10 ice sheet grounding events in the region since the Miocene (Fig. 5.8A). Time constraints were provided by DSDP Site 273. Isopach maps of units showed two discrete depocenters that reflect shifting ice streams during the Plio-Pleistocene. Bart (1998) was able to correlate glacial unconformities on the shelf with conformable surfaces within the trough mouth fan. The fan progrades across and downlaps the amalgamated RSU3-RSU2 unconformity, which is characterized by considerable relief, including submarine canyon heads (Fig. 5.8B, unconformity 1).

Eastern Basin. Hinz and Block (1984) conducted the first seismic stratigraphic analysis of the Eastern Basin. They recognized six seismic sequences, which have been subdivided further by Cooper and colleagues (1991a), Anderson and Bartek (1992), and Brancolini and colleagues (1995) (Table 5.2). Seismic profiles BGR 80-7 (Fig. 5.9A), PD90-30 (Fig. 5.10), and PD90-49 (Fig. 5.11) are used to illustrate the main seismic units of the Eastern Basin.

A regional unconformity (RSU6) separates deposits that fill more localized rift basins (RSS-1) from seismic units that are basinal in extent (Fig. 5.9A). Unconformity RSU6 appears to lie beneath the middle Oligocene glauconitic sand at DSDP Site 270 (Hinz and Block, 1984) (Fig. 5.5). Sequence RSS-2 (Unit 12 of Anderson and Bartek, 1992) (Table 5.2) is characterized by subhorizontal and concordant reflection patterns (Unit 12, Figs. 5.10 and 5.11), and higher-resolution records show strata that prograde from basement highs toward the flanking basins (De Santis et al., 1995) (Fig. 5.12). The contact between RSS-2 (Unit 12) and RSS-3 (Unit 11) sequences is locally conformable. Sequence RSS-3 is characterized by strong, subhorizontal reflections and by an aggradational stacking

pattern. A shelf-wide unconformity, RSU4A, separates RSS-3 and RSS-4 and marks a change from dominantly aggradation to dominantly progradation on the outer shelf (Cooper et al., 1991a) (Fig. 5.9A). Sequence RSS-4 is characterized by glacial erosion surfaces with relief similar to modern Ross Sea troughs, and is also characterized by massive sedimentary wedges that Anderson and Bartek (1992) interpreted as subglacial deposits. Later work by De Santis and colleagues (1995) came to the interpretation that these wedges grade basinward into proglacial fans deposited in an ice-proximal setting.

Sequence RSS-5 consists of strong, subhorizontal reflectors and is primarily a prograding sequence. It is separated from younger sequences by a shelf-wide unconformity (RSU-3) that marks a major basinward shift in deposition.

On the shelf, the sequences above RSU-3 are thinner than the older sequences and are best imaged using higher-resolution seismic data. Anderson and Bartek (1992) recognize seven seismic units above RSU-2 that stack in an aggradational manner (Fig. 5.10). Detailed analysis of these units by Alonso and colleagues (1992) showed that they are bounded by glacial unconformities and individual units show considerable seismic facies variability in both the strike and dip directions (Figs. 5.10 and 5.11). Higher-resolution data acquired during recent years are revealing more seismic units and bounding unconformities in the shallower part of the section than were previously recognized (Shipp et al., 1994).

Sequence Stratigraphic Interpretation

Cooper and colleagues (1991a) recognized a major shift in the style of Neogene stratal stacking patterns, changing upward in the section from aggradational (their Type IIA sequence) to progradational (their Type IA sequence). The RSU4A unconformity divides these sequences (Fig. 5.9A). They argued that this change records the onset of significant glaciation on the continental shelf when ice sheets began to deliver large quantities of sediment to the outer shelf. The chronostratigraphic control provided by DSDP Sites 270 and 272 indicates that this change occurred in the Early Miocene.

Bartek and colleagues (1991) argued that the overall stratal patterns seen in the Ross Sea are similar to those observed elsewhere in the world, which they call the "stratigraphic signature of the Neogene" (Fig. 5.9B). They contended that these stratal patterns are the product of global eustasy, which may or may not correspond to glacial events in the Ross Sea.

Anderson and Bartek (1992) studied intermediate-resolution seismic reflection data from the Ross Sea that provide greater stratigraphic resolution of the Neogene section than the MCS data set, enabling them to examine seismic facies and the nature of erosional surfaces. This higher-resolution data set was also used to examine seismic facies and map the thinner Plio-Pleistocene strata on the shelf (Alonso et al., 1992). Anderson and Bartek (1992) recognized 13 regional seismic units within

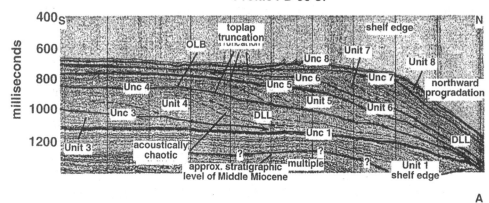

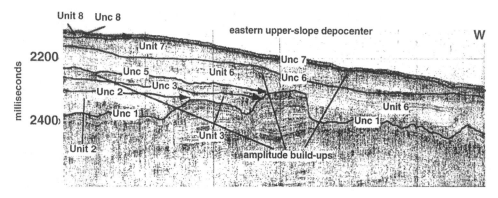

Figure 5.8. Interpreted dip-oriented seismic Profile PD90-37 showing Plio-Pleistocene seismic units of the Victoria Trough Mouth Fan and strike-oriented seismic Profile NBP95-4 showing downlap of fan units onto unconformity 1. (See Fig. 5.6 for profile locations.) (From Bart, 1998.)

the Eastern Basin (Figs. 5.10 and 5.11) whose ages were constrained using DSDP Leg 28 drill sites to develop a history of glaciation on the continental shelf (Fig. 5.13). They stressed that recognition of individual seismic units is possible only with combined dip-oriented and strike-oriented seismic profiles. Dip profiles show the onlapping and offlapping nature of seismic units (Fig. 5.10). Strike-oriented seismic profiles show the broad-scale erosional relief that is produced when ice sheets ground on the continental shelf (Figs. 5.11 and 5.12). Anderson and Bartek (1992) also stress that seismic facies analysis using higher-resolution data is possible in the Ross Sea, because glacial erosion has resulted in older units occurring near the seafloor where they can be imaged using high-frequency sound sources.

De Santis and colleagues (1995) integrated several different seismic data sets with varying degrees of resolution and used these data to conduct a seismic facies and stratigraphic analysis of the eastern Ross Sea, which included isopach maps of the major seismic units and structure contour maps of the bounding surfaces. Their results, along

with those of Anderson and Bartek (1992), provide the main sources for the following discussion of Neogene seismic stratigraphy.

The oldest possible glacial features on the eastern Ross Sea shelf are large V-shaped depressions that occur in the basement surface (Central High) in the central Ross Sea (Fig. 5.14). Busetti and Cooper (1994) interpreted these as half grabens; however, a detailed examination of these features by De Santis and colleagues (1995) led to the observation that they radiate in virtually all directions from the basement high. Internal seismic facies are chaotic and lobate sediment wedges, possibly trough mouth fans that occur at the mouths of these valleys. These observations led De Santis and colleagues (1995) to conclude that these are glacial valleys formed by ice caps that were grounded on the Central High. Note that these valleys and their fill occur below the U6 unconformity (Fig. 5.14) and below the Late Oligocene interval sampled at DSDP Site 270; therefore, the valleys are Late Oligocene or older in age. Although the unit containing the valley fill deposits can be traced for only a short distance before it dips below the water-bottom multiple, it shows no conspicuous evidence of glacial erosion or sedimentation, as the unit is characterized by strong and continuous reflectors. This implies that glaciation was confined to the basement high.

A   BGR 80-07

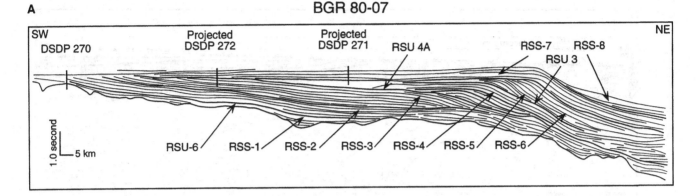

B

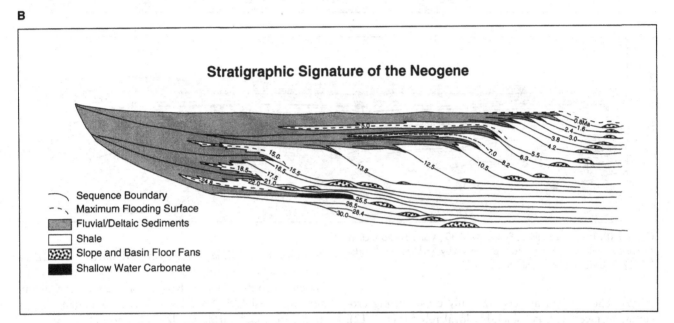

Stratigraphic Signature of the Neogene

Sequence Boundary
Maximum Flooding Surface
Fluvial/Deltaic Sediments
Shale
Slope and Basin Floor Fans
Shallow Water Carbonate

The existence of localized ice caps in the eastern Ross Sea during the Early Neogene is consistent with drilling results from CIROS-1. There, mid-Late Eocene to Early Oligocene deep-water mudstones with dropstones and Late Oligocene diamictite, interpreted as till, were sampled (Hambrey, 1993).

Sequence RSS-2, which is Late Oligocene to Early Miocene in age (Leckie and Webb, 1983), contains wedge-shaped bodies (Fig. 5.12) that are interpreted as proglacial fans by De Santis and colleagues (1995). DSDP cores from this stratigraphic interval sampled pebbly mudstones and massive diamictons. It is the oldest undisputed glacial unit in the eastern Ross Sea. Unfortunately, the extent of grounded ice on the continental shelf could not be determined from the PD-90 data set because RSS-2 dips below the water-bottom multiple a short distance from Site 270.

De Santis and colleagues (1995) concluded that the ice cap that deposited RSS-2 was confined to a basement high just west of Site 270. This basement high was surrounded by relatively deep basins that prevented the ice cap from expanding beyond the limits of the high. Cooper and colleagues (1991a) also argue that post-U6 Late Cenozoic rifting in the Ross Sea influenced the movement,

Figure 5.9. (A) The seismic sequences and bounding surfaces of the eastern Ross Sea are illustrated using this interpreted line (BGR80-07, Hinz and Block, 1984; modified from De Santis et al., 1995) (Table 5.2). (See Fig. 5.6 for profile location.) (B) The stratigraphic signature of the Neogene is shown for comparison. From Bartek et al., 1991. (Reprinted with permission of the American Geophysical Union, © 1995.)

and possibly the growth and decay, of large grounded ice sheets.

Seismic records indicate marine and glacial-marine sedimentation for RSS-3. RSS-3 is separated from younger units by a widespread erosional surface (RSU-4A) with considerable relief. This surface is an amalgamation of several unconformities, including unconformity U5 of Hinz and Block (1984) and U4A of Cooper and colleagues (1991a) (Fig. 5.9). The erosional surface is believed to record the onset of widespread expansion of the ice sheet onto the continental shelf (Cooper and colleagues, 1991a; Anderson and Bartek, 1992). Expansion of the ice sheet was possible only after the deep basins on the continental shelf were filled with sediment (De Santis et al., 1995). This episode of ice sheet expansion and sediment infilling of basins is manifested by thick wedges of strata prograding in all directions away from the highs (De Santis et al.,

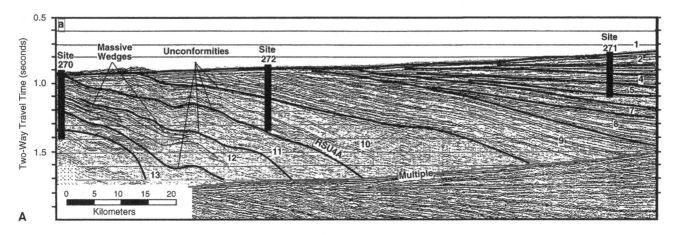

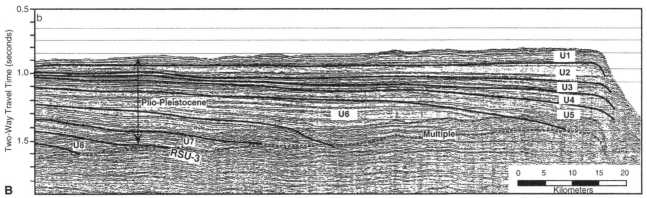

Figure 5.10. Interpreted seismic Profile PD90-30 (A and B) (dip section) showing seismic units recognized and mapped by Anderson and Bartek (1992). (See Fig. 5.6 for profile location.)

1995). These strata are cut by many cross-cutting erosional surfaces with considerable local relief (Fig. 5.12). Chronostratigraphic control on the age of RSS-4 from DSDP Leg 28 sites indicates that expansion of the ice caps occurred during the Early to Mid-Miocene (~ 21 to 16.2 Ma). This is approximately the same time that a major basinward shift in the ice sheet grounding line position occurred in the McMurdo Sound region (Bartek et al., 1996).

Using several different data sets, De Santis and colleagues (1995) characterized and mapped the mid-Miocene seismic units (RSS-5) of the Eastern Basin in greater detail than Anderson and Bartek (1992). The work of De Santis and colleagues (1995) revealed that Mid-Miocene seismic units are acoustically layered and quite continuous in their distribution on the shelf, implying that they are glacial-marine deposits. These results indicate that the ice sheet experienced a major recession from the Ross Sea continental shelf during the Mid-Miocene that began ~ 16.2 to 14.2 Ma and perhaps culminated in total retreat of the ice sheet from at least the eastern continental shelf by 13.8 Ma.

Late Miocene strata are virtually absent on the western Ross Sea continental shelf. These strata were removed by expanding and contracting ice sheets during

Plio-Pleistocene time (Anderson and Bartek, 1992). Ice sheet expansion during the Plio-Pleistocene carved a prominent unconformity, the Ross Sea Disconformity of Savage and Ciesielski (1983), which is an amalgamation of many erosional surfaces (U2, U3, and U4 of Hinz and Block (1984); RSU-2 in Table 5.2). Late Miocene strata are thicker on the eastern shelf.

Shipp and colleagues (1994) observed that Mid- to Late Miocene seismic units on the eastern shelf have a massive to laminated acoustic character and are strongly progradational at the shelf edge. In contrast, Plio-Pleistocene seismic units are separated by amalgamated erosional surfaces on the inner shelf and consist primarily of aggrading units on the middle and outer shelf, and seismic facies that are mainly massive and chaotic in character. Available lithologic information from DSDP Leg 28 sites that recovered Plio-Pleistocene material shows a greater proportion of massive units with more ice-rafted material than exists in the Miocene section (Hayes and Frakes, 1975). This change in depositional style between the Miocene and Plio-Pleistocene is interpreted as the change from temperate/subpolar interglacial conditions to polar interglacial conditions at some point during or just prior to the Early Pliocene (Shipp et al., 1994; Brancolini et al., 1995; De Santis et al., 1995).

The Plio-Pleistocene seismic stratigraphy of the eastern Ross Sea continental shelf was studied in detail by Alonso and colleagues (1992) and is the subject of an ongoing

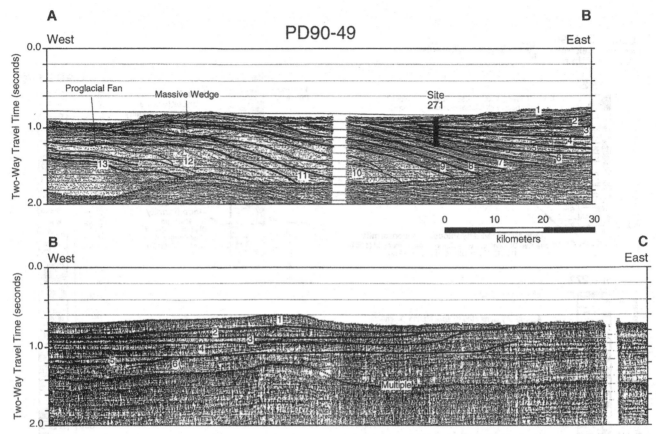

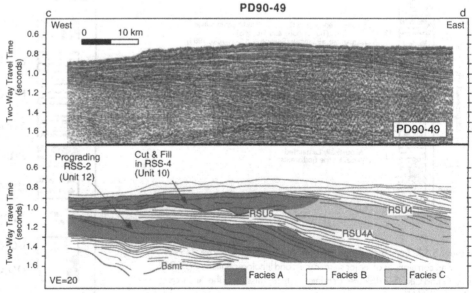

Figure 5.11. Interpreted seismic Profile PD90-49 (a strike section) illustrating unconformable surfaces between seismic units. These unconformities are more evident in this strike section than in the dip-oriented profiles (Fig. 5.10). They display broad-scale relief similar to that of the modern seafloor. (See Fig. 5.6 for profile location.)

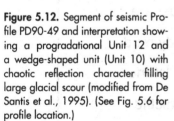

Figure 5.12. Segment of seismic Profile PD90-49 and interpretation showing a progradational Unit 12 and a wedge-shaped unit (Unit 10) with chaotic reflection character filling large glacial scour (modified from De Santis et al., 1995). (See Fig. 5.6 for profile location.)

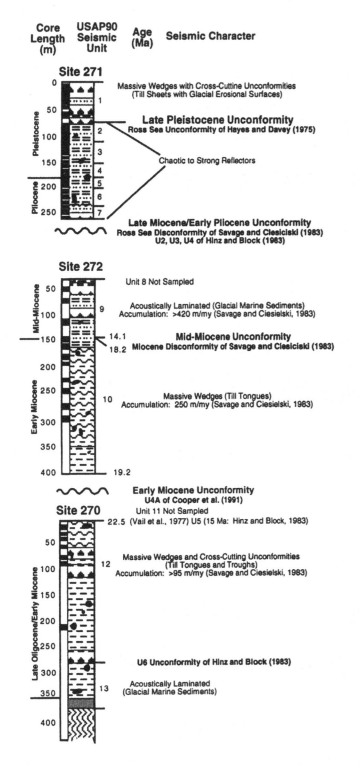

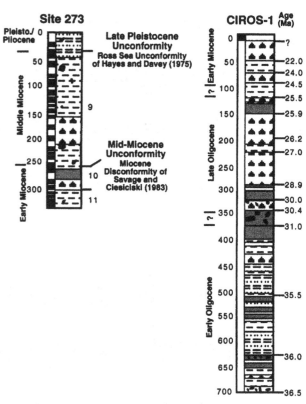

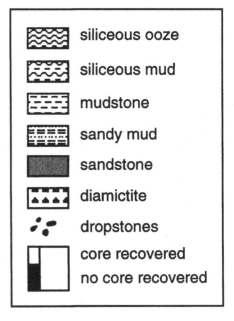

Figure 5.13. Summary diagram from Anderson and Bartek (1992) showing correlations of major seismic units and lithologic units from DSDP Leg 28 and CIROS-1 drill sites (reprinted with permission of the American Geophysical Union, © 1992).

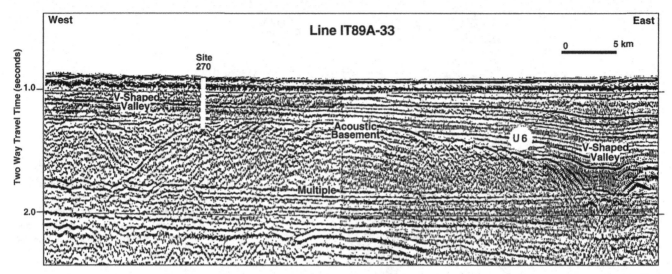

Figure 5.14. Enlargement of seismic Line IT89A-33 over DSDP Site 270 showing U6 onlapping acoustic basement about 7 km east of this site (modified from Busetti and Cooper, 1994). The two V-shaped valleys are interpreted by De Santis and colleagues (1995) as glacial troughs, whereas Busetti and Cooper (1994) interpret these features as half grabens.

study (Shipp, in progress). Alonso and colleagues (1992) recognize seven units (Fig. 5.10) that can be traced across the eastern Ross Sea, although these units thicken and thin abruptly due to cross-cutting glacial erosion surfaces. Shipp (in progress), working with a higher-resolution data set than Alonso and colleagues (1992), is finding that the Pleistocene section is composed of discontinuous till sheets separated by glacial unconformities. These combined results indicate that the Plio-Pleistocene was a time when ice sheets advanced and retreated from the continental shelf at a fairly high frequency, probably due to glacial-eustatic rise and fall caused by expanding and contracting Northern Hemisphere ice sheets (Alonso et al., 1992).

Bart (1998) identified 10 seismic units bounded by glacial unconformities in the Victoria Trough Mouth Fan, compared to seven glacial–interglacial cycles in the eastern Ross Sea. This is possibly due to the greater preservation potential of glacial units in the trough mouth fan than on the shelf of the eastern Ross Sea.

PACIFIC-ANTARCTIC MARGIN

The southeast Pacific sector of the Antarctic continental margin, hereafter referred to as the Pacific-Antarctic Margin, has had an active and complex tectonic history. The continental margin of the northern Antarctic Peninsula region, including Bransfield Basin, constitutes the only sector of the Pacific-Antarctic Margin not covered by sea ice throughout the year and the only part of the margin studied extensively to date. For this reason, this discussion of the Pacific-Antarctic Margin focuses on that region.

The seafloor extending from the Bransfield Basin to Marguerite Bay (Fig. 5.1) is typically covered by a relatively thin sediment package, especially on the inner shelf, so basement structures can be studied using relatively high-resolution seismic methods. This area provides an ideal setting in which to examine the impact of ridge-trench

collisions on continental margin evolution (Anderson, Pope, and Thomas, 1990; Larter and Barker, 1991a,b; Bart, 1993; Bart and Anderson, 1995; Larter et al., 1997). It is also a good setting in which to examine the interplay between tectonics and glaciation.

Geologic History and Tectonic Setting

Continental margin development has been influenced by a series of ridge-trench collisions between the Phoenix and Antarctic plates and involving the Aluk Ridge (Fig. 2.12). Ridge-trench collisions occurred progressively northeastward along the Antarctic Peninsula coast. Major fracture zones separate segments of the seafloor with different subduction histories (Herron and Tucholke, 1976; Barker, 1982) (Fig. 5.15). The Hero Fracture Zone marks the northern limit of ridge-trench collisions. The South Shetland Trench exists north of this feature, and seismic records show typical features associated with subduction, including a relatively thick accretionary prism (Gambôa and Maldonado, 1990; De Batist et al., 1994; Maldonado et al., 1994; Barker and Austin, 1994; Kim et al., 1994) (Fig. 5.16), although subduction is presently minimal. Seismic profiles that cross the margin just south of the Hero Fracture Zone show little evidence of subduction-related processes (Tucholke and Houtz, 1976; Larter and Barker, 1991b; Gambôa and Maldonado, 1990; De Batist et al., 1994; Kim et al., 1994).

Impact of Ridge-Trench Collisions on Continental Shelf Evolution

During the Late Cretaceous, the seafloor was being subducted beneath West Antarctica with fracture zones

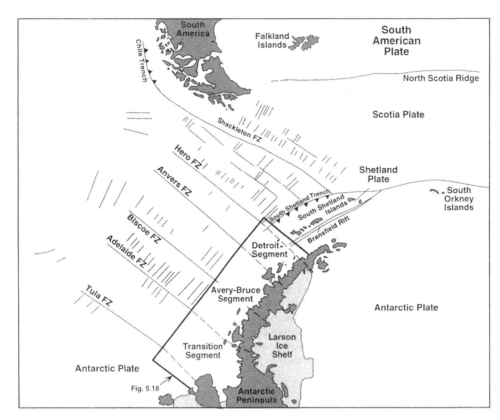

Figure 5.15. Major structural features of the southeast Pacific Ocean and Antarctic Peninsula (modified from Barker, 1982, and Cande and Mutter, 1982). Major structural segments of the Antarctic Peninsula are from Hawkes (1981). FZ = fracture zone.

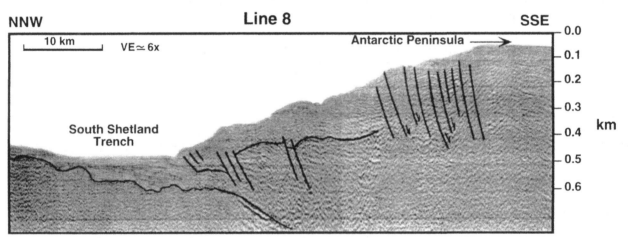

Figure 5.16. Interpreted seismic Line 8 across the South Shetland Trench showing accretionary prism and other subduction-related features (from Gambôa and Maldonado, 1990, reprinted with permission of the American Association of Petroleum Geologists).

trending parallel to the direction of subduction (Larter and Barker, 1991b) (Fig. 2.12). Thus, the boundaries between segments of different age seafloor remained in the same position on the margin. As each segment of the ridge approached the trench, magmatism in the corresponding arc segment ceased, and the arc and fore arc experienced tectonic uplift and erosion (Barker, 1982; Larter and Barker, 1989). Theoretical models predict several hundred meters of uplift of an arc during the subduction of the spreading center (DeLong, Schwartz, and Anderson,

1978). The models indicate that uplift culminates a few million years after collision and is followed by rapid thermal subsidence.

There is good evidence that the continental margin has experienced segmentation due to diachronous subduction of the buoyant ridge and associated uplift of margin segments (Hawkes, 1981; Barker, 1982; Garrett and Storey, 1987; Bart and Anderson, 1995; De Batist et al., 1994). De Batist and colleagues (1994) point out that the extension of the Hero Fracture Zone across the adjacent continental margin is marked by the Boyd Strait (Fig. 3.28). The Boyd Strait is the boundary between active rifting and volcanism of the Bransfield Basin and the now dormant Gerlache Strait. A seismic profile across the Boyd Strait

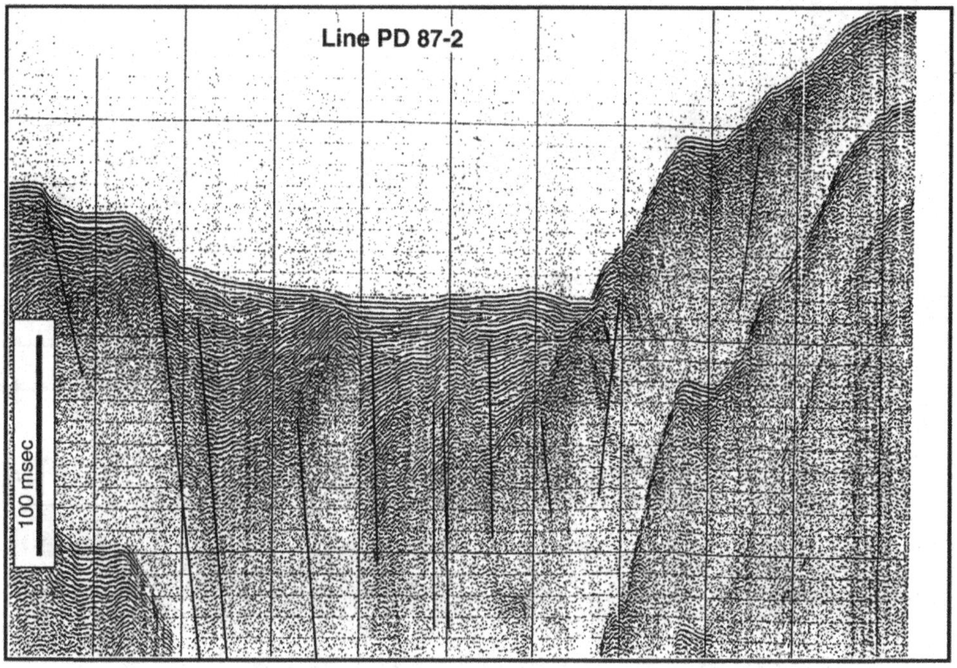

Line PD 87-2

Figure 5.17. Seismic Profile PD87-2 across the Boyd Strait (the landward extension of the Hero Fracture Zone) showing faults that offset even the youngest strata.

100 msec

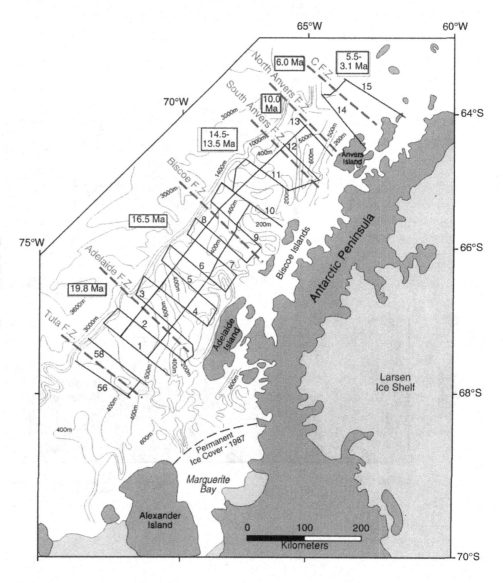

Figure 5.18. Tracklines for USAP-88 survey of Antarctic Peninsula continental shelf. Bold lines correspond to profiles described in text. Dashed lines are projections of fracture zones, and numbers in boxes are approximate ages for each segment of the seafloor according to Larter and Barker (1991a).

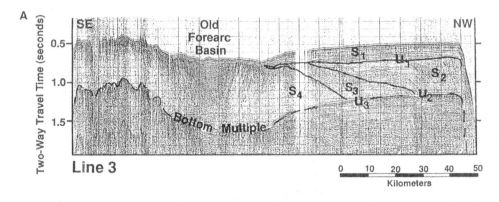

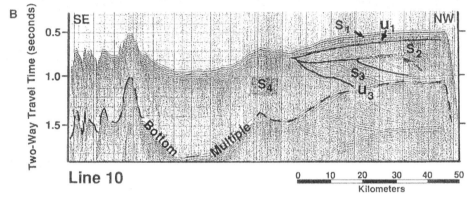

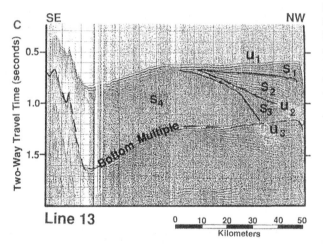

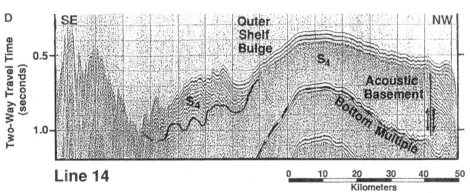

Figure 5.19. Representative USAP-88 seismic reflection profiles (A–F) from the northern Antarctic Peninsula continental shelf illustrating major sequences and sequence boundaries and changes in the thickness and width of the passive margin depositional wedge (S3–S1; from Bart and Anderson, 1995; reprinted with permission of the American Geophysical Union, © 1995). (See Fig. 5.18 for profile locations.)

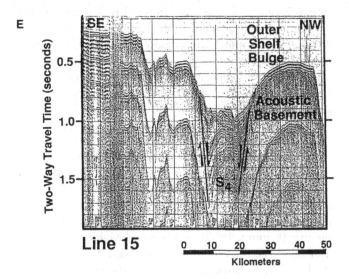

Line 15

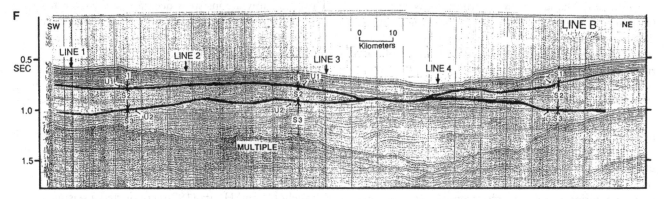

shows a series of half grabens with fault scarps extending to the seafloor (Fig. 5.17). De Batist and colleagues (1994) argue that a blueschist terrane on Smith Island, located adjacent to the Boyd Strait, is the product of subduction at the Hero Fracture Zone.

Transverse megafractures, which appear to correspond to fracture zones, tectonically segment the Antarctic Peninsula (Hawkes, 1981) (Fig. 5.15). Hawkes (1981) divides the Peninsula from south to north into four principal segments: the Transition, Avery-Bruce, Detroit, and Bransfield segments (Fig. 5.15). Each segment of the peninsula has a different history of magmatism, uplift and erosion, and post-orogenic sedimentation (Hawkes, 1981; Garrett and Storey, 1987; Parada, Orsini, and Ardila, 1992).

Anderson, Pope, and Thomas (1990) acquired seismic lines within each of the fracture zone–bounded segments between the Bransfield Basin and Marguerite Bay (Fig. 5.18). They noted major differences in the width and thickness of passive margin strata between major segments, as well as differences in the seismic stratigraphy of individual segments.

Figure 5.19 shows representative seismic lines from each of the fracture zone–bounded segments. The data show two main stratal packages. The oldest package

consists of folded and faulted strata (S4) deposited prior to ridge-trench collision (precollision accretionary prism of Larter and Barker, 1989). The magnitude of deformation of this package varies along the margin, as does the age of this package. Old fore-arc basins remain evident in some of the profiles (e.g., Fig. 5.19A, Profile PD88-3) and occur within those segments of the shelf bounded by the Adelaide and Biscoe fracture zones, as well as the South Anvers and North Anvers fracture zones (Fig. 5.18).

Passive margin strata (S1, S2, and S3) form a sedimentary wedge that overlies the tectonized package (Fig. 5.19). Onlap marks the unconformity between the passive margin wedge and the tectonized strata, as shown in Line PD88-3 (Fig. 5.19A). This is the collision unconformity. The angular discordance between S4 and S3 varies along the continental shelf, indicating variable rates of tectonic subsidence.

The shelf segments located between the North Anvers and Hero fracture zones are those segments where ridge-trench collision occurred most recently (within the past 6.0 Ma). Seismic lines from these segments (Fig. 5.19D,E, Lines 14 and 15) show a prominent basement high on the outer shelf where margin strata are virtually absent north of the C Fracture Zone. The only exceptions are strata that occur in a small fault-bounded shelf-margin basin in the area between the C Fracture Zone and the North Anvers Fracture Zone.

Comparison of seismic Lines 13 and 14 (Fig. 5.19C,D) illustrates the abrupt change in thickness and width of the passive margin wedge on either side of the projected North Anvers Fracture Zone. The maximum thickness of the wedge increases to the south, being virtually absent on Line 14 and ~ 1.5 sec thick on Line 13 (Fig. 5.19C,D). These two lines are separated by a distance of only 50 km. Between the North Anvers and Adelaide fracture zones, the passive margin wedge thickness varies only slightly (ranging from 1.5 to 1.8 sec; Fig. 5.20). This indicates that rates of subsidence dramatically decrease and are more or less uniform across the shelf once the margin becomes passive. The predominant change in the configuration of the shelf is a shift in the updip limit of the passive margin wedge, or hinge line for subsidence, from northeast to southwest. Changes in the width of the wedge appear to occur at the fracture zone projections (Fig. 5.20), indicating that the width of the passive margin is inherited partly from the early phase of rapid subsidence following ridge-trench collision.

Seismic Stratigraphy

Continental Shelf. The combined tectonic and glacial influences on continental shelf deposition have resulted in a complex stratigraphic setting where individual seismic units pinch and swell in all directions (Bart and Anderson, 1995). This stratigraphic complexity is not imaged in lower-resolution seismic profiles from the region (Larter and Barker, 1989). Bart and Anderson (1995) conducted a detailed seismic stratigraphic analysis using USAP-88 intermediate-resolution seismic data and grouped seismic units into stratal packages that contain two or more sequences (Fig. 5.19).

The oldest package is the deformed package (S4) formed prior to and during ridge-trench collision. North of the North Anvers Fracture Zone (Fig. 5.19D,E, Lines 14 and 15), S4 constitutes virtually the entire stratigraphic section on the shelf. Three stratal packages constitute the passive margin wedge south of the North Anvers Fracture Zone: an upper aggradational package (S1), a middle progradational package (S2), and a lower aggradational package (S3; Fig. 5.19A–C, Lines 3, 10, and 13). The lower aggradational package, S3, onlaps S4 and is generally characterized by strong, continuous reflections in the lower portion of the package and more discontinuous reflections in the upper part of the package.

The thickness of S2 varies widely across the shelf (Fig. 5.19). Discontinuous, often hyperbolic, reflectors and considerable cross-cutting of acoustic units (Fig. 5.19F) dominate the seismic expression of S2.

The youngest package is a dominantly draped sequence (S1) that thins landward (Fig. 5.19). It is separated from the underlying S2 sequence by a major unconformity, U1 (Fig. 5.19F). S1 strata primarily consist of horizontal to subhorizontal aggradational units. This sequence is distinct in its virtual absence of shelf-edge progradation. Individual units within S1 are usually bounded by landward-sloping unconformities and range in thickness from 20 msec to over 100 msec. They tend to be lens-shaped and are characterized by chaotic seismic facies and prograding foresets, which Bart and Anderson (1995, 1996) interpret as grounding zone wedges (Fig. 4.16B). The foreset strata of these wedges grade seaward into a seismic facies characterized by more continuous reflectors, which are interpreted as glacial-marine deposits.

Continental Slope and Rise. MCS profiles have been acquired from the continental slope, rise, and basin floor of the Pacific-Antarctic region by a number of different investigators (Kimura, 1982; Gambôa and Maldonado, 1990; Yamaguchi et al., 1988; Larter and Barker, 1989; Austin et al., 1991; Rebesco et al., 1994; Nitsche et al., 1997). The data show a consistent decrease in the overall thickness of continental rise strata from the Amundsen Sea (1.5 sec; Yamaguchi et al., 1988) northeastward to the northern Antarctic Peninsula continental rise (~ 0.6 sec. Larter and Barker, 1989). Seismic profiles acquired along the rise and parallel to the slope show a series of steps in the acoustic basement and associated thinning of strata toward the northeast (Fig. 5.21). These steps are associated with major fracture zones, and some are manifested as prominent ridges that extend to the base of the continental slope (Tucholke and Houtz, 1976; Larter and Cunningham, 1993; De Batist et al., 1994).

Tucholke and Houtz (1976) conducted a seismic facies and seismic stratigraphic analysis of the continental rise of the Bellingshausen Sea and the northern Antarctic Peninsula. They subdivided the continental rise into three seismic facies, based on differences in the near-surface strata. The upper rise is characterized by highly reflective, poorly layered strata that are interpreted as mostly sediment gravity flow deposits. The upper and middle rise is dominated by spectacular channel-levee complexes (Fig. 5.22A) and sediment drifts (Fig. 5.22B), while the lower rise is characterized by strong, parallel to subparallel reflections that are interpreted as mostly pelagic and hemipelagic sediments, as well as fine-grained turbidites. Individual levees are up to 1 km thick and show migration toward the northeast (Fig. 5.22A). Seismic profiles from the middle continental rise show middle-rise seismic facies resting rather sharply on lower-rise facies (Tucholke and Houtz, 1976; Gambôa and Maldonado, 1990; Larter and Cunningham, 1993; McGinnis and Hayes, 1995) and indicate a rapid phase of seaward growth of the rise (Tucholke and Houtz, 1976) (Fig. 5.22A).

Rebesco and colleagues (1994, 1997) mapped eight large depositional mounds (channel–levee and drift complexes) on the continental rise (Fig. 5.23). These mounds are interpreted as sediment drift deposits comprised of sediment that is transported basinward and toward the northeast by bottom currents. Several feeder channels for

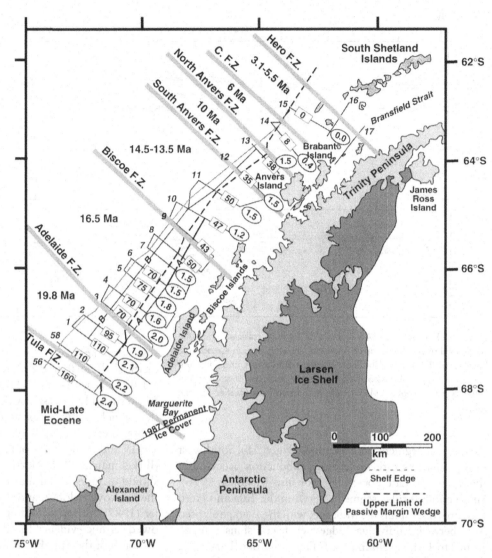

Figure 5.20. Changes in the width and thickness of the passive margin wedge. These changes appear to correspond to the extensions of major fracture zones (bold lines). Numbers in boxes are width of the wedges, and numbers in circles are maximum thickness (in two-way travel time) of the wedges, most of which are based on extrapolations of surfaces below the bottom multiple.

the mounds were imaged on GLORIA swath bathymetry records and show dendritic drainage patterns on the lower slope (Tomlinson et al., 1992).

Piston cores from channels on the continental slope and rise recovered disorganized gravel units and graded gravel and sand units (Wright, Anderson, and Fisco, 1983) (Fig. 4.65). Cores from the fans recovered sand and silt turbidites, and cores from the levees sampled interbedded and massive silts.

Sequence Stratigraphic Interpretation

Four sites were drilled on the Pacific-Antarctic shelf during ODP Leg 178, but only the preliminary results from this leg were available at the time this book was completed (Barker, Camerlenghi, and Acton, 1998). Unfortunately,

sample recovery at the four drill sites was poor, so the age and sedimentology of the major stratigraphic units on the shelf remain somewhat problematic.

In the absence of long and continuous core, collision unconformities continue to provide an alternate chronostratigraphic framework for seismic stratigraphic analysis on the Antarctic Peninsula continental shelf (Anderson, Pope, and Thomas, 1990; Larter and Barker, 1991a; Bart and Anderson, 1995, 1996).

Herron and Tucholke (1976), Barker (1982), and Larter and Barker (1991a) examined magnetic data from the southeast Pacific and demonstrated that progressively younger anomalies occur at the base of margin segments from south to north; hence, the youngest anomaly at the margin provides a date for cessation of subduction within

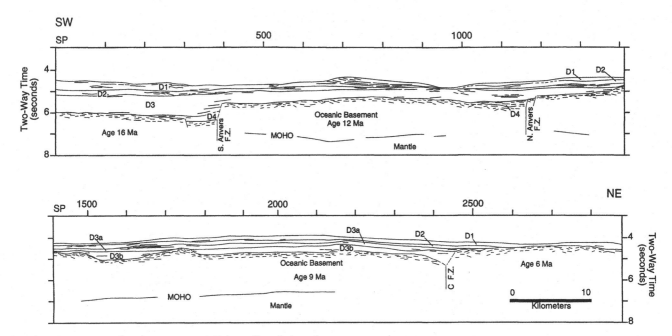

Figure 5.21. Interpreted BAS seismic Line 845-04, collected along the continental rise of the Antarctic Peninsula, showing steps in the acoustic basement in the vicinity of the South Anvers, North Anvers, and C fracture zones, as well as the associated changes in thickness of strata (from Larter and Cunningham, 1993, with permission from Elsevier Science Ltd.)

a particular segment of the seafloor (Fig. 5.18). Southwest of the Tula Fracture Zone, ridge subduction occurred in the Middle to Late Eocene (Herron and Tucholke, 1976). Subduction of ridge segments between the Tula and Hero fracture zones spans the Miocene, with the average elapsed time between individual ridge-trench collisions ranging from 2 to 4 Ma, excluding the C Fracture Zone (Larter and Barker, 1991a). The segment north of the Hero Fracture Zone constitutes the sole surviving segment of the spreading ridge and subduction zone (Fig. 5.16). This segment apparently stopped spreading ~4 Ma before the ridge reached the margin (Barker, 1982).

Bart and Anderson (1995) identified six collision unconformities on the shelf. From southwest to northeast, individual collision unconformities rise in stratigraphic position, and the number of unconformities increases. The assigned maximum ages of the collision unconformities were taken from the latest results of Larter and Barker (1991a): CU6 – 19.8 Ma; CU5 – 16.5 Ma; CU4 – 14.5 Ma; CU3 – 10 Ma; and CU2 – 6.0 Ma. The CU1 unconformity is a seafloor unconformity (within the bubble pulse) that occurs only in the northeastern part of the study area (Fig. 5.19D,E). It is believed to be related to the latest collision event, which occurred between 5.5 and 3.1 Ma (Larter and Barker, 1991a).

The oldest seismic package on the shelf (S4) consists of folded and faulted strata deposited prior to ridge-trench collision. The age of S4 decreases from southwest to north-

east along the margin. S4 is situated at or near the seafloor on the shelf north of the North Anvers Fracture Zone (Fig. 5.19D,E, Lines 14 and 15). The sediments that compose S4 presumably consist of volcaniclastic material similar to that which presently exists on the seafloor north of the South Shetland Islands (Jeffers, Anderson, and Lawver, 1991) and in Tertiary outcrops on King George Island (Birkenmajer, 1987). North of the Tula Fracture Zone, S4 should include glacial-marine sediments deposited during the Krakow, Polonez, and Legru glaciations of Birkenmajer (1987), although folding and faulting have overshadowed seismic evidence for glaciation (Chapter 2).

South of the Tula Fracture Zone, the basal strata of S3 may be as old as Eocene (Herron and Tucholke, 1976) and span that interval of time when temperate glaciers existed on the Peninsula in harmony with temperate forests (Birkenmajer, 1987; Askin, 1992). The maritime setting at that time probably was similar to that of the present Gulf of Alaska. Anderson and Molnia (1989) emphasize that rapid rates of terrigenous sedimentation characterize temperate glacial-marine settings, such as the Gulf of Alaska. In contrast, slow rates of sedimentation characterize subpolar and polar glacial-marine settings, such as Antarctica. Thus, thinning of S3 from southwest to northeast (Fig. 5.19) may also reflect a change in the climate and reduced terrigenous sedimentation on the shelf with time.

North of the Tula Fracture Zone, the S3 sequence includes the CU4-CU6 collision unconformities and, therefore, is believed to span the early–middle Miocene. An aggradational stacking pattern for this time period is consistent with seismic stratigraphic interpretations from the Ross Sea (Anderson and Bartek, 1992) (Fig. 5.9A) and other areas of the globe (Bartek et al., 1991) (Fig. 5.9B).

A

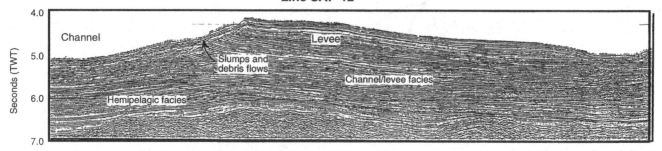

Line SAP-12

B

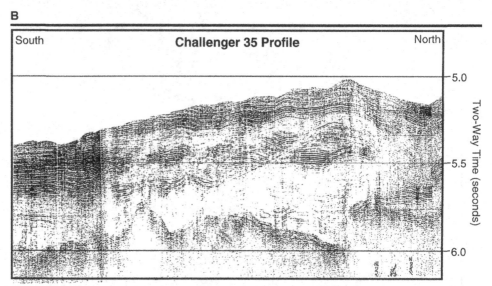

Figure 5.22. (A) Seismic Profile SAP-12 from the northern Antarctic Peninsula continental rise showing channel–levee system (from McGinnis and Hayes, 1995; reprinted with permission of the American Geophysical Union, © 1995); (B) segment of Challenger 35 profile from the Bellingshausen Sea continental rise showing sediment waves (from Tucholke and Houtz, 1976).

Larter and Barker (1989) and Bart and Anderson (1995) attempted to establish the onset of glaciation on the Antarctic Peninsula continental shelf using seismic reflection profiles from the region. Larter and Barker (1989) argued that prograding strata on the shelf that are bounded by major unconformities were produced by ice sheets that advanced across the shelf. Later, Bart and Anderson acquired a higher-resolution, more dense seismic grid from the region (Fig. 5.18) that imaged a total of 19 glacial unconformities. In strike-oriented profiles, these unconformities show cross-sectional profiles similar to present-day glacial troughs on the shelf and indicate episodes of glacial erosion in which many tens to hundreds of meters of erosion occurred (Fig. 5.19F). The data also showed massive and chaotic seismic facies between these unconformities and grounding zone wedges in the upper part of the stratigraphic section. These combined data led Bart and Anderson to conclude that ice sheets had grounded on the continental shelf on several occasions in the past. The main difference between Larter and Barker's (1989) and Bart and Anderson's (1995) interpretations concerned the timing of these grounding episodes; both groups were hindered by a lack of chronostratigraphic data from drill cores. Bart and Anderson (1995) concluded that ice sheets

first grounded on the shelf during the Miocene, whereas Larter and Barker (1989) argued that ice sheets did not advance onto the shelf until the Plio-Pleistocene.

Recent drilling on the shelf, during ODP leg 178, has yielded some new information with which to constrain the timing of glaciation on the Antarctic Peninsula shelf, however, none of the sites on the shelf succeeded in recovering material from the older part of the glacial section. Preliminary results from Site 1097, located within the Marguerite Trough, showed that the Plio-Pleistocene section contains glacial deposits (Barker, Camerlenghi, and Acton, 1998), but the base of the glacial section was not sampled. Only preliminary results from the Leg 178 data had been released at the time this book was being published.

The change from progradation (S2) to aggradation (S1) and associated decrease in unit thickness and depth of glacial erosion surfaces observed on the Antarctic

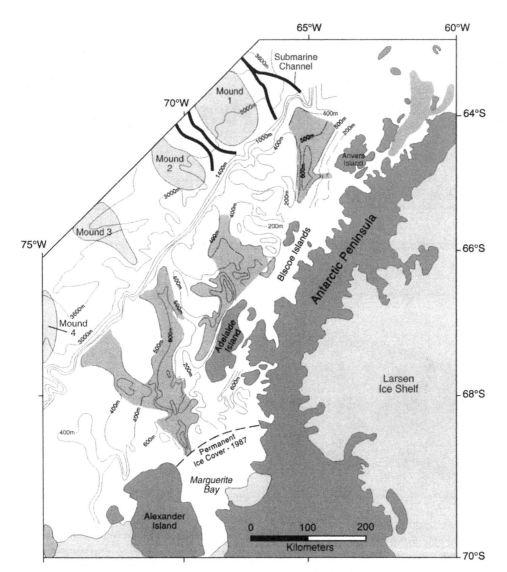

Figure 5.23. Map showing distribution of glacial troughs (dark shaded), submarine canyon channels (heavy black lines), and sediment mounds (light shaded) on the Antarctic Peninsula continental margin (compiled from Pope and Anderson, 1992, and Rebesco et al., 1994).

Peninsula continental shelf (Fig. 5.19A,B) is similar to stratigraphic changes that occur on the Ross Sea continental shelf (Fig. 5.9). There, these changes are attributed to shortened duration of ice sheet grounding events caused by high-frequency sea-level fluctuations of the Plio-Pleistocene (Alonso et al., 1992). A similar interpretation seems logical for the Antarctic Peninsula region (Bart and Anderson, 1996).

Tucholke and Houtz (1976) point out that development of the continental rise has progressively occurred from southwest to northeast as successive episodes of ridge-trench collision led to the trench becoming passive and filled with sediment. This, in turn, allowed terrigenous sediment transport onto the continental rise. Thus, the stratigraphic boundaries between continental rise facies should be diachronous along the margin. This stratigraphic

pattern may be complicated even more by the shifting of fans and associated drift deposits (Tucholke and Houtz, 1976). The observed stratigraphic change in seismic facies may also record the onset of glacial conditions on the Antarctic Peninsula and associated mass wasting of the continent and continental shelf, but the tectonic and climatic signals are difficult to deconvolve. It is noteworthy that the observed change from distal to more proximal seismic facies on the middle continental rise of the northern Antarctic Peninsula continental margin occurs sometime in the Early to Middle Miocene, based on extrapolation of seismic surfaces to DSDP Leg 35 drill sites (Tucholke and Houtz, 1976; McGinnis and Hayes, 1995). The oldest glacial unconformity on the adjacent continental shelf was interpreted as a Middle Miocene feature by Bart and Anderson (1995). This is consistent with interpretations of seismic records from the Amundsen Sea continental rise that show an upward change from a strongly layered seismic facies, believed to be hemipelagic and pelagic deposits, to a seismic facies with turbidite features.

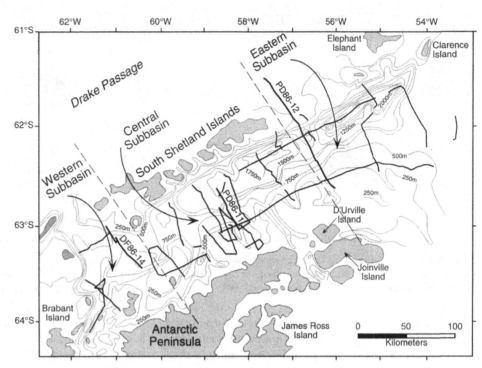

Figure 5.24. Bathymetry and locations of single-channel high-resolution seismic lines in the Bransfield Basin used by Jeffers and Anderson (1990) to conduct seismic stratigraphic analysis of the basin.

The boundary between these seismic facies is inferred from DSDP Site 324 as being Middle Miocene in age (Yamaguchi et al., 1988).

BRANSFIELD BASIN

Bransfield Basin is an extensional basin separating the South Shetland Islands from the northernmost Antarctic Peninsula (Fig. 5.24). It is the latest in a series of convergent margin basins that have developed throughout the history of the Pacific margins of Antarctica and South America. The basin is ∼100 km wide and lies between the Hero and Shackleton fracture zones (Fig. 5.15).

Rifting within the basin is believed to have begun either at ∼2 Ma (Weaver, Saunders, and Tarney, 1982; González-Ferrán, 1985, 1991), or possibly in the Early Pliocene (Jeffers, Anderson, and Lawver, 1991). This is a relatively young basin. It differs from the well-studied back-arc basins of the western Pacific in its tectonic evolution and its sedimentary character (Barker and Dalziel, 1983; Jeffers, Anderson, and Lawver, 1991; Lawver et al., 1996). There is no strong evidence for normal seafloor spreading within the basin (Lawver et al., 1996). Also, it is associated with the waning stages of Pacific-Antarctic subduction (Barker, 1982; Barker and Dalziel, 1983), rather than with ongoing subduction and arc magmatism. Glacial–marine processes and their associated lithofacies have dominated sedimentation throughout the history of the basin (Jeffers and Anderson, 1990; Birkenmajer, 1992). Although unique in the present, the Bransfield Basin may be analogous to several deformed ancient basins of the Pacific-Antarctic sector of West Antarctica.

Geologic History and Tectonic Setting

Seismic refraction data reveal the general crustal structure of the Bransfield Basin (Ashcroft, 1974; Guterch et al., 1985; Guterch et al., 1985; Barker and Austin, 1994). Similarity of crustal structures led Ashcroft (1974) to propose that the South Shetland Islands and the Antarctic Peninsula originally were joined prior to a Tertiary rifting event. The South Shetland Islands, a Late Mesozoic magmatic arc, sit atop a 30- to 32-km-thick block of igneous and metamorphic continental lithosphere (Guterch et al., 1985, 1991; Birkenmajer, 1992). The Antarctic Peninsula is essentially a Mesozoic magmatic arc. Crustal thickness beneath the Peninsula ranges between 38 and 44 km (Guterch et al., 1985, 1991). The crust beneath the axis of the Bransfield Basin is much thinner (20 to 25 km), resembling a slightly thickened oceanic section with physical properties similar to a continental ridge (Guterch et al., 1984). Barker and Austin (1994) conclude that the basin is primarily composed of attenuated pre-existing crust.

A Bouguer gravity high over the center of the Bransfield Basin supports the interpretation of the presence of dense, quasi-oceanic crust beneath a thin sedimentary cover (Renner, Sturgeon, and Garrett, 1985). The crust generally thickens to the northeast along strike, both along the South Shetland Islands and within the basin (Guterch et al., 1991). Northeast-southwest-trending normal faults, associated with Pliocene to recent volcanics exposed on the South Shetland Islands (Barton, 1965; Weaver et al., 1979;

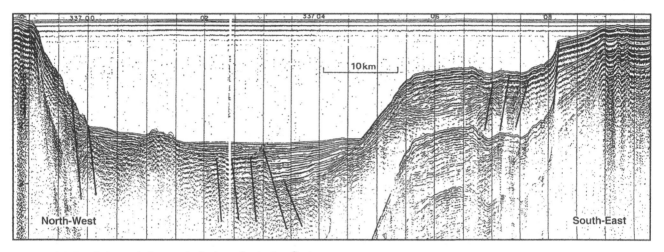

Figure 5.25. British Antarctic Survey single-channel seismic reflection profile across the Bransfield Basin showing steep, fault-bounded basin margins (modified from Barker, Dalziel, and Storey, 1991).

González-Ferrán, 1991), also occur offshore, forming the steep northwestern margin of the Bransfield Basin (Fig. 5.25). Similar faults occur on the Trinity Peninsula shelf (Fig. 5.25).

Recently acquired swath bathymetry records show linear volcanic features roughly aligned along the basin axis (Canals et al., 1994; Grácia et al., 1996; Lawver et al., 1996) (Fig. 5.26). Presently, volcanic activity is concentrated along the northern edge of the basin on Deception, Penguin, and Bridgeman islands (Weaver et al., 1979; Saunders and Tarney, 1982; González-Ferrán, 1991) (Fig. 3.28). There also are considerable submarine volcanism and hydrothermal activity in the Bransfield Basin (Han and Suess, 1987). Nagihara and Lawver (1989) acquired 22 heat flow measurements in the central part of the Bransfield Basin. The highest values were measured in the central part of the basin and along the southwest and northeast edges of the basin. They also noted large differences in heat flow values from closely spaced stations, and suggested that this might indicate hydrothermal activity. Seismic activity occurs along the axis of the basin (Forsyth, 1975; Pelayo and Wiens, 1989). High-resolution seismic reflection profiles from the basin show numerous seafloor scarps associated with faults (Jeffers and Anderson, 1990). The volcanic, hydrothermal, tectonic, and seismic activity imply that extension within the Bransfield Basin continues, although the mechanism of extension remains problematic (Lawver et al., 1996).

González-Ferrán (1985) viewed the extensional volcanism of the Bransfield Basin within the larger framework of the entire northern Antarctic Peninsula. In addition to Bridgeman, Deception, and Penguin islands, he discussed two other Pliocene-Recent extensional volcanic centers, both located under Larsen Ice Shelf. He named them the Prince Gustav Rift (Paulet Islands) and the Larsen Rift (Coley, Argo, and Seal nunataks). González-Ferrán (1985) considered all these components to be part of one large intraplate "fan rift" system. In general, the Pliocene-Recent volcanic rocks on Deception, Penguin, and Bridgeman islands are transitional between ocean floor basalts and island arc volcanics (Weaver et al., 1979). Analyses of basalts and basaltic andesites dredged from seamounts show $^{87}Sr/^{86}Sr$ and $^{143}Nd/^{144}Nd$ ratios typical of back-arc basin basalts (White and Cheatham, 1987; Keller and Fisk, 1987; Keller, Fisk, and White, 1991).

Basin Evolution

Several different mechanisms for the formation of the Bransfield Basin have been proposed. Barker (1982) and Barker and Dalziel (1983) attributed the rifting to continued sinking of the subducted slab after cessation of Aluk-Antarctic Plate subduction, resulting in northwest migration of the trench and extension in the Bransfield Basin. Seafloor spreading at the Aluk-Antarctic ridge in the Drake Passage occurred rapidly from ~ 20 to 4 Ma and apparently ceased thereafter (Barker and Dalziel, 1983). The Bransfield Basin extension began at approximately the same time that oceanic seafloor spreading ended.

Interpretation of magnetic anomalies as possible indicators of the seafloor spreading history in the Bransfield Basin has proven unreliable for estimating the age of the basin. Barker (1976) interpreted a single magnetic profile across the Bransfield Basin to show no magnetic reversals and suggested that the seafloor in the Bransfield Basin formed during a single epoch of uniform normal magnetic polarity. Roach (1978), however, modeled the same profile by inserting narrow, reversely magnetized strips of seafloor on the flanks of the positively magnetized spreading center and speculated that seafloor spreading has been occurring for the past ~ 1.3 Ma. More recently, Kim, Chung, and Nam (1992) measured several anomalies that display an irregular distribution pattern, which they attribute to changes in the regional stress field of plate motion. Because of the short history and the uncertain processes that formed the seafloor of the Bransfield Basin, age determination from magnetic profiles is not possible (Lawver et al., 1996).

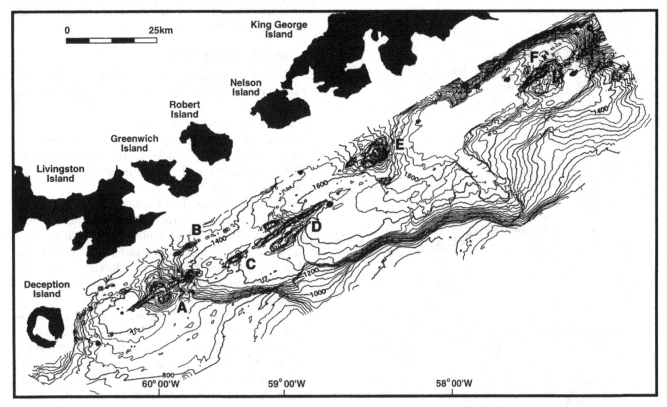

Figure 5.26. Swath bathymetry map of the central subbasin of the Bransfield Basin showing volcanic ridges roughly aligned with the basin axis (provided courtesy of M. Canals). Contours in meters.

Tokarski (1987) suggested that propagation of stress from the eastern Scotia Sea to the Bransfield Basin along the South Scotia Ridge may have been important in initiating rifting. He observed a transition from strike-slip deformation to extension on King George Island from the Late Miocene to present, and a concomitant clockwise rotation of the principal stress axis from east-southeast to west-southwest. Tokarski (1987) attributed the change in stress to the cessation of subduction at the South Shetland Trench and the onset of east-west seafloor spreading in the eastern Scotia Sea.

Birkenmajer (1982) and Birkenmajer and colleagues (1986) recognized four phases of Late Cretaceous to Cenozoic deformation of King George Island. Northeast-southwest-trending right-lateral strike-slip faults and associated folds, mappable across the entire length of King George Island, formed in a strike-slip regime during the Cretaceous to Early Miocene (Birkenmajer et al., 1986). Numerous shorter strike-slip faults spaced 5 to 15 km apart offset the earlier faults and reflect the transition from compression to extension during the Late Miocene. Normal faulting related to rifting began in the Pliocene. Faulting continues along with volcanism into the present.

Barker and Austin (1994) examined MCS profiles across Bransfield Basin and discovered complex fan-shaped faulting patterns along the Antarctic Peninsula margin. They interpreted these as evidence that extension occurred along separate diffuse zones, rather than being along a central spreading center.

Jeffers, Anderson, and Lawver (1991) re-examined the various geologic and geophysical data from the Bransfield Basin and presented a model for basin evolution. Their model is illustrated in Figure 5.27. In the initial phase of basin development, the Brabant segment of the Aluk-Antarctic ridge crest collided with the trench at anomaly 4-3 time (6 to 4 Ma), causing spreading and subduction to cease as this part of the Pacific Plate became rigidly locked to the Antarctic Plate (Fig. 5.27A). Spreading may have continued slightly longer on the Shetland segment, with continued convergence at the South Shetland Trench and consequent dextral shear along arc-parallel faults during this phase. Continued strike-slip motion in the Shetland segment resulted in extension normal to the segment boundary and the formation of the Boyd Strait (Fig. 5.27B).

Extension in the Bransfield Basin is believed to have begun in response to the cooling and sinking of the subducted oceanic slab (Barker, 1982; Barker and Dalziel, 1983). Seismic records show that extension was highly segmented (Jeffers and Anderson, 1990; Prieto et al., 1998). On land, the transition from a compressional regime to an extensional regime probably correlates with the development of the "Admiralty phase" strike-slip faults in the Late Miocene (Birkenmajer, 1986). The subsequent Bransfield phase normal faults (Late Plio-Pleistocene) reflect the early rifting of the Bransfield Basin (Fig. 5.27B). Crustal

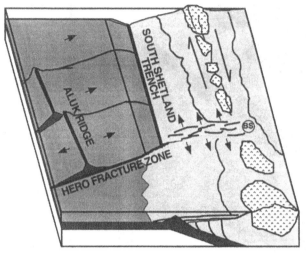

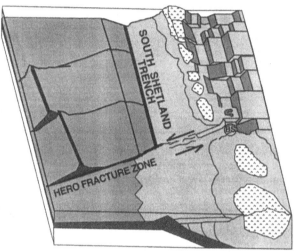

Figure 5.27. Proposed evolution model for Bransfield Basin (modified from Jeffers, Anderson, and Lawver, 1991): (A) Basin development commenced 6 to 4 Ma when the Brabant segment of the Aluk-Antarctic ridge crest collided with the trench at anomaly 4-3 time; (B) continued convergence at the South Shetland trench and consequent dextral shear along the arc-parallel faults during this phase was completed with back-arc extension. BS = Boyd Strait.

failure may have occurred along pre-existing arc-parallel strike-slip faults. As extension and crustal thinning continued, dikes and sills were injected into the sedimentary strata, creating thick interbeds of volcanic and terrigenous sedimentary units. This style of emplacement of igneous material explains the lack of well-defined magnetic anomalies in the basin (Lawver and Hawkins, 1978).

The regime of back-arc extension in the Bransfield Basin appears to extend southward through the Croker Passage and Gerlache Strait (Storey and Garrett, 1985; González-Ferrán, 1985). This is supported by the regional Bouguer anomaly map, which shows a gravity high extending through these straits, indicating that they are underlain by anomalously dense material (Davey, 1972). Onshore, large faults flank these straits and are believed to be products of extension (Goldring, 1962; Hooper, 1962; Curtis, 1966; Dewar, 1970).

Seismic Stratigraphy

Seismic data indicate that the three subbasins composing the Bransfield Basin (Chapter 3) represent discrete segments with different tectonic and depositional histories owing to different sinking rates of the subducted slab (Jeffers, 1988; Jeffers, Anderson, and Lawver, 1991; Jeffers and Anderson, 1990) (Fig. 3.25). The seismic stratigraphy of the three subbasins is discussed independently.

Western Subbasin. The bottom topography of the western subbasin is extremely rugged at depths shallower than 500 m. There is little evidence of substantial sediment accumulation at water depths shallower than this. An axial ridge did not develop in this subbasin, indicating that it is the least mature subbasin.

Seismic profiles from the western subbasin show that it is a narrow fault-bounded feature (Fig. 5.28A, DF86-14). These faults are typically manifested at the seafloor by scarps with tens of meters of relief. Hummocky and incoherent subbottom reflectors on the southern slope of the basin indicate active slumping.

Three straits enter the western subbasin from the adjoining platforms, including the Orléans Strait, Croker Passage, and Boyd Strait (Fig. 3.25). These straits cut through relatively flat-topped banks and are deeply incised into acoustic basement (Fig. 5.29). The banks generally are devoid of acoustically penetrable sediment, whereas the troughs are characterized by hummocky seafloor, a chaotic seismic facies in their updip portions, and a highly faulted, acoustically layered seismic facies in their more basinward portions (Fig. 5.29). Seafloor scarps associated with these faults indicate neotectonic activity. Faults appear to trend northeast-southwest, or parallel to the main axis of the troughs.

The western subbasin is the repository for sediment that has been eroded from the peninsula and archipelago and then transported westward from the Gerlache Strait through the Croker Passage and Orléans Strait, as well as sediments eroded from the southern portion of the South Shetland platform. Seismic profiles show that the sediment of the western subbasin increases in thickness toward the center of the basin from 0.1 sec to at least 0.4 sec (Jeffers and Anderson, 1990).

Central Subbasin. An abrupt deepening of the basin floor marks the boundary between the western and central subbasins (Jeffers and Anderson, 1990). Prominent features of the central subbasin include volcanic ridges that extend for a few kilometers parallel to the axis of the central subbasin (Grácia et al., 1996; Lawver et al., 1996) (Figs. 5.26 and 5.28B). These ridges formed along the flanks of half grabens that are offset along the basin axis (Prieto et al., 1997) (Fig. 5.27B).

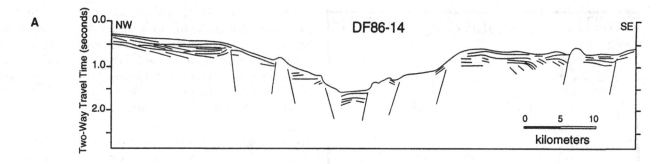

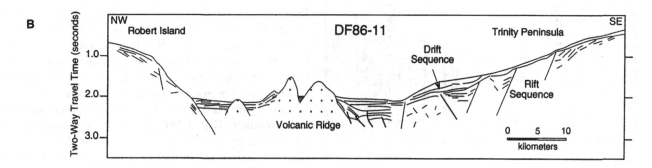

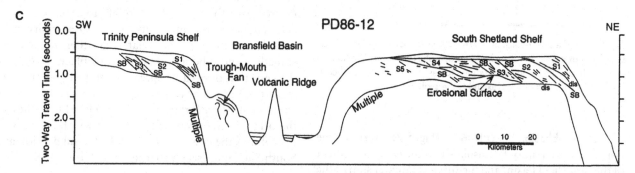

Figure 5.28. Interpreted high-resolution seismic reflection profiles (A) DF86-14, (B) DF86-11, and (C) PD86-12 from the western, central, and eastern subbasins, respectively, of the Bransfield Basin. These profiles illustrate the different structural style and stratigraphy of these subbasins. (See Fig. 5.24 for locations of the profiles.)

Gambôa and Maldonado (1990) and Bialas and GRAPE- team (1990) interpreted MCS profiles that cross the central subbasin to show two main stratigraphic packages, a 1.2-sec-thick drift sequence and an older faulted rift sequence exceeding 1.0 sec thick. Intermediate-resolution seismic profiles show complex stratal patterns within the upper sequence and highly variable thickness along strike. Cross-cutting unconformities and chaotic seismic facies indicate a glacial origin for these strata (Banfield and Anderson, 1995).

Stratal patterns on the Trinity Peninsula platform indicate basinward transport of sediment through several deeply incised troughs that dissect the Trinity Peninsula platform (Fig. 5.24). A prograding complex of coalescing trough mouth fans exists on the lower slope and adjacent basin floor (Jeffers and Anderson, 1990; Banfield and Anderson, 1995; Grácia et al., 1996). Seismic records that cross the trough mouth fans show a series of onlapping wedges prograding into the basin (Fig. 5.30A). A steep (~ 9°) slope separates the wedges from the 1500-m-deep basin floor. As much as 0.7 sec of sediment covers the adjacent basin floor, represented by coherent continuous reflectors divisible into two general types. Strong continuous

reflectors constitute draping units of nearly constant thickness. Packages of weaker, somewhat discontinuous reflectors onlap and thin toward the slope (Fig. 5.30B). The geometry and character of these reflectors indicate cycles of dominantly pelagic (draping) and dominantly terrigenous (turbiditic) deposition. Line PD86-11 (Fig. 5.30B) shows that two downward shifts in onlap occur in the prograding wedge, with two episodes of predominantly turbidite deposition on the basin floor. The inferred processes indicate that these packages represent two complete glacial eustatic cycles (Jeffers and Anderson, 1990). The turbidites represent distal equivalents of the prograding complex (Jeffers and Anderson, 1990; Banfield and Anderson, 1995). When the troughs are inactive, the supply of terrigenous sediment to the basin floor is relatively low, and hemipelagic sedimentation predominates. All the basinal strata eventually lap out against the axial ridge (Fig. 5.28B).

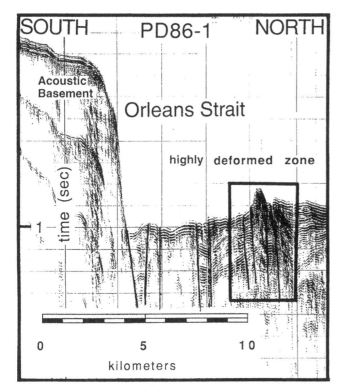

Figure 5.29. Seismic reflection Profile PD86-1 across the Orleans Strait showing that this trough is incised into acoustic basement. The floor of the trough contains a layered seismic facies that is cut by northeast-southwest-trending faults that extend to the seafloor.

Eastern Subbasin. Line PD86-12 (Fig. 5.28C) crosses the northernmost Trinity Peninsula shelf, the eastern segment of the Bransfield Basin, and the submarine extension of the South Shetland shelf northeast of King George Island (Fig. 5.24). Two sets of prograding shelf sequences are seen: one on the Trinity Peninsula margin that prograges into the Bransfield Basin and another that prograges northward from the South Shetland shelf over the inner slope of the South Shetland Trench. Four distinct seismic sequences are identified on the South Shetland margin and three on the Trinity Peninsula margin by basinward shifts in onlap. These sequences represent relative sea-level falls, possible glacial advances, and development of "lowstand deposits" (Jeffers and Anderson, 1990).

On the South Shetland shelf, the youngest sequence, S1, is bounded at the top by a rugged sea bottom interpreted as a glacial erosion surface formed during the last glacial maximum. A downward shift in onlap delimits its lower boundary. Below it, three sequences are identified by downward shifts in onlap. The upper parts of these sequences display erosional truncation near the seafloor (Fig. 5.28C). The basal surface of S4 is a very strong reflector, suggesting that it is the product of prolonged exposure and/or erosion. Strata beneath this surface could be considerably older. The three sequences on the Trinity Peninsula margin are similar in geometry to the youngest three sequences on the South Shetland shelf. Sequence S1 forms a "wedge" at the shelf margin, clearly showing a downward shift in onlap. Below, two older sequences onlap acoustically opaque basement.

The three youngest sequences (S1 through S3) on the Trinity Peninsula and South Shetland margins are similar in geometry (Fig. 5.28C) and are probably correlative. These three sequences represent sedimentation as the two margins evolved in parallel after rifting of the Bransfield Basin (Jeffers and Anderson, 1990). The basal surface in the fore arc (base of S4) probably represents a tectonically enhanced unconformity. The unconformity reflects subsidence of the fore arc related to either the slowing of spreading and subduction, which took place ~ 6 Ma, or to the cessation of oceanic spreading and ridge-trench collision at ~ 4 Ma. The relative ages of the seismic sequences are consistent with the subduction-related hypothesis for the origin of the Bransfield Basin, in which back-arc rifting follows the cessation of Aluk Ridge subduction at the South Shetland Trench (Barker and Dalziel, 1983).

Sequence Stratigraphic Interpretation

Haq, Hardenbol, and Vail (1987) identified five third-order eustatic cycles in the Plio-Pleistocene, bracketed by sequence boundaries occurring at 0.8 Ma, 1.6 Ma, 2.4 Ma, 3.0 Ma, and 3.8 Ma. Although the depositional

Figure 5.30. High-resolution seismic profiles (A) PD91-21 and (B) PD86-11 across the Trinity Peninsula platform and adjacent central subbasin illustrate stratal patterns on the shelf, slope, and basin floor. The profiles extend down the axis of a glacial trough and across the adjacent trough mouth fan. (See Fig. 5.24 for profile location.)

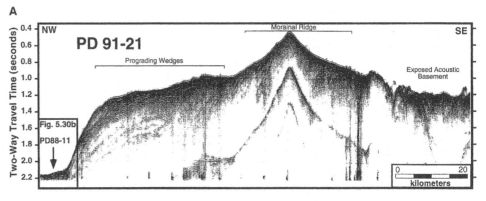

sequences in the Bransfield Basin have not been directly dated, if they are correlative with the global cycles of Haq and colleagues (1987), the chronology of the physical stratigraphic record of the Bransfield Basin is consistent with cessation of Aluk-Antarctic spreading at ~4 Ma and the formation of the basin between ~3 and 2 Ma (Jeffers and Anderson, 1990). Deposition of the Bransfield Basin sequence S4 would have occurred 2.4 to 3.0 Ma; the period of initial fore-arc subsidence after ridge-driven subduction stopped. The first sequence deposited on the rifted back-arc margin, S3, would have been deposited 1.6 to 2.4 Ma. Sequence S2, the first sequence deposited onto quasi-oceanic crust, would be younger than 1.6 Ma. Regardless of their absolute ages, the relative ages of the Bransfield Basin depositional sequences suggest that a period of fore-arc subsidence, possibly due to the cessation of ridge-driven subduction at the South Shetland Trench, predated the back-arc rifting (Jeffers and Anderson, 1990).

Because the Bransfield Basin is a relatively recent feature, deposition of the stratigraphic sequences that fill the basin occurred after ice caps supposedly formed in the region (Chapter 6). Figures 5.31 and 5.32 illustrate the two

Figure 5.30. (*Continued*)

B

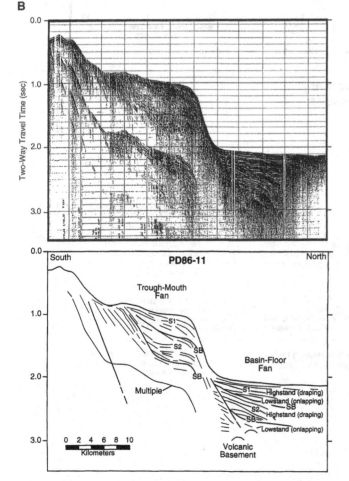

depositional systems active during the history of the Bransfield Basin.

Figure 5.31 shows the major depositional systems in the modern Bransfield Basin. They have been compiled by combining the surface sediment distribution pattern observed in piston cores and bottom grab samples with results from seismic facies analyses (Jeffers and Anderson, 1990; Banfield and Anderson, 1995). In total, these depositional systems compose a representative time-slice of the highstand/interglacial systems tract. During this interval, the basin is blanketed with siliceous hemipelagic muds, except on the shallow portions of the platforms and banks where strong bottom currents prevent deposition of these sediments. Sediment gravity flow processes dominate sedimentation on steep slopes.

The sedimentary regime in the Bransfield Basin changes dramatically during eustatic lowstands and glacial advances (Fig. 5.32). Marine ice sheets ground on the platforms and banks, eroding sediment of the previous highstand (Banfield and Anderson, 1995). Prominent grounding line ridges mark the position of the most recent grounding line (Figs. 4.18 and 5.30A). Troughs serve as flow conduits for ice streams. Troughs are first to experience erosion, and later, tills are deposited in the floors of these troughs (Banfield and Anderson, 1995). The sediment that is delivered to the troughs is transported downslope and accumulates in trough mouth fans and aprons at and near the base of the slope (Fig. 5.30B).

WESTERN WEDDELL SEA MARGIN

The continental shelf of the western Weddell Sea is one of the widest in Antarctica. It is bounded on the west by the mountains of the Antarctic Peninsula and to the east by the Weddell Abyssal Plain. Typically, sea ice covers the continental shelf throughout most of the year. Until recently, sea ice has prevented marine geophysical work (Anderson, Shipp, and Siringan, 1992b; Sloan, Lawver, and Anderson, 1995; Fechner and Jokat, 1995).

Geologic History and Tectonic Setting

The western Weddell Sea has experienced 175 myr of tectonic activity (Lawver et al., 1991). Prior to the breakup of Gondwana, the Antarctic Peninsula was on the leading edge of a subduction zone. The Weddell Sea was not in existence; the peninsula was situated offshore of the coast of Queen Maud Land and Coats Land (Fig. 2.5). Rifting between east and west Gondwana was marked partially by extension between the Antarctic Peninsula and East Antarctica, in addition to extensive volcanic intrusions in the TAM during the Jurassic. Counterclockwise rotational movement of the peninsula with respect to East Antarctica between 127 and 118 Ma resulted in separation of India and East Antarctica (Lawver et al., 1991).

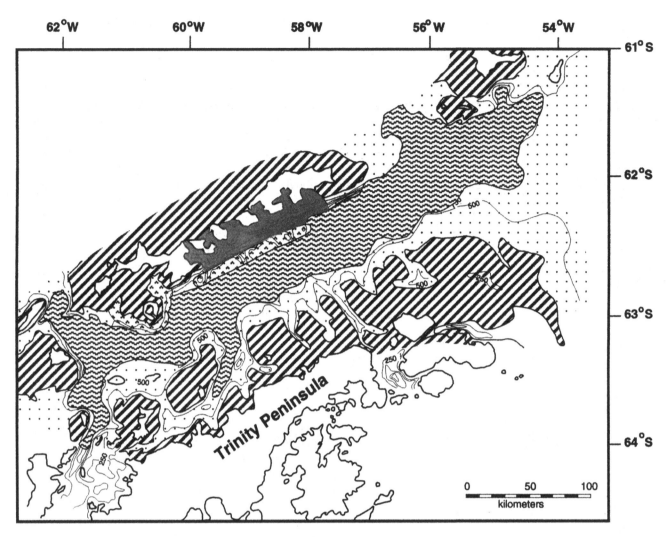

Sedimentary Environments	Sedimentary Processes	Lithofacies
Shallow Banks (<250m)	Intensely scoured by marine currents. (Current-related transport diminishes with depth).	Coarse gravel lags (residual glacial marine).
Slopes	Slumps, debris flows, turbidites and glacial marine muds are ubiquitous on the steeper slopes.	Muddy sands and sandy muds, with an overall fining offshore. Disorganized conglomerates and crudely graded gravels and sands on steeper slopes
Bays, canyons, and associated fan-like lobes	Glacially eroded sediment from the South Shetland Islands delivered to the bays by glacial meltwater is transported offshore and downslope to depositional lobes on the basin floor.	Terrigenous muds with occasional sandy horizons; basin floor sediments are enriched in biogenic material.
Central volcanic ridge	Vocaniclastic material is delivered directly downslope by lava flows and by passive settling through the watercolumn. Secondary downslope transport is by sediment gravity flow processes. Pelagic sedimentation dominates between eruptions.	Graded volcanic ash units interbedded with diatomaceous muds and oozes.
Basin floor	Biosiliceous sedimentation with increasing terrigenous influence near the basin margins and volcanic influence near the central ridge.	Ash-bearing diatomaceous muds and oozes. Total organic carbon may exceed 2%.

Figure 5.31. Depositional model for highstand–interglacial interval based on seismic stratigraphic analysis of high-resolution seismic reflection profiles and piston cores from the Bransfield Basin (from Jeffers and Anderson, 1990; reprinted with permission of the American Association of Petroleum Geologists). Bathymetry in meters.

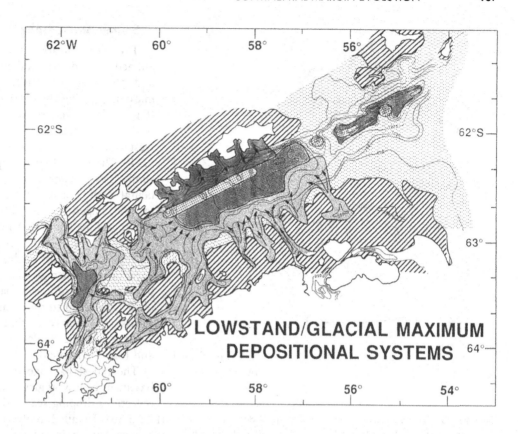

Figure 5.32. Depositional model for lowstand–glacial interval based on seismic stratigraphic analysis of high-resolution seismic data from the Bransfield Basin (from Jeffers and Anderson, 1990; reprinted with permission of the American Association of Petroleum Geologists).

	Sedimentary environments	Sedimentary Processes
	Shallow banks	Glacial erosion and deposition of basal tills.
	Slopes	Sediment eroded from the shelf progrades basinward where canyons or troughs are not present.
	Bays, canyons, and associated fan-like lobes	Sediment from the bays is glacially eroded and transported offshore and downslope to depositional lobes on the basin floor.
	Central volcanic ridge	Juvenile material is delivered directly by lava flows and by passive settling through the watercolumn. Secondary downslope transport is by sediment gravity flow processes. Pelagic sedimentation dominates between eruptions
	Troughs and associated prograding wedges	Material eroded from the continent and shelf is carried through the troughs to prograding wedges on the slope.
	Basin floor	Biogenic sedimentation is sharply reduced. Turbidites on the basin floor are sourced from the trough-wedge complex.

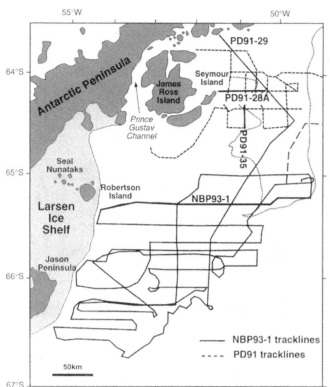

Figure 5.33. Geographic map of the northwestern Weddell Sea showing locations of seismic profiles used to conduct a seismic stratigraphic investigation of the area. Seismic lines illustrated in Figures 5.34, 5.35, 5.36, and 5.37 are indicated.

Throughout much of its geologic history, the Antarctic Peninsula was a paleomagmatic arc fronted on its western margin by accretionary complexes (Storey and Garrett, 1985) (Fig. 2.22). The arc, while in existence prior to the breakup of Gondwana, shows evidence of expansion during the Late Jurassic to Early Cretaceous. Continued ridge-crest and trench collisions during the Tertiary essentially terminated subduction (Herron and Tucholke, 1976; Barker, 1982; Macdonald and Butterworth, 1990). On the Weddell Sea side of the peninsula, a series of back-arc basins existed throughout the Late Cretaceous and Early Tertiary (Fig. 2.22). The sequence of geologic events in the region is recorded in the Larsen Basin, located on the northern portion of the shelf (del Valle, Elliot, and Macdonald, 1992), and the Lataday Basin, located on the southern portion of the shelf, which contain several kilometers of sedimentary strata (Traube and Rybnikov, 1990; Pirrie, Whitham, and Ineson, 1991).

There are no drill sites in the region, so the stratigraphy of the Larsen and Lataday basins is inferred from outcrops along the Lassiter, Orville, and Black coasts, in addition to exposures on James Ross, Seymour, and adjacent islands. In general, these outcrops consist of deep marine fan facies and slope-apron facies overlain by shallow marine shelf-deltaic facies (Macdonald and Butterworth, 1990; Pirrie, Whitham, and Ineson, 1991; Chapter 2).

Seismic Stratigraphy

The seismic stratigraphy of the Larsen Basin was examined by Anderson and colleagues (1992b) and Sloan and colleagues (1995) using intermediate-resolution seismic data collected during two separate cruises to the area (Fig. 5.33). Seismic records show a seaward-thickening wedge of sedimentary deposits prograding from northwest to southeast and onlapping acoustic basement (Fig. 5.34). The overall stratigraphic package shows broad similarities to packages on the Pacific-Antarctic continental shelf. For example, compare the stratal architecture imaged in Line NBP93-1 (Fig. 5.34) with that shown in Line USAP88-13 (Fig. 5.19C).

Highly tectonized acoustic basement subcrops throughout the inner shelf. The basement is poorly imaged, but in some regions the upper 100 to 200 msec displays a chaotic reflection pattern (Fig. 5.35). The basement is believed to represent the seaward extension of Jurassic volcanic rocks that are exposed on adjacent islands (e.g., Eagle, Andersson, Jonassen, Paulet, Dundee, and Joinville islands).

The oldest package of strata (S5) onlaps acoustic basement at low angles and is characterized by strong reflections dipping seaward to the southeast at 3 to 4° (Fig. 5.36). Detailed analysis has shown that S5 can be subdivided into at least five sequences (Sloan, Lawver, and Anderson, 1995). There is no seismic expression of glaciation within S5.

Units S5 and S4 are separated by a northeast-southwest-trending fault zone (Fig. 5.36). Individual faults are nearly vertical, and flower structures occur in the northern part of the survey area. The fault zone may extend along the entire western margin of the Weddell Sea, as indicated by a similar structural feature on the southwestern continental shelf (Jokat, Fechner, and Studinger, 1995). Sloan and colleagues (1995) interpret this as a strike-slip fault zone; however, there is evidence for considerable vertical displacement along this zone. Note that S5 strata are not deformed, which supports a strike-slip versus compressional origin for the fault zone. The seismic data show sediment wedges that onlap the faulted strata (Fig. 5.34), indicating that these are syntectonic deposits, probably fan-delta systems shed from the emerging Antarctic Peninsula. These fan-deltas occur within the lower portion of Unit S4. Unit S4 shows pronounced progradation on the continental shelf. Given the deformed nature of S4, it is not possible to discern glacial features within this package.

Units S4 and S3 are separated by a regional unconformity (U3) that possibly shows a landward-sloping profile (Fig. 5.34). If so, this is the oldest evidence of glaciation on the shelf (Anderson, Shipp, and Siringan, 1992b). Unit S3 is characterized by extensive seaward-dipping reflectors that display mainly aggradational geometries. This package is characterized by broad-scale cut-and-fill structure interpreted to indicate glacial erosion (Anderson, Shipp, and Siringan, 1992b; Sloan, Lawver, and Anderson, 1995).

NBP 93-1

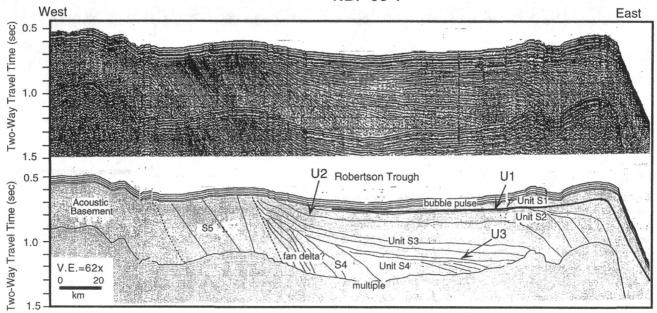

Figure 5.34. Uninterpreted and interpreted seismic Profile NBP93-1 showing seismic units on the western Weddell Sea continental shelf (modified from Sloan, Lawver, and Anderson, 1995). (See Fig. 5.33 for profile location.)

Troughs with widths on the order of tens to hundreds of meters and depths of tens of meters are common features within Unit S3. The troughs are infilled by seismically massive or chaotic deposits with clinoforms that indicate dominant progradation toward the southeast (Fig. 5.37). These features, taken in conjunction with the volume of sediment composing this sequence, support the interpretation that S3 is a glacial sequence.

The boundary between Units S3 and S2 is marked by a change from dominantly aggradation to dominantly progradation (Fig. 5.34). Otherwise, Units S3 and S2 show similar seismic facies and glacial erosion surfaces. The youngest widespread glacial unconformity on the shelf is designated U1 (Fig. 5.37). It represents a period of deep (hundreds of meters) erosion on the shelf, which contributed to the present overdeepened topography of the shelf. Unit S1 overlies U1 and contains prominent subglacial features, including glacial troughs and grounding zone wedges (Fig. 4.16C). On the inner shelf, massive and chaotic reflector patterns characterize the individual units (Fig. 5.37). These grade seaward into more stratified units on the middle and outer shelves, interpreted to be glacial-marine deposits (Anderson, Shipp, and Siringan, 1992b).

Sequence Stratigraphic Interpretation

The oldest stratigraphic package in the Larsen Basin (Unit S5) shows seismic attributes that are indicative of fine-grained sediment accumulation on a subsiding shelf. The probable age of this package, based on along-strike correlation with strata on Seymour Island, is Middle Eocene through Middle Oligocene (Anderson, Shipp, and Siringan, 1992b). There is no seismic evidence of glaciation on the shelf during S5 time, which is consistent with

outcrop data from Seymour Island, indicating a relatively warm, temperate climate at this time (Askin, 1992).

Deposition of S5 strata culminated with an episode of tectonic activity and associated fan-delta deposition (Unit S4) within the western portion of the Larsen Basin. The timing of this tectonic event is uncertain, but it most likely occurred during the Oligocene. The S4 and S3 boundary is a regional unconformity (U3; Fig. 5.34) that is tentatively interpreted as a glacial erosion surface. This unconformity was perhaps formed during the Middle Miocene, when ice sheets first advanced on the northern Antarctic Peninsula continental shelf (Bart and Anderson, 1995). This was followed by numerous advances of the ice sheet onto the

Figure 5.35. Representative segment of seismic Profile PD91-29 illustrating seismic character of acoustic basement, which is inferred to be Jurassic and younger volcanic and volcaniclastic deposits (modified from Anderson et al., 1992). (See Fig. 5.33 for profile location.)

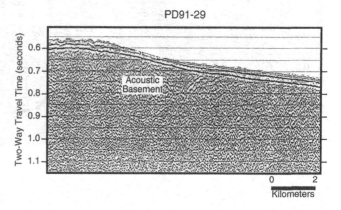

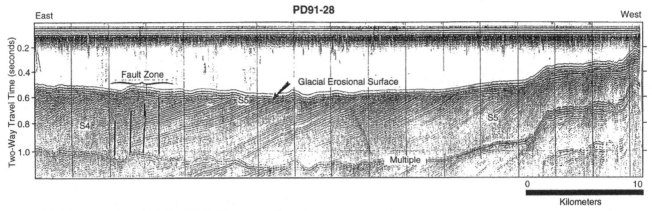

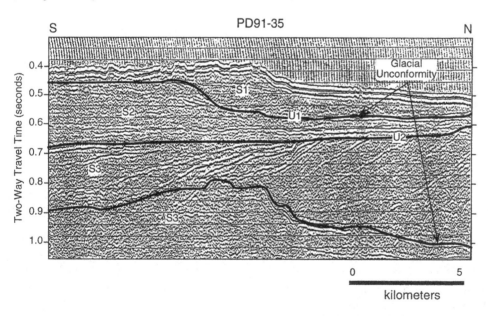

Figure 5.36. Representative seismic line (PD91-28) segment illustrating seismic Units S5 and S4 of the Larsen Basin. Sequence S5 is characterized by strong subhorizontal reflections that dip seaward at 3 to 4°. Its approximate age is Middle Eocene through middle Oligocene. This line segment also shows the fault zone that bounds Units S5 and S4 (modified from Anderson et al., 1992b). (See Fig. 5.33 for profile location.)

continental shelf. A shift from dominantly progradational strata to dominantly aggradational strata in the upper part of the section is correlated to a similar change in stratal geometry on the northern Antarctic Peninsula shelf (Fig. 5.19). There, this change from progradation to aggradation is attributed to higher-frequency sea-level oscillations during the Plio-Pleistocene (Bart and Anderson, 1995).

Only a few seismic profiles have been acquired in the southern Weddell Sea. These profiles show a relatively

Figure 5.37. Representative segment of seismic profile (PD91-35) showing broad-scale cut-and-fill structure of the continental shelf that is indicative of glacial erosion and deposition. The clinoforms are interpreted as grounding zone wedges that prograded oblique to the plane of the seismic section toward the southeast (modified from Anderson et al., 1992b). (See Fig. 5.33 for profile location.)

flatlying stratigraphic succession on the south-central shelf and a highly tectonized section on the southwestern shelf (Fechner and Jokat, 1995).

EASTERN WEDDELL SEA MARGIN

The Fimbul and Riiser-Larsen ice shelves rim virtually all of the eastern Weddell Sea margin and cover most of the narrow continental shelf. A discontinuous chain of mountains follows the trend of the Queen Maud Land (also referred to as Dronning Maud Land) coast ~100 km inland (the Ritscher Upland, Mühlig-Hofmann, Drygalski, Wohlthat, Humboldt, and Ør Rondane mountains).

Geologic History and Tectonic Setting

Prior to the breakup of Gondwana, the eastern margin of India/Sri Lanka and the southeastern margin of Africa were adjacent to the eastern Weddell Sea and Queen Maud Land (Fig. 2.5). Interaction among these plates was complicated, but the large-scale pre-breakup geology of the Antarctic margin resembles that of its Indian/Sri Lankan and South African counterparts. Post-breakup stratigraphy reflects increasing dissimilarities as the continents drifted farther apart (Chapter 2).

Separation of India and Antarctica began between 130 and 118 Ma (Johnson, Powell, and Veevers, 1976; Katz, 1978; Lawver and Scotese, 1987). Determination of the timing is difficult, because the magnetic anomalies off the Queen Maud Land margin remain unidentified (Lawver et al., 1991). In the Cauvery, Palar, and Godavari-Krishna basins of eastern India/Sri Lanka, pre- and early rift phase deposits consist of non-marine to marine sandstones, shales, and limestones (Sahni, 1982). Following separation, these marginal basins filled with non-marine to shallow marine sandstones, shales, and limestones (Sahni, 1982). By the Late Cretaceous, the continents were sufficiently separated, and sedimentation on their margins differed significantly.

Initiation of separation between eastern and western Gondwana began in the Permian with the development of a system of normal faults along the eastern margin of Africa (de Wit et al., 1988). Eventually, this fault system developed into a rift, and the separation of Africa and Madagascar/Antarctica began. The Karoo series, composed of thick non-marine sediment, was deposited in the north-trending basins created along the east African margin. Rocks of similar age and composition crop out in the Ellsworth and Pensacola mountains of Antarctica (Chapter 2). The margins of Africa and Antarctica underwent transtensional shearing in the Mid-Jurassic to Earliest Cretaceous (170 to 130 Ma; Fig. 2.6). By Late Jurassic time, the eastern Weddell Sea was more than 400 km wide (Leitchenkov, Miller, and Zatzepin, 1996). By the Early Cretaceous (115 Ma), oceanic crust formed between Antarctica and its conjugate margins, virtually completing the separation of Africa and Antarctica. The eastern Weddell Sea margin experienced passive margin conditions as the continents drifted apart.

The oceanic crust that underlies the northern part of the Weddell Sea is characterized by northwest-southeast- and northeast-southwest-oriented gravity anomalies that are interpreted as short offset fracture zones resulting from the separation of Antarctica and South America (Bell et al., 1990; Livermore and Woollett, 1993) (Fig. 2.11). Seismic records from the northern Weddell Sea show a thick acoustically laminated stratigraphic package that thins abruptly to the north, passing first into a region of isolated seamounts and then onto the rugged flanks of the America-Antarctic Ridge (Fig. 5.38A,B). The southern Weddell Sea opened as the Ellsworth-Whitmore Mountains block rotated away from the Antarctic Peninsula between 175 and 155 Ma (Grunow, 1993). The southern Weddell Sea was subsequently subjected to subduction beneath the Antarctic Peninsula and Thurston Island blocks.

At ~75° S, the Weddell Sea continental shelf break bends sharply to the west (Fig. 5.38). There are dramatic differences in the structure and stratigraphy of the margin north and south of this boundary. To the north, the continental shelf is narrow, the continental slope is quite steep, and the continental rise ends abruptly at the Explora Escarpment (Hinz and Krause, 1982) (Fig. 5.38). Haugland and Kristoffersen (1986) proposed that the Explora Escarpment marks the plate boundary between the Antarctic continent and the Weddell Sea. This portion of the continental margin is believed to be a sheared margin related to rotation of West Antarctica away from East Antarctica and the opening of the Weddell Basin (Chapter 2).

South of ~75° S, crystalline basement, the roots of the TAM, occupies the inner shelf (Fig. 5.38), and seismic records show an abrupt thickening of strata that occur along the rifted shoulder of the TAM in the vicinity of the Thiel Trough (Roquelpo-Brouillet, 1982; Elverhøi and Maisey, 1983; Kamenev and Ivanov, 1983; Haugland, Kristoffersen, and Velde, 1985; Miller et al., 1990a). The boundary between these continental margin provinces is marked by a narrow, deep basin that Kamenev and Ivanov (1983) call the Brunt Megatrough (Fig. 5.38). This feature is undoubtedly related to the early rifting phase of the Weddell Basin evolution.

Lithostratigraphy

ODP Leg 113 drill sites provide chronostratigraphic and lithostratigraphic information about the seismic units observed on the eastern Weddell Sea continental margin and abyssal floor (Fig. 5.39). Core recovered from Site 692 consisted of Early Cretaceous (Aptian to Albian?) nannofossil claystone with organic carbon content averaging 8.6% (Barker, Kennett, and Shipboard Scientific Party, 1988a). Shipboard analysis of these claystones indicates a strong resemblance to similar-age deposits of the Falkland Plateau and further indicates deposition under weakly aerobic conditions in water depths of 500 to 1000 m. The Cretaceous deposits are overlain by gravels of unknown age (canyon deposits) and Miocene to Pleistocene muds. Diatoms from ODP Site 692 place the age of canyon-cutting at this site as early Late Miocene or earlier (Barker, Kennett, and Scientific Party, 1988a).

Core from Site 693 penetrated 398 m of Pleistocene to lower Oligocene hemipelagic muds resting unconformably on lower Cretaceous claystone, mudstone, diatomite, and silicified sandstone (Barker et al., 1988a). The results of the shipboard analyses indicate that the lower Cretaceous mudstones are the oxidized equivalents of the Aptian mudstones recovered at Site 692.

Site 694 was drilled on the central abyssal plain of the Weddell Sea and contains a record of Weddell Fan evolution. Sandy turbidites dominate the Plio-Pleistocene portion of the section and are interbedded with hemipelagic sediments of early Late Miocene to Early Pliocene age. The fan did not reach this portion of the abyssal plain until sometime in the Middle Miocene.

The only information concerning the probable age of the deposits underlying the continental shelf comes from

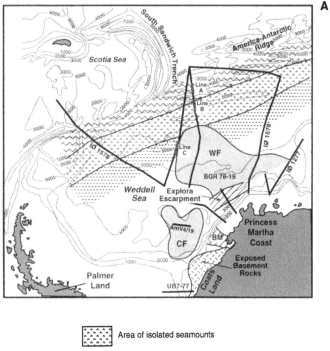

A

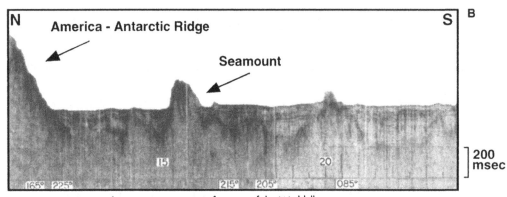

IO 1578 - A

B

Figure 5.38. (A) Map showing major tectonic features of the Weddell Sea. Also shown are locations of Islas Orcdas 1277 and 1578 seismic profiles, which are used to map basement features and seismic Profiles BGR78-19, ANTV/4-19, and UB7-77, which are used to illustrate continental margin structure and stratigraphy. WF = Weddell Fan; CF = Crary Fan; BM = Brunt Megatrough. (B) Islas Orcadas 1578 seismic profile illustrating the change from seamounts and ponded strata in the northern Weddell Sea to the rugged basement exposures along the southern flank of the America-Antarctic Ridge.

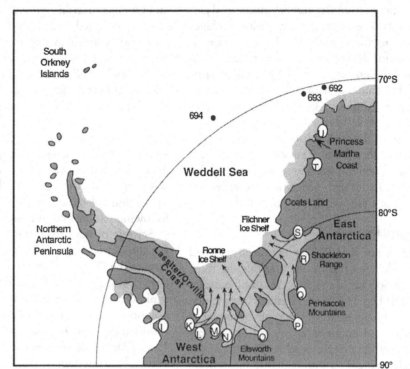

Figure 5.39. Core lithologies from ODP Sites 692, 693, and 694 from the Weddell Sea. (modified from Barker, Kennett, and Shipboard Scientific Party, 1988a).

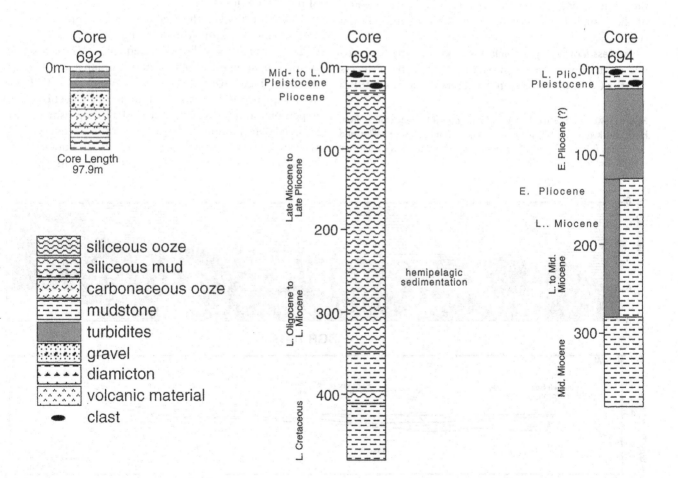

reworked palynomorphs from Quaternary glacial and glacial-marine sediments. These deposits contain assemblages dominated by marine dinoflagellates, which indicate an origin in mostly marine deposits of Aptian through Cenomanian age (Truswell and Anderson, 1984). These same deposits contain abundant sedimentary rock clasts that include quartz arenites (Andrews, 1984). The sedimentary clasts are believed to have been eroded from the continent and deposited on the shelf during expansion of the ice sheet (Anderson et al., 1991a). The occurrence of these clasts indicates the presence of sedimentary basins, similar to the Karoo Basin of southern Africa, located either on Queen Maud Land or the adjacent continental shelf.

Seismic Stratigraphy

More than 30,000 km of MCS data have been acquired in the eastern Weddell Sea region, making this the most thoroughly studied portion of the East Antarctic margin. Unfortunately, ODP Leg 113 to the Weddell Sea yielded only meager chronostratigraphic information for sequence stratigraphic analysis of continental margin strata. Because the character of the margin north and south of the Brunt Megatrough changes so dramatically, the seismic stratigraphy of these areas is discussed separately. The area north of the Brunt Megatrough is hereafter referred to as the Princess Martha Coast Margin, and the area south of the trough is referred to as the Coats Land Margin (Fig. 5.39).

Princess Martha Coast Continental Margin. The first seismic stratigraphic investigations offshore of the Princess Martha Coast were conducted by Hinz and Krause (1982)

Figure 5.40. (A) Seismic Line BGR78-19 and (B) interpretation from Hinz and Krause (1982) illustrating major seismic units of Miller and colleagues (1990b; Table 5.3) of the eastern Weddell Sea continental margin. (Location given in Fig. 5.37.)

and Okuda and colleagues (1983). Hinz and Kristoffersen (1987) examined seismic data acquired on different German and Norwegian expeditions to the eastern Weddell Sea margin. They mapped several major unconformities on the shelf. Miller and colleagues (1990b) conducted a seismic stratigraphic analysis that included higher-resolution data and led to the recognition of more seismic units than previously identified by Hinz and Kristoffersen (1987). Miller and colleagues (1990b) revised Hinz and Krause's (1982) seismic stratigraphic interpretation of sequences based on results from ODP Leg 113 drilling on this part of the continental margin. The interpretations of Miller and colleagues (1990b) are briefly described in the following paragraphs and are tabulated in Table 5.3.

Seismic profiles from the Princess Martha Coast margin show a package of seaward-dipping reflectors (W1), termed the "Explora Wedge" by Hinz and Krause (1982) (Fig. 5.40). Sequence W1 is interpreted to consist of volcanic deposits of the early rifting stage, based on extrapolation to similar features on other margins and drilling results (Hinz and Krause, 1982; Hinz and Kristoffersen, 1987). The wedge extends into the southern Weddell Sea and eventually connects with a 150-km-wide rift zone that approximately parallels the eastern Weddell Sea shoreline. Volcanics associated with the rift zone yielded age dates of the Middle Jurassic.

Unconformity U2, the Weddell Sea Unconformity (Fig. 5.40), separates sequences W1 and W2 and is believed to be a Late Jurassic horizon. Sequence W2 consists of continuous subparallel reflectors that onlap in a landward direction (Fig. 5.40). Unconformity U3 separates W2 from W3. Sequence W3 consists of hummocky reflections and hyperbolae and shows evidence of shale diapirism and associated deformation (Miller et al., 1990b). Sequence W4 is characterized by weak, subparallel discontinuous reflections and many diffractions. The seismic units situated

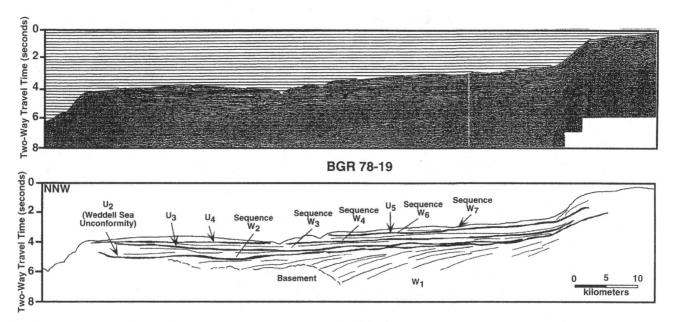

above U5 are very different in character from the older seismic units. Sequences W5, W6, and W7 display continuous, parallel reflections. Sequences W5 and W6 are truncated by unconformity U7 and draped by W7.

Coats Land Continental Margin. Unfortunately, only generalized line drawings and short line segments have been published from the area offshore of Coats Land (Elverhøi and Maisey, 1983; Haugland, Kristoffersen, and Velde, 1985; Miller et al., 1990b). To my knowledge, no attempt has been made to work out the seismic stratigraphy of the shelf using these data. The data show several undeformed progradational sequences that lap onto acoustic basement in an onshore direction (Haugland, Kristoffersen, and Velde, 1985). Seaward of the area of exposed acoustic basement lies a thick westward-dipping sequence of strata that is partially exposed in the Crary Trough. Seismic reflection profiles acquired across the southern part of the Crary Trough show this sequence (Elverhøi and Maisey, 1983; Ivanov, 1983; Haugland, Kristoffersen, and Velde, 1985; Jokat, Fechner, and Studinger 1995) (Fig. 5.41).

Table 5.3. Seismic stratigraphic units of the eastern Weddell Sea continental margin

Unit	Unconformity	Age	Interpretation
W 7		Plio-Pleistocene	Hemipelagic and glacial-marine
	U7		
W 6		Late Miocene-Pliocene	Hemipelagic and glacial-marine
	U6		
W 5		Late Miocene	Diatomaceous mudstone and nannofossil-rich mudstone
	U5		
W 4		Early to Late Oligocene to Early to Mid-Miocene	Hemipelagic sediments
	U4		
W 3		Early Cretaceous	Carbonaceous mudstone
	U3		
W 2		Late Jurassic to Early Cretaceous	
	U2		
W1		Late Jurassic	
	U1		Weddell Sea Continental Margin Unconformity

Seismic stratigraphic units from Miller et al. (1990).

Two sonobuoy measurements from the trough indicate sediment thickness of ~2 km near the axis of the trough and 4.5 km in its western flank (Haugland, 1982). According to Ivanov (1983), this westward-dipping sedimentary sequence may approach 10 km in total thickness west of the Crary Trough.

Haugland and colleagues (1985) found that there is little evidence of large tectonic features in the Crary Trough, which indicates that the strata within the trough postdate the final phase of uplift of the TAM in this region. The younger strata on the shelf are cut by large-scale erosion surfaces, believed to be glacial troughs, but the age of these surfaces is unknown.

Fairly detailed seismic stratigraphic work has been conducted on the continental slope and rise offshore of the Crary Trough; this work has focused on the evolution of the Crary Fan (Roquelpo-Brouillet, 1982; Haugland, Kristoffersen, and Velde, 1985; Miller et al., 1990b; Kuvaas and Kristoffersen, 1991; Moons et al., 1992; De Batist et al., 1994; Bart, De Batist, and Miller, 1994). Miller and colleagues (1990b) recognized four major fan sequences, all of which downlap or onlap the U4 unconformity. A more recent seismic stratigraphic and facies analysis of combined Norwegian and German-Belgian data by De Batist and colleagues (1994) and Bart and colleagues (1994) has led to the recognition of three major channel–levee complexes within the Crary Fan and mapping of feeder canyons on the Coats Land Margin. The canyons show multiple stages of fill (Fig. 5.42A). The fan consists of an axial channel with widths of several tens of kilometers and levees that are up to 1500 m thick (Fig. 5.42B). These systems show several stages of fill and gradual seaward progradation of the fan with time. The early stages of channel fill were characterized by deposition of thick, acoustically chaotic facies, interpreted as being dominantly composed of debris flow deposits (Bart, 1998). This stage of fan evolution is believed to mark widespread mass wasting on the continental shelf by ice sheets. The latter stage of fan evolution is characterized by channel–levee systems that display seismic attributes identical to other fan systems of the world. The younger channel–levee complex reflects more point-sourced sediment supply relative to the older chaotic units. This change toward more focused sedimentation may have resulted from the evolution of the Crary Trough and associated ice streams.

The base of the oldest channel–levee complex has not been dated, but the base of the middle complex was tentatively assigned a Middle to Early Miocene age (De Batist et al., 1994). The upper channel–levee complex, which includes the Deutschland Canyon, rests on the U5 unconformity of Miller and colleagues (1990b) and implies that this episode of fan development began in the Late Miocene.

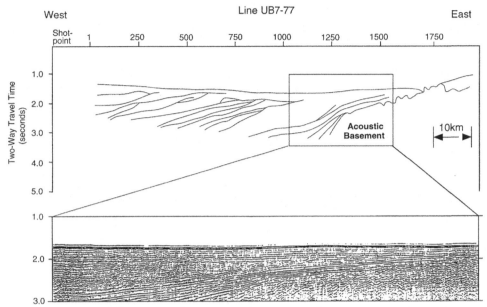

Figure 5.41. Seismic Profile UB7-77 and interpretation showing exposed basement in the eastern flank of the Crary Trough and westward-dipping units situated above the basement (modified from Haugland, Kristoffersen, and Velde, 1985). (Location shown in Fig. 5.37.)

Sequence Stratigraphic Interpretation

Sequences W1 through W3 represent the early evolution of the Weddell Basin, from Late Jurassic through Early Cretaceous time (Hinz and Krause, 1982; Miller et al., 1990b). Sequences W1 and W2 are believed to be composed of volcanic and volcaniclastic deposits shed from the continent during its separation from Africa and probably show similarities to Jurassic and Early Cretaceous strata of South Africa (Chapter 2). The top of W3 was sampled at ODP Site 693 and consisted of Early Cretaceous (Aptian to Albian) organic mudstone (Fig. 5.39). A prominent unconformity (U4) separates W3 and W4; the missing section spans Early Cretaceous through Early Oligocene time. Sequence W4 consists of nannofossil-bearing clayey mudstone and diatomaceous mudstone and is Early to Late Oligocene to Early to Middle Miocene in age (Barker et al., 1988a,b).

Unconformity U5 represents a period of erosion on the continental rise that spans much of the middle Miocene and corresponds to the initial influx of sandy turbidites at Site 694 on the Weddell Abyssal Plain (Fig. 5.39). This was probably the time when the Wegener Canyon and other canyons on the eastern Weddell Sea margin were formed. Directly above the U5 unconformity at Site 693 on the continental rise are nannofossil and diatomaceous mudstones of Late Miocene age (W5). Sequence W6 is believed to correspond to the Late Miocene to Early Pliocene hemipelagic sediments recovered in the upper half of Site 693 (Miller et al., 1990b) (Fig. 5.39). During the Early Pliocene, the more proximal (mid-fan) portion of the Weddell Fan shifted into the location of Site 694.

The Crary Fan is believed to have evolved as a result of the glacial erosion of the Crary Trough (Haugland, Kristoffersen, and Velde, 1985). This implies that glacial erosion was occurring on this part of the continental shelf sometime during the Late Oligocene to Early Miocene (Haugland, Kristoffersen, and Velde, 1985). The Crary Fan experienced episodes of major growth during the Middle to Late Miocene (De Batist et al., 1994), which was probably a time of deep glacial erosion on the continental shelf in the Weddell Sea.

WILKES LAND

The Wilkes Land margin is bounded entirely by the EAIS, which is grounded primarily at the coast (Fig. 5.1). A number of small ice shelves and several ice tongues dot the coastline. This portion of the continent is primarily composed of Precambrian cratonic rocks, although volcanic and sedimentary strata locally outcrop along the coast (Mawson, 1942) (Fig. 5.43).

Geologic History and Tectonic Setting

Prior to the breakup of Gondwana, the Australian and Antarctic continents were joined along what is now the Wilkes Land continental margin of Antarctica (Fig. 2.5). The general character of Mesozoic and Early Cenozoic strata that lie beneath the Wilkes Land shelf is inferred from descriptions of correlative strata in south Australian basins, and from rare grab samples and sediment cores that have sampled older strata on the Wilkes Land continental shelf (Chapter 2).

Initial studies of magnetic data from the Indian Ocean by Weissel and Hayes (1972) led to the conclusion that seafloor spreading between Australia and Antarctica began 55 Ma. Cande and Mutter (1982) reinterpreted

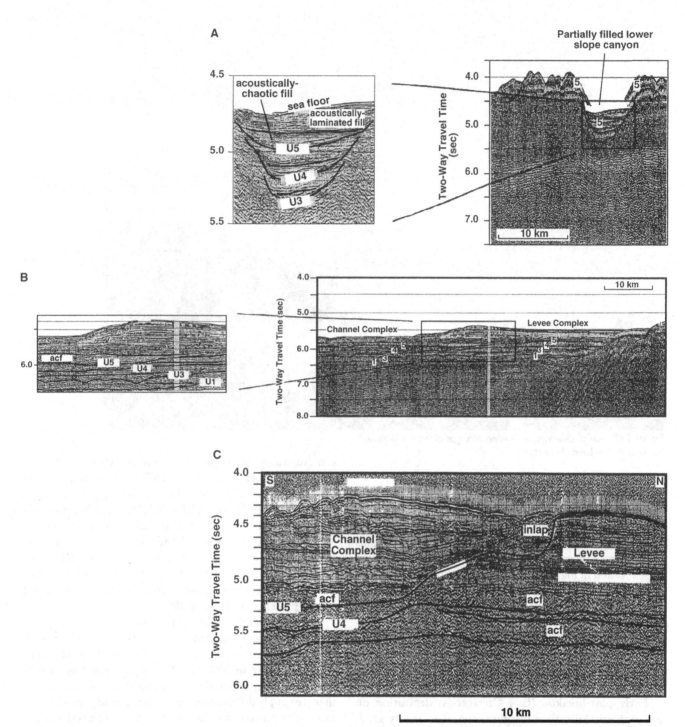

Figure 5.42. Portions of seismic profiles and interpretations from the Crary Fan illustrating (A) canyon with multiple fill, (B) channel-levee complex, and (C) seismic facies within the channel-levee complex (from Bart, 1998). (Data provided courtesy of Dr. H. Miller.)

these data and concluded that initial rifting between the two continents began between 110 and 90 Ma, whereas Veevers (1986) constrained the time of breakup to 95 ± 5 Ma. Actual breakup was preceded by a long period of extension and rifting, which Falvey (1974) refers to as the "Rift Valley Phase." This phase probably began in the Mid-Jurassic (Veevers, 1987a,b). During this time, rift basins formed in the vicinity of the incipient breakup axis. Initially, these basins were broad troughs that then developed into narrow fault-bounded basins as rifting progressed (Falvey and Mutter, 1981).

Subsidence rates on Australia's southern margin were high (0.1 km/myr average) during the rifting phase and slowed appreciably (0.01 km/myr average) during the post-breakup phase (Falvey and Mutter, 1981). Eittreim and Smith (1987) studied the subsidence history for the Wilkes Land margin. They employed the McKenzie (1978) crustal stretching and thinning model, backtracked the

Figure 5.43. Aerial photograph showing outcrops along the George V Coast of Wilkes Land, Antarctica.

rift-onset and rift unconformities to sea level, corrected for sediment compaction, and assumed a breakup time of 95 Ma. During the rifting phase, a thick sequence of predominantly continental deposits accumulated in the basins of southern Australia. Periodic marine invasions from the west resulted in marginal marine deposition in basins that now occupy the southwestern portion of the margin (e.g., Eucla Basin) during Aptian/Albian time (Deighton, Falvey, and Taylor, 1976). To the east, rift basins filled with continental deposits consisting predominantly of sandstone and mudstone. Up to 9 km of pre-breakup deposits occur in the Great Australian Bight Basin (Falvey and Mutter, 1981).

Early post-breakup (Late Cretaceous) deposition on the southern Australian margin is characterized by gradual marine deepening with deposition of a transgressive-regressive series of sands, silts, and muds (Deighton, Falvey, and Taylor, 1976). Deighton and colleagues (1976) present a paleogeographic model for post-breakup development of the Australian margin based on core and magnetic anomaly patterns (Fig. 5.44). Although their model uses 55 Ma as the initiation of breakup of the Antarctic and Australian continents based on the early work of Falvey (1974) and Weissel and Hayes (1972), it is still useful in the visualization of depositional environments through time. During the Paleocene and Early Eocene, marine transgressions were more widespread in south-

ern Australia, although these marine deposits are relatively thin (Boeuf and Doust, 1975). There is no record of carbonate or evaporite deposition during the early post-breakup interval. Truswell (1982) concluded that reworked Late Cretaceous–Eocene palynomorphs from the Wilkes Land and George V Land continental shelves indicate a setting similar to that of southern Australia at this time, characterized by predominantly continental deposition with occasional shallow marine incursions indicated by the presence of dinoflagellates.

The close parallel between the Antarctic and Australian continental margins ended with the onset of Middle Eocene to Early Oligocene carbonate deposition on the southern Australian margin (Falvey, 1974). By Eocene time, a narrow seaway separated the continents (Fig. 5.44), resulting in increasingly different climatic regimes and providing the moisture necessary for the formation of ice caps (Bartek et al., 1991). In Antarctica, the Eocene was a period of pronounced cooling and the initial development of an ice sheet on the continent (Chapter 6). It was also a time of destruction of land vegetation, indicated by decreased diversity of reworked palynomorphs of this age (Truswell, 1982).

Lithostratigraphy

Although the Wilkes Land continental shelf has not been drilled, there is evidence that Cretaceous deposits similar to those of the Southern Australian margin underlie the present shelf. Truswell (1982) conducted a palynologic analysis of samples acquired from the Wilkes

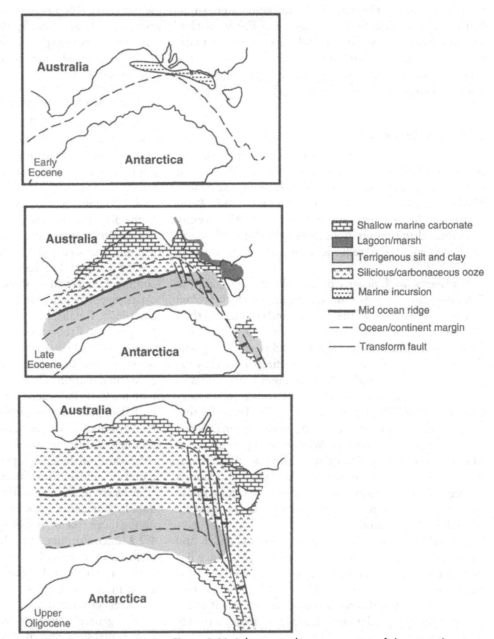

Figure 5.44. Paleogeographic reconstruction of depositional environments during breakup and dispersal of Antarctica and Australia (modified from Deighton, Falvey, and Taylor, 1976).

Land continental shelf by Douglas Mawson during the Australasian Expedition of 1911 to 1914. These samples include lignites that contain abundant pollen and spores ranging from Early Cretaceous to Early Tertiary in age. A Deep Freeze 79 piston core from the Mertz Trough, located offshore of George V Land, penetrated an outcrop of mudstone containing abundant pollen and spores of Aptian age, but lacking marine microfossils (Domack, Fairchild, and Anderson, 1980). Detailed analysis of the pollen and spore assemblage showed it to be identical to that of Early Cretaceous deposits of the Otway Basin of southern Australia (Domack, Fairchild, and Anderson, 1980). Glacial sediment sampled in piston cores from the northern flank of the Mertz Trough contain abundant quartz, stable heavy minerals, and sandstone clasts that

are probably situated at or near the seafloor on the inner continental shelf (Domack, 1987).

The only drill sites in the vicinity of the Wilkes Land continental margin are DSDP Sites 268 and 269 on the lower continental rise (Fig. 5.45). These sites yielded a record of hemipelagic sedimentation that was occasionally interrupted by turbidite deposition from the Oligocene through the end of the Miocene. During the Plio-Pleistocene, gravelly to sandy turbidites were deposited at DSDP Site 269 on the continental rise (Hayes and Frakes, 1975) (Fig. 5.45). These sands and gravels are similar in composition to glacial sediments acquired in piston cores on the adjacent shelf. This implies the turbidites were supplied to canyons directly by ice streams (Wright, Anderson, and Fisco, 1983). At Site 268, there was a dramatic increase in siliceous biogenic sedimentation during the Plio-Pleistocene, which implies cooling surface waters and more vigorous surface circulation (Hayes and Frakes, 1975).

Seismic Stratigraphy

Several seismic stratigraphic studies have been conducted on the Wilkes Land continental margin (Wannesson et al., 1985; Tanahashi et al., 1987; Eittreim and Smith, 1987; Hampton, Eittriem, and Richmond, 1987; Wannesson, 1990). Eittreim and colleagues (1995) recently re-examined the results from these different studies and correlated the major sequences and bounding surfaces (Table 5.4).

Continental Shelf. Seismic profiles show more than 6 to 7 km of sedimentary deposits resting on the block-faulted continental basement of the Wilkes Land continental shelf (Wannesson et al., 1985) (Fig. 5.46; Table 5.4). The oldest sequence (E) consists of syn-rift and post-rift deposits that are offset by faults. Sequence D directly overlies the WL4 unconformity and infills fault-bound minibasins in the WL4 surface. Sequence D attains a maximum thickness of 2.5 km (Wannesson et al., 1985). The reflectors are moderately discontinuous. The sequence pinches out in a landward direction onto WL4. The upper surface of this sequence is bounded by unconformity WL3. Overlying WL3 and pinching out in a landward direction is sequence C. This sequence also reaches a maximum thickness of 2.5 km. Internal reflectors are subparallel in nature and relatively continuous.

Unconformity WL2 separates sequences C and B and shows 300 to 600 m of erosion on the shelf. Sequence B is dominantly a progradational sequence. WL1 is restricted to the outer shelf and represents 350 to 700 m of erosion. Sequence A is also restricted to the outer shelf and is dominantly a progradational sequence, probably a trough mouth fan.

Continental Slope and Rise. Hampton, Eittreim, and Richmond (1987) conducted a detailed seismic facies and stratigraphic analysis of seismic profiles from the continental slope and rise off Wilkes Land. Their seismic stratigraphy was later modified by Eittreim, Cooper, and Wannesson (1995); however, the seismic facies interpretations of Hampton and his colleagues remains one of the most detailed analyses of continental slope and rise strata in Antarctica.

The most recent sequence on the continental slope and rise consists of massive slumps and debris flows interbedded with contourites. These grade seaward into channel-levee complexes. The upper sequence is between 1.5 and 2.0 km thick. Unconformity WL2 can be traced from the continental shelf to the continental rise, where it forms the base of this upper unit (Eittreim, Cooper, and Wannesson, 1995) (Fig. 5.46). Below the WL2 unconformity is an acoustically laminated sequence that is interpreted as being composed of hemipelagic sediments and fine-grained turbidites (Hampton, Eittreim, and Richmond, 1987).

Sequence Stratigraphic Interpretation

A comparison of Wilkes Land margin development with that of the Australian margin reveals that while the general structural and stratigraphic framework of each margin resembles the other, some differences exist (Boeuf and Doust, 1975; Falvey and Mutter, 1981; Mutter et al., 1985). The majority of these differences can be attributed to the different climatic regimes experienced by each continent. Both regions display evidence of relative tectonic stability and a period of Neogene shelf progradation following breakup. The Tertiary sequence of the Wilkes Land continental margin is thicker and contains more siliciclastic deposits than the Australian margin (Tanahashi et al., 1987). Tanahashi and colleagues (1987) note that the Early to Late Cretaceous sequences, interpreted as deposition in a broad rift basin and subsiding rift basin, respectively, are less deformed than Cretaceous deposits of the Australian margin. They correlate the WL3 unconformity with the Early Paleogene breakup unconformity described by Falvey and Mutter (1981) off the Australian margin. No carbonate buildups occur on the early Antarctic shelf (Hampton, Eittreim, and Richmond, 1987), while the post-rift Australian shelf displays numerous carbonate buildups dated as younger than Eocene (Boeuf and Doust, 1975).

A recent analysis of combined French and U.S. seismic data by Eittreim, Cooper, and Wannesson (1995) focused on the more recent stratigraphic section resting above the WL2 unconformity. The WL2 unconformity marks a change in the style of stratal geometry to greater progradation and a general absence of preserved topsets, which Eittreim and colleagues (1995) interpret to be caused by ice sheets grounding on the shelf.

The general seismic stratigraphy of the Wilkes Land continental slope and rise is similar to that of the southeast Pacific-Antarctic margin in that more proximal sediment gravity flow deposits appear to rest directly on more distal hemipelagic deposits. These stratigraphic relationships imply a dramatic seaward shift in more proximal facies that was probably forced by glaciation and mass wasting on the adjacent continent and continental shelf. The combined data indicate that the WL2 unconformity marks the onset of glaciation and mass-wasting on the Wilkes Land continental shelf at some time during the Oligocene (Eittreim, Cooper, and Wannesson, 1995; Escutia, Eittreim, and Cooper, in press).

There is a need for drill sites on the Wilkes Land continental margin to constrain the age of events seen in seismic profiles from the region. Also lacking from the region are intermediate- and high-resolution seismic data that would allow more detailed seismic stratigraphic analysis of the

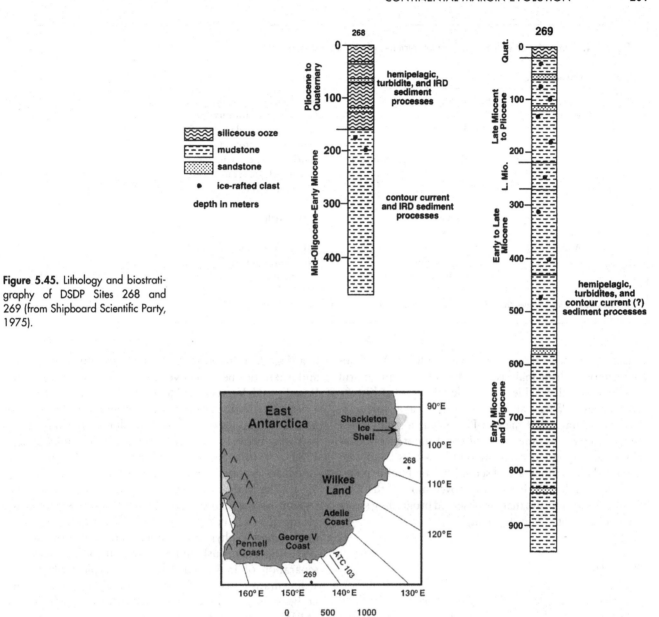

Figure 5.45. Lithology and biostratigraphy of DSDP Sites 268 and 269 (from Shipboard Scientific Party, 1975).

Figure 5.46. (below). Interpreted seismic reflection Line ATC103 showing major sequences and bounding unconformities (Table 5.4) of the Wilkes Land continental margin (modified from Eittreim, Cooper, and Wannesson, 1995).

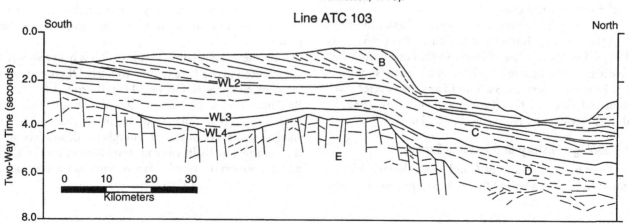

Line ATC 103

Table 5.4. Summary of Wilkes Land continental margin seismic stratigraphy

Unit	Unconformity	Age	Interpretation
A		Holocene	Subglacial and glacial
	WL1		Glacial Unconformity
B		Late Eocene to Mid Miocene	Subglacial and glacial marine
	WL2		Glacial Unconformity
C		Mid Eocene to Early Miocene	Marine on shelf, fan, and hemipelagic sediment on continental rise
	Wl3		
D		Mid Cretaceous to Mid Eocene	Fluvial deltaic
	WL4		Rift Unconformity
E		Early Cretaceous	Pre-rift terrigenous sediment and coal

Seismic stratigraphy compiled from Hampton et al. (1987), Wannesson (1991), and Eittreim et al. (1995).

glacial and glacial-marine strata that overlie the WL2 unconformity. Such an analysis is needed for selection of drill sites that would provide a more detailed glacial history for the region. What can be said given the available low-resolution data is that the Plio-Pleistocene aggradational strata observed on other parts of the Antarctic continental shelf are apparently missing on the Wilkes Land continental shelf, or they are obscured by the bubble pulse. This suggests that the ice sheet removed most of the Plio-Pleistocene section and that the observed prograding strata are predominantly Miocene in age.

PRYDZ BAY

Prydz Bay forms a deep embayment between Wilkes Land and Queen Maud Land (Fig. 5.1). The Ingrid Christensen Coast, with its numerous glaciers, bounds the southeastern side of the bay (Fig. 5.47). MacRobertson Land and Amery Ice Shelf, the fourth largest ice shelf of the continent, bound the southwestern edge. The Amery Ice Shelf is fed primarily by Lambert Glacier, the largest ice stream outlet in East Antarctica.

Prydz Bay lies within a sedimentary basin, the Prydz Bay Basin (Stagg, Ramsay, and Whitworth, 1983; Stagg, 1985). This basin is the offshore portion of a more extensive feature, the Lambert Graben, which extends some 700 km beneath Amery Ice Shelf and Lambert Glacier (Wellman and Tingey, 1976; Kurinin and Grikurov, 1982; Fedorov, Ravich, and Hofmann, 1982) (Fig. 5.47).

Geologic History and Tectonic Setting

Prior to the breakup of Gondwana, the eastern margin of India was adjacent to the Amery region of Antarctica (Fig. 2.5). Initial separation between India/Sri Lanka and Antarctica began between 130 and 118 Ma (Johnson, Powell, and Veevers, 1976; Katz, 1978; Lawver et al., 1991). The northeast-southwest orientation of the Prydz Bay Basin implies that the later evolution was influenced by rifting between Antarctica and India/Sri Lanka (Stagg, 1985). Grabens normal to the shelf edge (Mahanadi and Godavari grabens) underlie two of India's basins (Fig. 5.47). The Mahanadi Graben perhaps represents the extension of the Lambert Graben (Fedorov, Ravich, and Hofman, 1982). Formation of the Indian and Amery structures, whether physically connected or not, probably occurred simultaneously in response to the initial phases of separation of India and Antarctica. Typical rift-margin sedimentation occurred within the pull-apart basins and grabens. Early basin sedimentation consisted predominantly of Permian to Late Jurassic or Early Cretaceous continental sediments, based on studies of the Mahanadi Graben (Chapter 2).

Following the initial rifting phase, shallow marine clastics and carbonates were deposited on the Indian margin. By the Early Cretaceous (115 Ma), separation of India and Antarctica was virtually complete (Fig. 2.7), marking the inception of independent stratigraphies for these continents.

Continental glaciation in this region is believed to have commenced in the Late Eocene to Early Oligocene, based on core analyses (ODP Leg 119; Barron, Larsen, and Baldauf, 1989). Numerous glacial episodes have occurred since that time, resulting in large-scale erosion, stacked sequences of glacial–interglacial deposits beneath the continental shelf, canyon development on the slope, and deposition of complex fan systems on the continental rise (Cooper et al., 1990a).

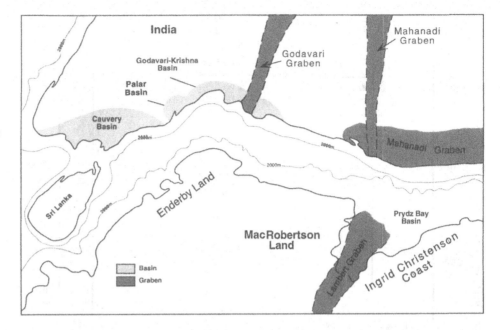

Figure 5.47. Basins and grabens of the India and Prydz Bay region (reprinted from Stagg, 1985, with permission from Elsevier Science Ltd.)

Lithostratigraphy

During ODP Leg 119, several sites were drilled on the continental shelf in Prydz Bay with the objective of establishing the onset of glaciation in this part of East Antarctica. Figure 5.48 provides lithologic descriptions and ages of the units sampled at these sites.

Site 740, drilled on the inner shelf, penetrated Mesozoic and possibly older alluvial deposits unconformably overlain by a glacial-marine sequence (Barron, Larsen, and Shipboard Scientific Party, 1989). The younger deposits consist of sandstone, siltstone, mudstone, and coal beds with abundant palynomorphs that yielded an Early Cretaceous (Albian) age (Turner and Padley, 1991). Diatomaceous muds and oozes cap the sequence. Site 741 recovered carbonaceous Eocene(?) deposits unconformably overlain by diamicton, thought to be till (Hambrey, Ehrmann, and Larsen, 1991) (Fig. 5.48). The cores from Site 742 contain an older sequence of glacial fluvial–lacustrine deposits of unknown age, which grade upward into diamictites and pebbly mudstones of subglacial and glacial-marine origin. The diamictons date back to the Middle(?)–Late Eocene (Barron, Larsen, and Shipboard Scientific Party, 1989b). The base of the glacial sequence was not sampled at Site 742 (Barron, Larsen, and Shipboard Scientific Party, 1989). Cooper and colleagues (1990) believe that it lies ~ 100 m below core bottom. This sequence is capped by conglomerates, sandstones, and mudstones, which presumably reflect marine sedimentation under less extreme climatic conditions. Till and glacial-marine deposits of Early–Late Pliocene age overlie this sequence and mark a return to more severe glacial/climatic conditions.

On the outer shelf, coring operations at Site 739 penetrated an Early Oligocene through Quaternary sequence almost 500 m thick, which consists of subglacial and glacial-marine deposits (Hambrey, Ehrmann, and Larsen, 1991). Site 743, located on the upper continental slope, recovered pebbly mudstone. The entire section is believed to be Quaternary in age.

Seismic Stratigraphy

The most extensively studied data set in the Amery region is the 1982 Australian Bureau of Mineral Resources MCS survey (Stagg, Ramsay, and Whitworth, 1983; Stagg, 1985). Leitchenkov, Shelestov, Gandjuhin, and Butsenko (1990) interpreted a multichannel data set gathered by the Soviet Antarctic Expedition between 1985 and 1988.

Stagg (1985) subdivides the shelf and slope suite into an Upper Series (PS1 and PS2) and a Lower Series (PS3, PS4, and PS5; Fig. 5.49). PS5 is at least 750 m thick and displays few reflectors. It is capped by an erosional surface, over which lie concordant reflectors of Unit PS4. This sequence is a minimum of 1,000 m thick and has parallel to slightly divergent geometries that display minor faulting. Its upper bounding surface is erosional. PS3 concordantly overlies or downlaps onto this surface. PS3 displays subparallel reflectors and minor faulting. Its upper boundary is an erosional surface. The Upper Series is divided into PS1 and PS2 (Fig. 5.49). PS2 displays complex sigmoid-oblique prograding reflectors near the shelf edge, is at least 750 m thick, and is bounded at its upper surface by a strong erosional truncation. Sequence PS1 is composed of stacked seaward-thickening strata that reach a maximum thickness of 300 m (Cooper et al., 1990a).

On the continental rise, sequence PD5 constitutes acoustic basement. Its internal reflectors are discontinuous and normally faulted. PD4 overlies the erosional surface between these two sequences. PD4 occurs in faulted

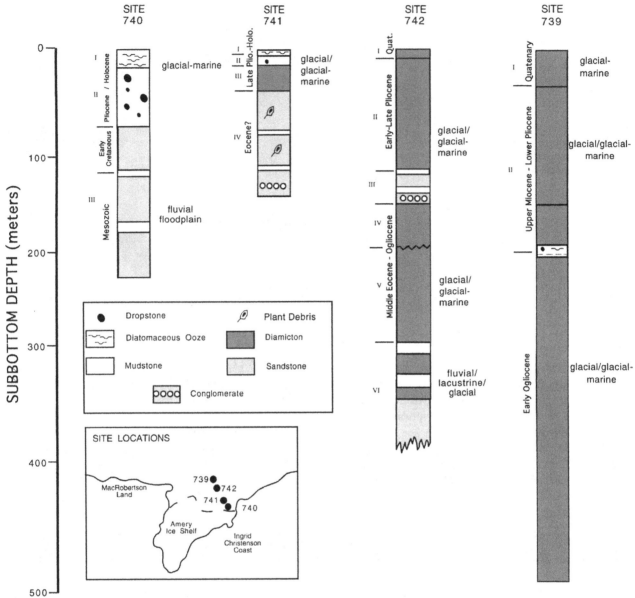

Figure 5.48. Core lithologies from ODP Sites 739, 740, 741, and 742 in Prydz Bay (from Shipboard Scientific Party, 1989).

depressions within PD5 and displays discontinuous reflections. Locally, it reaches a thickness of 500 m. Sequence PD3, a package of flat-lying reflectors, overlies PD4. Sequence PD2 displays poor reflector continuity and reaches a thickness of 1,500 m. An unconformity separates sequences PD2 and PD1. Sequence PD1 displays continuous reflections with evidence of slumping, erosion, and growth faulting. It is up to 1250 m thick.

Using intermediate-resolution seismic data collected during ODP Leg 119, Cooper, Stagg, and Geist (1991c) examined the younger sequence of the area. The overall stacking pattern of Neogene strata on the shelf is one that shifts from progradation to aggradation (Fig. 5.50). Cooper and colleagues (1991c) subdivide PS2 into two units, PS2A and PS2B. Unit PS2A is a seaward-dipping

unit that is relatively thin (400 m) in the mid-shelf region and thickens beneath the outer shelf to 4000 m.

Sequence Stratigraphic Interpretation

The probable stratigraphic succession of the Prydz Bay Basin can be inferred from information presented in Chapter 2. On the shelf and slope, Stagg (1985) interpreted a basal unit (PS6) to represent Cambrian and older cratonic metamorphosed basement. PS5 is a pre-breakup sedimentary package composed of tillite (Late Paleozoic glacial deposits), conglomerate, sandstone, siltstone, and coal. This equates to the lower unit of Leitchenkov and colleagues (1990). The erosional unconformity between PS5 and PS4 may represent the rift-onset unconformity initiated in the Early Permian-Triassic (Stagg, 1985). PS4 and PS3 were deposited during the period between rift onset in Early Permian and breakup of the margin at some point in the Early Cretaceous. The middle group of

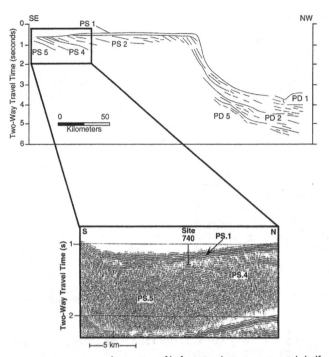

Figure 5.49. Interpreted seismic profile from Prydz Bay continental shelf and slope (reprinted from Stagg, 1985, with permission from Elsevier Science Ltd., and seismic line segment illustrating seismic Units PS5 and PS4 (from Cooper, Stagg, and Geist, 1991c).

Leitchencov and colleagues (1990) has a similar interpretation.

Cooper, Stagg, and Geist (1991c) interpreted the upper portion of PS2 as a glacial sequence (Unit PS2A). The reflectors change from gently dipping on the inner shelf to steeply dipping foresets on the outer shelf (Fig. 5.50). The age of PS2 ranges from Middle(?)–Late Eocene to Early–Middle Miocene and corresponds to the oldest lithologic evidence for glaciation on the continental shelf, Middle(?)–Late Eocene tills from Site 742.

The upper Miocene–lower Pliocene interval of ODP Site 739 must include portions of seismic unit PS1. These correlations indicate that the shift from dominantly progradation to aggradation of shelf strata occurred toward the end of the Miocene or beginning of the Pliocene (Cooper, Stagg, and Geist, 1991c); this record is similar to the overall stratigraphic signature observed elsewhere on the Antarctic continental shelf (Bartek et al., 1991).

The surface of the basal deep water unit of Stagg (1985), PD5, is thought to represent rift-onset (Early Permian–Triassic(?)) or, alternatively, margin breakup (Early Cretaceous). PD4 includes Triassic to Jurassic continental sediment if the upper surface of PD5 represents rift onset. If the top of PD5 defines the margin breakup, PD4 contains shallow water marine sediments with deposition initiated in the Early Cretaceous. PD3 may consist of volcanics emplaced as a relatively flat-lying sill in the Early Cretaceous. Units PD2 and PD1 represent marine sediments passing from shallow to deep water facies as sub-

sidence occurred along the margin between the Cretaceous and Recent. PD2 was assigned a Cretaceous age by Cooper, Stagg, and Geist (1991c). The unconformity between PD2 and PD1 was traced to ODP Site 739, thus assigning it an age of Early Oligocene to Late Miocene (Cooper et al., 1991c). Cooper and colleagues (1991c) attributed the formation of this unconformity to events that accompanied complete separation between Australia and Antarctica as well as the establishment of the Circum-Antarctic Current, thus refining the age further to the Middle Oligocene. They saw no evidence of canyon formation prior to development of the Oligocene unconformity and suggested canyon development commenced in the Late Oligocene to Early Miocene. PD1 displays a distinct fan morphology and extensive faulting, which indicates rapid deposition. Seismic records acquired during the Soviet Antarctic Expedition show excellent examples of large channel–levee complexes and sediment drifts (Kuvaas and Leitchenkov, 1992). Kuvaas and Leitchenkov (1992) suggest that the initiation of turbidite sedimentation occurred in response to glaciation and mass wasting on the continental shelf.

SUMMARY

The Antarctic continental margin shows a range of tectonic settings and evolutionary histories. Seismic profiles record dramatic changes in the style of continental margin stratal architecture that are associated with the onset of glaciation on the continent and continental shelf. The most notable evidence for glaciation are glacial unconformities on the continental shelf that record hundreds of meters of erosion and associated progradation of the shelf margin. This mass wasting on the continental shelf is manifested on the continental slope and rise as a dramatic change from dominantly hemipelagic sedimentation to dominantly sediment mass movement in the form of huge slumps, debris flows, and submarine fans, which, virtually everywhere around the continent, constitute a stratigraphic section that is over 1 km thick and represent many tens to a few hundred kilometers of continental rise progradation. In West Antarctica, this glacial influence on continental margin development began in the Miocene, mainly in the Middle Miocene, whereas it began in Middle(?)–Late Eocene to Early Oligocene time in East Antarctica. There was an episode of deep erosion of the continental shelf during the Middle Miocene that influenced both East and West Antarctica. More rapid advance and retreat of the ice sheets onto the conti- nental shelves during the Plio-Pleistocene are recorded in virtually every seismic record from the continental shelf.

There is a dire need for more core data from the continental margin to help constrain the timing of ice sheet grounding episodes. The cores need not be long, because glacial ice has cut deeply into the shelf, thereby exposing

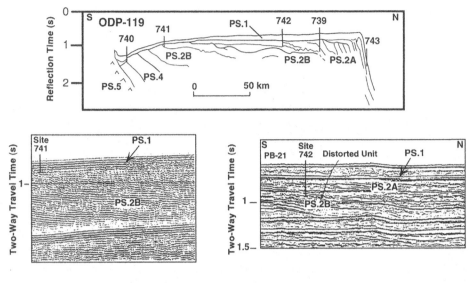

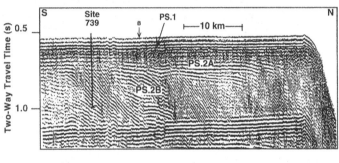

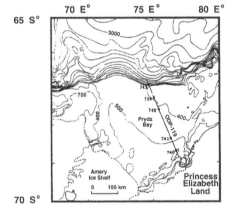

Figure 5.50. ODP-119 seismic reflection profiles illustrating seismic character of Units PS2B and PS2A and change from dominantly progradation in Unit PS2B to aggradation in Unit PS2A (from Cooper, Stagg, and Geist, 1991c).

older strata. Recovering relatively complete stratigraphic sections of Cenozoic strata is possible in some areas, but the advancing and retreating ice sheets have resulted in a cut-and-fill structure on the shelf and poor stratigraphic continuity. Careful seismic stratigraphic analysis using the best possible stratigraphic resolution is needed to piece together the stratigraphic record on the shelf and is essential to optimize future drilling efforts.

CHAPTER SIX

Antarctica's Glacial History

To many geologists, the term "Ice Ages" conjures up information learned in undergraduate classes about Northern Hemisphere glaciation during the Pleistocene. Yet, the Antarctic Ice Sheet appears to have been in existence at least an order of magnitude longer than the Northern Hemisphere ice sheets, as far back as the Mid–Late Eocene. Throughout most of the late Cenozoic, the Antarctic Ice Sheet has driven global eustasy and deep ocean circulation. It has played a crucial role in regulating global climate, particularly in the Southern Hemisphere (Barrett, 1991; Denton, Prentice, and Burckle, 1991; Flower and Kennett, 1994). Only in more recent geologic time has the Antarctic Ice Sheet passed into a secondary role in the global cryosphere. Equilibrated with the polar climate, it responds to the fall and rise of sea level resulting from waxing and waning Northern Hemisphere ice sheets. Still, if there is to be any significant increase in global sea level over the next several centuries, it will likely come from Antarctica.

The first explorers to Antarctica recorded signs that the ice sheet has a history of expansion and contraction. The Polish explorer Arctowski (1901) was struck by the spectacular glacial landforms as he sailed along the inland passage of the Antarctic Peninsula. He stated, "We must not imagine that the Antarctic lands are at the present day as heavily loaded with glaciers as they might be, for traces of a wider extension, dating doubtless from the glacial epoch, are still preserved" (Arctowski, 1901; p. 372). Later, Scott (1905) suggested that Ross Ice Shelf had once stood at the edge of the continental shelf.

Over a half century passed between these observations and the first efforts focused on unraveling Antarctica's complex glacial history. Even now, many unsolved mysteries about the history of the Antarctic Ice Sheet remain. When and where did it first form? When, if ever, did it achieve stability? What have been the driving forces causing it to expand and recede? How have fluctuations in the ice sheet impacted global eustasy, oceanography, climate, and life?

The intent of this chapter is to provide the reader with a review of what we know and do not know about Antarctica's glacial history. The literature is filled with contradictory information, and the evidence must be carefully weighed before reaching conclusions. To assist the reader in assessing the literature, the initial part of this chapter offers a brief description of the marine geologic and geophysical criteria used to interpret Antarctica's glacial history. A review of pertinent literature discussing the subject of Antarctic glacial history follows, with my assessment of these results. This exercise produces a reasonably coherent history of glaciation on the continent.

USING "PROXY" CRITERIA TO DEFINE ANTARCTICA'S GLACIAL HISTORY

Geologists have attempted to decipher Antarctica's glacial history from the marine sedimentary record since the first sediment cores were collected from the Subantarctic seafloor. The approach used has not proceeded in an ideal fashion. Most sediment cores collected during early marine geologic expeditions were acquired by ships with only limited ice-breaking capabilities; thus, coring efforts were typically conducted far from the continent. By far, the majority of these sediment cores (literally thousands) were collected by the USNS research vessel *Eltanin* (Fig. 6.1). As a result of this coring strategy, scientists were faced with interpreting the deep-sea ("proxy") record of glaciation without information about sedimentary processes occurring near the continent.

To complicate matters, drilling on the Antarctic seafloor has not always occurred in the best locations. The drill ships used, the *Glomar Challenger* and later the *Glomar Explorer*, were poorly suited for work in ice-covered waters. Southern Ocean drilling legs were often planned based on the passage of the drill ship through a region en route to lower-latitude site locations. Furthermore, sites were often planned using sparse and poor-quality seismic records, with a poor understanding of the stratigraphy of a region, and with little information on which to develop biostratigraphic zonations. To date, ten expeditions have drilled core from the Southern Ocean: DSDP Legs 28, 29, 35, 36, and 71, and ODP Legs 113, 114, 119, 120, and 178 (Fig. 6.2). Only three of these cruises, DSDP Leg 28, ODP Leg 119, and ODP Leg 178, recovered core from the Antarctic continental shelf. Thus, the sampling

Figure 6.1. U.S. research vessel *Eltanin*.

strategy in the Southern Ocean has profoundly influenced scientific research; there has been far greater emphasis on the deep-sea record than on the continental margin record.

Some of the criteria used to interpret the glacial record from deep-sea sediments include downcore changes in: (1) oxygen isotope (δ^{18}O) concentrations; (2) ice-rafted detritus; (3) sediment type; (4) clay mineralogy; (5) microfossil assemblages; and (6) the occurrence of hiatuses. The literature generated since the 1970s on the proxy record of Antarctic glaciation and paleoceanography is voluminous; no attempt is made here to summarize it in its entirety. What should be noted is that the results of these studies often are contradictory. Contradictions stem from the different methods used, different assumptions made, and inadequate chronostratigraphic control. Kennett and Hodell (1995) offer a recent review of the proxy evidence for Antarctic glacial history.

Oxygen Isotope Record

The most widely used proxy indicator of ice volume change is the oxygen isotope record, specifically, changes in the δ^{18}O signal with time. The problem with this method is that enrichment of δ^{18}O occurs in response to either an increase in ice volume or a decrease in seawater temperatures, and the two signals are not separated easily.

Salinity also influences the δ^{18}O concentration of seawater. A linear relationship exists between salinity and the isotopic composition of water, with a 1% salinity change resulting in 0.5% change in δ^{18}O of seawater (Broecker, 1989). Salinity changes might be significant in high-latitude regions where freshwater runoff varies significantly between temperate and polar settings (Poore and Matthews, 1984; Wei and Wise, 1990; Wise et al., 1992). Problems related to dissolution and diagenesis are not presented in this book; see Killingley (1983) for further discussion on these problems.

One possible way to circumvent the temperature versus ice volume dilemma is to use oxygen isotopic data derived from benthonic foraminifera. The assumption is that bottom water temperatures and salinities have not varied significantly; therefore, the ice volume effect dominates changes in δ^{18}O concentrations. Prentice and Matthews (1991), however, argue that deep water temperature variations have been sufficient to mask the ice volume signal because of opposite signatures for deep water temperature changes and ice volume changes. Another approach is to use planktonic foraminifera from tropical oceans. Matthews and Poore (1980) argue that reasonable ice volume information can be derived from tropical planktonic

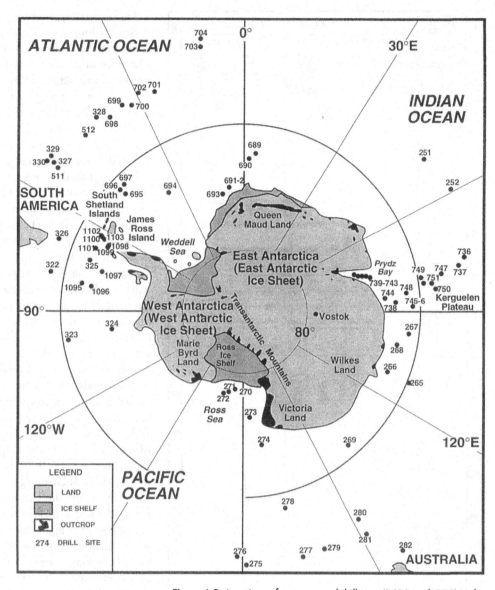

Figure 6.2. Locations of outcrops and drill sites (DSDP and ODP) in the Antarctic region.

curves due to the greater stability of tropical sea surface temperatures away from upwelling regions. This assumption has been supported by theoretical and empirical evidence.

Over the years, both methods have been employed in many drill sites from around the globe. The result is a composite oxygen isotope curve (Fig. 6.3) that, according to some authors (Matthews and Poore, 1980; Miller, Fairbanks, and Mountain, 1987b; Denton, Prentice, and Burckle, 1991; Wise et al., 1992; Abreu and Anderson, 1998), is a reasonable approximation of ice volume on earth. Although there is still some difference of opinion as to how the oxygen isotope record should be interpreted (the range of uncertainty spans nearly 30 Ma), there is a growing consensus that significant volumes of ice existed on Antarctica at or near the Eocene–Oligocene boundary (Fig. 6.4).

It is important to recognize that most Tertiary isotope curves lack sufficient resolution to show high-frequency variations such as the 41,000- and 19,000-year orbital periodicities (Denton, Prentice, and Burckle, 1991). This low resolution predominantly results from imprecise chronostratigraphic control, but the evolution of improved instrumentation and chronostratigraphic methodologies is resolving the issue. A reasonably good isotopic curve has been developed for the last 450,000 years (Fig. 6.5, Curve A). This curve is being increasingly accepted as a proxy ice volume curve for the Late Pleistocene. The curve shows good agreement with the $\delta^{18}O$ curve from Vostok ice cores of East Antarctica (Fig. 6.5, Curve B).

Global Sea-level Record

Sea-level changes provide the most direct measure of ice volume changes on earth. Contributions from Northern Hemisphere ice sheets to global eustasy commenced in the last 3 to 2 Ma and are manifested as high-frequency

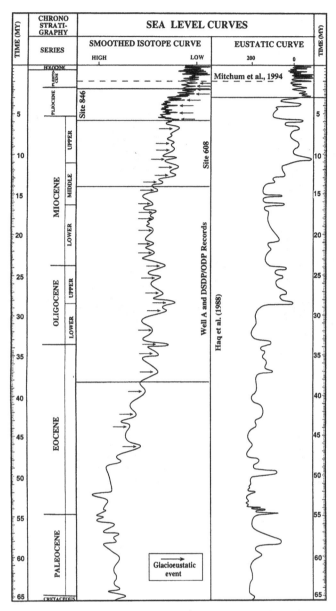

Figure 6.3. Composite, smoothed oxygen isotope (ice volume) curve based on measurements of benthic foraminifera from the Atlantic and Pacific oceans (from Abreu and Anderson, 1998) and sea-level curve of Haq and colleagues (1987) and Mitchum and colleagues (1994).

Unfortunately, methods used to reconstruct sea-level curves for longer time spans (beyond ~ 20,000 years) are imprecise. In addition, sea-level curves do not indicate where ice masses existed. By far, the best sea-level curve is that for the last glacial eustatic rise, which is accurate to within a few meters per year (Fairbanks, 1989) (Fig. 6.5, Curve C). Although we know far more about the distribution of ice on Earth during this period than other intervals of geologic time, debate focuses on which ice sheets contributed to sea-level changes of the last 20,000 years and when the contributions were made.

The ice volume component of the oxygen isotope record has been calculated to be 0.11 ppm for each 10 m of equivalent sea-level change (Fairbanks and Matthews, 1978), assuming the present isotopic composition of the Antarctic Ice Sheet. Using this calibration, the oxygen isotope curve (Fig. 6.5, Curve A) has been used as a proxy to sea-level changes extending back through several glacial–interglacial cycles. However, we must keep in mind that this approach does not account for temperature and salinity effects. Calibration of the oxygen isotope curve and sea level is still needed.

Detailed analysis of seismic records from around the world reveals what is believed to be a long-term eustatic record manifested in the onlap (coastal onlap), and off-lap of strata on continental shelves (Vail et al., 1977; Vail and Hardenbol, 1979; Haq, Hardenbol, and Vail, 1987; Mitchum et al., 1994) (Fig. 6.3). Lowstands are manifested by seaward shifts in deposition that typically result in thick shelf-margin deltas and fluvial erosion on the continental shelf to create sequence boundaries. As sea level rises, deposition shifts landward (coastal onlap) and the shelf is again subjected to erosion – this time by waves (shoreface ravinement). The end result is an overall stratal stacking pattern that records eustatic change but is an inaccurate gauge of the magnitude of eustatic rise and fall. The latter point is made clear by examination of the eustatic curve shown in Figure 6.3. The magnitude of sea-level change inferred from major Oligocene and Miocene sequence boundaries ranges up to 250 m, calibrated to the sea-level rise of the past 20,000 years (Fig. 6.5, Curve C); this calls for a doubling of global ice volume of the last glacial maximum. This is unlikely. Abreu and Anderson (1998) argue that $\delta^{18}O$ records more likely indicate sea level changes during the late Cenozoic in the range of 25 to 55 m. They further argue that the $\delta^{18}O$ curve is a more reasonable estimate of the magnitude of sea-level change than is derived from seismic stratigraphic analyses, and provide some reasons as to why the two methods of estimating global ice volumes differ. Perhaps the most significant problem with estimating sea-level change from seismic records has to do with the poor stratigraphic resolution of the seismic data.

Remember that the stratigraphic resolution of seismic data decreases with depth in the subsurface. This leads to

oscillations. Thus, the eustatic changes that occurred prior to this time reflect the period of Antarctic glaciation, provided the glacial-eustatic component of sea-level change can be distinguished from other eustatic controls, such as tectonics. It is generally assumed that eustatic changes of tens of meters are caused by either variations in the volume of ice sheets or volumetric changes in large-scale tectonic features on the seafloor (most important, ocean ridges). Glacial-eustatic changes of this magnitude can occur in tens of thousands of years, whereas tectono-eustatic changes of this magnitude require millions of years to occur.

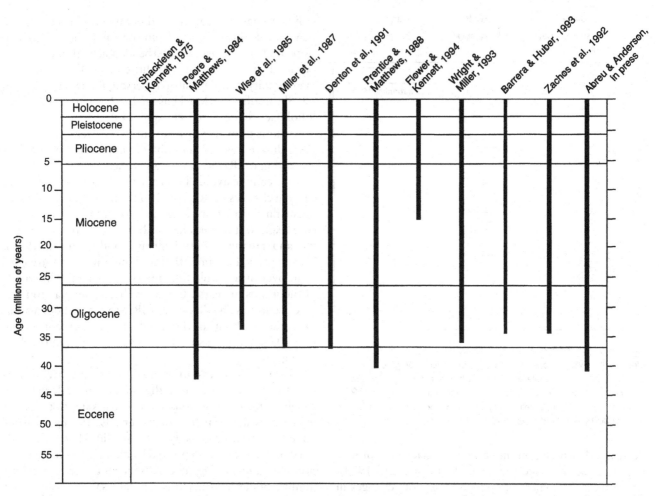

Figure 6.4. Interpretations of the oxygen isotope records vary widely with regard to Antarctica's glacial history. This figure shows the age of the Antarctic Ice Sheet based on the isotopic record from various investigators. There is an emerging consensus that the East Antarctic Ice Sheet reached continental proportions by the Latest Eocene-Early Oligocene.

poor estimation as to the exact degree of coastal onlap. Indeed, calibration of the $\delta^{18}O$ curves and high-resolution seismic data yields different results in different areas, due mainly to autocyclic effects (Thomas and Anderson, 1994). Furthermore, high-frequency eustatic events, such as those that occurred during the Late Pleistocene (Fig. 6.5, Curve A), produce changes in stratal stacking patterns that are beyond the limits of resolution of conventional seismic data. Therefore, high-frequency glacial eustatic cycles are not likely to be recorded in seismic records if they occur below the upper few hundred meters in the stratigraphic column.

Given the constraints described above, there still should be reasonable chronostratigraphic correlation between sea-level curves derived from seismic stratigraphic studies and ice volume curves derived from $\delta^{18}O$ analyses, but the magnitudes of sea-level changes that are inferred from these two methods differ. In general, there is reasonable agreement between the $\delta^{18}O$ curve and the sea-level curve at the third-order level (millions of years) for the Oligocene to Present (Miller, Fairbanks, and Mountain, 1987b, 1996; Abreu and Anderson, 1998) (Fig. 6.3). This implies that the combined isotopic and seismic stratigraphic

records of eustasy can be taken as a reasonable proxy for ice sheet evolution at this time scale.

Ice-Rafting Record

The concentration of ice-rafted debris (IRD) in deep-sea sediment has been used extensively to examine the glacial history of the Antarctic continent. A variety of assumptions have been made in these studies of the complex relationship between IRD concentration and ice expansion and contraction. Some authors contend that increased ice rafting occurs during glacial maxima (Conolly and Ewing, 1965; Goodell et al., 1968; Margolis and Kennett, 1971; Ledbetter and Watkins, 1978; Watkins, Ledbetter, and Huang, 1982). Others argue that increased IRD concentrations may reflect glacial minima (Frakes, 1983; Drewry and Cooper, 1981; Bornhold, 1983; Anderson, 1986; Kellogg, 1987). Some have argued that, during any single glacial event, ice rafting may either increase or

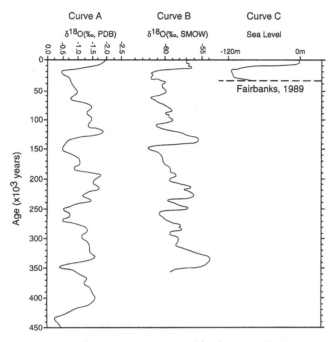

Figure 6.5. (A) The oxygen isotope record for the past 450,000 years (Prell et al., 1986) is believed to be a good proxy for global ice volume and shows reasonable correlation to the (B) oxygen isotope record obtained from ice cores at Vostok, East Antarctica (Lorius et al., 1985, Jouzel et al., 1987, and Petit et al., 1997). (C) Sea-level curve for the past 20,000 years (Fairbanks, 1989).

decrease, depending on the severity of glacial conditions and distance from the continent (Watkins et al., 1974). Given these differences of opinion, downcore changes in the concentration of IRD have the advantage of being excellent supporting data for any given researcher's paleoglacial scenario.

Another drawback to the IRD approach is that most deep-sea sediments, particularly the hemipelagic sediment that occurs around Antarctica, consist of a number of different components with flux rates to the seafloor that are virtually impossible to determine. Biogenic and fine-grained terrigenous sediments are most abundant; authigenic and IRD usually constitute only a small percentage (by volume) of the total sediment (generally < 10% by volume and commonly < 3%). A change in the relative concentration of any single component changes the relative concentration of the others, including the ice-rafted component. For example, variation in IRD within biogenic deposits may reflect changes in the flux of biogenic material to the seafloor, and this may be unrelated to ice volume on Antarctica. Unfortunately, IRD curves often are presented without information concerning lithologic changes in a core. Consequently, it is not possible to assess possible sedimentologic influences on IRD concentration.

The uncertainty in the flux rate of different sediment components to the deep seafloor increases closer to the continent. The amount of fine-grained terrigenous sediment bypassing the shelf and reaching the deep seafloor responds strongly to changes in glacial maritime setting,

sediment mass movement, and oceanographic circulation (Chapter 4). Piston cores from the Antarctic continental slope and rise recovered interbedded glacial-marine pebbly mud, laminated silt with little or no IRD, and sediment gravity flow deposits (Anderson, Kurtz, and Weaver, 1979b; Grobe and Mackensen, 1992; Chapter 4). These lithologic changes reflect significant differences in the downslope transport of sediment to the deep seafloor, and these differences are undoubtedly manifested in the concentration of IRD in deep-sea sediments.

One cannot avoid the problem of variable flux rates causing changes in IRD concentrations by calculating sediment flux rates from averaged accumulation rates in a core (e.g., Watkins et al., 1974; 1982; Watkins, Ledbetter, and Huang, 1982; Ledbetter and Watkins, 1978). This approach assumes that accumulation rates are constant over measurable time intervals (generally hundreds of thousands of years). In reality, there is no way to measure accurately the flux rate of IRD to the seafloor within time frames of tens of thousands of years, except for the past 30,000 years or so – the range of reliable radiocarbon dating.

The approach most often used to measure IRD in sediment cores is to wet-sieve the sample and calculate the weight percent terrigenous sand. The assumption that all terrigenous material that does not pass through a 63-μ screen is ice-rafted is likely to be invalid. However, sand-sized material can be deposited by alternative sedimentary processes and concentrated subsequent to deposition by a number of mechanisms. It is generally safe to assume that the grain size distribution of till on the continental shelf, which tends to be unsorted (Fig. 4.7), is indicative of the material being rafted by icebergs.

I have conducted grain size analyses on sand-sized terrigenous material from piston cores in the Weddell Abyssal Plain. The sand is composed of three distinct size distributions: (1) well-sorted, fine- to very fine-grained sand; (2) negatively skewed sand with the very fine-grained sand, silt, and clay components missing; and (3) poorly sorted sand. These different components are almost equally represented in any given core and reflect a combination of sedimentary processes, including iceberg rafting, eolian transport onto sea ice and dispersal by that means, and current winnowing and transport.

Even given a reliable estimate of IRD supply, there is considerable uncertainty of how variations in the flux rate relate to glacial conditions on the continent (Drewry, 1986; Anderson, 1986). One important factor concerns changes in the drift tracks of icebergs with time (Cooke and Hays, 1982; Wise et al., 1991). Chapter 1 contains a detailed discussion of the distribution of icebergs around Antarctica (Figs. 1.19–1.21). Clearly, the concentration of icebergs at any given distance from Antarctica varies around the continent and is related to major oceanographic and climatic fronts. Just how the drift tracks varied in response to changes in sea-ice distribution, water mass structure, and oceanographic and atmospheric circulation through time is unclear, but these variations have occurred.

Figure 6.6 is a map of the concentration of terrigenous sand and gravel in surface sediments of the southwest Pacific-South Atlantic region. No grain size analyses were conducted on these samples. Note that the concentration of coarse terrigenous sediment varies significantly on a regional scale. Only in the southeast Pacific is there a crude correlation to iceberg concentrations and distance from the continent (Fig. 1.19) or iceberg drift tracks (Fig. 1.21). The strongest correlation is with sediment type (Fig. 4.49). The highest IRD concentrations are associated with the South Pacific Scour Zone (SPSZ) (Fig. 4.66) and are therefore interpreted as lag deposits.

Given the problems associated with using downcore IRD concentrations to reconstruct glacial conditions on the continent, the results of previous studies employing this method should be viewed with caution. Detailed sedimentologic investigations are necessary; all sedimentary components should be examined and downcore variations in the components should be tied to $\delta^{18}O$ records and other chronostratigraphic data. Grobe and Mackensen (1992) successfully utilized this approach. Their analyses of continental margin sediment in the eastern Weddell Sea provides a credible Late Quaternary record of climate cycles for this region. One of their observations is that ice rafting in this region is greatest during the transition from an interval of peak warmth to a glacial episode. Also, Anderson and Andrews (in press) have demonstrated that the IRD concentrations in the western Weddell Sea are highest at times when glacial-marine sedimentation was occurring on the adjacent continental shelf, not when ice sheets were grounded on the shelf.

IRD has also been used to indicate onset of continental glaciation by its first occurrence in deep-sea cores. This criterion has been used widely to interpret DSDP and ODP sedimentary records. The trick is to agree on the first "significant" appearance of IRD in cores; history has shown that one individual's significant occurrence is another person's sparse occurrence. In addition, the first occurrence of IRD typically does not correspond to the earliest evidence for glaciation based on other criteria (as discussed in the following section). Figure 6.7 shows the occurrence of the earliest IRD in DSDP and ODP cores from around Antarctica. The first occurrence varies by millions of years around the continent and even within the same region.

Increased Bottom Current Activity

The deep-sea sedimentary record contains major widespread hiatuses, which are typically recognized when magnetostratigraphic and biostratigraphic data are combined. These hiatuses often are interpreted as episodes of increased bottom water production and flow velocity (Watkins and Kennett, 1972, 1977; Ciesielski and Wise, 1977;

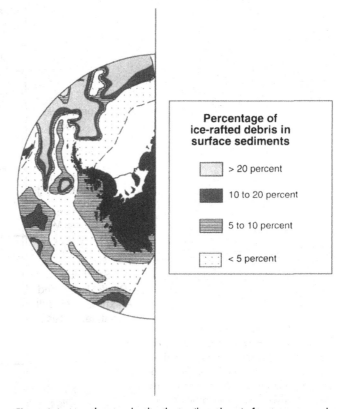

Figure 6.6. Map showing the distribution (by volume) of terrigenous sand and gravel (ice-rafted debris; IRD) in surface sediments of the South Pacific and South Atlantic sectors of the Southern Ocean.

Ciesielski et al., 1978; Kidd and Davies, 1978; Ledbetter and Huang, 1978; Ledbetter, 1979, 1981; Ciesielski, Ledbetter, and Ellwood, 1982; Ledbetter and Ciesielski, 1982, 1986; Ledbetter et al., 1983; Wright and Miller, 1993). These studies have also interpreted stratigraphic changes in sediment grain size and particle alignment as evidence for variations in bottom current velocity. One of the important arguments made in these studies is that the unconformities and sedimentologic changes are correlative over vast areas. Denton, Prentice, and Burckle (1991) questioned the synchroneity of many of the deep-sea hiatuses described in the literature, citing poor biostratigraphic control of the surfaces. Other problems exist as well.

Silt grain size was used as a gauge for bottom current velocity (Ledbetter and Johnson, 1976; Huang and Watkins, 1977; Ledbetter, 1979; Ledbetter and Ellwood, 1980; Ledbetter, 1981; Ledbetter and Ciesielski, 1982, 1986; Ledbetter, 1986). Anderson and Kurtz (1985) critiqued the use of mean grain size of fine-grained sediments as a paleovelocity gauge; they list a number of factors that influence the size distribution of fine-grained abyssal sediment. Among these are deep-sea storm events, winnowing during settling, variations in the rate of bioturbation (which renders bottom current winnowing more effective), and variations in the flux of different fine-grained components to the seafloor. Still, paleovelocity data from some areas, such as the Argentine Basin (Ledbetter, 1986) and

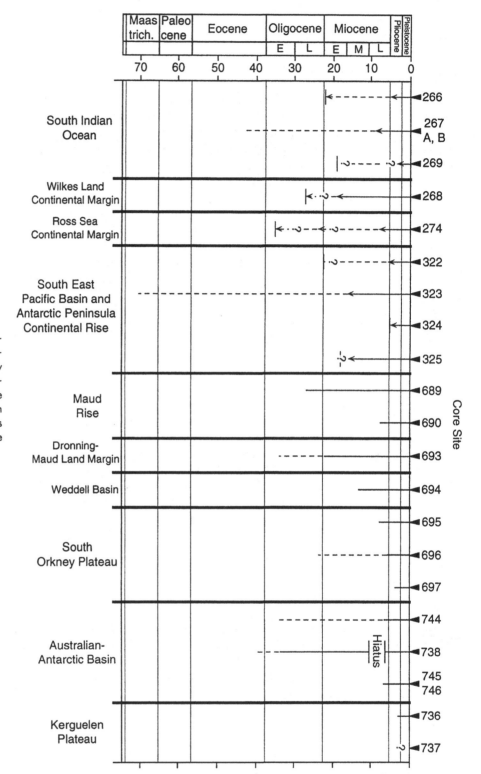

Figure 6.7. Earliest occurrence of ice-rafted debris in DSDP and ODP Southern Ocean drill sites. Dashed lines show minor occurrences, solid lines show "significant" occurrences. The first occurrence varies widely around the continent, even within the same areas. These variations are due to geographic location, distance from the continent, and other factors.

the northern Weddell Sea (Pudsey, Barker, and Hamilton, 1988), have yielded meaningful results. The question is, how are these paleovelocity changes related to glacial conditions on Antarctica?

It is important to be aware of the assumptions that are commonly made in relating bottom current paleovelocity changes to glacial conditions on the continent. These include assumptions that: (1) hiatuses in the deep-sea sediment column are caused by high-velocity bottom currents; (2) increased bottom current velocity translates to increased bottom water production rates; (3) the bottom water mass responsible for the hiatuses on the abyssal floor is AABW; (4) freezing at the underside of ice shelves produces the high-salinity shelf water needed to make AABW;

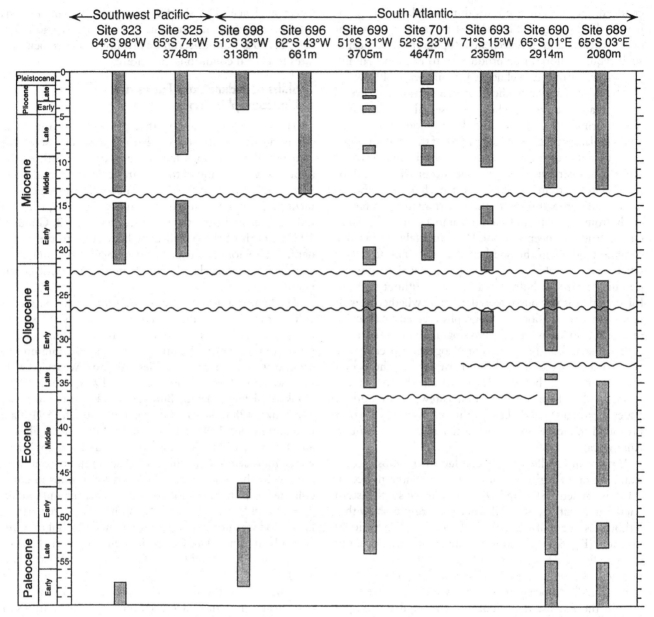

Figure 6.8. Stratigraphic occurrence of Cenozoic hiatuses in the Southern Ocean. Shaded areas indicate recovered core (modified from Wright and Miller, 1993).

therefore, increased production of AABW implies the presence of large ice shelves somewhere in Antarctica (this includes the added inference that the climate was polar); and (5) large ice shelves are associated only with the WAIS.

The conclusion reached via this chain of assumptions is that deep-sea hiatuses in the Southern Ocean mark periods of expansion of the WAIS, an interpretation that has been challenged by several authors (Anderson, 1986; Corliss, Martinson, and Keffer, 1986; MacDonald and Anderson, 1986).

The assumption that AABW production requires the presence of ice shelves is poorly supported by physical oceanographic studies. Gill (1973) and Foster and Carmack (1976) have argued that seasonal sea-ice production creates most of the brine needed to form saline shelf water, an important ingredient for AABW production (Chapter 1). This implies that there may, in fact, be an

inverse relationship between AABW formation and the extent of glacial ice or perennial pack ice on the continental shelf. Ice shelves and perennial pack ice occupy space on the continental shelf where seasonal sea-ice production might otherwise occur (Anderson, 1986). An analysis of fine-grained sediment on the Weddell Abyssal Floor by Pudsey, Barker, and Hamilton (1988) led to the conclusion that bottom currents are weaker during glacial periods, which seems to be consistent with the current mode of AABW formation.

On a long-term scale, it is important to recall that ice shelves are polar features and may not have been present throughout much of Antarctica's glacial history. Furthermore, the process of AABW formation has undoubtedly changed, and production rates diminished, as the continental shelf was glacially scoured to create the present overdeepened and foredeepened topography. A deeper shelf requires more brine production to raise the salinity of shelf water to the level needed to mix with CPDW.

The deep seafloor of the Southern Ocean has several widespread scour zones where deep-sea hiatuses are presently forming (Fig. 4.64). The distribution of the scour zones indicates that the physiography of the seafloor, especially the relief of fracture zones, is a dominant control of scour zone distribution; scour zones occur where seafloor features restrict current flow. Clearly, changes in the relief of seafloor features with time exert a strong influence on bottom currents and have an impact on their capacity to erode the deep seafloor. Unfortunately, knowledge of plate motions in the Southern Ocean is not at a state where we can reliably reconstruct the relief of physiographic barriers to bottom water flow. Without such information, it is not possible to determine whether ice volume changes or seafloor physiography created a particular hiatus. We do know that the physiography of the seafloor changed significantly during the Neogene, especially the South Pacific sector of the Southern Ocean, the Scotia Sea, and the area between Tasmania and Victoria Land (Chapter 2). These are important gateways for bottom water circulation, and their evolution must have had a profound influence on bottom water circulation around Antarctica.

Wright and Miller (1993) conducted the most recent examination of regional hiatuses in the Southern Ocean. They examined DSDP and ODP sites in the southeastern Pacific and southwestern Atlantic. Their results show that although some hiatuses are localized, four are regional in extent (Fig. 6.8). The most prominent hiatuses occur in the Middle–Late Eocene, at the Eocene–Oligocene boundary, near the Oligocene-Miocene boundary, and near the Early–Middle Miocene boundary. A more localized surface in the Early–Late Oligocene is recognized. Wright and Miller (1993) point out that these surfaces correspond to times of $\delta^{18}O$ increases, which implies that the hiatuses formed when ice sheets expanded. They attribute hiatus formation to increased bottom water production and stronger bottom water flow during glaciations. Conversely, Corliss and colleagues (1986) conclude that AABW circulation "does not show a simple relationship with paleoclimatic oscillations, indicating that changes in oceanographic conditions in the Southern Ocean had little effect on AABW formation" (p. 1106).

In summary, studies of the paleovelocity of deep-sea currents have produced questionable results. This is perhaps why Denton, Prentice, and Burckle (1991) argue that paleovelocity data should be rejected when evaluating Antarctica's glacial history. I believe that widespread hiatuses, such as those identified by Wright and Miller (1993), may be related to glaciation on the continent; however, these data should be considered in light of other, more direct evidence. The shorter-term paleovelocity records were omitted from my analysis of Antarctica's glacial history, because I am uncertain as to how paleovelocity relates to glacial conditions on Antarctica.

Shifts in Sedimentary Facies and Biogeographic Zones

The cold surface waters that presently surround the Antarctic continent are recorded in the surface sediment as a broad biosiliceous facies that extends north to the approximate position of the Antarctic Polar Frontal Zone (Fig. 4.49). North of this zone, calcareous biogenic sediment dominates at water depths above the CCD, whereas red clay occurs below the CCD (Lisitsyn, 1972; Goodell, 1973). South of the biosiliceous belt, close to the continent, is a region of terrigenous sediment, predominantly silty diatomaceous clay derived from the Antarctic continent (Chapter 4).

The main planktonic microfossil groups (diatoms, coccoliths, radiolarians, and planktonic forams) show strong biogeographic zonation around Antarctica. Many studies illustrate the temporal shifts in microfossil communities around the continent (e.g., Ciesielski and Weaver, 1974; Lozano and Hays, 1976; Kennett, 1978; Haq, 1980; Cooke and Hays, 1982; Burckle, 1984; Wise, Gombos, and Muza, 1985; Barron, Larsen, and Baldauf, 1991; Barrera and Huber, 1993; Lazarus and Caulet, 1993; Flower and Kennett, 1994). Radiolarians and diatoms appear to be the most sensitive to shifting oceanographic fronts and tend to be the best preserved microfossil groups in cores collected near the Antarctic continent (Lazarus and Caulet, 1993; Burckle et al., 1996). Coccolithophores are particularly good sea surface temperature indicators, but they are typically absent in late Cenozoic sediments near the continental margin (Burckle and Pokras, 1991; Wise et al., 1991).

Variations in the terrigenous sediment-biogenic silica belts near Antarctica have been related to fluctuations in sea-ice cover (e.g., Lazarus and Caulet, 1993). However, the effect of variable flux of terrigenous sediment from the continent and associated masking of biogenic material also should be considered. The latter effect is clear from modern distribution patterns of abyssal sediment around Antarctica showing major deviations between the sedimentary facies boundaries and the average position of the ice front (Fig. 4.49).

Cooke (1978) and Cooke and Hays (1982) demonstrate that the boundary between biosiliceous and terrigenous facies corresponds approximately with the average limit of summer sea ice around Antarctica. Lozano and Hays (1976), Cooke (1978), and Cooke and Hays (1982) used these sedimentologic facies to demonstrate that the Polar Front shifted northward a maximum of 6° during the Stage 2 glacial maximum. Burckle, Robinson, and Cooke (1982) argue that the sedimentary facies boundary between siliceous ooze and silty diatomaceous clay more

accurately reflects the spring limit of sea-ice cover. Burckle (1984) and Burckle and Cirilli (1987) determined that
the spring sea-ice limits in the Southern Ocean during the
last glacial maximum were 3 to 8° farther north than the
present limits. While these authors disagree as to whether
sea ice or surface circulation is the dominant influence
on distribution of biosiliceous sediments in the Southern
Ocean, their combined results indicate that biosiliceous
sedimentation shifts northward when the Antarctic climate becomes more severe, but not necessarily when the
ice sheet expands.

Ciesielski and Weaver (1974) observed that the siliceous
biogenic facies shifted northward during the Late Neogene, indicating progressive cooling of surface waters. Results from more recent ODP legs in the Southern Ocean
indicate that the northward shift of biosiliceous sediments
began in the Latest Eocene to Earliest Oligocene (Barron,
Larsen, and Baldauf, 1991; Kennett and Barker, 1990;
Barrera and Huber, 1993; Lazarus and Caulet, 1993).
Again, these authors caution that this change in the style
of deep-sea sedimentation does not necessarily indicate
significant buildup of ice on Antarctica.

In summary, latitudinal shifts in sedimentary facies and
biogeographic zones with time reflect changing surface
water temperatures and oscillations in the position of major oceanic fronts. These shifts do not provide direct evidence for Antarctic ice volume change. Wise and colleagues (1985), Lazarus, Pallant, and Hays (1987), and
Burckle and colleagues (1996) offer more thorough reviews of the subject.

CONTINENTAL AND CONTINENTAL SHELF RECORD OF ICE SHEET EVOLUTION

The most direct evidence for Antarctic glaciation exists on
the continent and continental shelf. The continental record
of Antarctic glaciation includes raised trimlines, erratics,
drift deposits, striated bedrock, polar desert pavements,
tills, raised marine terraces, and subglacial volcanic deposits. Denton and colleagues (1991) review these works.
The continental shelf record holds much promise because
it contains direct evidence for subglacial sedimentation
and erosion in the form of glacial unconformities, tills,
and grounding zone–proximal geomorphic features (e.g.,
grounding zone wedges). These deposits and features are
interbedded with glacial-marine strata that contain fossils
for biostratigraphic analysis. Our ability to obtain ages for
these deposits is improving. Unfortunately, few drill sites
targeted the Antarctic continental shelf (Fig. 6.2). To date,
only the Ross Sea, Prydz Bay, and Antarctic Peninsula
continental shelves have been drilled. Yet, these shelf site
areas provide valuable records that dramatically altered
prior conceptions of the extent and timing of glaciation
on Antarctica. Our understanding of Antarctica's glacial
history grows tremendously with each properly located
drill site on the continental shelf; additional drill sites are
necessary for expanding our understanding.

Sedimentologic Evidence of Glaciation

Glaciation of the Antarctic continent must have occurred in a series of steps as the climate shifted from temperate to subpolar and eventually to polar. This climatic
evolution was undoubtedly punctuated by shorter-term
climatic events (Milankovich Cycles). Climate has a strong
impact on sedimentation on the continental shelf; therefore, sediment facies can be used to reconstruct climate
history. Anderson and Molnia (1989) and Anderson and
Ashley (1991) listed the main differences between temperate and polar–subpolar glacial-marine environments and
deposits. The most significant difference is that polar and
subpolar ice sheets and glaciers ablate almost exclusively
by iceberg calving, whereas temperate glaciers ablate almost exclusively by melting. This difference is manifested
in the sedimentary record. Antarctica contains only localized meltwater deposits in Quaternary strata. In contrast,
there is a predominance of meltwater-derived sediment in
Quaternary strata of temperate glacial settings (Anderson
and Molnia, 1989).

Till versus Glacial-Marine Sediment. The most compelling
evidence for glaciation is the occurrence of diamicton;
however, key questions always arise concerning such deposits. Are the deposits till, are they glacial-marine in origin, or are they debris flows? How extensive was the
glaciation? Controversial evidence for Paleogene glaciation includes diamictons of Eocene and Oligocene age that
occur on King George Island (Birkenmajer, 1988), in drill
cores from Prydz Bay (Barron, Larsen, and Baldauf, 1991;
Hambrey, Ehrmann, and Larsen, 1991), and on the Ross
Sea continental shelf (Hayes and Frakes, 1975; Barrett
et al., 1989). Careful sedimentologic analysis coupled with
seismic work can provide proper interpretation of these
deposits.

In past years, considerable effort was focused on developing criteria to distinguish subglacial diamicton from
glacial-marine diamicton on the Antarctic continental shelf
(Chapter 4). The results of these works show that such
distinctions can be made only with careful sedimentologic
analyses, such as grain size, grain and pebble shape and
fabric, mineral and rock composition, and fossil and carbon content analyses. Chapter 4 offers a discussion of criteria used to distinguish subglacial and glacial-marine sediments.

Interglacial Deposits. Till provides the best evidence for
grounded ice, but it provides little or no information about
climatic conditions. Ice sheets exist under a wide range
of climatic regimes, from polar to temperate. Temperate
ice sheets did ground on the continental shelves of North
America and Europe (i.e., Miller, 1953; King and Fader,
1986; Vorren et al., 1989; Stoker, 1990; Belknap and
Shipp, 1991), so marine ice sheets are not unique to polar
settings.

The interglacial deposits interbedded with till provide important climatic information. A wide range of criteria exists for drawing paleoclimatic interpretations from glacial-marine sediment. For example, temperate interglacial sequences are unique in the predominance of meltwater deposits and geomorphic features that indicate running water (Anderson and Ashley, 1991). In contrast, modern Antarctic glacial-marine sediments, which reflect the current polar climate, contain significant amounts of biogenic material (Chapters 4 and 5). Unfortunately, detailed sedimentologic and paleontologic analyses of interglacial deposits on the Antarctic continent and continental shelf have progressed slowly. More work is necessary to resolve key questions concerning the nature of the climatic transition from temperate to polar and possible major shifts in climate.

Other Paleoclimatic and Glacial Evidence. Some of the most informative data concerning the progressive cooling of the Antarctic climate comes from palynologic analyses of exposures in the northern Antarctic Peninsula region (Birkenmajer, 1991; Askin, 1992) and analyses of recycled palynomorphs in glacial and glacial-marine sediment all around the continent (Truswell, 1982, 1983; Truswell and Anderson 1984; Truswell and Drewry, 1984; Hill and Truswell, 1993). One important aspect of the palynologic record is the virtual disappearance of palynomorphs by the end of Oligocene time nearly everywhere in Antarctica, except for what appear to be stunted *Nothofagus* plants, which may have inhabited the continent as recently as Pliocene to early(?) Pleistocene time (Webb and Harwood, 1991; McKelvey et al., 1991), although this is highly controversial.

The presence of hyaloclastites (subglacially erupted volcanic rocks) also serves as a good glacial indicator (Stump, 1980; Drewry, 1981; LeMasurier et al., 1979; LeMasurier and Rex, 1982). Equally important, hyaloclastites can be dated using the K-Ar techniques. Proper identification of hyaloclastites requires careful field observations.

Seismic Expression of Glacial Erosion and Deposition

Seismic profiles provide an important data set for examining the stratigraphic record of glaciation in Antarctica. Direct evidence for glaciation on the continental shelf exists in the form of glacial erosion surfaces as well as subglacial and ice-proximal seismic facies (Chapter 4). Indirect evidence of glaciation on the continental shelf includes mass flow deposits that reside on the continental slope and rise, representing the products of glacial erosion on the shelf.

Seismic methods can demonstrate whether a till unit encountered at isolated drill sites is associated with a widespread glacial unconformity on the continental shelf, thus providing evidence for (or against) continent-scale glaciation (Anderson and Bartek, 1991, 1992). This correlation requires relatively high-resolution seismic data and

velocity data for calculating subbottom depths (Bartek, Anderson, and Henrys, 1994; De Santis et al., 1995). Anderson and Bartek (1992) demonstrated that diamictons in the Early Miocene section of Ross Sea DSDP Leg 28 drill sites occur at approximately the same stratigraphic levels as regional glacial erosion surfaces. The origin of these tills (subglacial or glacial-marine) previously had been in question (Barrett, 1975; Balshaw, 1981; Hambrey, 1993). More recent analysis of seismic records across CIROS-1 in McMurdo Sound provided evidence that the Late Oligocene diamicton correlates to regional unconformities in the western Ross Sea (Henrys et al., 1994; Barrett et al., 1995; Bartek et al., 1996). This information lends support to the interpretation that these are major glacial episodes, probably expansions of the EAIS into the Ross Sea rather than merely mountain glaciations.

The identification of seismic facies is based on the intensity of acoustic contrasts, morphology of bounding surfaces, and the geometry, intensity, and frequency of internal reflections. This procedure, adapted from Vail and colleagues (1977), has been modified for smaller-scale, higher-resolution data from high-latitude margins by Belknap and Shipp (1991). Seismic facies analysis, like other sedimentologic analyses used for critical paleoenvironmental interpretation, should rely on stratigraphic and lateral facies relationships; no single seismic reflection character is unique to any particular environment or climatic setting. A large number of detailed investigations into the seismic signature of glacial deposits have been conducted in the Northern Hemisphere (e.g., Barrie and Piper, 1982; Belknap and Shipp, 1991; King et al., 1991), and one study was conducted in the inland passage of Chile (DaSilva, Anderson, and Stravers, 1997). These studies provide an important basis for characterizing seismic facies in temperate and subpolar glacial settings. See Chapter 4 for further discussion.

Outcrop studies of glacial-marine deposits indicate that the diagnostic facies and geomorphic features are typically on the order of meters to a few tens of meters thick and display considerable lateral variability (Anderson, 1983a). Low-resolution seismic data, in which a single reflector represents several tens of meters of the stratigraphic section, are not appropriate for seismic facies analysis (Fig. 6.9). In general, high-resolution reflection profiling requires high-frequency sound sources (> 500 Hz), and the depth of penetration of these sources is restricted to a few hundred meters (Fig. 6.9). Most facies and geomorphic features can be imaged using intermediate-resolution sources (50 to 500 Hz), which can penetrate to depths of over a kilometer (Anderson and Bartek, 1992; De Santis et al., 1995) (Fig. 6.9).

The thickness of stratigraphic units and the facies that compose them is predominantly controlled by fluctuations in sea level. The rate of sea-level change increased during the late Cenozoic (Fig. 6.3), resulting in a corresponding

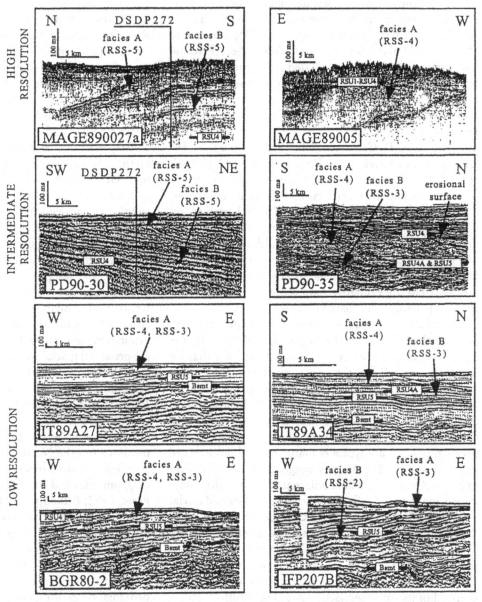

Figure 6.9. Examples of high- (sparker), intermediate- (generator-injector air gun), and low-resolution (air gun) seismic lines acquired in approximately the same location illustrating different levels of stratigraphic resolution (De Santis et al., 1995). The RSU4 unconformity provides a good point of reference for comparing the different records. The high-resolution profile resolves seismic facies, but penetration is limited to a few hundred milliseconds. The intermediate-resolution record resolves seismic facies and erosion surfaces to subsurface depths of up to 1 sec. Note that the lower-resolution record does not resolve the acoustically layered glacial-marine facies (Unit C1) from overlying ice-proximal facies (chaotic reflection pattern) or glacial erosion surfaces separating these units (from De Santis et al., 1995; reprinted with permission of the American Geophysical Union).

decrease in the thickness of sequences (Vail et al., 1977; Mitchum et al., 1994). This also is true in Antarctica because sea level controls the duration of ice sheet grounding events on the continental shelf (Bartek et al., 1991; Chapter 5). Thus, high-resolution seismic sources can be used to image the younger, thinner sequences, and intermediate-resolution tools can be used to image older sequences. In addition, the ice sheets have in many areas eroded deeply into the stratigraphic section, leaving older sequences just below the seafloor and, therefore, subject to analysis using high-resolution seismic tools. The Ross Sea is a case in point. The eastern Ross Sea has a relatively thick Plio-Pleistocene section. Alonso and colleagues (1992) investigated these deposits using intermediate-resolution seismic sources. In the western Ross Sea, little Plio-Pleistocene strata exist on the inner shelf and Miocene strata occur at or near the seafloor. These Miocene deposits can

be imaged with high-resolution seismic sources (Shipp et al., 1994). Thus, direct comparison of Miocene and Plio-Pleistocene seismic facies is possible using high- and intermediate-resolution seismic records (Anderson and Bartek, 1992; Shipp et al., 1994; De Santis et al., 1995; Bartek et al., 1997).

The first high-resolution seismic data acquired from the Antarctic continental shelf came from the Ross Sea (Houtz and Meijer, 1970; Houtz and Davey, 1973; Hayes and Davey, 1975) and later from the Weddell Sea (Elverhøi and Maisey, 1983). A concentrated effort toward acquiring grids of high- to intermediate-resolution seismic reflection data specifically for seismic facies and stratigraphic analysis subsequently followed in the Bransfield Basin (Jeffers and Anderson, 1990; Canals et al., 1994; Banfield and Anderson, 1995; Prieto et al., 1998), in the Ross Sea (Anderson and Bartek, 1991; De Santis et al., 1995; Bartek et al., 1998; Shipp et al., in press), on the Antarctic Peninsula continental shelf (Anderson, Bartek, and Thomas, 1991b; Bart and Anderson, 1995, 1996; Larter et al., 1997), in the northwestern Weddell Sea (Anderson, Bartek, and Thomas, 1991b; Sloan, Lawver, and Anderson, 1995), and in Prydz Bay region (O'Brien et al., 1995; O'Brien and Harris, 1996; O'Brien and Leitchenkov, 1997). Other data sets exist but are primarily scattered in distribution. The following discussion is based on these studies.

Glacial Unconformities. Anderson (1991) and Anderson and Bartek (1992) stress that regional, landward-dipping erosional surfaces provide the strongest evidence that ice sheets grounded on the Antarctic continental shelf (Chapter 3). These surfaces are best imaged on high- to intermediate-resolution records (Fig. 6.9). In strike view, considerable relief (often mimicking the present-day seafloor) characterizes the landward-dipping surfaces, and individual troughs are typically many tens of kilometers in width (Fig. 3.5). Depths may range from 100 to 200 m (measured from the seafloor adjacent to the trough). These dimensions dramatically differ from those of incised fluvial valleys and fluvial channels, which rarely exceed 100 m in depth and are seldom more than a few tens of kilometers in width. Glacial unconformities typically amalgamate on the inner shelf (Fig. 3.4). In many regions, such as along the northern Antarctic Peninsula, the inner shelf has been stripped of its entire sediment cover (Fig. 3.4B).

Erosional surfaces have been traced across large areas of the continental shelves of the Ross Sea (Alonso et al., 1992; Anderson and Bartek, 1992; Shipp et al., 1994; De Santis et al., 1995; Bart, 1998), the northwestern Weddell Sea (Anderson, Bartek, and Thomas, 1991b; Sloan, Lawver, and Anderson, 1995), and the Antarctic Peninsula (Anderson, Pope, and Thomas, 1990; Bart and Anderson, 1995) to demonstrate that these erosional surfaces do reflect grounding by large ice sheets. In all these areas, petrographic studies confirmed that modern glacial troughs correspond to the former positions of ice streams (Anderson, Brake, and Myers, 1984, Kennedy and Anderson, 1989, Anderson et al., 1991b). Therefore, these erosional surfaces are interpreted to be troughs carved by relatively rapidly flowing ice streams within the ice sheet.

Another scale of glacial erosion exists on the Antarctic continental shelf. The erosional features are smaller in lateral extent than the surfaces described above. Typically, the cross-sections display a U-shaped profile averaging 2 to 10 km wide and tens of meters deep. In the western Ross Sea, these smaller erosional features are interpreted to have been formed by localized erosion beneath small ice streams and outlet glaciers flowing through the TAM. On the Antarctic Peninsula shelf and in the Bransfield Basin, these troughs are similar in scale to modern troughs that extend offshore of outlet glaciers (Bart and Anderson, 1995; Banfield and Anderson, 1995).

Subglacial and Grounding Zone Facies and Features. On all the continental shelf areas where higher-resolution seismic data have been gathered to date (Ross Sea, Weddell Sea, Antarctic Peninsula shelf, and Bransfield Basin), subglacial and ice-proximal seismic facies dominate the Plio-Pleistocene section (Anderson, Bartek, and Thomas, 1991b, 1992a; Alonso et al., 1992; Anderson and Bartek, 1992; Banfield and Anderson, 1995; Bart and Anderson, 1995, 1996; Shipp et al., in press). Two subglacial seismic facies have been recognized, till sheets (Fig. 4.4) and ice stream boundary ridges (Fig. 4.5). These subglacial facies are bounded by glacial erosion surfaces. Within the grounding zone, there is a close association of subglacial, ice-proximal glacial-marine and sediment gravity flow facies. Two prominent grounding zone facies are recognized: grounding zone wedges (Fig. 4.16) and grounding line ridges (Fig. 4.18). Chapter 4 provides a detailed description of these seismic facies.

GLACIAL-MARINE FACIES. Studies of high-resolution seismic profiles from the Northern Hemisphere glaciated shelves revealed a variety of glacial-marine seismic facies based on geometries and internal characteristics (e.g., Barrie and Piper, 1982; Belknap and Shipp, 1991; King et al., 1991). Seismic character provides clues to the proximity of the deposits to the grounding zone. These studies benefit from on-land continuations of the units identified in seismic profiles and/or cores that directly sample the units.

At this point, it is not always possible to distinguish proximal from distal glacial-marine seismic facies in Antarctica. Seismic facies that display a relatively continuous, layered internal character are interpreted to be glacial-marine in origin (Fig. 6.9). They display the same kinds of stratal geometries (e.g., draping, onlap, offlap, etc.) that are observed in lower-latitude marine strata. The thickness of this facies varies from a few tens to hundreds of meters; generally, the thickness increases with depth in the stratigraphic column. Shipp and colleagues (1994) suggest that the greater thickness of glacial-marine facies in the Miocene stratigraphic section of the Ross Sea continental shelf, compared to the Plio-Pleistocene units, possibly marks the onset of Pliocene polar and subpolar conditions. The increasing polar conditions resulted in decreased meltwater sedimentation during interglacial periods.

TROUGH MOUTH FANS. Trough mouth fans are accumulations of progradational sediment packages that occur at the continental shelf edge offshore of large glacial troughs. They typically display a concave-up erosion surface that cuts into the prograding, internally stratified units (Fig. 3.15), indicating that the ice sheet advanced across the fan and that the sediment that composes the clinoforms is ice-proximal in origin. The trough mouth fan complex displays a generally well-layered seismic character (Fig. 3.15). On the outer shelf, prograding clinoforms downlap onto older glacial erosion surfaces (Bart and Anderson, 1995, 1996; Bart, 1998). Some units display toplap (rollover). Downdip, many of the units show evidence of slumping. The entire fan complex can build across the shelf edge many kilometers during a single grounding event. Figure 3.15 illustrates the complex stratal geometries that result from several episodes of grounding and associated delta progradation. Trough mouth fans have been identified in the western Ross Sea (Hayes and Davey, 1975; Bart, 1998), in the Weddell Sea (Crary Fan – Kuvaas and Kristoffersen, 1991; Moons et al., 1992; De Batist et al., 1994), in the Bransfield Basin (Jeffers and Anderson, 1990; Banfield and Anderson, 1995; Canals et al., 1994; Prieto et al., 1998), on the Antarctic Peninsula shelf (Bart and Anderson, 1996), offshore Prydz Bay (Kuvaas and Leitchenkov, 1992; O'Brien, 1994), and on the Wilkes Land continental margin (Eittreim, Cooper, and Wannesson, 1995). All these authors interpret the trough mouth fans as being products of glacial erosion on the continental shelf and progradation during the glacial maxima.

Bart and Anderson (1996) and Bart (1998) demonstrated that glacial unconformities on the continental shelf in the northwestern Ross Sea (Victoria Fan) and on the Antarctic Peninsula shelf extend seaward as conformable surfaces separating downlapping glacial-marine strata within the upper slope portion of the fan. Thus, trough mouth fans provide direct evidence of glacial erosion (ice sheet expansion) on the shelf and glacial-marine strata for dating these events. To date, no trough mouth fans have been drilled in Antarctica.

Missing Features and Seismic Facies. In this discussion of geomorphic features and seismic facies, it is important to consider those features observed in more temperate settings of the Northern Hemisphere that are not seen in Antarctica. The absence of incised fluvial valleys, transgressive shoreface ravinement surfaces, carbonate mounds, and fluvial deltas comes as no surprise, given the polar climate and great depth of the Antarctic continental shelf. But other features that occur on glaciated shelves of the Northern Hemisphere and in the Chilean Inland Passage also are absent or greatly restricted in their distribution. These include: (1) anastomosing tunnel valley networks; (2) proglacial deltas, fans, and aprons; (3) till tongues; and (4) lift-off moraines. The intermediate- to high-resolution seismic data from Antarctica should image these features where present, but none have been observed.

Proglacial deltas, fans, and aprons are associated with more temperate glacial settings where copious meltwater debouches from the glacier terminus. Subglacial tunnel

valleys occur in the Northern Hemisphere glacial settings (Ashley, Shaw, and Smith, 1985; Boyd, Scott, and Douma, 1988; Piotroswki, 1994), and they imply that subglacial meltwater was abundant and concentrated within these channels. Till tongues are associated with relatively stable ice margins (King and Fader, 1986; King et al., 1991; King, 1993), whereas lift-off moraines imply highly concentrated basal debris zones at the ice margin (Ashley, Shaw, and Smith, 1985; King, 1993). The absence of these features implies a paucity of subglacial meltwater and thin basal debris zones, both of which are consistent with polar marine ice sheets.

High- to intermediate-resolution seismic records image the Eocene through Miocene strata in areas such as the Ross and northwestern Weddell seas, where these strata occur in the shallow subsurface (Anderson, Shipp, and Siringan, 1992b; Anderson and Bartek, 1992; Bartek et al., 1997). Future studies should focus on characterizing seismic facies and identifying geomorphic features in older strata that might record the climatic transition in Antarctica from temperate to polar.

Stratal Stacking Patterns. During the late 1980s and early 1990s there was a significant increase in marine geophysical research around the Antarctic continental margin. Data acquisition included many tens of thousands of kilometers of multichannel seismic reflection profiles (Behrendt, 1990; Anderson, 1991; Cooper et al., 1994). These data were acquired for purposes of examining the deep crustal structure and overlying stratigraphic package of the continental margin. Such data have provided a wealth of information in that regard (Chapter 5). In recent years there has been a re-examination of these data to see if they can provide information about the glacial history of Antarctica. The stratigraphic resolution of these data is not sufficient to allow direct observation of most glacial erosional and depositional features (Fig. 6.9). Therefore, these studies have focused on determining if the overall stacking patterns of strata provide clues to ice sheet grounding events and deposition on the continental shelf (Larter and Barker, 1989, 1991b; Cooper et al., 1991a,b; Kuvaas and Kristoffersen, 1991; Cooper et al., 1993; Larter and Cunningham, 1993; Eittreim, Cooper, and Wannesson, 1995).

Larter and Barker (1989) argued that, given its great depth and foredeepened topography, the Antarctic shelf is not subject to many of the marine agents (specifically shoreface ravinement and stream valley incision) that create sequence boundaries and transgressive ravinement surfaces. Larter and Barker (1989, 1991b) and Cooper and colleagues (1991a,b) further suggested that the unique geometry of shelf margin strata in Antarctica results from glaciation on the continent. Their models do not differ significantly; both argue for progradation of the continental shelf during repeated advances of the ice sheet. These two-dimensional models are based solely on dip lines. They assume that sediment delivery to the shelf break and beyond

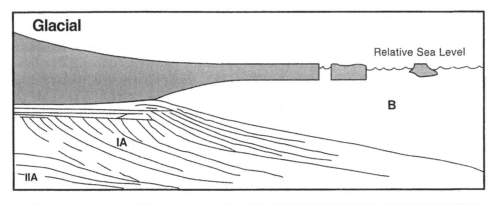

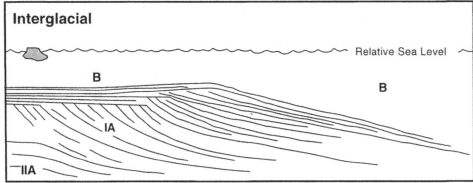

Figure 6.10. Seismic stratigraphic model of Cooper and colleagues (1991a) showing Type IA and Type IIA sequences.

by ice sheets as opposed to fluvial-deltaic and/or glacial-marine (temperate) processes.

Cooper and colleagues (1991a) distinguish two main types of stratal geometries. Type IA sequences consist of strongly prograding strata that have relatively thin topset beds and steeply dipping foreset beds; the topset beds typically are eroded on the inner shelf (Fig. 6.10). Relatively gently dipping and continuous strata stacked in an aggradational fashion characterize Type IIA sequences. Type IA strata are believed to reflect an interval of shelf development when grounded ice sheets advanced across the shelf and delivered sediment directly to the shelf break. Type IIA sequences are interpreted to be pre-glacial strata. An underlying assumption in the models of Larter and Barker (1989) and Cooper and colleagues (1991a) is that ice sheets are required to prograde sediment to the shelf edge. This further implies that the glacial setting during shelf progradation is polar because temperate glacial-marine settings, such as the Gulf of Alaska continental shelf, have progradational strata primarily composed of melt-water-derived glacial-marine deposits.

Bartek and colleagues (1991) conducted a sequence stratigraphic analysis of the Ross Sea and concluded that the overall stacking pattern is broadly similar to those in other parts of the world (Fig. 5.9B). They suggest that sea-level changes are the primary control on the sequence stratigraphy recorded by the strata of the Ross Sea continental shelf. They further argue that the strong tie between

sea level and sedimentation on the Antarctic continental shelf is due to the influence of rising or falling sea level on ice sheet growth and stability. When sea level falls, ice sheets expand onto the shelf, whereas a sea-level rise forces the ice to retreat.

Larter and Cunningham (1993) and Cooper and colleagues (1991a) believe that sediment is supplied to the shelf break as a line source and that the line source is indicative of sediment supply from ice sheets rather than from rivers. Bart and Anderson (1996) took a different stance, and used isopach maps of individual glacial units to demonstrate that glacial deposition on the continental shelf is point-sourced, not line-sourced. These are not trivial differences in interpretation. Non-linear sedimentation on the outer shelf results in greater stratigraphic complexity than is indicated by the models proposed by Larter and Barker (1989) and Cooper and colleagues (1991a).

Seismic Profile PD88-B (Fig. 5.19F), a strike-oriented profile collected along the outer shelf of the Antarctic Peninsula, is key to understanding the long-term depositional patterns on the shelf and resultant seismic stratigraphy. Profile PD88-B shows broad-scale cut-and-fill structures and poor lateral continuity of units. Paleotroughs are of approximately the same dimensions as modern troughs in that area. Note also that the troughs have shifted laterally on the shelf, which implies that deposition on the upper slope has also shifted. Isopach maps of individual units support shifting depocenters (Bart and Anderson, 1996). These combined data demonstrate that sedimentation on the outer shelf and upper slope of the Antarctic Peninsula is point-sourced. Detailed seismic stratigraphic analysis in the Weddell Sea (Miller et al., 1990b; Kuvaas and Kristoffersen, 1991; Moons et al., 1992; De Batist et al., 1994; Sloan, Lawver, and Anderson, 1995), Ross Sea (Alonso et al., 1992; Bart, 1998), Wilkes Land (Eittreim, Cooper, and Wannesson, 1995), and Prydz Bay (Kuvaas and Leitchenkov, 1992; O'Brien and Leitchenkov, 1997) continental margins has also led to the conclusion that glacial sedimentation of these margins is non-linear and focused primarily near trough mouths.

Another point to be made using the PD88 data set is that any single dip-oriented seismic profile provides only a partial record of glacial erosion and deposition on the shelf. Using the strike-oriented seismic profile (PD88-B), imagine the different glacial reconstructions that would result from an analysis of drill sites placed along Profile PD88-1 versus Profile PD88-4 (Fig. 5.19F). The stratigraphic complexity shown in Profile PD88-B is characteristic of those areas where high- and intermediate-resolution seismic data have been collected. It also serves to illustrate the underlying problem of reconstructing ice sheet grounding history on the continental shelf using seismic data that lack the stratigraphic resolution needed to image the relatively thin glacial units and bounding glacial unconformities seen in Profile PD88-B.

Dip-oriented, intermediate-resolution seismic data from the Ross Sea and Antarctic Peninsula continental shelves show more complex stratal patterns than the simple shift from aggradation to progradation observed in lower-resolution seismic data from these areas; they also show that glacial erosion surfaces with associated ice-proximal geomorphic features and seismic facies occur in both prograding and aggrading sequences (Anderson and Bartek, 1992; Alonso et al., 1992; Bart and Anderson, 1995, 1996). Indeed, the PD88 data from the Antarctic Peninsula continental shelf record good examples where the same seismic units are aggradational in one locality and progradational in another (Bart and Anderson, 1995, 1996). Intermediate-resolution seismic data from the Ross Sea and Antarctic Peninsula shelves were used to construct models that attempt to illustrate how multiple grounding events result in the stratal packaging observed on these shelves (Bartek et al., 1991; Bart and Anderson, 1995, 1996).

Continental Slope and Rise Facies

Seismic records from around the Antarctic continental slope and rise show a consistent stratigraphic change in seismic facies, with more stratified units in the deeper part of the section overlain by a sequence with large channel–levee complexes (Figs. 5.22 and 5.42). This change is interpreted to represent the onset of glaciation on the continental shelf and associated glacial erosion (Tucholke and Houtz, 1976; Kuvaas and Leitchenkov, 1992; Miller et al., 1990b; Moons et al., 1992; McGinnis and Hayes, 1995; Rebesco et al., 1994; Eittreim, Cooper, and Wannesson, 1995).

Summary

Seismic data provide strong evidence for grounding of ice sheets on the continental shelf. The most widespread and diagnostic features are glacial erosion surfaces, which typically show landward slopes and cross-sectional profiles that mimic the relief of troughs on the modern seafloor. High- to intermediate-resolution data offer the vertical resolution appropriate for imaging these features and for seismic facies analysis. Repeated episodes of glacial erosion on the continental shelf have resulted in broad-scale cut-and-fill stratal patterns, resulting in poor lateral continuity of units. For this reason, seismic stratigraphic studies require closely spaced seismic lines with good stratigraphic resolution. The practice of using overall stratal stacking patterns observed in low-resolution seismic data to decipher the glacial history of the continental shelf is similar to interpreting the depositional environment of a sandstone layer several kilometers away through binoculars; it is worth a drive to the outcrop.

HISTORY OF THE ANTARCTIC CRYOSPHERE

Triggering Mechanisms for Cryosphere Development

Conventional theory holds that the EAIS initially formed by the coalescence of temperate glaciers flowing from centers on the Gamburtsev Subglacial Mountains (Denton, Armstrong, and Suiver, 1971) and the TAM (Calkin, 1964; Mercer, 1978; Drewry, 1975). Bentley and Ostenso (1961) argued that the WAIS formed when ice shelves thickened and grounded on the continental shelf. These ice shelves created a protective barrier behind which the ice sheet formed (Mercer, 1978). Because ice shelves are considered to be polar features, this implies that development of the WAIS occurred in a polar climate, and therefore lagged behind the formation of the EAIS.

Denton, Prentice, and Burkle (1991) point out that a problem with these models is that they assume little change in the topography of the continent and do not account for tectonic uplift. Webb (1990) and Behrendt and Cooper (1991) concur and argue that uplift of the TAM represents one of the more significant structural events in the development of the Antarctic Ice Sheet. The ancestral TAM may have been only a minor obstacle to the spread of the early EAIS into West Antarctica; although, their uplift history remains problematic (Webb and Harwood, 1991; Fitzgerald, 1994). Equally important is the fact that glacial erosion has created much of the low-lying topography on Antarctica. When the first ice sheets formed, they advanced across a low-lying West Antarctic continent with shallow inland seas (Abreu and Anderson, 1998). Evidence for this comes from recycled Paleogene palynomorphs in glacial sediment from the Ross Sea and Weddell Sea continental shelves (Truswell and Drewry,

1984; Anderson et al., 1991a). These palynomorphs originated in the interior portions of West Antarctica where deep subglacial basins now exist. Glauconitic sandstone of Oligocene age at DSDP Leg 28 Site 270 (Hayes and Frakes, 1975) indicates a shallow marine shelf setting; this also is supported by the presence of shallow water benthic foraminifera in the sand (Leckie and Webb, 1983). De Santis and colleagues (1995) cite seismic evidence that glaciation of the Ross Sea continental shelf began with separate ice caps centered over islands. These ice caps eventually coalesced as the rift basins separating the islands filled with sediment. Ultimately, uplift of the TAM and erosion of the Ross and Weddell sea continental shelves resulted in partitioning of the EAIS and WAIS, and contributed to the very different ice sheet settings.

Kennett (1977) stressed the importance of the breakup and separation of Gondwana continents and development of ocean passages around Antarctica in cryosphere development. He contended that the development of circumpolar circulation resulted in thermal isolation of Antarctica and eventual glaciation of the continent. Prior to separation of the continents, meridional ocean circulation would have allowed subtropical currents to reach far south, resulting in a more temperate marine climate around Antarctica (Kvasov and Verbitsky, 1981). The onset of circumpolar circulation around the continent is thought to have begun when the Tasman Rise subsided in the Early to Mid-Oligocene and when the Drake Passage opened approximately at the Oligocene–Miocene boundary (Barker and Burrell, 1977; Exon, Hill, and Royer, 1995). Schnitker (1980) also stressed the role of paleoceanographic events in triggering Antarctic Ice Sheet development. He proposed that upwelling WDW around the continent resulted in increased moisture supply to the continent and triggered development of the EAIS.

Robin (1988) constructed a glaciologic model for the development of the Antarctic Ice Sheet constrained by geologic data. His model suggests that extensive glaciation of East Antarctica began in the Eocene (after 40 Ma), with polar conditions established by 10 Ma, resulting in the formation of the WAIS. Climatic modeling by Bartek, Sloan, Anderson, and Ross (1992) suggests that development of the Antarctic Ice Sheet was in part triggered by separation of Antarctica and Australia in the Middle Eocene, which resulted in the establishment of large low-pressure systems over East Antarctica. This condition was conducive to prevailing onshore flow of moist air, leading to increased precipitation over the continent and buildup of the ice sheet. One important aspect of glaciologic models is that the transition from temperate to polar climatic conditions probably resulted in decreased ice volume on Antarctica because of decreased precipitation over the continent (e.g., Pattyn, Decleir, and Huybrechts, 1992; Huybrechts, 1992, 1993; Budd et al., 1994).

Two schools of thought have emerged concerning the long-term stability of the Antarctic Ice Sheet. One school contends that Antarctica's climate has been polar since at least the Pliocene and that the ice sheets have been relatively stable since that time (Kennett, 1978; Denton et al., 1991; Kennett and Hodell, 1995). This argument hinges in large part on the contention that the factors leading to glaciation (e.g., tectonic isolation of the continent and development of circumpolar circulation) are irreversible (Kennett, 1978). The other school maintains that warm or subpolar ice sheets did not become polar until the Late Pliocene to Early Pleistocene, almost 20 million years after isolation of the continent and development of the circumpolar current (Webb and Harwood, 1991). The issue of when and if the Antarctic Ice Sheet achieved stability is yet to be resolved and may best be addressed using the marine sedimentary record.

Stability of the Antarctic Ice Sheet

Hollin (1962) was the first to suggest that the stability of the Antarctic Ice Sheet is strongly linked to global sea level, because large sectors of the ice sheet are grounded below sea level. Clark and Lingle (1977) and Thomas and Bentley (1978) later developed glaciologic models that demonstrated the feedback between sea level and ice sheet stability. These models indicated that the WAIS, in particular, is capable of rapid retreat when buoyed by rising sea level. Anderson and Thomas (1991) argued that this mechanism of rapid ice sheet retreat was the most likely cause of rapid sea-level events that occurred during the most recent, Late Pleistocene through Holocene, transgression.

Recent results from glaciologic modeling have led to the conclusion that the WAIS is inherently unstable and undergoes episodes of growth and collapse at time scales of hundreds of thousands of years (MacAyeal, 1992). Some of the controlling factors on ice sheet growth and decay, such as deformation of the bed on which the ice sheet rests, may behave somewhat independently from climatic cycles, or at least have a delayed response to climate change (Alley, 1990). Another mechanism of ice sheet instability is undermelting of ice shelves by warm water masses impinging onto the shelf (Potter and Paren, 1985; Jenkins and Doake, 1991; Jacobs et al., 1992; Jenkins et al., 1996). The response time of the ice sheet to this mechanism is on the order of decades to centuries, especially if undermelting works in concert with tidal pumping within the grounding zone environment (MacAyeal, 1992; Anandakrishnan and Alley, 1997).

Ice Sheet History: Cretaceous to Pleistocene

The following discussion weighs the geologic evidence for cryosphere development in Antarctica and draws upon several recently published reviews on the subject (Denton et al., 1991; Moriwaki, Yoshida, and Harwood, 1992; Wise et al., 1991; Hambrey, 1993; Barrett, 1997; Abreu and Anderson, 1998). Courageous students who tackle the vast literature on the subject will undoubtedly find themselves confused and frustrated by the contradictions that exist. To resolve some of this confusion, I have attempted to weigh the evidence in light of the problems described in the previous section. Evidence for Antarctic cryosphere development is presented in chronologic fashion, beginning with the Cretaceous.

Cretaceous. Matthews and Poore's (1980) interpretation of the composite $\delta^{18}O$ curve indicates possible ice buildup in Antarctica during the Cretaceous. Geologic evidence for widespread glaciation on the continent at that time is lacking, but it is important to recognize that there are few Cretaceous exposures on the continent. Some insight into Antarctica's climate during the Cretaceous is afforded by examining the stratigraphic records of neighboring Gondwana continents, particularly those attached to Antarctica during the Cretaceous (Chapter 2).

Cretaceous deposits from Australia, Africa, and South America yield no evidence of glaciation, except for ice-rafted material in Early Cretaceous deposits of southern Australia. There, the presence of IRD has been cited as evidence of glaciers in Antarctica by Frakes and Francis (1988). However, a single piston core into Early Cretaceous deposits on the adjacent Antarctic margin off Wilkes Land yielded a rich pollen-spore assemblage indicative of a relatively warm temperate climate (Domack, Fairchild, and Anderson, 1980). Early Cretaceous (Middle Albian) deposits were also sampled at ODP Site 740 in Prydz Bay (Fig. 6.2), including sandstone, mudstone, and coal. These deposits show no signs of glaciation in the region at that time (Turner and Padley, 1991). Recently, Ditchfield, Marshall, and Pirrie (1994) conducted $\delta^{18}O$ analyses on molluscan macrofossils in Late Jurassic to Eocene strata of James Ross and Alexander islands. They concluded that cold temperatures or subpolar conditions were established as early as Albian time in the region.

Late Cretaceous exposures are restricted to the Antarctic Peninsula region (Fig. 2.24). These include good exposures of the Latady and Fossil Bluff formations on James Ross and Seymour islands. These deposits contain a rich pollen-spore assemblage, including conifers, cycads, ginkgoes, and some angiosperms, indicating a conifer-dominated rain forest inhabited the region at that time (Askin, 1992). By latest Cretaceous, the flora of the Antarctic Peninsula region had diversified. The most significant change occurred in angiosperms, including many endemic species that evolved at the expense of cryptogams and gymnosperms (Askin, 1992).

ODP Leg 113, Sites 689 and 690 on the Queen Maud Land margin (Fig. 6.2) penetrated Late Cretaceous deposits. Studies of $\delta^{18}O$ concentrations, clay mineralogies, and planktonic microfossil assemblages indicate temperate to cool subtropical climatic conditions in the region (Kennett and Barker, 1990).

In summary, the evidence does not favor the existence of a large ice sheet on Antarctica during the Cretaceous, although ice caps may have inhabited the central portions of the continent, and some evidence for tidewater glacial activity exists in Australia.

Paleocene. Paleocene strata are confined to the northern Antarctic Peninsula region; none of these deposits contain ice-rafted stones (Chapter 2). Palynologic data from exposures in James Ross Basin indicate that Late Cretaceous–Paleocene climates in the Antarctic Peninsula region were warm to cool temperate, with high rainfall (Askin, 1992).

The relatively warm, maritime climate of the Antarctic Peninsula region is consistent with results from ODP Leg 113, which sampled diverse calcareous planktonic and benthonic microfossil assemblages and clay mineralogies indicative of warm climatic conditions in the more northern sector of the Weddell Sea (Robert and Kennett, 1994). The $\delta^{18}O$ data from these sites also indicate relatively warm surface and bottom waters in the Weddell Sea throughout most of the Paleocene. Oxygen isotopes from Maud Rise suggest surface water temperatures of ~ 9 to 10°C (Kennett and Stott, 1990).

Stott, Kennett, Shackleton, and Corfield (1990) argue that oxygen isotope data indicate the Paleocene–Eocene boundary was possibly the warmest time during the Cenozoic in Antarctica. This extreme warming event may have been short-lived, perhaps lasting only 20 kyr when surface temperatures reached 18 to 22°C. The most recent sequence stratigraphic chart (Hardenbol et al., in press) shows approximately 500,000-year eustatic cycles during the Late Paleocene through Early Eocene, which implies ice sheet fluctuations in Antarctica during this time interval.

There is no direct evidence that significant volumes of ice occurred on Antarctica during the Paleocene. However, outcrops of this age are restricted to the northern Antarctic Peninsula region. The proxy record of paleoclimates in the Antarctic Peninsula and Weddell Sea region indicates temperatures that were too warm for significant glaciation in that region during the Paleocene. The eustatic record indicates ice sheet fluctuations somewhere in Antarctica, perhaps interior East Antarctica, during the late Paleocene (Fig. 6.3). Thus, the record of Paleocene glaciation in Antarctica remains unclear.

Eocene

DEEP-SEA RECORD. Shackleton and Boersma (1981) argued that polar surface waters were 10 to 15°C warmer than the present during the Early Eocene. The oxygen isotope record shows a more or less steady increase in $\delta^{18}O$ during the Eocene (Fig. 6.3). Prentice and Matthews (1988) and Poore and Matthews (1988) suggested that this $\delta^{18}O$ increase resulted from ice volume increases and that there was a significant buildup of ice on Antarctica during the latest middle Eocene, perhaps reaching present proportions by 42 Ma. More recently, Zachos, Stott, and Lohmann (1994) reviewed the $\delta^{18}O$ data from Southern Ocean ODP sites and concluded that these data indicate a progressive decline in bottom water temperatures from ~10°C in the early Eocene to near 0°C at the end of the Eocene. This is supported by Kennett and Stott (1990), who observed a step-like increase in $\delta^{18}O$ in the later part of the Eocene at Site 689 on Maud Rise, which they attribute to a temperature decrease. Abreu and Anderson (1998) interpret the composite oxygen curve as indicating initial ice sheet growth in Antarctica during the Middle through Late Eocene, although the eustatic record shows only minor signs of sea-level lowering during this time interval (Fig. 6.3).

Kennett and Barker (1990) cite clay mineralogic (smectite-rich clays; Diester-Haas, Robert, and Chamley, 1993) and paleobotanical evidence (*Nothofagus*-dominated pollen assemblage on the South Orkney Plateau) in support of their model for an ice-free continent during the Eocene, although neither line of evidence is inconsistent with the existence of ice sheets elsewhere in Antarctica. Wise and colleagues (1991, 1992) summarized the known occurrences of Paleogene IRD, including a number of Eocene cases. They concluded that none of these occurrences of IRD are suitable documentation for widespread glaciation on Antarctica at the time.

At the Eocene–Oligocene boundary, a major shift in $\delta^{18}O$ occurs at sites on the Campbell Plateau (Kennett and Shackleton, 1976) and at some lower-latitude locations (Douglas and Savin, 1973; Wright and Miller, 1993) (Fig. 6.3). There is also a significant eustatic fall at the Eocene–Oligocene boundary (Vail and Hardenbol, 1979; Haq, Hardenbol, and Vail, 1987) (Fig. 6.3), and deep-sea hiatuses mark the Eocene–Oligocene boundary in portions of the South Atlantic (Wright and Miller, 1993) (Fig. 6.8, 40-36 Ma).

CONTINENTAL SHELF RECORD. In Prydz Bay, Site 742 of ODP Leg 119 sampled direct evidence of Eocene glaciation on the East Antarctic continental shelf. The drill core recovered massive diamictons that were interpreted as a waterlain till (Hambrey, Ehrmann, and Larsen, 1991) and are believed to be ~36 to 40 Ma in age; however, the age is somewhat problematic and these diamictons may be as young as Early Oligocene (Barron, Larsen, and Baldauf, 1991; Hambrey, Ehrmann, and Larsen, 1991). Seismic records show that this till lies close to a major unconformity on the continental shelf that marks a change in seismic facies from gently dipping layers (PS2B) on the inner shelf to steeply dipping foresets (PS2A) on the outer shelf (Fig. 5.50). The age of PS2 ranges from Middle(?)–Late Eocene to Early Miocene (Cooper et al., 1987a). The till and associated unconformity provide strong evidence for an ice sheet on East Antarctica by this time (Barron, Larsen, and Baldauf, 1991; Hambrey, Ehrmann, and Larsen, 1991).

Glacial-marine sediments in the lower part of CIROS-1 in western Ross Sea were dated as middle Eocene in age using fossil dinoflagellates (Hannah, 1994), although more recent paleomagnetic work has led to the reinterpretation of these sediments as being of latest Eocene age (Barrett, 1997; Wilson et al., 1998).

Seismic profiles acquired in Larsen Basin of the western Weddell Sea imaged probable Eocene and Oligocene strata, based on extrapolation of seismic units to onshore exposures on Seymour Island (Anderson, Shipp, and Siringan, 1992b). These records show no signs of glacial erosion or deposition on the continental shelf during this time interval (Anderson, Shipp, and Siringan, 1992b; Sloan, Lawver, and Anderson, 1995).

CONTINENTAL RECORD. Early–Middle Eocene glacial deposits occur in small outcrops on King George Island. These deposits indicate glacial conditions at that time and are designated as part of the Krakow Glaciation of Birkenmajer (1988, 1991). The age of these deposits (< 49.4 Ma) is derived from radiometric dating of basalts that overlie them. Coccoliths within a unit just below the basalt indicate a Late Paleocene to Early Eocene age (Birkenmajer, 1991). There is little evidence that these glacial strata indicate more than local alpine glaciations. The Krakow Glaciation was followed by the Arctowski Interglacial (Middle Eocene–Early Oligocene, 49 to 32 Ma), when a relatively warm moist climate and a diverse vascular plant community existed in the South Shetland and Seymour islands. Eocene floras of Seymour Island, located to the south of King George Island, indicate relatively warm conditions, with a subsequent shift toward *Nothofagus*-dominated flora indicating cooling (Askin, 1992). There is no sedimentologic evidence of glaciation in the form of ice-rafted clasts in Eocene deposits on Seymour Island (Zinsmeister, 1982).

Recent studies of erratics from the McMurdo Sound area have yielded fossil remains of crocodiles, sharks, and pelican-like birds (Harwood, personal communication). These erratics range in age from Middle to Upper Eocene. Detailed examination of these erratics has yielded a wealth of information about the Eocene climate of the region, and little evidence for significant glaciation at that time. These results will be published in a forthcoming volume of the American Geophysical Union's Antarctic Research Series.

SUMMARY. The deep-sea record of glaciation in the Eocene is still being debated, but the $\delta^{18}O$ records are consistent with large ice sheets in Antarctica during the Middle–Late Eocene. Drilling on the continental shelf in Prydz Bay has yielded convincing evidence for glaciation in East Antarctica by Mid(?)-Late Eocene time. A widespread unconformity on the Prydz Bay shelf supports extensive ice sheet grounding as opposed to more localized glaciation at this time. There is no direct evidence for significant glaciation on West Antarctica in the Eocene. Late Eocene glacial-marine sediment recovered at CIROS-1 in the western Ross Sea likely records an advance of the EAIS into the region. Outcrops on King George Island include glacial deposits, but there is no evidence to suggest that these record anything other than alpine glaciation.

Oligocene

DEEP-SEA RECORD. The oxygen isotope record of cryosphere development during the Oligocene has been debated widely. Initially, Shackleton and Kennett (1975) and Kennett and Shackleton (1976) argued that the $\delta^{18}O$ increase in the Oligocene (Fig. 6.3) was due to the introduction of cold bottom waters. They suggested that a permanent ice sheet did not form until the middle Miocene, although widespread mountain glaciation probably was occurring. Wise, Gombos, and Muza (1985) challenged this interpretation based on analysis of mixed benthic foraminiferal samples from DSDP Site 511 on the Falkland Plateau. They argued that Shackleton and Kennett's (1975) temperature estimates for Early Oligocene bottom waters were too cold and that part of the $\delta^{18}O$ increase was indicative of ice buildup on Antarctica. Shackleton (1986) later revised his estimates of bottom water temperatures, based on additional data, and agreed that some ice had to exist on the continent in the early Oligocene. Miller, Fairbanks, and Mountain (1987b) and Wright and Miller (1992, 1993) suggest an increase in Antarctic ice volume throughout the Oligocene based on the initiation of high rates of change in $\delta^{18}O$ in the oceans at this time. Zachos, Breza, and Wise (1992) interpreted the Early Oligocene $\delta^{18}O$ record of ODP sites from the Kerguelen Plateau to indicate the existence of ice sheets similar in volume to the present Antarctic Ice Sheet. Zachos, Stott, and Lohmann (1994) later revised the estimate of ice volumes on Antarctica to be 60% of the present volume in the Early Oligocene and 130% of the present volume in the late Oligocene.

Denton, Prentice, and Burckle (1991) and Abreu and Anderson (1998) conclude that the $\delta^{18}O$ record suggests significant ice volumes in Antarctica during most of the Oligocene. Relative maxima in ice volume are indicated for the latest Oligocene to Early Miocene (Fig. 6.3). Wright and Miller (1992) measured high-frequency $\delta^{18}O$ variations of ~ 1 Ma duration during the Oligocene and attributed these variations to ice sheet volume changes. They recognized distinct episodes of ice sheet growth at $\sim$ 36, 32, and 28 Ma.

The 28-Ma isotopic event corresponds to a major eustatic fall at this time (Fig. 6.3). Indeed, Miller and colleagues (1996) have made a strong argument that major sequence boundaries on the New Jersey continental shelf correspond to shifts in the isotopic record and that ice volume was the primary driving force controlling eustasy during the Oligocene.

In the Weddell Sea, the clay mineralogy at ODP Site 689 shows a shift within the Oligocene section from smectite-dominated clay mineral assemblages to illite-dominated assemblages. A similar change in clay mineralogy occurs within the Oligocene section at ODP sites on the Kerguelen Plateau (Ehrmann, 1991) and at CIROS-1 in the western Ross Sea (Ehrmann, 1995). The change from smectite- to illite-dominated clay assemblages reflects intense weathering and implies a predominance of glacial erosion over chemical erosion at that time (Grobe, Fuetterer, and Spiess, 1990; Robert and Maillot, 1990; Ehrmann, 1991; Diester-Haas, Robert, and Chamley, 1993). These mineralogic changes may, in fact, be a delayed response to

continental glaciation; initial weathering of the continent by ice probably reworked large quantities of smectite-rich soil (Ehrmann, 1991).

Seismic profiles from the continental slope and rise off Wilkes Land and Prydz Bay show prominent unconformities separating more distal (hemipelagic deposits) seismic facies below from more proximal (fan deposits) above. Offshore Wilkes Land, this unconformity (WL2) spans much of the Oligocene and Early Miocene (Eittreim et al., 1995). Offshore of Prydz Bay, the unconformity (PD2) spans the Early Oligocene to late Miocene (Cooper, Stagg, and Geist, 1991c). These unconformities are interpreted as being caused by intensified bottom current flow associated with the early phases of glaciation on the continent (Hampton, Eittreim, and Riehmond, 1987; Cooper, Stagg, and Geist, 1991c; Eittreim, Cooper, and Wannesson, 1995). The subsequent basinward shift in more proximal rise facies (fan growth) is interpreted as the culmination of glacial erosion and mass wasting on these continental shelves (Eittreim, Cooper, and Wannesson, 1995; Kuvaas and Leitchenkov, 1992).

By the Late Oligocene, cold water species replaced warm water silicoflagellates on the South Tasman Rise (Ciesielski, 1975), an increase in the percentage of cold water calcareous nannofossils occurred at the Kerguelen Plateau (Wei and Wise, 1992; Wise et al., 1992), the biosiliceous belt expanded around Antarctica, and a reorganization of planktonic foraminiferal biogeographic provinces occurred. These changes indicate a significant cooling in the Southern Ocean (Haq, Lohmann, and Wise, 1977; Kennett, 1978; Wei and Wise, 1992) and are associated with the first occurrence of IRD at several drill sites around the continent (Fig. 6.7).

CONTINENTAL SHELF RECORD. In Prydz Bay, ODP Site 739 penetrated Early Oligocene diamictite interpreted as till (Hambrey, Ehrmann, and Larsen, 1991). The presence of till indicates that a grounded ice sheet extended to the continental shelf break (Hambrey, Ehrmann, and Larsen, 1991). Early Oligocene glacial-marine sediment was also recovered at CIROS-1 in the western Ross Sea and at DSDP Site 270 in the eastern Ross Sea (Hambrey, 1993).

CIROS-1 in the western Ross Sea also recovered abundant evidence for glaciation in the area in Late Oligocene time. The core recorded no less than seven grounding events (Hambrey, 1993) indicated by glacial erosion surfaces and tills. The revised magnetobiostratigraphy for CIROS-1 indicates that the EAIS did not overflow the TAM and ground on the continental shelf until the Late Oligocene (Wilson et al., 1998). A major unconformity in the CIROS-1 core separates lower Oligocene from upper Oligocene-Lower Miocene strata. This unconformity perhaps corresponds to a widespread unconformity on the Ross Sea continental shelf (Bartek et al., 1991, 1996). The Late Oligocene section at CIROS-1 consists of alternating till and glacial-marine deposits and interglacial units that include fluvial to deep-water mudstone facies

(Hambrey, 1993). The fluvial deposits, in conjunction with the occurrence of a temperate pollen-spore assemblage in the mudstones (Mildenhall, 1989), reflect cool temperate interglacial conditions (Hambrey, 1993).

The glacial episodes recorded by CIROS-1 show reasonable correlation to the global sea-level curve (Hambrey, 1993; Barrett, 1997). However, such a correlation is expected for tidewater glaciers due to the influence of sea level on glacier grounding line positions and does not necessarily imply that these glacial deposits and unconformities were caused by expansion of the EAIS.

Recent analysis of seismic reflection records from the Ross Sea by De Santis and colleagues (1995) suggests that during the Oligocene, large areas of the Ross Sea continental shelf consisted of basement highs with low relief. These banks and islands were exposed and eroded during eustatic lowstands. Erosional surfaces on the banks include deep U-shaped valleys that De Santis and colleagues (1995) interpret as glacial valleys. Possible deltas located at the valley heads indicate that these were temperate glaciers. The isopach maps of De Santis and colleagues (1995) show that rift basins surround the basement highs. The basins potentially prevented the spread of ice caps from the banks across the shelf.

On the Wilkes Land continental margin, the oldest major unconformity on the shelf interpreted to be a glacial erosion surface (WL2) separates Oligocene and Early Miocene strata, based on extrapolation from DSDP Site 269 on the adjacent continental rise (Eittreim, Cooper, and Wannesson, 1995).

CONTINENTAL RECORD. Birkenmajer (1988) described till and glacial-marine deposits of the Polonez Cove Formation and the South Shetland Islands, and concluded that they were deposited by an ice cap that extended across the northern Antarctic Peninsula. Basalts with K-Ar ages of 32.8 and 30 Ma overlie the glacial deposits. The ages are supported by calcareous nannofossil ages (Birkenmajer and Gazdzicki, 1986), although Wise and colleagues (1992) caution that the nannofossil specimens illustrated by Birkenmajer and Gazdzicki (1986) appear to have been reworked. Birkenmajer (1988) refers to the glaciation as the Polonez Glaciation. The Polonez Glaciation was followed by the Wesele Interglacial (Middle Miocene). A late Oligocene (29.5 to 25.7 Ma) glaciation, the Legru Glaciation, occurred between 30 and 26 Ma. There are no equivalent age deposits on Seymour and James Ross islands, so the extent of the Polonez and Legru glaciations is uncertain. Vegetation in the Antarctic Peninsula region consisted of sparse *Nothofagus*-fern communities (Askin, 1992).

Other evidence for Oligocene glaciation in West Antarctica occurs in the form of 25-Ma-old hyaloclastites in Marie Byrd Land (LeMasurier and Rex, 1982, 1983). Recently, these results have been questioned by Wilch (1997), who argued that although the oldest glaciovolcanic deposits at Mt. Petras in West Antarctica are Middle Oli-

gocene (29 to 27 Ma) in age, these deposits reflect the presence of a thin, local ice cap.

SUMMARY. There is good evidence for the existence of an ice sheet on East Antarctica during the Oligocene. Localized glaciation, predominantly in the form of mountain glaciers and ice caps, occurred in West Antarctica, although the existence of a larger WAIS in the region beneath the present ice sheet cannot be ruled out. By the Mid-Oligocene, the ice sheet spread into the western Ross Sea.

Miocene

DEEP-SEA RECORD. Savin, Douglas, and Stehli (1975), Shackleton and Kennett (1975), Woodruff, Savin, and Douglas (1981), and Flower and Kennett (1994) argued that the $\delta^{18}O$ enrichment at ~ 15 Ma (Fig. 6.3) indicates considerable growth of the EAIS. The majority of investigators, however, contend that significant cryosphere development occurred earlier (Fig. 6.4). Miller, Wright, and Fairbanks (1991) recognize $\delta^{18}O$ increases at 24, 22, 20, 18, 16, 13.5, 10, and 8 Ma. They argue that these represent step-like buildups of ice on the continent.

The results of drilling in the southeast Pacific Basin (DSDP Leg 35) led to the conclusion that significant glaciation occurred in West Antarctica by the Early–Middle Miocene (Tucholke et al., 1976). This is based on the first significant occurrence of IRD at Site 325 on the Bellingshausen continental rise (Fig. 6.7). Also during the Early–Middle Miocene, there was a change from distal seismic facies (pelagic and hemipelagic sedimentation) to more proximal facies (predominance of sediment mass movement) on the continental rise of the Bellingshausen Sea. This change is interpreted as being due to glacial erosion on the Antarctic Peninsula continental shelf (Tucholke and Houtz, 1976; McGinnis and Hayes, 1995).

During the Middle Miocene, a widespread deep-sea hiatus formed in the Southern Ocean (Fig. 6.8). Formation of the hiatus is attributed to more vigorous bottom water circulation (Moore et al., 1978; Wright and Miller, 1993).

On the South Orkney Plateau (Site 696), the first significant occurrence of IRD occurs in the latest Miocene section (Fig. 6.7). This depositional event corresponds to a change in dominant clay mineralogy from smectite to illite, the appearance of hemipelagic sediment, and the rapid deposition of turbidites on the Weddell Abyssal Plain at Site 694 (Kennett and Barker, 1990). Kennett and Barker (1990) employed these lines of evidence to argue that development of the WAIS did not occur until the Latest Miocene–Early Pliocene.

The Kennett and Barker (1990) scenario for glaciation involves a number of assumptions that do not hold up under further scrutiny. Their argument is partially based on the occurrence of displaced benthic and neritic planktonic diatoms. These microfossils are abundant in Middle- and Early–Late Miocene portions of the Weddell Fan turbidite sequence drilled at Site 694. Kennett and Barker (1990) argue that the diatom tests were probably derived from

West Antarctica and that they imply an absence or near absence of ice cover over the shallow portions of the West Antarctic continental shelf during this time, thus negating the presence of the WAIS prior to the Late Miocene. The argument excludes the possibility that the diatoms originated in interglacial deposits in either East or West Antarctica. Diatomaceous mud and ooze are the primary sediment constituents occurring in modern shelf basins (Anderson et al., 1983a), and these materials typically contain benthic diatoms (Harwood, personal communication).

Kennett and Barker (1990) further contend that cessation of turbidite deposition at Site 694 in the Early Pliocene (~ 4.8 Ma) is evidence for the development of the WAIS. There are two flaws in this logic. First, Anderson, Wright, and Andrews (1986) presented petrographic data in favor of an East Antarctic, not West Antarctic, source for turbidites of the Weddell Fan. Second, turbidite deposition on the Weddell Abyssal Plain has remained active until recent time; turbidites have been recovered at or near the seafloor in piston cores (Anderson, Wright, and Andrews, 1986; Chapter 4).

More recent work on the Weddell Fan indicates that it is linked to the Crary Fan via a large submarine channel that extends along the base of the eastern Weddell Sea continental slope (Bart, De Batist, and Miller, 1994; Chapter 4). Detailed sequence stratigraphic analysis of the Crary Fan (Moons et al., 1992; Kaul, 1992; Bart et al., 1994; De Batist et al., 1994) illustrates construction of a minimum of five channel–levee units, each representing a period of significant fan development. De Batist and colleagues (1994) argue that episodes of fan development correlate with ice sheet grounding events. Detailed seismic stratigraphic work in the area (Moons et al., 1992; Bart, De Batist, and Miller, 1994; De Batist et al., 1994) shows that the channel–levee systems occur between two prominent unconformities, U4 and U5, of Miller and colleagues (1990b). Based on correlation to ODP Leg 113 Site 693, unconformities U4 and U5 are believed to be Oligocene and middle Miocene in age, respectively (Miller et al., 1990b). Based on these observations, De Batist and colleagues (1994) concluded that at least five long-term glacial expansions onto the continental shelf, and no less than 14 smaller-scale expansions, occurred in the Weddell Sea region since the Middle Miocene.

CONTINENTAL SHELF RECORD. In Prydz Bay, major hiatuses span the Late Oligocene to Middle Miocene. The deposits beneath these hiatuses are overconsolidated, indicating erosion by glacial ice (Solheim, Forsberg, and Pittenger, 1991). During the Late Miocene to Middle Pliocene, the ice sheet retreated from the Prydz Bay area, and glacial-marine sedimentation dominated (Hambrey, Ehrmann, and Larsen, 1991). A major Mid-Miocene(?) unconformity that is attributed to ice sheet grounding on the shelf also occurs on the Wilkes Land continental shelf (Eittreim, Cooper, and Wannesson, 1995).

Early marine geologic and geophysic investigations on the Ross Sea continental shelf led to the discovery of a shelf-wide unconformity, the Ross Sea Unconformity (Houtz and Meijer, 1970). Hayes and Frakes (1975) argued that the unconformity marked the first major advance of the ice sheet across the continental shelf ~ 5.0 Ma. Later, Savage and Ciesielski (1983) re-examined DSDP Sites 272 (eastern Ross Sea) and 273 (western Ross Sea) and interpreted the Ross Sea Unconformity as being created between 13.8 and 4.0 Ma.

Cooper and colleagues (1991a) and Anderson and Bartek (1992) argue that glacial topography with relief similar to that of the present characterized the Ross Sea continental shelf during the Early Miocene. Also, diamicton in early Miocene sections of CIROS-1 and DSDP Site 270 has been reinterpreted as till (Hambrey, 1993). Early Miocene glacial deposits contain mineral and rock compositions indicative of a source area in the interior portion of West Antarctica (Barrett, 1975). De Santis and colleagues (1995) argue that the ice caps were restricted to islands and banks on the Ross Sea continental shelf during the Early Miocene. At this time, rift basins on the shelf gradually filled with sediments, setting the stage for ice sheets to advance across the entire continental shelf by the Mid-Miocene. This widespread grounding event is marked by a shelf-wide Early–Middle Miocene (21 to 16.2 Ma) unconformity that occurs in seismic records (Anderson and Bartek, 1992; De Santis et al., 1995) and by a basinward shift in glacial unconformities in the McMurdo Sound region (Bartek et al., 1996). At DSDP Sites 272 in the eastern Ross Sea and 273 in the western Ross Sea, glacial-marine deposits dominate the Late Miocene (16.2 to 13.8 Ma) section that rests on top of the Early–Middle Miocene unconformity (Hayes and Frakes, 1975; Anderson and Bartek, 1992; De Santis et al., 1995).

Miocene glacial and glacial-marine strata recovered at DSDP sites on the Ross Sea continental shelf and at CIROS drill sites in the westernmost Ross Sea are interbedded with meltwater deposits and diatomaceous oozes, indicating shifts from temperate to subpolar or polar interglacial climates throughout the Miocene.

Larter and Barker (1989, 1991) conclude that ice sheets did not ground on the Antarctic Peninsula continental shelf until Latest Miocene–Early Pliocene time. Their argument is based on a change in the overall stacking pattern of continental margin deposits seen in low-resolution seismic records from the Antarctic Peninsula shelf. Larter and Barker (1989, 1991) derive their estimate of the age of ice grounding from the stratigraphic position of interpreted grounding line deposits resting above a collision unconformity. The age of the unconformity is approximated from seafloor magnetic data and a tentative tie with DSDP Site 325 on the Bellingshausen continental rise. Bart and Anderson (1995) examined a more extensive, higher-resolution data set from the same region. The data set included at least one dip line from each of the major tectonic segments of the margin between Marguerite Bay and the Bransfield Basin (Figs. 5.18 and 5.19). The grid provided better resolution of glacial erosion surfaces, glacial seismic facies, and geomorphic features. Bart and Anderson (1998) used the same method used by Larter and Barker (1991a) to establish their chronostratigraphy. The oldest widespread glacial erosion surface (U2 unconformity) seen in the Bart and Anderson (1995) data was interpreted to be of Middle Miocene age, although recent results from ODP Leg 178 Site 1097 would now place the U2 unconformity in the Lower Pliocene. However, the site also sampled ice-proximal deposits beneath the U2 unconformity (Barker et al., 1998). More detailed biostratigraphic analysis of this site will hopefully produce more definitive time constraints on the glaciation of the shelf. Evidence exists for several grounding events during the Pliocene (Bart and Anderson, 1995).

CONTINENTAL RECORD. The continental record of glaciation in West Antarctica during the Miocene includes subglacially erupted volcanics at the head of Scott Glacier; radiometric dates provide ages of 15 to 9 Ma (Stump et al., 1980). Craddock, Anderson, and Webers (1964) obtained a 22 ± 12 Ma date from a volcanic flow deposit resting on tillite in the Jones Mountains of West Antarctica. Rutford, Craddock, and Bastien (1968) argued that glaciation in the Jones Mountains occurred before 12 to 7 Ma. Further evidence for glaciation in West Antarctica exists in the form of volcanic flows overlying glacially striated surfaces. The volcanics yielded radiometric dates of 10 to 7 Ma (Rutford, Craddock, White, and Armstrong, 1972). Hyaloclastites in Victoria Land have been dated to be 7.3 to 5.4 Ma old (Drewry, 1981), and glaciovolcanic sequences at Mount Murphy record glaciations at 7 and 9 Ma (Wilch, 1997). Laser-fusion ^{40}Ar/^{39}Ar feldspar dates from volcanic ashes, interbedded with glacial drifts and overlying scoured terrain, indicate that ice sheet overriding events occurred in the TAM (Mayewski, 1975; Denton et al., 1984) prior to the late Miocene (Marchant and Denton, 1996). On King George Island, fossiliferous glacial-marine strata record an Early Miocene glacial episode, the Melville Glaciation (Birkenmajer, 1988). Pirrie et al (1997) studied Miocene (? Late Miocene) strata on James Ross Island, the Hobbs Glacier Formation, which contain diamictons that they interpret as "grounding line proximal" deposits.

Mercer and Sutter (1982) developed a glacial history for southern South America from interbedded tillite and basaltic flows. They concluded that glacial episodes were probably in phase with those occurring in West Antarctica, possibly dating back to 7 Ma.

SUMMARY. Solid evidence exists for the presence of large ice sheets on East and West Antarctica during the Miocene. In my opinion, much of the proxy evidence against the existence of the WAIS during the Miocene can be refuted. Indeed, Denton and colleagues (1991) believe that ice volume on Antarctica was equal to roughly one half of the present volume by 15 Ma and ice volumes perhaps exceeded those of the present by 12 Ma. There is strong evidence for significant growth and development of the WAIS during the Early–Middle Miocene. Expansion of the ice sheet across the continental shelf during the Middle and Late Miocene is corroborated by major isotopic and eustatic events (Fig. 6.3). Widespread glacial unconformities on the Antarctic Peninsula continental shelf indicate that ice caps in the region advanced onto the shelf, perhaps as early as Middle Miocene time. There is also onshore evidence for large ice caps in the Antarctic Peninsula during the Miocene. During the Miocene the WAIS was probably grounded mostly above sea level, unlike the modern WAIS, and therefore contributed to the observed eustatic fall of the Mid-Miocene (Abreu and Anderson, 1998).

Pliocene

DEEP-SEA RECORD. Miller, Wright, and Fairbanks (1991) argue that an δ^{18}O increase at ∼ 5.5 to 4.5 Ma is related to growth of the WAIS. Prentice and Fastook (1990) and Denton and colleagues (1991) conclude that isotopic data for the period between 6.0 and 4.4 Ma (Fig. 6.3) suggest high-frequency ice volume changes with little evidence of extensive deglaciation. Kennett and Hodell (1993) argue that oxygen isotope data from the Subantarctic region imply warming of surface water of less than ∼ 3°C for the warmest interval of the Pliocene (4.8 to 3.2 Ma). Webb (1990) and Webb and Harwood (1991) contend that changes in the latest Miocene-Pliocene oxygen isotope records, in addition to other "proxy" indicators, are not inconsistent with expanding and contracting ice sheets in Antarctica during that time. Supporting evidence for warmer than present interglacial conditions during the Pliocene comes from coastal deposits of the eastern United States that indicate sea-level highstands higher than present during the Pliocene (Dowsett and Cronin, 1990; Krantz, 1991).

In general, the biogeographic record from the Southern Ocean lends support to the concept of a highly variable climatic setting in Antarctica during the Pliocene. Ciesielski and Weaver (1974) present evidence for a major warm interval 4.3 to 3.95 Ma. Kemp, Frakes, and Hayes (1975) record a northward shift of the siliceous ooze–calcareous ooze boundary between ∼ 5.5 and 3.5 Ma. Abelmann, Gersonde, and Spiess (1990) report the incursion of warm water–related microfossils into the Weddell Sea 4.8 to 4.4 Ma. Burckle et al. (1996) argues that a significant cooling event occurred between 4.6 and 4.5 Ma. Recent micropaleontologic work by Bohaty and Harwood (1995) on ODP Sites 748B and 751B from the southern Kerguelen Plateau has provided evidence for warm events at ∼ 4.4, 4.3, 4.2, and 3.6 Ma. Barron (1996) suggests that the Polar Front migrated as much as 6° of latitude south of its present position in the southeastern Atlantic and Indian oceans between 3.1 and 3.0 Ma based on diatom and nannofossil distribution patterns. Other micropaleontologic investigations have yielded evidence for warm interglacial events in the Pliocene (Hays and Opdyke, 1967; Bandy, Casey, and Wright, 1971; Blank and Margolis, 1975; Kennett and Vella, 1975; Keany, 1978; Ciesielski and Grimstead, 1986; Flemming and Barron, 1996).

Burckle and Pokras (1991) reviewed the micropaleontologic evidence for an early Pliocene warming event in the Southern Ocean. They conclude that surface water temperatures during the austral summer were at most a few degrees higher than at present. Kennett and Hodell (1993) argue that sea surface temperatures during the Pliocene were never warm enough to cause displacement of Antarctic by Subantarctic planktonic assemblages. Burckle and colleagues (1996) (Dowsett, Barron and Poole, 1996) provide reviews of the micropaleontologic and isotopic evidence for warming episodes in the Pliocene. Wise, Gombos, and Muza (1985) note a significant northward shift in the position of the polar front ~ 2.7 to 2.6 Ma, which they argue was the climatic threshold into the modern glacial regime. Burckle and Abrams (1986) recorded an increase in ice-related diatoms in the Southern Ocean at ~ 2.4 Ma. The modern sea-ice related diatom assemblage of the Ross Sea appeared at roughly this same time (Harwood, 1991).

CONTINENTAL SHELF RECORD. Alonso and colleagues (1992) recognized a number of shelf-wide unconformities bounding till sheets within the Plio-Pleistocene section of the eastern Ross Sea. They argued that an increase in the frequency of grounding events in the Plio-Pleistocene relative to the Miocene was caused by higher-frequency sea-level changes linked to Northern Hemisphere ice sheet expansion and contraction (Fig. 6.3). Seismic records from the Antarctic Peninsula continental shelf also indicate high-frequency grounding events during the Plio-Pleistocene (Bart and Anderson, 1995). Seismic facies analysis of Plio-Pleistocene strata in the Ross Sea, the Antarctic Peninsula shelf, and the northwestern Weddell Sea shelf are still in progress; however, these studies have yielded no good evidence to date for significant meltwater discharge (e.g., incised fluvial valleys or deltas) from the continent during the Pliocene.

Evidence for temperate interglacial climates during the Pliocene comes from CIROS-2 (Ferrar Fjord) and DVDP-11 (Taylor Valley) in the westernmost Ross Sea. Diamicton in the cores is interbedded with mud, sand, and gravel facies interpreted to be meltwater deposits (Hambrey, 1993; Barrett et al., 1992).

CONTINENTAL RECORD. Mayewski (1975) argued that at some time during the Pliocene an ice sheet covered the TAM. He referred to this as the Queen Maud Glaciation. Denton and colleagues (1991) cite other lines of evidence for a thicker ice sheet having deposited the Sirius Tillite at that time.

The most interesting and controversial record of Pliocene glaciation in Antarctica centers around the Sirius Group, first described by Mercer (1972, 1981) and Mayewski (1975). The Sirius Group predominantly consists

of tillite and generally occurs above 2000 m elevation in the TAM. The strata contain paleosols and plant remains, including in situ roots of *Nothofagus* at 1800 m elevation near the Beardmore Glacier (Webb and Harwood, 1987, 1991; Carlquist, 1987; McKelvey et al., 1991). The most surprising discovery was the presence of reworked diatoms of supposedly Late Pliocene–(?)Early Pleistocene age within the strata. This indicates that land vegetation persisted in Antarctica until this time (Webb and Harwood, 1987). This is supported by the occurrence of *Nothofagus* palynomorphs in the Mid-Pliocene section of DSDP Site 274 (Flemming and Barron, 1996), situated north of the Ross Sea (Fig. 2.2).

Francis (1995) likens the *Nothofagus* plants of the Sirius Formation to the willows that inhabit the Arctic today, which implies mean annual temperatures below freezing, frozen and nutrient-poor soils, and periods of glacial melting. Hill and Truswell (1993) argue that the presence of the stunted, impoverished plant community implies temperatures of ~8°C for a minimum of three months each year, in addition to liquid precipitation. The interpretation implies that interglacial temperatures during the Late Pliocene–(?)Early Pleistocene were never as cold as today. Further, this leads to the conclusion that prior glacial events were subpolar in nature and a polar ice sheet did not emerge until after 2.5 Ma (Webb and Harwood, 1991).

Even more controversial interpretations from Sirius deposits center on the origin of diatoms found within these deposits. It has been proposed that the marine diatom tests in the Sirius Group originated from interior basins in East Antarctica, such as the Wilkes Subglacial Basin (Harwood, 1983, 1986; Webb and Harwood, 1991; Barrett et al., 1992). This implies that these basins were ice-free on several occasions during the Pliocene–(?)Early Pleistocene. According to Huybrechts' (1993) glaciologic modeling studies, deglaciation of the interior basins would require temperatures much warmer (> 15°C above present temperatures) than those required to sustain *Nothofagus* vegetation. Huybrechts' (1993) temperature estimates may be high because they assume an ice sheet and bedrock configuration in East Antarctica similar to that of the present and because the model ignores the effect of reduced sea ice and ice shelf mantling of the continental shelf (Wilson, 1995). Alternatively, major uplift of the TAM subsequent to deposition of the Sirius Group would not require greatly increased temperatures (Webb, Leckie, and Ward, 1986; Webb and Harwood, 1991). Webb and Harwood (1987) suggest that 1000 to 3000 m of uplift may have occurred. This is, however, inconsistent with fission-track thermochronology results (Fitzgerald, 1992; Chapter 2) and ^{10}Be exposure ages (Ivy-Ochs et al., 1995), although Webb (1991) counters that it is reasonable that different mountain blocks have different uplift histories. Fitzgerald (1992) agrees that uplift may have been episodic and that individual blocks probably do have different uplift histories. Behrendt, LeMasurier, and Cooper (1992) propose that uplift rates since the Pliocene occasionally may have reached 1 km/Myr. Wilch, Lux, Denton, and McIntosh (1993) argue that Plio-Pleistocene uplift rates in the Mc-Murdo Dry Valleys region were minimal.

In support of their model for a temperate climate throughout the Pliocene, Webb and Harwood (1991) and Barrett and colleagues (1992) cite examples of other areas in East Antarctica where deglaciations occurred during the Pliocene. These regions include the Vestfold Hills, where major ice sheet recession occurred in the Early to Mid-Pliocene (Pickard et al., 1988; Quilty, 1991; McKelvey and Stephenson, 1990), and the Prince Charles Mountains, where marine sedimentation occurred inland of the present ice front at some time since the Late Pliocene (Bardin, 1982; McKelvey and Stevenson, 1990; McKelvey et al., 1991). In Prydz Bay, Site 742 records Early Pliocene glacial recession of the ice sheet (Barron, Larsen, and Baldauf, 1991). Deglaciation also occurred in the Sør Rondane Mountains during the Middle Pliocene (Moriwaki et al., 1992).

The argument for a temperate climate and dynamic EAIS in the Pliocene contradicts the "accepted wisdom" and is still being received critically (Clapperton and Sugden, 1990; Denton, Prentice, and Burckle, 1991; Denton et al., 1993; Sugden, Marchant, and Denton, 1993; Marchant and Denton, 1996). In particular, Denton and his colleagues (1993) contend that evidence from the McMurdo Dry Valleys region indicates a persistent cold desert setting throughout the Pliocene with no evidence for meltdown of the EAIS. The evidence for sustained subpolar to polar climatic conditions in Victoria Land extending back to at least the beginning of the Late Miocene is as follows: (1) Desert pavements and thermal contraction cracks are covered by isotopically dated Pliocene ash falls (Marchant et al., 1993; Marchant et al., 1996). (2) There is a general absence of Pliocene landforms associated with temperate glaciers and associated outwash environments (e.g., outwash fans, kame terraces, and deltas) above 800 to 1000 m elevation in the western McMurdo Dry Valleys region (Marchant et al., 1993; Marchant and Denton, 1996). (3) There is strong evidence that uplift in the McMurdo Dry Valleys region of the TAM during the Plio-Pleistocene was less than 300 m, based on ages and elevations of subaerial cinder cone deposits in the Taylor Valley (Wilch et al., 1989). (4) There is no evidence for a major glacial advance of alpine glaciers and of the Taylor Glacier during the Pliocene, which is inconsistent with major expansion of the EAIS (Denton, Prentice, and Burckle, 1991). (5) Lower-elevation exposures of the Sirius Till commonly occur at the edges of deep glacial valleys where they are predicted to occur (Denton, Prentice, and Burckle, 1991). (6) The Peleus Till in the Wright Valley, which is at an elevation between 1150 and 1100 m (Prentice, 1985; Prentice et al., 1987), contains recycled marine diatoms that yield a maximum age of 4.0 Ma (Denton, Prentice, and Burckle, 1991). The till records a glacial event in which ice from East Antarctica virtually filled the Wright Valley. There is no evidence that the ice sheet flowed through the valley since that time, nor is there evidence for major melting events since then (Hall et al., 1997). (7) Striation patterns, till provenance, and [10]Be dating indicate that a local ice sheet existed in northern Victoria Land up to the Early Pliocene. Subsequently, the present drainage systems with deeply incised glacial valleys evolved (Van der Wateren and Verbers, 1992).

Denton and colleagues (1991, 1993) and Sugden and colleagues (1993) envision the glacial setting at the time of deposition of the lower-elevation Sirius Till as one where the ice sheet was essentially confined to East Antarctica and drained by glaciers flowing through the TAM. This configuration was theoretically achieved many times in the Pliocene.

Others have challenged the concept of a warm Pliocene. Recently published results from sedimentologic analyses of Sirius deposits at Mount Fleming and Table Mountain show that the clay minerals at both locations do not indicate chemical weathering (Helfer and Schluchter, 1995). Microfossils of possible East Antarctic origin were not found in either location (Helfer and Schluchter, 1995). These results indicate a cold dry climate since at least 5 Ma (Ivy-Ochs et al., 1995). Also, Stroeven, Prentice, and Kleman (1996) interpret the Sirius deposits of Mount Fleming as having been deposited in an alpine glacial setting, which is incompatible with diatoms in these deposits having been recycled from East Antarctica.

Dispute also focuses on the age of the Sirius Formation. Burckle and Pokras (1991) argued that the diatoms of the Sirius Group could be much older than Pliocene–(?)Early Pleistocene, perhaps as old as Oligocene. The Pliocene–Early Pleistocene age is supported by radiometric dating of deposits that are possibly correlative to the Sirius Group in a drill core from the Ferrar Fjord. Dating yielded an age of 2.8 ± 0.3 Ma (Barrett et al., 1992). However, exposure age dating of sirius tillites at Mount Fleming and Table Mountain have yielded ages of 6.5 Ma and 6.0 Ma, respectively (Bruno et al., 1997).

Burckle and Potter (1995) reported Pliocene marine diatoms in cracks in the Beacon Heights Orthoquartzite, which they view as evidence that diatoms in the Sirius Group may have been transported there by wind. This is supported by the work of Kellogg and Kellogg (1996), who observed terrestrial and small marine benthic diatoms in snow samples from interior areas in Antarctica, including the South Pole. Furthermore, recent studies have shown that diatoms are concentrated mainly in the upper weathered layer of the Sirius Till at Mount Fleming (Stroeven, Prentice, and Kleman, 1996). Harwood and Webb (1998) argue that the diatoms that occur within the Sirrus Group include larger tests than those found at the South Pole and in cracks of ancient strata (eolian-derived), and also includes clasts with diatoms. This indicates that the sirius diatoms are in situ, not wind-derived.

They further argue that the diatom assembly found within cracks in Pre-Cenozoic rocks are more abundant and diverse and derived from an open-ocean setting, unlike the Sirius assemblage.

Recent work on fossil diatoms in the Sirius Group by Burckle and colleagues (1996) and Stroven et al., (1997) focused on the diversity of diatoms in these strata and compared diatom assemblages within the Sirius Group to those in tills of Sweden and Finland. The results showed a much more diverse diatom assemblage in the Fennoscandian tills, including planktic and benthic marine, brackish, freshwater, and terrestrial species. The Sirius Group diatom assemblage consists dominantly of planktic and a few freshwater species. The diatom tests are dispersed throughout the Fennoscandian Tills, but are restricted to the surface and near-surface deposits in the Sirius Group. The debate about extreme warming events and associated deglaciation in East Antarctica continues.

SUMMARY. The deep-sea paleontologic record indicates that the Pliocene was a time of significant sea surface temperature fluctuations around Antarctica, but this does not imply corresponding ice volume changes. The oxygen isotope and eustatic records, however, do indicate significant ice volume fluctuations during the Pliocene. Several glacial unconformities occur in the Pliocene section on the continental shelves of the Ross Sea and the Antarctic Peninsula and provide evidence of waxing and waning marine ice sheets in these areas. No evidence of significant meltwater production exists on these shelves. The continental record is being debated. How cold was Antarctica in the Pliocene? Did the ice sheet undergo massive changes in volume? Paleobotanic evidence suggests that it was at least as cold as the high Arctic today, which is compatible with the existence of an impoverished *Nothofagus* vegetation living on or near frozen ground. During glacial episodes, the ice sheets advanced to the edge of the continental shelf, but refugia of *Nothofagus* may have remained in isolated coastal regions. The greatest controversy centers around evidence of major deglaciation in East Antarctica, in the form of recycled diatom fossils in Sirius tillites. The debate about the implications of the Sirius Formation continues, but the weight of the evidence seems to be against extreme deglaciation in the Pliocene (Barrett, 1997).

Pleistocene. There is one climatic event for which all records seem to agree: a major cooling event took place sometime around the Plio-Pleistocene boundary. Webb and Harwood (1991) argue that the climatic cooling at the end of the Pliocene was the threshold between Tertiary unipolar, temperate, and cyclic glaciations to bipolar ice sheets. Prior to the Pliocene cooling, climates were too warm to support ice sheets in mid-latitudes. More or less continuous glaciation took place in southern South America after the late Pliocene (Mercer, 1978, 1983). The Pleistocene was also a time of significant cooling in New Zealand (Kennett, Watkins, and Vella, 1971; Mildenhall, 1980). The global oxygen isotope record and the sea-level record

indicate a shift to higher-frequency (~100,000-year frequency) ice volume changes during the Pliocene and Pleistocene (Fig. 6.3), but debate still focuses on Antarctica's contribution to these proxy signals. Oerlemans (1982) argued that the Antarctic Ice Sheet could expand during global warming episodes because of the increased flow of relatively warm, moist air onto the continent. Hollin (1962) and Thomas and Bentley (1978) argue that the Antarctic Ice Sheet expands and contracts in harmony with the Northern Hemisphere ice sheets because of its sensitivity to sea-level rise and fall.

Denton and colleagues (1991) summarize the available evidence from the Ross Sea region and support an in-phase relationship between glacial cycles in the Northern and Southern hemispheres. However, they also point out that there may be significant lag times built into the complex glacial systems of both hemispheres. In Antarctica, the response of the marine-based ice sheet to sea-level rise may be almost immediate (Thomas and Bentley, 1978), whereas the response of the interior ice sheet to accumulation changes may be on the order of 10,000 years or more (Robin, 1983; MacAyeal, 1992).

Since Pewé (1960), Hollin (1962), and Nicols (1964) published their classic papers describing the state of understanding of Antarctica's Late Pleistocene glacial history, steady progress has been made toward reconstructing ice sheet configurations in both East and West Antarctica during the last few glacial–interglacial cycles. This research is driven largely by the need to examine the response of the ice sheets to global (bipolar) forcing mechanisms, such as eustasy and climate change, versus internal forcing mechanisms such as ice sheet-ice bed interaction. We still have a great deal of work to do before we can resolve these issues.

ICE CORE RECORD. The ice core (oxygen isotope) record from Vostok, near the center of the EAIS (Fig. 6.2), provides a reasonable proxy of ice volume changes for the past 450,000 years (Petit et al., 1997) (Fig. 6.5B). The Vostok record shows strong correlation with the global isotopic record (Fig. 6.5A), which indicates linkages between Northern and Southern Hemisphere ice sheets (Lorius et al., 1985; Jouzel et al., 1987; Petit et al., 1997). However, different factors may regulate the activity of the ice sheets and complicate the phase relationships between the ice masses of the two hemispheres. For example, the importance of the interaction between the ice sheet and the bed on which it rests is appreciable; these processes may operate more or less independently of climate (Alley, 1990; MacAyeal, 1992). Other important questions concern: (1) the configuration of the EAIS and WAIS during glacial maxima and minima; (2) the rate of change from a glacial to an interglacial configuration; and (3) the forcing agents that cause the ice sheet to expand and contract. Answers to questions of ice sheet activity lie in the Pleistocene geologic record of glaciation, and the most complete stratigraphic record exists offshore.

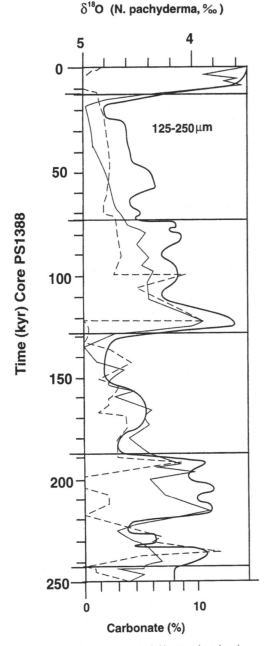

$\delta^{18}O$ (N. pachyderma, ‰)

125-250 µm

Time (kyr) Core PS1388

Carbonate (%)

Figure 6.11. Oxygen isotope curve (solid line) and total carbonate curve (dashed line) for sediment Core PS 1388 from the eastern Weddell Sea continental slope. The bold curve is the $\delta^{18}O$ chronostratigraphic curve from Martinson and colleagues (1987; modified from Grobe and Mackensen, 1992).

DEEP-SEA RECORD. Denton and colleagues (1991) note that the Late Quaternary pattern of variability in the oxygen isotope record shows two important events. Planktonic $\delta^{18}O$ amplitude was less during the interval between 850,000 and 450,000 BP than since 450,000 BP, indicating that Antarctic ice volume was less variable during interglacials and always larger than present. These same data indicate that the Antarctic Ice Sheet fluctuated quasi-periodically during the past 450,000 years (since Stage 12), and that ice volumes have never been much greater

than during Stage 2 and never significantly less than at present.

The literature describing the results of sedimentologic studies of deep-sea sediments recovered in piston cores from around Antarctica is voluminous and, unfortunately, filled with contradictory results. The reasons for this are poor chronostratigraphic resolution and a poor understanding of complex sedimentologic processes that occur on the deep seafloor. Researchers during the late 1960s through the 1970s relied heavily on paleomagnetic and biostratigraphic data for chronostratigraphic information. A general paucity of calcium carbonate in sediment cores collected near the continent (Fig. 4.49) precluded oxygen isotope and radiocarbon analyses. This situation has improved with the advent of better instrumentation capable of yielding good results with small sample sizes, but reexamination of the deep-sea sedimentary record of Antarctic climatic and paleoceanographic changes has progressed slowly during the past decade, despite the fact that promising results have been achieved.

One of the most thorough studies was conducted by Mackensen and colleagues (1989) and Grobe and Mackensen (1992), who examined sediment cores from the Weddell Sea continental slope and rise. One of these cores (PS 1388) yielded a particularly good isotope curve, which correlates well with the global isotope record (Fig. 6.11). They note that biogenic carbonate and IRD sedimentation increased during interglacials and suggest correlation between glacial cycles in the Weddell Sea and those of the Northern Hemisphere.

CONTINENTAL SHELF RECORD. High- and intermediate-resolution seismic profiles from the Ross Sea, northwestern Weddell Sea, and the Antarctic Peninsula show that the Pleistocene section on the continental shelf is characterized by numerous glacial erosion surfaces and discontinuous units (Chapter 4). The seismic facies are dominated by stacked till sheets and interbedded glacial-marine deposits that are thin and discontinuous. The data indicate that the Pleistocene sections on these shelves are dominated by subglacial and grounding zone-proximal deposits (Shipp et al., 1994; Bart and Anderson, 1995). This interpretation is supported by drill core from the Ross Sea, Prydz Bay, and the Antarctic Penisula shelf. Furthermore, Sites DVDP-10 and DVDP-11 in the McMurdo Sound region penetrated thick diamicton with abundant clasts, indicative of more severe glacial conditions in the Pleistocene than in the Pliocene (McKelvey, 1991).

CONTINENTAL RECORD. The Pleistocene record of glaciation in East Antarctica is fragmentary; Pleistocene strata generally are confined to the TAM and widely scattered coastal outcrops. The best documented exposures occur in northern Victoria Land. The Latest Pliocene-Earliest Pleistocene was a period of major cooling in the TAM (Mercer, 1972; Denton, Anderson, and Conway, 1986a; Denton et al., 1986b; Denton et al., 1989). Abundant sedimentologic evidence (Sirius, Peleus, and Prospect Mesa drift deposits) and geomorphologic evidence (high-elevation trimlines and striations) exist to indicate that the EAIS overrode the TAM, perhaps on several occasions during the Pleistocene. The timing of these events, however, remains problematic (Denton, Prentice, and Burckle, 1991).

The continental record of Pleistocene glaciation in West Antarctica is also fragmentary. Hyaloclastites of Pleistocene age (1.18 Ma) on Ross Island and in McMurdo Sound indicate that the ice sheet grounded on the continental shelf and was hundreds of meters thicker than at present (Kyle, 1981). Glaciovolcanic tuffs at Sechrist Peak, Mount Murphy, record an ice sheet highstand of greater than 550 m during the Middle Pleistocene (590 ± 15 ka; Wilch, 1997).

SUMMARY. Polar conditions have prevailed in Antarctica throughout the Pleistocene. Both the EAIS and WAIS have been much thicker than present and advanced to the edge of the continental shelf on several occasions during the Pleistocene. The Vostok ice core isotopic record and isotopic curves obtained from sediment cores from the Weddell Sea suggest that the Antarctic Ice Sheet advanced and retreated in harmony with Northern Hemispheric ice sheets, but additional work is needed to establish these bipolar linkages and to determine the driving forces (e.g., climatic, eustatic, etc.) causing ice sheets to advance and retreat in both hemispheres.

LATE QUATERNARY ICE SHEET ACTIVITY: CASE STUDIES

As relatively low latitude inhabitants, our perception of glacial-interglacial cycles (120,000 yr) is that interglacials dominate. However, the Antarctic has not experienced true interglacial conditions (e.g., when significant ice shelf melting occurs) since the Miocene. Hence the terms "glacial maxima" and "glacial minima," not "glacial" and "interglacial," are commonly used. Analysis of physical and chemical properties of the Vostok ice cores indicates that as many as three glacial maxima occurred during the interval between 120,000 and 13,000 yr BP (Barkov, Korotkevich, and Petrov, 1991) (Fig. 6.5). Denton and colleagues (1991) describe a drift sheet that is stratigraphically older than the Stage 2 Drift in the McMurdo Sound area. They believe it represents a glacial maximum with ice configurations in the area similar to those of the Stage 2 glacial maximum. Dagel (1985) obtained uranium-series dates on calcium carbonate from lacustrine beds within this drift unit that range from 185,000 to 130,000 yr BP, indicating that this drift records the Stage 6 glacial maximum.

The Stage 5e interglacial may have been characterized by a smaller global ice volume than exists today; sea level was ~ 5 m higher (Emiliani, 1969; Bloom et al., 1974). A 6-m rise could result from either complete melting of the WAIS (Mercer, 1968), complete melting of the Greenland Ice Sheet (Emiliani, 1969), or partial melting of northern and southern hemisphere ice sheets (Denton, Armstrong, and Suiver, 1971; Robin, 1983; Hughes, 1987). Scherer (1998) interpreted marine diatoms in sediment sampled beneath Ice Stream B as supporting evidence for a WAIS grounding line positioned inland of its present location during the Late Pleistocene. Burckle (1993) refuted the Late Pleistocene age of these microfossils; however, more recent work by Scherer et al. (1998) showed high concentrations of beryllium[10] in sediments containing diatoms. This is strong evidence for retreat of the ice sheet during the Quaternary.

Mercer (1968) argued for the virtual collapse of the WAIS during Pleistocene interglacials based on an interpretation of summer warming of 7 to 10°C relative to present-day temperatures. He provided geologic support for this from the Reedy Glacier area. Denton and colleagues (1991), however, point out that such high temperatures are not supported by the CLIMAP sea surface temperature reconstructions (Stuiver et al., 1981) or the ice-core $\delta^{18}O$ records from Vostok. These records indicate that summer temperatures were only ~ 2°C warmer during Stage 5e than present.

Denton and colleagues (1991) offer two alternative mechanisms for ice sheet collapse. One calls for grounding line recession in Pine Island Bay in the Antarctic Peninsula region. Currently, this region is fed by an unbuttressed ice stream (Hughes, 1977). The other proposed mechanism calls for collapse of the marine portions of the WAIS in response to the global rise in sea level (Hollin, 1964; Thomas and Bentley, 1978). A third alternative is that the ice sheet grew thinner due to accelerated flow velocities caused by sliding over a deforming bed (Alley et al., 1989).

Huybrechts (1990) constructed a three-dimensional thermo-mechanical ice sheet model for the entire Antarctic Ice Sheet for the last glacial–interglacial cycle. His model calls for retreat of the ice sheet caused by rising sea level that takes 6000 years to complete.

The configuration of the Antarctic Ice Sheet during the Last Glacial Maximum (LGM) was modeled by Stuiver and colleagues (1981; CLIMAP reconstruction) and later revised by Denton and colleagues (1991). In the CLIMAP reconstruction, flowlines of the ice sheet on the continental shelf are based on extensions of present-day flowlines, which are made to correspond to glacial troughs on the shelf. The grounding line was placed near the continental shelf edge based on an ice surface profile reconstructed from the Ross Sea (Stuiver et al., 1981). A later reconstruction by Denton and colleagues (1991) has a much greater East Antarctic contribution of glacial drainage into the Ross Sea region (Fig. 6.12). Another difference between these two models is that the more recent model of Denton and colleagues (1991) shows little interior surface elevation change but considerable thickening of peripheral ice in East Antarctica. This is consistent with oxygen isotope data from the interior Vostok and Dome Circle locations

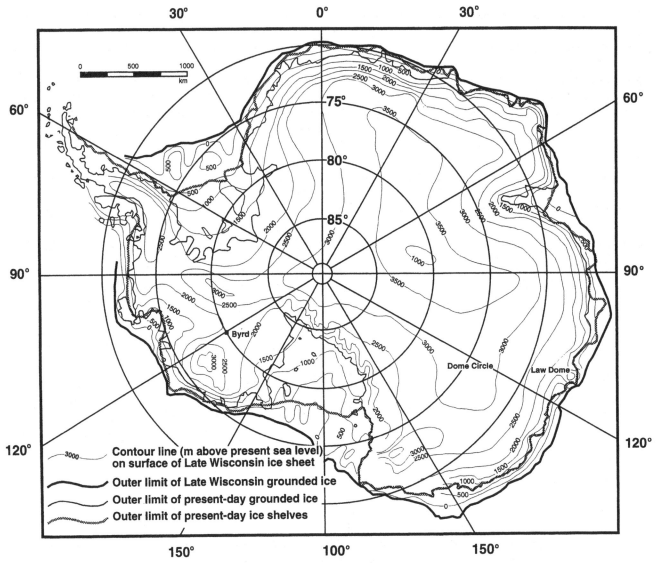

Figure 6.12. Ice sheet reconstruction of Denton and colleagues (1991) for the last glacial maximum.

as well as the peripheral Law Dome and Terre Adelie sites (Robin, 1983; Lorius et al., 1984, 1985).

Drewry's (1979) model for the Ross Sea requires only minimum expansion of the WAIS during the LGM. The model predicts a vast ice shelf pinned on shelf banks and covering most of the continental shelf. His model fits ice core data (Byrd Ice Core; Stuiver et al., 1981), but not geologic data that calls for ice at higher elevations (Denton, Prentice, and Burckle, 1991).

Continental Record

Stuiver and colleagues (1981) and Denton and colleagues (1991) review the evidence that ice sheet surface elevations in East Antarctica were greater during the Late Pleistocene than at present. Included in this evidence are lateral moraines in the TAM that occur well above the Ross Ice Shelf. These are dated by many radiocarbon measurements as "Late Wisconsin" features (see Denton,

Prentice, and Burckle, 1991 for a review). In the Ross Sea region, the most widespread high-elevation unit is the Ross Sea Drift; it is perched 240 to 610 m above present sea level (Stuiver et al., 1981). These moraines merge with the inland ice plateau near glacier heads, indicating that the inland EAIS stood at about the same elevation as today (Denton, Prentice, and Burckle, 1991). This is supported by ice core data from interior East Antarctica (Lorius et al., 1984, 1985). Evidence for expansion and thickening of the EAIS in Victoria Land exists as perched erratics, moraines, and striations along the coast between Cape Adare and Rennick Glacier.

Perched erratics on Beaufort and Franklin Islands in the southwestern Ross Sea occur at elevations of up to 320 m above sea level. In addition, there is evidence that as much as 1325 m of ice filled McMurdo Sound during the LGM (Stuiver et al., 1981). The western Ross Sea region has also produced the greatest number of radiocarbon dates on Late Pleistocene deposits, and these dates indicate the LGM occurred between 21,200 and 17,000 yr BP (Stuiver et al., 1981). Glaciovolcanic deposits at ~350 m above the present ice surface at Mount Takahe yielded ^{40}Ar/^{39}Ar ages of 29,000 ± 12,000 yr BP and record the maximum glaciation in this region (Wilch, 1997). A parasitic cinder cone dated at 34,000 ± 8000 yr BP on the west flank of Mount Frakes records ice elevations ∼ 100 to 150 m above the present ice sheet.

Denton and colleagues (1991) present numerous radiocarbon dates for the completion of the Holocene grounding line recession between ∼ 7000 and 5000 yr BP in the western Ross Sea region. These ages are consistent with those from ice cores in the interior portions of the EAIS (Law Dome) that indicate ice surface lowering prior to 8000 yr BP (Robin, 1983). The ages also concur with ice cores from coastal regions of both East and West Antarctica that range from 8400 to 6000 yr BP (Denton, Prentice, and Burckle, 1991).

Colhoun, Mabin, Adamson, and Kirk (1992) reexamined the implications of raised beaches around Antarctica, including the Ross Sea, and argue that deglaciation all around Antarctica was well advanced by 10,000 BP and was complete by 6000 yr BP. Terra Nova Bay was deglaciated prior to 7065 ± 250 yr BP, based on the occurrence of penguin rookeries of that age (Baroni and Orombelli, 1991).

Denton and colleagues (1992) surveyed trimline elevations and striations in the Ellsworth Mountains and observed a former trimline at about 400 to 650 m above the present ice surface on the landward side of the mountains and up to 1900 m higher on the seaward side of the mountains. The age of these features remains problematic. Carrara (1981) conducted field studies in the Orville Coast of the Antarctic Peninsula region and presented evidence that the ice sheet had once stood 450 m higher than at present, possibly during the LGM. In the northern Pensacola Mountains, glacial erratics occur 400 m above the present ice sheet surface. Waitt (1983) examined nunataks in the Lassiter Coast of Southern Palmer Land and noted striations and erratics up to 500 m higher than the present ice elevation. He projected the ice sheet grounding line nearly 800 km north of its present position in the Weddell embayment. The age of the observed feature is unknown. Sugden and Clapperton (1977), Clapperton and Sugden (1982), and Hall (1983, 1984) presented evidence that glacial expansion occurred on Subantarctic islands during the "Late Wisconsin." Bentley and Anderson (1998) offer a recent summary of the onshore and offshore records of glaciation in the Weddell Sea–Antarctic Penninsula region during the LGM.

Evidence for higher ice elevations in coastal East Antarctica during the late Pleistocene is presented by Young and colleagues (1984), who argue for ice elevations of ∼ 400 m above the present 250 km inland from the coast. A greater ice thickness near the coast is indicated by high-elevation (relative to present ice surface elevations) drift, glacial erratics and striated bedrock, and raised Holocene marine deposits at Windmill Islands (Goodwin, 1993) and

in the Vestfold Hills (Adamson and Pickard, 1983, 1986). Hirakawa and Moriwaki (1990) presented geomorphic evidence that the EAIS was at least 400 m higher, possibly during the LGM, than at present in the central Sør Rondane Mountains. The ice sheet was as much as 700 m thicker than at present in the Vestfjella and Heimefrontfjella nunataks of western Queen Maud Land (Hirvas, Lintinen, and Nenonen, 1994). Mabin (1992) believes moraines at elevations ∼100 m above present levels in the Lambert Glacier area might mark ice surface elevations during the LGM.

Expansion of the EAIS during the LGM does not appear to have affected all areas. Mabin (1992) examined raised beaches in the Bunger Hills of East Antarctica and concluded that the EAIS was thinner and less extensive during the LGM than was previously believed. This has been supported by the work of Burgess, Spate, and Shevlin (1994), who provide evidence that the Larsemann Hills were partly ice-free during the LGM. Igarashi, Harada, and Moriwaki (1995) presented evidence from raised beaches around Lützow-Holm Bay that the EAIS was only slightly larger than at present during the LGM. Colhoun (1991) summarizes other lines of evidence for a thinner (< 300 m) EAIS in coastal areas than had been previously hypothesized. Furthermore, the results of analyses of gas content in bubbles in the Vostok and Byrd ice cores indicate ice surface elevations that were actually lower than at present during the LGM (Raynaud and Label, 1979; Raynaud and Whillans, 1982). Thus, the size and extent of the EAIS and WAIS during the LGM remain problematic, largely because of poor age control on glacial features. In view of this uncertainty, marine deposits may yield the best record of ice sheet size and dimension during the most recent glacial maximum.

Results of Marine Geologic Investigations

Detailed marine geologic investigations, including till provenance analyses aimed at reconstructing the ice sheet configuration during the Late Pleistocene, have been conducted in two areas of East Antarctica: off Wilkes Land (Anderson et al., 1980b; Domack, 1982; Domack et al., 1989), and on the eastern Weddell Sea continental shelf (Anderson et al., 1980b, 1991a) (Fig. 6.13). In Prydz Bay, O'Brien and Harris (1996), O'Brien and Leitchenkov (1997), and Domack et al. (1998) used bottom profiler data, seismic data and sediment cores to reconstruct the Late Pleistocene ice sheet configuration. Domack and colleagues (1991b) examined ODP Leg 119 drill core from the area to constrain the timing of ice sheet retreat from the shelf. Similar studies have been conducted on the West Antarctic continental shelf in the Ross Sea (Anderson et al. 1980b; Anderson, Brake, and Myers, 1984; Anderson et al., 1992a; Jahns, 1994; Shipp et al., in press; Domack et al., in press; Licht et al., 1996), in the Antarctic Peninsula region (Kennedy and Anderson, 1989; Herron and Anderson, 1990; Anderson, Bartek, and Thomas, 1991b;

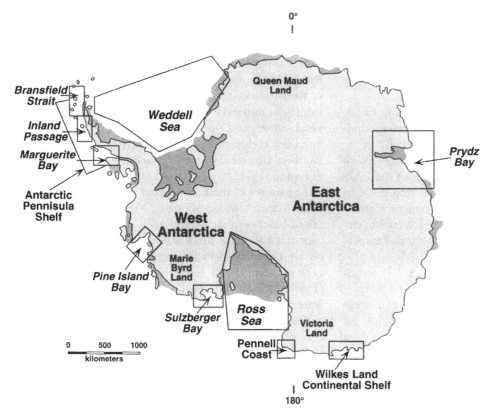

Figure 6.13. Areas of the Antarctic continental shelf where marine geologic and geophysical surveys have been conducted in order to reconstruct the glacial setting during the Late Pleistocene.

Pope and Anderson, 1992; Pudsey, Barker, and Larter, 1994), and on the continental shelf offshore of Pine Island Bay (Kellogg and Kellogg, 1987, unpublished results). The evidence for ice sheets having grounded on these continental shelves during the Late Pleistocene is irrefutable. In all these areas, the common stratigraphy seen in piston cores consists of tills overlain by glacial-marine sediments. Seismic records from these areas also provide strong evidence of recent grounding on the continental shelf in the form of subglacial facies and geomorphic features resting on top of regional glacial erosion surfaces (Kennedy and Anderson, 1989; Herron and Anderson, 1990; Anderson, Bartek, and Thomas, 1991b, 1992a,b; Pope and Anderson, 1992; Shipp and Anderson, 1995; Shipp et al., in press). Glacial striations have been observed on the continental shelf off Wilkes Land (Barnes, 1987) and on the Antarctic Peninsula continental shelf (Pudsey, Barker, and Larter, 1994). Detailed multibeam surveys of the Ross Sea continental shelf have revealed some of the most spectacular evidence for ice sheet grounding on the shelf in the form of megascale glacial lineations, drumlins, and a number of other subglacial and ice-proximal geomorphic features (Chapter 3).

EAIS Reconstructions

Wilkes Land Continental Shelf. The glacial reconstruction for the Wilkes Land continental shelf of East Antarctica

is based on a till provenance study by Domack (1982), side-scan sonar data of Barnes (1987), and seismic stratigraphic analyses by Eittreim and colleagues (1995) (Fig. 6.14). This reconstruction places the ice sheet grounding line at the edge of the continental shelf and calls for the existence of several large ice streams. AMS (tandem accelerator mass spectrometer) radiocarbon dates of glacial-marine sediment that directly overlies till indicate that the transition from subglacial to glacial-marine sedimentation occurred prior to ~9000 yr BP, and that the retreat of the grounding line to its present coastal position was complete by ~2000 yr BP (Domack et al., 1986; Domack et al., 1991a). Deglaciation of the Windmill Islands, west of the George V Coast between 110°E and 111°E, occurred between 8000 and 5000 yr BP (Goodwin, 1993).

Weddell Sea. The glacial reconstruction for the eastern Weddell Sea is illustrated in Figure 6.15. It is based on sedimentologic work by Anderson and colleagues (1980b) and Elverhøi (1981) and on a till provenance study by Andrews (1984) and Anderson and colleagues (1991a). This configuration is consistent with the reconstruction by Denton and colleagues (1991) and shows ice streams from both East and West Antarctica extending to the outer continental shelf. The largest glacial feature on the shelf is the Crary Trough. Tills occur at water depths just over 1000 m in this trough and imply a minimum of 1200 m of grounded ice (Bentley and Anderson, 1998). Only a few piston cores were collected west of the Crary Trough, so

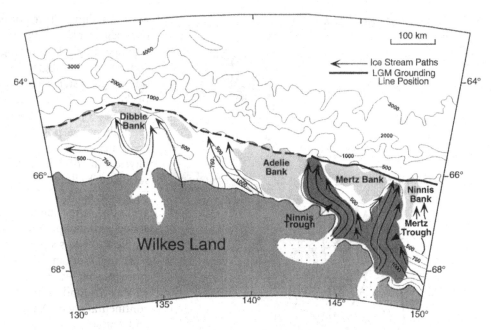

Figure 6.14. Late Pleistocene ice sheet reconstruction for the Wilkes Land continental shelf based on a till petrography study by Domack (1982), side-scan sonar investigation by Barnes (1987), and seismic stratigraphic analysis by Eittreim and colleagues (1995). Bathymetry is from Chase (1987). Dark shading represents areas (troughs) where large ice streams existed. Arrows designate paleoflow lines (modified from Eittreim, Cooper, and Wannesson, 1995).

determining the extent of the ice sheet on this portion of the continental shelf remains problematic.

Age control on these glacial deposits in the Weddell Sea is limited. Elverhøi (1981) obtained radiocarbon dates ranging from 37,830 to 21,240 yr BP on shell hash beds that overlie till and glacial-marine sediments on the outer shelf and upper slope. Elverhøi and Maisey (1983) acquired high-resolution seismic profiles from the area that show a mid-shelf grounding zone wedge resting above a glacial unconformity that extends to the shelf break. Bentley and Anderson (1998) obtained radiocarbon ages for two samples from 205 and 400 cm depth in core 3-17-1 from the outer shelf offshore of the Princess Martha Coast of 14,940 and 23,870 yr BP respectively. These samples were both taken from a glacial-marine unit resting on till. They also obtained radiocarbon ages from glacial-marine sediments in core 3-7-1 that range back to 26,660 yr BP. These combined data indicate that the EAIS advanced across the continental shelf sometime prior to 37,830 yr BP and retreated from the shelf by about 26,000 yr BP.

Mackensen and colleagues (1989) acquired isotopic records from both planktic and benthic foraminifera in sediment cores from the continental slope and rise offshore of Queen Maud Land. Detailed analysis of these cores by Grobe and Mackensen (1992) showed that the last glacial-interglacial cycle corresponded to that of the Northern Hemisphere (Fig. 6.11). Their data also show smaller concentrations of planktic foraminifera and IRD in surface sediments compared to sediments from the last (Stage 5e) interglacial. Analysis of gravity cores from the Weddell Sea continental slope, rise, and abyssal floor west of 40° W have shown planktic foraminifera within interglacial, IRD-rich sediments range from 26,000 to beyond the limits of radiocarbon dating and suggest that the LGM

occurred after 26,000 yr BP (Anderson and Andrews, in press). These data suggest that the present glacial setting is more severe than that which existed during Stage 5e.

Prydz Bay. O'Brien (1994) and O'Brien and Harris (1996) used bottom profiler records to characterize the bottom morphology of Prydz Bay. Their map of the seafloor shows several troughs, and bottom profiler records show features that are interpreted as being megaflutes. Seismic profiles from the shelf show prograding clinoforms and reflect subglacial sediment transport across the shelf during the Plio-Pleistocene (O'Brien and Harris, 1996). These combined data were used to infer that the ice sheet was grounded to the shelf edge during the late Pleistocene (Fig. 6.16). Later work on sediment cores from Prydz Bay showed that the ice sheet was not grounded in the deeper portions of troughs during the LGM (Domack et al., 1998).

The ODP Leg 119 sites in Prydz Bay penetrated diamicton just below the seafloor, which was interpreted as late Pleistocene till (Barron, Larsen, and Baldauf, 1991). Domack and colleagues (1991b) acquired radiocarbon ages from glacial-marine sediments resting above the Late Pleistocene till. These dates indicate that the ice sheet retreated from Prydz Bay sometime around 11,500 yr BP. A Mid-Holocene re-advance of the Lambert/Amery system occurred between 7000 and 4000 yr BP. Goodwin (1996) recorded a similar Mid-Late Holocene readvance of the EAIS at Law Dome, along the Budd Coast.

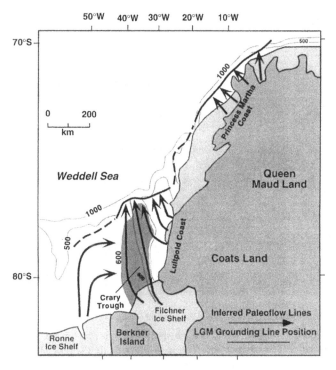

Figure 6.15. Late Pleistocene glacial reconstruction for the eastern Weddell Sea based on a till provenance study by Andrews (1984) and Anderson and colleagues (1991a). (See Fig. 4.13 for data.) Arrows show paleoflow lines. Ice streams from both East and West Antarctica were grounded in the Crary Trough. The extent of the grounding line west of the Crary Trough is inferred; no tills were recovered from this part of the continental shelf, but only short cores were collected in the region.

WAIS Reconstructions

Ross Sea. Scott (1905) was the first to speculate that the Ross Sea once was covered by an extensive ice sheet, the "Great Ice Sheet," which he believed was grounded on the continental shelf as far north as Cape Adare. It turns out that Scott was not far off in his speculation. The presence of a grounded ice sheet in the Ross Sea during the Late Pleistocene is strongly supported by marine geophysical and geologic data.

High-resolution seismic profiles show a series of regional unconformities on the continental shelf with surface relief that mimics the modern glacial topography of the shelf (Alonso et al., 1992; Chapter 3). The youngest of these unconformities occurs at or near the seafloor. The unit that rests directly on this unconformity includes an extensive till sheet (Fig. 4.4), but other subglacial facies and geomorphic features also occur above this unconformity, including ice stream boundary ridges (Fig. 4.5) and grounding zone wedges (Anderson et al., 1992a; Fig. 4.16A). Deep-tow side-scan and multibeam surveys reveal subglacial features including striations, moraines and mega-scale glacial lineations, (Fig. 3.6) on virtually every part of the shelf surveyed as well as drumlins on one part of the shelf (Fig. 3.8).

Piston cores from the Ross Sea continental shelf have sampled a general stratigraphy of diatomaceous glacial-marine sediment/transitional glacial-marine sediment/till (Kellogg et al., 1979; Anderson et al., 1980). Detailed sedimentologic and petrographic analyses were used to confirm these subglacial deposits and reconstruct paleodrainage systems (Anderson et al., 1980b; Anderson, Brake, and Myers, 1982a; Anderson et al., 1992a; Myers, 1984; Jahns, 1994; Domack et al., in press). The grounding line positions and ice stream boundaries reconstructed from the combined sedimentologic and geophysical data (Shipp et al., in press) along with recent results from Kellogg, Hughes, and Kellogg (1996) are shown in Figure 6.17. These data show that the ice sheet was grounded at the shelf break, and the grounding line associated with that portion of the ice sheet that was nourished by the EAIS was situated near Coulman Island.

Recent work in the Ross Sea has yielded the best chronostratigraphic record, with over 100 radiocarbon dates, of ice sheet grounding and retreat on any part of the Antarctic continental shelf (Anderson et al., 1992a; Taviani, Reid, and Anderson, 1993; Licht et al., 1996, in press; Domack et al., 1995; Domack et al., in press). These dates were acquired by AMS radiocarbon dating of shell material and total organic carbon. The carbon reservoir correction that should be applied to organic carbon dates remains problematic, but is somewhere between 1500 and 3000 yr (Domack, personal communication), so these dates are less reliable than carbonate dates.

Radiocarbon dates on shell material from an outer shelf carbonate bank in the western Ross Sea range from greater

Figure 6.16. Late Pleistocene glacial reconstruction for Prydz Bay based on seafloor bathymetry and geomorphic data from O'Brien (1994) and O'Brien and Harris (1996).

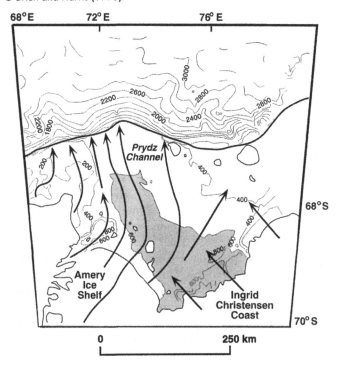

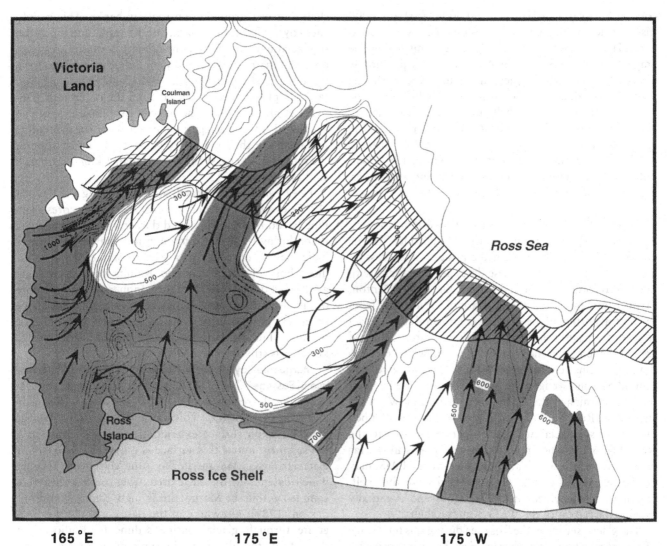

165°E   175°E   175°W

Figure 6.17. Late Pleistocene glacial reconstruction for the Ross Sea. Compiled from orientations of megascale glacial lineations (Fig. 3.6) and drumlins (Fig. 3.8), locations of ice stream boundary ridges (Fig. 4.5) and troughs, and results from till provenance studies. Shaded areas are where major ice streams existed, and the cross-hatched area designates the location of the LGM grounding zone (from Myers, 1982a; Anderson et al., 1984b, 1992a; Jahns, 1995; Shipp and Anderson, 1995; Shipp et al., in press).

than 34,000 yr BP to 22,700 yr BP and indicate open-marine conditions on this portion of the continental shelf during isotope Stage 3 (Taviani, Reid, and Anderson, 1995). The absence of an ice sheet grounded on the outer continental shelf during isotope Stage 3 is supported by radiocarbon dates on calcareous fossils from the Terra Nova drift in northern Victoria Land, which yielded ages older than 25,300 yr BP (Orombelli, Baroni, and Denton, 1991) and glacial-marine sediments in the western Ross Sea that have yielded AMS ages of 27,000 to 22,000 yr BP (Licht et al., 1996). Radiocarbon analyses have focused on the oldest glacial-marine sediment directly overlying till and transitional glacial-marine sediment. Cores collected in close proximity to the grounding zone wedge adjacent to Coulman Island yielded ages from ~29,000 to ~20,000 yr BP (Licht et al., 1996). Cores collected south of Coulman Island grounding zone yielded ages that range from ~18,000 to ~11,000 yr BP (Anderson et al., 1992a; Licht et al., 1996).

The combined results from these recent studies indicate an LGM ice sheet configuration with the ice sheet grounded near the shelf break in the eastern Ross Sea and south of Coulman Island in the western Ross Sea (Fig. 6.17). This is a smaller ice sheet than the one that advanced to the edge of the continental shelf prior to 34,000 yr BP, creating the widespread glacial unconformity and shelf-edge grounding zone wedges and trough mouth fans seen in seismic records (Shipp and Anderson, 1995) (Fig. 3.15). The older grounding event may record the marine extension of the ice sheet during oxygen isotope Stage 6, which is recorded by a 185,000 to 130,000 yr BP–aged drift deposit in the TAM (Dagel, 1985; Denton, Prentice, and Burckle, 1991). Alternatively, this grounding event may have occurred during the 60,000 to 50,000 yr BP (Stage 4) glacial maximum that is recorded by ice cores at Vostok (Fig. 6.5) and by 60,000 (K/Ar) to 25,000 (fission track) yr BP-aged striations on the coast of Gaussberg (Tingey, 1982). The Vostok records imply that the Stage 4 glacial maximum in Antarctica was potentially as severe as the Stage 2 glacial maximum (Fig. 6.5). The Stage 4 grounding event seems most probable, given the radiocarbon dates on the shelf, and is consistent with results from coastal studies in East Antarctica that indicate retreat of the ice sheet prior to ~ 30,000 yr BP (Yoshida, 1983, 1989; Orombelli, Baroni, and Denton, 1991; Colhoun, 1991). These data are also consistent with observations in the Weddell Sea, discussed earlier in this section.

By 11,500 yr BP, the ice sheet in the western Ross Sea retreated south to the vicinity of the Drygalski Ice Tongue and, by ~ 7000 yr BP, it reached a position near Ross Island (Licht et al., 1996). These dates are consistent with onshore records (Baroni and Orombelli, 1991; Denton, Prentice, and Burckle, 1991). Retreat of the ice sheet from the central portion of the Ross Sea continental shelf, which receives drainage exclusively from West Antarctica, lagged behind the retreat in the western Ross Sea, which receives drainage exclusively from the EAIS. This is indicated by a grounding zone wedge that prograded from Iselin Bank (the paleodrainage divide between EAIS and WAIS ice streams in Ross Sea) toward the west and downlapped onto older subglacial and glacial-marine deposits to the west (Fig. 6.18). This implies that ice streams in the far western Ross Sea, which are nourished by the EAIS, may be the "weak links" that first retreated from the shelf, creating an embayment in the ice front and eventually causing the retreat of the WAIS from the shelf.

The glacial setting in the Ross Sea during this final stage of retreat may have been characterized by an extensive ice shelf that was grounded on shallow banks. This is indicated by the flat-topped nature of the banks and pinch-out of glacial units onto the banks (Shipp et al., in press), an apparent hiatus of ~ 15,000 yr in radiocarbon dates from the outer shelf (Licht et al., 1996), and by sedimentary facies in piston cores collected from the outer shelf (Domack et al., in press). The paucity of sediment cover on subglacial features suggests that retreat of the grounding line from the shelf may have been rapid, but the rate of the retreat of the ice shelf was less than 1 km/yr (Domack et al., in press).

Pine Island Bay and Amundsen Shelf. Kellogg and Kellogg (1987) examined seismic records and piston cores from Pine Island Bay and suggested that an ice sheet grounded there during the LGM, although no detailed sedimentologic analyses of the cores have been conducted and no radiocarbon dates were acquired. Results from a petrologic analysis of diamictons from the continental shelf offshore of Pine Island Bay were inconclusive as to whether an ice sheet was grounded there during the LGM (unpublished results). Thus, the Late Pleistocene glacial history of this key area is still unresolved.

Marguerite Bay, Antarctic Peninsula Continental Shelf, and Inland Passage. One of the first areas studied in the Antarctic Peninsula region was the Marguerite Bay (Kennedy and Anderson, 1989). This work revealed a prominent glacial erosion surface extending across the entire bay; the sediment above this surface is quite thin. The largest glacial trough in Marguerite Bay is the Marguerite Trough, which extends to the shelf break (Fig. 6.19). Tills were sampled within the Marguerite Trough (Kennedy and Anderson, 1989). Elsewhere in the bay, cores did not penetrate through glacial-marine sediments. Radiocarbon dates from the glacial-marine unit in two piston cores yielded surface ages in the range of 4100 ± 150 yr BP to 4260 ± 160 yr BP (Harden, DeMaster, and Nittrouer, 1992), indicating the presence of significant old carbon in these cores. The corrected dates for the oldest

Figure 6.18. Seismic profile from the western Ross Sea showing westward-prograding strata associated with Paleo-Ice Stream B advancing into a trough that was previously occupied by Paleo-Ice Stream A.

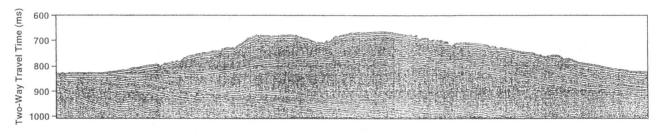

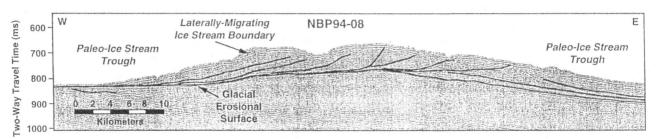

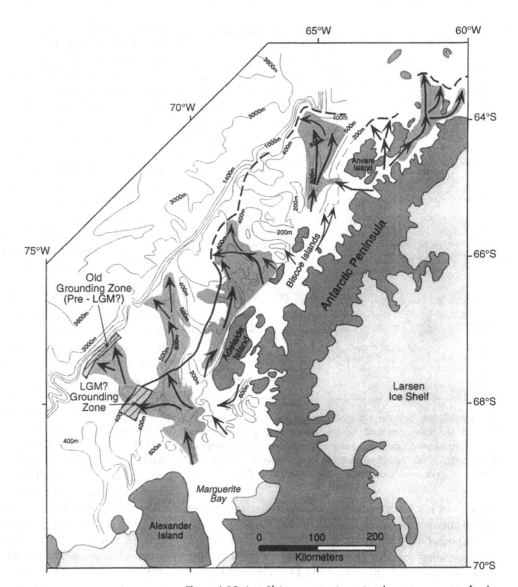

Figure 6.19. Late Pleistocene maximum ice sheet reconstruction for the Antarctic Peninsula continental shelf. Arrows show paleoflow lines, and the heavy line marks the approximate location of a prominent grounding zone wedge (cross-hatched area) on the inner shelf. The heavy line is the inferred maximum grounding line during the LGM (compiled from Pope and Anderson, 1992, Bart and Anderson, 1995, and Pudsey et al., 1994).

glacial-marine sediments are in the order of 6000 to 6700 yr BP, so the ice sheet had retreated from the central part of Marguerite Bay by this time. This agrees with the results of Clapperton and Sugden (1982), who suggested that George VI Sound was ice-free by 6000 yr BP, based on pelecypod shells of that age in a moraine on Alexander Island.

Examinations of high-resolution seismic data, side-scan sonar records, bottom profiler records, and piston cores from the Antarctic Peninsula shelf have yielded somewhat controversial results concerning the extent of grounded ice on the continental shelf during the LGM (Pope and Anderson, 1992; Pudsey, Barker, and Larter, 1994). Pope and Anderson (1992) concluded that the sediment recovered in piston cores on the shelf indicate that the LGM grounding line was situated on the inner shelf and that a fringing ice shelf extended across portions of the middle shelf (Fig. 6.19). They based this conclusion on the fact that ridges within troughs, interpreted to be megaflutes, exhibit greater relief on the inner shelf and appear to be

draped by glacial-marine sediment on the outer shelf, and none of the cores they acquired from the outer shelf sampled tills. Pudsey and colleagues (1994) examined side-scan sonar records and four sediment cores from a portion of the continental shelf offshore of Anvers Island. Their records show glacial flutes, thus providing indisputable evidence of the ice sheet having grounded on the inner shelf (Fig. 3.7). They reasoned that the grounding line of the ice sheet was situated at the shelf break during the LGM, and later work by Larter and Vanneste (1995) showed a grounding wedge on the outer shelf in this region. Bart and Anderson (1996) later showed a prominent glacial unconformity extending to the shelf break and a

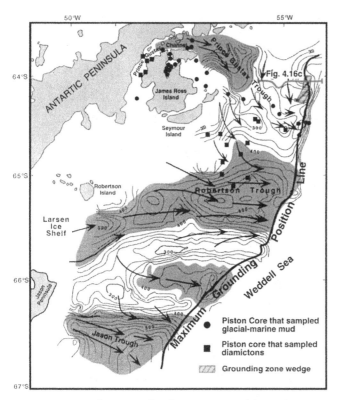

Figure 6.20. Late Pleistocene glacial reconstruction of the northwestern Weddell Sea continental shelf. Arrows show inferred paleoflow lines. Cross-hatched area designates the grounding zone wedge (compiled from Smith, 1985, Anderson et al., 1991c, Anderson, Shipp, and Siringan, 1992b, and Sloan, Lawver, and Anderson, 1995).

number of stacked grounding zone wedges resting above this unconformity on the inner shelf (Fig. 4.16B).

Radiocarbon dates from glacial-marine sediment above the grounding zone proximal facies in two cores yielded corrected ages of 11,481 to 12,430±140 yr BP and 10,317 to 11,335 ± 110 yr BP, respectively (Pope and Anderson, 1992). Pudsey, Barker, and Larter (1994) acquired age dates younger than 11,000 yr BP on glacial-marine sediment on the continental shelf offshore of Anvers Island. These combined data indicate that a glacial-maritime setting similar to that of the present was established by 12,430 yr BP.

As in Marguerite Bay, high-resolution seismic records from the Gerlache Strait show that this portion of the inland passage virtually has been scraped clean of sediment (Fig. 3.26), except for a thin layer of post-glacial diatomaceous glacial-marine sediment (Griffith and Anderson, 1989). Radiocarbon ages from a core collected in the central part of the Gerlache Strait indicate that glacial-marine sedimentation began sometime after 8000 yr BP (Harden, DeMaster, and Nittrouer, 1992), and radiocarbon dates from fjords that surround the Gerlache Strait indicate that ice retreat from these fjords by 8000 yr BP (Shevenell et al., 1998).

Weddell Sea. Seismic records from the continental shelf offshore of James Ross and Seymour islands in the

northwestern Weddell Sea show widespread glacial unconformities and subglacial seismic facies (Anderson, Shipp, and Siringan, 1992b; Sloan, Lawver, and Anderson, 1995) (Figs. 4.16C and 5.37). As seen on the Pacific side of the Peninsula, there is a prominent glacial erosion surface located near the seafloor that extends to the shelf-break, and a grounding zone wedge rests atop this unconformity on the mid-shelf (Fig. 4.16C). Piston cores from this area did not penetrate till. Detailed analysis of these piston cores by Smith (1985) included petrographic work that showed strong provinciality of transitional glacial-marine deposits. This was taken as evidence for deposition beneath a fringing ice shelf that extended to the shelf edge (Anderson et al., 1991c), but these diamictons could also be deformation till. Radiocarbon ages have not been acquired from these deposits.

High-resolution seismic records and sediment cores collected within Prince Gustav Channel and fjords along the channel show that these areas have been virtually scraped clean of sediment, except for a thin layer of diatomaceous glacial-marine mud that drapes acoustic basement within fjords and the Gustav Trough. Onshore studies in the area indicate that deglaciation onshore occurred prior to 7400 yr BP (Hjort, Ingólfsson, and Björk, 1992; Hjort et al., 1997). Figure 6.20 shows the late Pleistocene glacial reconstruction for this region.

The extent of the grounded ice sheet on the southern Weddell Sea continental shelf is poorly constrained. Piston cores from the Crary Trough sampled till indicating that an ice sheet was grounded in the trough; however, attempts to radiocarbon date the glacial-marine sediments that overlie these tills have been hampered by a lack of carbonate material (Bentley and Anderson, 1998). Short sediment cores collected there during a 1968 glacier cruise did not penetrate basal tills (Anderson et al., 1980b). Longer cores acquired on several expeditions of the R/V *Polarstern* did penetrate "stiff" diamictons, interpreted as till, along the western front of the Ronne Ice Shelf (Fütterer and Melles, 1990). A 3.5-kHz subbottom profiler record collected along the front of the ice shelf showed hummocky seafloor topography, typical of glaciated seafloor and identical to seafloor morphology in the Ross Sea where megascale glacial lineations exist (Fig. 3.7).

Bransfield Strait. The late Quaternary glacial history of the Bransfield Strait was the subject of study by Banfield (1994) and Banfield and Anderson (1995), who examined high-resolution seismic reflection profiles and piston cores acquired from the basin during several marine geologic expeditions. Their results show that the ice sheet was grounded at the edge of the continental shelf at present-day water depths of up to 700 m during the Late Pleistocene (Fig. 6.21); the grounding line position is marked by grounding line ridges (Fig. 4.18). Radiocarbon dates from ice-proximal deposits in piston Core DF 82-48, using both bulk organic carbon and foraminiferal tests, indicate that the ice sheet retreated from the outer continental shelf

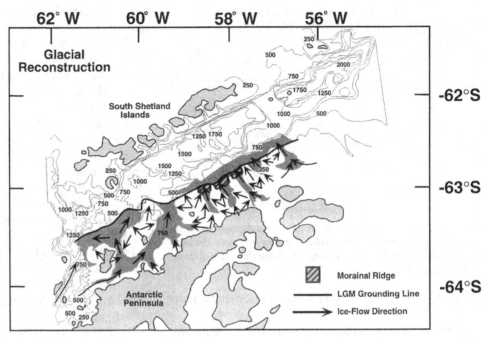

Figure 6.21. Late Pleistocene ice sheet reconstruction for the Bransfield Strait (from Banfield and Anderson, 1995).

Figure 6.22. Late Pleistocene ice sheet reconstruction for the South Orkney Plateau based on distributions of the younger glacial unconformities and occurrence of tills on the shelf. Heavy line is the approximate grounding line position (compiled from Herron and Anderson, 1990).

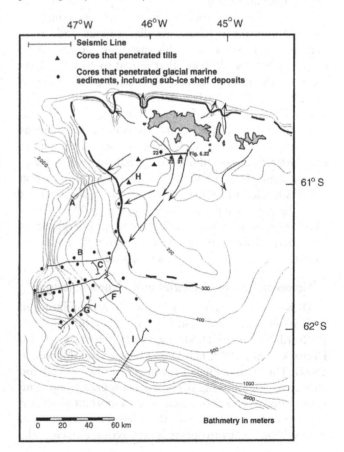

prior to ~ 13,000 to 14,000 yr BP (Banfield and Anderson, 1995). Piston cores from the central basin did not penetrate deep enough to determine if a permanent floating ice canopy covered the basin during the glacial maximum. Radiocarbon ages for diatomaceous glacial-marine sediment in five piston cores taken from the basin floor yielded maximum ages extending back only to the Mid-Holocene (Harden et al., 1992), so the basin was free of a permanent ice canopy by that time. Sediment cores from Yanou Lake, King George Island, indicate that deglaciation of the island started prior to 8000 yr BP (Yang and Harwood, 1995). Piston cores from the Central Basin show little variation in biogenic silica versus terrigenous sediment concentrations during the past 6000 years (Singer, 1987).

South Orkney Plateau. Herron (1988) and Herron and Anderson (1990) conducted a detailed study of high-resolution seismic data and piston cores from the South Orkney Plateau. This plateau is an important "dip stick" for the level at which small, relatively low latitude ice caps grounded in response to sea level fall during the Late Pleistocene. Several troughs occur offshore of the islands and indicate locations of major outlet glaciers (Sugden and Clapperton, 1977) (Fig. 6.22). A seismic profile across one of these troughs shows the glacial unconformity draped in acoustically laminated glacial-marine deposits (Fig. 6.23). Piston cores from the plateau penetrated till overlain by, and is in sharp contact with, compound (diatomaceous) glacial-marine sediments (Fig. 6.23). Radiocarbon dates from articulated pelecypods found within compound glacial-marine sediments in Core DF85-22 (Fig. 6.23) indicate that the ice cap retreated from the plateau prior to 6000 to 7000 yr BP (Herron and Anderson, 1990). Cores collected at water depths deeper than 250 m recovered glacial marine diamicton, interpreted as sub-ice

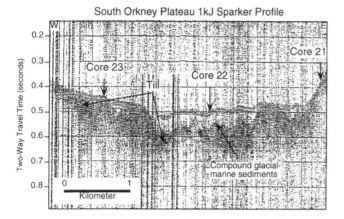

South Orkney Plateau 1kJ Sparker Profile

Figure 6.23. Seismic profile showing the glacial erosion surface on the South Orkney Plateau and overlying glacial-marine sediments. The locations of piston Cores 85-21, 85-22, and 85-23 are also shown (Herron and Anderson, 1990).

shelf deposits, beneath compound glacial-marine sediments.

Summary. There is abundant evidence that both the East and West Antarctic ice sheets advanced onto the continental shelf during the LGM, but this was not the most widespread glacial grounding event in the Late Pleistocene. Indeed, some coastal areas of East Antarctica appear to have remained unaffected by this glacial episode, and marine geologic data from several areas of the continental shelf support more extensive earlier grounding events during the Late Pleistocene.

Retreat of the ice sheet from the Ross Sea continental shelf appears to have been diachronous, indicating that certain ice streams, in this case those nourished by the EAIS, are "weak links" that retreat to create embayments in the grounding line, thus setting up retreat of the ice sheet. Radiocarbon dates are desperately needed from all areas of the shelf, but the few dates that are available indicate that deglaciation of the continental shelf was relatively complete by Mid-Holocene time. This implies that Antarctica's contribution to global sea-level rise was largely complete by the Mid-Holocene. More radiocarbon work is needed to resolve the timing of ice sheet retreat from the continental shelf following the LGM. There is some evidence to indicate that the present interglacial is characterized by more extensive ice cover than was the case during the last (Stage 5e) interglacial.

Holocene Record of Glacial and Climate Change

Diatomaceous sediment is presently accumulating in shelf basins and fjords and bays at relatively high rates (Ledford-Hoffman, DeMaster, and Nittrouer, 1986; Domack et al., 1989; Harden, DeMaster, and Nittrouer, 1992). This material may provide a record of glacial and climatic changes during the past few thousand years and offers information about some very important problems concerning feedback between the lithosphere, cryosphere, atmosphere, and hydrosphere in Antarctica, as well as the

response of the Antarctic cryosphere to different forcing agents.

This research was not possible before the advent of AMS radiocarbon dating technology because there previously was no way of dating these sediments, which contain only small amounts of organic carbon and rarely any calcareous material. With the advent of this new technology, a whole new field of Antarctic marine geologic research opened. To date, only a few papers describing the results of investigations of Holocene sediments have been published, but the results have been significant. One of the greatest challenges in this area of research is that of understanding the carbon reservoir effect, which causes radiocarbon dates to be older, in some cases by as much as a few thousand years.

Ice cores show an Early Holocene climatic optimum between 10,000 and 6000 yr BP (Ciais et al., 1992). This appears to have been a time of ice sheet retreat from the continental shelf all around Antarctica. Domack and colleagues (1991b) present evidence for a Mid-Holocene re-advance of the EAIS in Prydz Bay and offshore of Wilkes Land. Goodwin (1996) presented evidence for a re-advance of the EAIS at Law Dome, Budd Coast, during the Mid–Late Holocene. These authors suggest that this re-advance may have been a direct consequence of climate warming during an Antarctic "hypsithermal." Denton and colleagues (1991) also present evidence that there was expansion of alpine glaciers in the TAM during the Holocene, and Clapperton and Sugden (1982) report evidence for Holocene re-advance of glaciers in Marguerite Bay. Expansion of both the EAIS and WAIS during the hypsithermal implies that the response of these ice sheets to future warming will be to expand, not contract.

Domack and colleagues (1995) document a Holocene re-advance of the Müller Ice Shelf (Anvers Island, Antarctic Peninsula) during the Little Ice Age that was followed by retreat of its calving line in recent time. Yet, the Siple ice core record indicates warmer temperatures during the Little Ice Age compared to today (Mosley-Thompson, 1992). Domack and colleagues (1995) suggest that the apparent discrepancy between the ice core record and the marine sedimentary record is likely due to the fact that either the ice core record is a high-elevation record that is insensitive to maritime influences or there is a lag between climatic and glacial events. They attribute these changes to recent warming trends in the Antarctic Peninsula region (Chapter 1).

Leventer, Dunbar, and DeMaster (1993) examined diatomaceous sediment in the McMurdo Sound region for a century-scale record of climatic change. Their results indicate a complex relationship between climate and sea-ice cover; decreased sea-ice cover can result from either increased warming or from stronger katabatic winds during colder periods.

Domack, Mashiotta, and Burkley (1993), Leventer and colleagues (1996), and Shevenell and colleagues (1996) have documented multi-century variation in organic matter preservation and biogenic silica accumulation rates in Antarctic fjord sediments, which they believe records changes in productivity and meltwater sedimentation. These multicentury variations are attributed to paleoclimatic and/or oceanographic changes in the Peninsula region, most notably changes in sea-ice extent. These events are not recorded in the Siple ice core record. More recently, Domack and Mayewski (personal communication) have demonstrated a strong correlation between paleoproductivity events in fjords of the Antarctic Peninsula region and paleotemperature changes recorded in Greenland ice cores (Fig. 6.24), which suggests a bipolar and possibly global-forcing mechanism. Recently published ice core records from East Antarctica show similar multicentury climatic events (Mayewski et al., 1996; Mosley-Thompson, 1996) and thus indicate that the Antarctic climate changes more rapidly than previously thought.

Students searching for exciting marine geologic research projects in Antarctica are well advised to consider the high-resolution sedimentary record of climatic, glacial, and oceanographic change to be found in shelf basins, bays, and fjords. These high-frequency climatic events must be studied in greater detail if we are going to be able to assess the shorter-term (historic) record of climate change in Antarctica.

SUMMARY

During the past two decades, the proxy record of ice volume change derived from oxygen isotopes and the sea-level records derived from sequence stratigraphic analyses have steadily improved. These records support the existence of large ice sheets during much of the Cenozoic. Concurrently, the geologic record of glaciation on Antarctica has improved considerably. The result of geologic studies has been extension of the age of the ice sheet further back in time so that the geologic record of ice sheet development is now more consistent with the eustatic and oxygen isotope records. The geologic data also indicate a complex history of ice sheet growth and decay.

Current evidence does not favor the existence of a large ice sheet in Antarctica during the Cretaceous through the Paleocene, although ice caps may have inhabited the central portions of the continent; Cretaceous and Paleocene exposures are restricted to the northern Antarctic Peninsula region.

There is good evidence that the EAIS was present by Middle(?)–Late Eocene time and that it experienced significant volume changes during the Late Eocene and Oligocene. This early phase of ice sheet evolution occurred at a time when all paleoclimatic data and climate models indicate a cold temperate climate, at least in coastal regions. So, the early ice sheet was more sensitive to climatic changes than it is today and perhaps waxed and

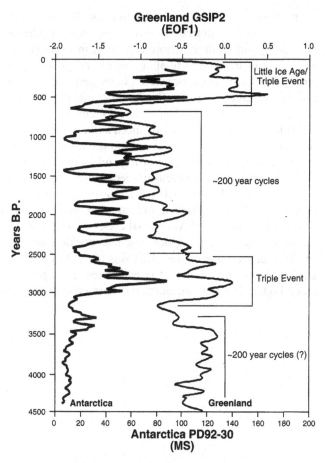

Figure 6.24. Comparison of Holocene climate record from Greenland ice core (GISP2) and magnetic susceptibility record for piston core (PD 92-30) from the Palmer Deep in the Antarctic Peninsula region (courtesy of Eugene Domack).

waned in response to high-frequency (Milankovitch-scale) climatic oscillations. During the Oligocene, the EAIS advanced onto the continental shelf all around East Antarctica, with the possible exception of the Queen Maud Land region. The EAIS also advanced into the western Ross Sea across what may have been a much lower TAM chain.

Initial glaciation of West Antarctica probably began with a number of different ice caps situated on small continental blocks and islands separated by seaways that occupied rift basins. With time, these seaways were filled with glacial-marine sediments, allowing the WAIS to spread across the West Antarctic continent and adjacent continental shelf. The early WAIS was grounded nearer to sea level than at present. This implies that the WAIS contribution to global sea level was greater during its early development than is the case today. Possible uplift of the TAM created a greater physiographic barrier to EAIS drainage into West Antarctica, resulting in the segregation of the EAIS and WAIS.

The oxygen isotope and deep-sea paleontologic records indicate significant ice volume fluctuations in Antarctica during the Pliocene. Several glacial unconformities occur in the Pliocene section on the continental shelves of the Ross Sea and the Antarctic Peninsula, providing evidence of waxing and waning marine ice sheets in these areas. There is still much debate about when the present polar conditions on Antarctica were established, but there is general agreement that this important climatic threshold was complete by the Plio-Pleistocene boundary.

Both the EAIS and WAIS advanced and retreated from the continental shelf on several occasions during the Pleistocene, perhaps in response to rising and falling sea level, implying an in-phase relationship between Northern and Southern Hemisphere glacial cycles. However, existing geologic data, although limited, indicate that the EAIS may be out of phase with WAIS glacial cycles. There is some evidence that the EAIS grounding line position on the continental shelf during the LGM was located in a more landward position relative to an earlier (either Stage 6 or Stage 4) glacial maximum, and that the present interglacial setting is characterized by a more extensive ice edge than existed during the last (Stage 5e) interglacial.

Future marine geologic research should focus on obtaining a better stratigraphic record of ice sheet expansion and retreat from the continental shelf and on those mechanisms influencing ice sheet advance and retreat. Also, recent results from investigations of glacial-marine sediments in fjords and shelf basins have yielded exciting results concerning Antarctica's climatic record during the Holocene and indicate century-scale climate cycles. This research must be continued if we are going to determine whether ongoing changes in the Antarctic climate and glacial setting are due to anthropogenically induced global warming or part of natural climate cycles. There is also a need to revisit the deep-sea (proxy) sedimentary record of Antarctica's glacial and climatic history, given improved age dating techniques and a better understanding of marine sedimentation around the continent.

Bibliography

Aario, R. 1977. Classification and terminology of morainic landforms in Finland. *Boreas*, vol. 6, pp. 87–100.

Abelmann, A., R. Gersonde, and V. Spiess. 1990. Pliocene-Pleistocene paleoceanography in the Weddell Sea – siliceous microfossil evidence. In U. Bleil and J. Thiede, eds., *Geological History of the Polar Oceans: Arctic Versus Antarctic*. Kluwer Academic Publishers, Boston, pp. 729–759.

Abreu, V. S., and J. B. Anderson. In press. Glacial eustasy during the Cenozoic: sequence stratigraphic implications. *Am. Assoc. Petrol. Geologists Bull.*

Abreu, V. S., R. Savini, and S. Barrocas. 1992. Paleoceanography and microbiostratigraphy of the Moby Dick Group, Melville Peninsula, Northern King George Island, Western Antarctica. Fourth International Conference on Paleo Oceanography, Siena, Italy, p. 48.

Abreu, V. S., and G. A. Haddad. In press. Glacioeustatic fluctuations: the mechanism linking stable isotope events and sequence stratigraphy from the early Oligocene to middle Miocene time. In P. C. DeGraciansky, J. Hardenbol, T. Jacquin, P. R. Vail, and M. B. Farley, eds., *Sequence Stratigraphy of the European Basins*. Society of Economic Paleontologists and Mineralogists Special Publication No. 57.

Ackley, S. F. 1979. Mass balance aspects of Weddell Sea pack ice. *J. Glaciol.*, vol. 24, pp. 391–405.

Adams, C. J. D., J. E. Gabites, and G. W. Grindley. 1982. Orogenic history of the central Transantarctic Mountains: new K-Ar age data on the Precambrian–Lower Paleozoic basement. In C. Craddock, ed., *Antarctic Geoscience*. University of Wisconsin Press, Madison, pp. 817–826.

Adamson, D., and J. Pickard. 1983. Late Quaternary ice movement across the Vestfold Hills, East Antarctica. In R. L. Oliver, P. R. James, and J. B. Jago, eds., *Antarctic Earth Science*. Cambridge University Press, New York, pp. 465–469.

Adamson, D. A., and J. Pickard. 1986. Cainozoic history of the Vestfold Hills. In J. Pickard, ed., *Antarctic Oasis – Terrestrial Environments and History of the Vestfold Hills*. Academic Press, New York, pp. 63–97.

Adie, R. J. 1957. The Petrology of Graham Land. III. Metamorphic Rocks of the Trinity Peninsula Series. Falkland Islands Dependencies Survey Scientific Reports, no. 20. Her Majesty's Stationary Office, London, England, 26 pp.

Adie, R. J. 1972. Evolution of volcanism in the Antarctic Peninsula. In R. J. Adie, ed., *Antarctic Geology and Geophysics*.

International Union of Geological Sciences, Series B, vol. 1. Universitetsforlaget, Oslo, Norway, pp. 137–141.

Ahmad, N. 1981. Late Palaeozoic Talchir tillites of Peninsular India. In M. J. Hambrey and W. B. Harland, eds., *Earth's Pre-Pleistocene Glacial Record*. Cambridge University Press, New York, pp. 326–330.

Aitchison, J. C., M. A. Bradshaw, and J. Newman. 1988. Lithofacies and origin of the Buckeye Formation: Late Paleozoic glacial and glaciomarine sediments, Ohio Range, Transantarctic Mountains, Antarctica. *Palaeogeog. Palaeoclimatol. Palaeoecol.* vol. 64, pp. 93–104.

Aitkenhead, N. 1965. The Geology of the Duse Bay-Larsen Inlet Area, North-East Graham Land (with Particular Reference to the Trinity Peninsula Series). British Antarctic Survey Scientific Reports, No. 51, 62 pp.

Alley, R. B. 1990. West Antarctic collapse – how likely? *Episodes*, vol. 13, pp. 231–238.

Alley, R. B., S. Anandakrishan, C. R. Bentley, and N. Lord. 1994. A water-piracy hypothesis for the stagnation of Ice Stream C, Antarctica. *Ann. Glaciol.* vol. 20, pp. 187–194.

Alley, R. B., D. D. Blankenship, C. R. Bentley, and S. T. Rooney. 1987a. Till beneath ice stream B, 3. till deformation: evidence and implications. *J. Geophys. Res.*, vol. 92, pp. 8921–8929.

1987b. Till beneath ice stream B, 4. a coupled ice-till flow model. J. Geophys. Res., vol. 92, pp. 8931–8940.

1989. Sedimentation beneath ice shelves – the view from ice stream B. *Marine Geol.*, vol. 85, pp. 101–120.

Alley, R. B., and I. M. Whillans. 1991. Changes in the West Antarctic Ice Sheet. *Science*, vol. 254, pp. 959–963.

Allison, I. 1979. The mass budget of the Lambert Glacier drainage basin, Antarctica. *J. Glaciol.*, vol. 22, pp. 223–234.

Alonso, B., J. B. Anderson, J. I. Díaz, and L. R. Bartek. 1992. Pliocene-Pleistocene seismic stratigraphy of the Ross Sea: evidence for multiple ice sheet grounding episodes. In D. H. Elliot, ed., *Contributions to Antarctic Research III: Antarctic Research Series*, vol. 57: American Geophysical Union, Washington, D.C., pp. 93–103.

Anandakrishnan, S., and R. B. Alley. 1997. Tidal forcing of basal seismicity of Ice Stream C, West Antarctica, observed far inland. *J. Geophys. Res.*, vol. 102, pp. 15,183–15,196.

Anderson, J. B. 1971. Marine origin of sands in the Weddell Sea. *Ant. J. U.S.*, vol. 6, pp. 168–169.

1972a. The Marine Geology of the Weddell Sea. PhD Dissertation, Florida State University, Tallahassee, 222 pp.

1972b. Nearshore glacial-marine deposition from modern sediments of the Weddell Sea. *Nat. Phys. Sci.*, vol. 240, pp. 189–192.

1975a. Ecology and distribution of foraminifera in the Weddell Sea of Antarctica. *Micropaleontology*, vol. 21, pp. 69–96.

1975b. Factors controlling CaCO₃ dissolution in the Weddell Sea from foraminiferal distribution patterns. *Marine Geol.*, vol. 19, pp. 315–332.

1983a. Ancient glacial-marine deposits: their spatial and temporal distribution. In B. F. Molnia, ed., *Glacial-Marine Sedimentation*. Plenum Press, New York, pp. 3–92.

1983b. Preliminary results of the 1982-1983 piston-coring program. *Ant. J. U.S.*, vol. 18, pp. 157–161.

1985. Antarctic Glacial Marine Sedimentation: A Core Workshop. Geological Society of America, Boulder, Colorado, 66 pp.

1986. Critical evaluation of some criteria used to infer Antarctica's glacial and climatic history from deep-sea sediments. *S. Afr. J. Sci.*, vol. 82, pp. 503–505.

1990. Sediments. In G. P. Glasby, ed., *Antarctic Sector of the Pacific*. Elsevier Oceanography Series vol. 51, New York, pp. 187–206.

1991. The Antarctic continental shelf: results from marine geological and geophysical investigations. In R. J. Tingey, ed., *The Geology of Antarctica*. Clarendon Press, Oxford, pp. 285–334.

Anderson, J. B., and J. T. Andrews. 1999. Radiocarbon constraints on ice sheet advance and retreat in the Weddell Sea, Antarctica. *Geology*.

Anderson, J. B., B. A. Andrews, L. Bartek, and E. M. Truswell. 1991a. Petrology and palynology of Weddell Sea glacial sediments: implications for subglacial geology. In M. R. A. Thomson, J. A. Crame, and J. W. Thomson, eds., *Geological Evolution of Antarctica*. Cambridge University Press, New York, pp. 231–235.

Anderson, J. B., and G. M. Ashley. 1991. Glacial marine sedimentation; paleoclimatic significance; a discussion. In J. B. Anderson and G. M. Ashley, eds., *Glacial Marine Sedimentation; Paleoclimatic Significance*. Geological Society of America Special Paper 261, Boulder, Colorado, pp. 223–226.

Anderson, J. B., K. Balshaw, E. Domack, D. Kurtz, R. Milam, and R. Wright. 1979a. Geologic survey of the East Antarctic continental margin aboard USCGC Glacier. *Ant. J. U.S.*, vol. 14, pp. 142–144.

Anderson, J. B., and L. R. Bartek. 1991. High resolution seismic record of Antarctic glacial history. *Kor. J. Polar Res.*, vol. 2, pp. 71–78.

Anderson, J. B., and L. R. Bartek. 1992. Cenozoic glacial history of the Ross Sea revealed by intermediate resolution seismic reflection data combined with drill site information. In J. P. Kennett and D. A. Warnke, eds., *The Antarctic Paleoenvironment: A Perspective on Global Change I*. Antarctic Research Series, vol. 56. American Geophysical Union, Washington, D.C., pp. 231–263.

Anderson, J. B., L. R. Bartek, and M. A. Thomas. 1991b. Seismic and sedimentologic record of glacial events on the Antarctic Peninsula shelf. In M. R. A. Thomson, J. A. Crame, and J. W. Thomson, eds., *Geological Evolution of Antarctica*. Cambridge University Press, New York, pp. 687–691.

Anderson, J. B., C. Brake, E. W. Domack, N. Myers, and J. Singer. 1983a. Sedimentary dynamics of the Antarctic continental shelf. In R. L. Oliver, P. R. James, and J. B. Jago, eds., *Antarctic Earth Science*. Cambridge University Press, New York, pp. 387–389.

Anderson, J. B., C. Brake, E. Domack, N. Myers, and R. Wright. 1983b. Development of a polar glacial-marine sedimentation model from Antarctic Quaternary deposits and glaciological information. In B. F. Molnia, ed., *Glacial-Marine Sedimentation*. Plenum Press, New York, pp. 233–264.

Anderson, J. B., C. F. Brake, and N. C. Myers. 1984. Sedimentation on the Ross Sea continental shelf, Antarctica. *Marine Geol.*, vol. 57, pp. 295–333.

Anderson, J. B., E. W. Domack, and D. D. Kurtz. 1980a. Observations of sediment-laden icebergs in Antarctic waters: implications to glacial erosion and transport. *J. Glaciol.*, vol. 25, pp. 387–396.

Anderson, J. B., D. S. Kennedy, M. J. Smith, and E. W. Domack. 1991c. Sedimentary facies associated with Antarctica's floating ice masses. In J. B. Anderson and G. M. Ashley, eds., *Glacial Marine Sedimentation; Paleoclimatic Significance*. Geological Society of America Special Paper 261, Boulder, Colorado, pp. 1–25.

Anderson, J. B., and D. D. Kurtz. 1980. The scientific program: USCGC Glacier Deep Freeze 80 expedition. *Ant. J. U.S.*, vol. 15, pp. 114–117.

Anderson, J. B., and D. Kurtz. 1985. The use of silt grain size parameters as a paleovelocity gauge: a critical review and case study. *Geo-Marine Lett.*, vol. 5, pp. 55–59.

Anderson, J. B., D. D. Kurtz, E. W. Domack, and K. M. Balshaw. 1980b. Glacial and glacial marine sediments of the Antarctic continental shelves. *J. Geol.*, vol. 88, pp. 399–414.

Anderson, J. B., D. D. Kurtz, and F. M. Weaver. 1979b. Sedimentation on the Antarctic continental slope. In L. J. Doyle and O. Pilkey, eds., *Geology of Continental Slopes*. Society of Economic Paleontologists and Mineralogists Special Publication No. 27, pp. 265–283.

Anderson, J. B., and B. F. Molnia. 1989. *Glacial-Marine Sedimentation*. Short Course in Geology, vol. 9. American Geophysical Union, Washington, D.C., 127 pp.

Anderson, J. B., P. G. Pope, and M. A. Thomas. 1990. Evolution and hydrocarbon potential of the northern Antarctic Peninsula continental shelf. In B. St. John, ed., *Antarctica as an Exploration Frontier – Hydrocarbon Potential, Geology, and Hazards*. American Association of Petroleum Geologists Studies in Geology, No. 31, pp. 1–12.

Anderson, J. B., D. E. Reid, and M. Taviani. 1991d. Compositional and geochemical characteristics of recent biogenic carbonates from the Ross Sea (Antarctica). In C. A. Ricci, ed., *Earth Science Investigations in Antarctica*. Memorie della Societa Geologica Italiana, vol. 46, pp. 561–562.

Anderson, J. B., S. S. Shipp, L. R. Bartek, and D. E. Reid. 1992a. Evidence for a grounded ice sheet on the Ross Sea continental shelf during the late Pleistocene and preliminary paleodrainage reconstruction. In D. H. Elliot, ed., *Contributions to Antarctic Research III*. Antarctic Research

Series, vol. 57. American Geophysical Union, Washington, D.C., pp. 39–42.

Anderson, J. B., S. S. Shipp, and F. P. Siringan, 1992b. Preliminary seismic stratigraphy of the northwestern Weddell Sea continental shelf. In Y. Yoshida, K. Kaminuma, and K. Shiraishi, eds., *Recent Progress in Antarctic Earth Science*. Terra Scientific Publishing Co., Tokyo, Japan, pp. 603–612.

Anderson, J. B., and M. J. Smith. 1989. Formation of modern sand-rich facies by marine currents on the Antarctic continental shelf. In R. Morton and D. Nummedal, eds., *Shelf Sedimentation, Shelf Sequences, and Related Hydrocarbon Accumulation*. Gulf Coast Section of the Society of Economic Paleontologists and Mineralogists 7th Annual Research Conference Proceedings, vol. 85, pp. 41–52.

Anderson, J. B., and M. A. Thomas. 1991. Marine ice-sheet decoupling as a mechanism for rapid, episodic sea-level change: the record of such events and their influence on sedimentation. *Sediment. Geol.*, vol. 70, pp. 87–104.

Anderson, J. B., R. Wright, and B. Andrews. 1986. Weddell Fan and associated abyssal plain, Antarctica: morphology, sediment processes, and factors influencing sediment supply. *Geo-Marine Lett.*, vol. 6, pp. 121–129.

Andersson, J. G. 1906. On the geology of Graham Land. *Uppsala University Geology Institute Bulletin*, vol. 7, pp. 19–71.

Andrews, B. A. 1984. A Petrologic Study of Weddell Sea Sediments: Implications for Provenance and Glacial History. MA Thesis, Rice University, Houston, Texas, 203 pp.

Andrews, J. T. 1968. Postglacial rebound in Arctic Canada: similarity and prediction of uplift curves. *Can. J. Earth Sci.*, vol. 5, pp. 39–47.

Angino, E. E. and R. S. Andrews. 1968. Trace element chemistry, heavy minerals, and sediment statistics of Weddell Sea sediments. *J. Sediment. Petrol.*, vol. 38, pp. 634–642.

Arctowski, H. 1901a. Exploration of Antarctic lands. *Geog. J.*, vol. 17, pp. 150–180.

Ashcroft, W. A.. 1974. Crustal Structure of the South Shetland Islands and Bransfield Strait. British Antarctic Survey Scientific Reports, No. 66, 43 pp.

Ashley, G. M., J. Shaw, and N. D. Smith (eds.). 1985. Glacial Sedimentary Environments. Short Course No. 16. Society of Paleontologists and Mineralogists, Tulsa, Oklahoma, 246 pp.

Askin, R. A. 1988a. Campanian to Paleocene palynological succession of Seymour and adjacent islands, northeastern Antarctic Peninsula. In R. M. Feldman and M. O. Woodburne, eds., *Geology and Paleontology of Seymour Island, Antarctic Peninsula*. Geological Society of America Memoir 169, pp. 131–153.

1988b. The palynological record across the Cretaceous/Tertiary transition on Seymour Island, Antarctica. In R. M. Feldman and M. O. Woodburne, eds., *Geology and Paleontology of Seymour Island, Antarctic Peninsula*. Geological Society of America Memoir 169, Boulder, Colorado, pp. 155–162.

1992. Late Cretaceous–early Tertiary Antarctic outcrop evidence for past vegetation and climates. In J. P. Kennett, and D. A. Warnke, eds., *The Antarctic Paleoenvironment: A Perspective on Global Change I*. Antarctic Research

Series, vol. 56. American Geophysical Union, Washington, D.C., pp. 61–73.

Austin, J. A. Jr., T. H. Shipley, L. A. Lawver, D. E. Hayes, J. C. Mutter, and J. P. McGinnis. 1991. Initial multichannel seismic results from the northern west Antarctic Peninsula: Bransfield Strait, a nascent backarc basin. EOS (Transactions of the American Geophysical Union), vol. 72, pp. 4.

Baegi, M. 1985. Turbidite Deposition on the Bellingshausen Abyssal Plain: Sedimentologic Implications. MA Thesis, Rice University, Houston, Texas, 109 pp.

Balshaw, K. M. 1981. Antarctic glacial chronology reflected in the Oligocene through Pliocene sedimentary section in the Ross Sea. PhD dissertation, Rice University, Houston, Texas, 140 pp.

Bandy, O. L., R. E. Casey, and R. C. Wright. 1971. Late Neogene planktonic zonation, magnetic reversals, and radiometric dates, Antarctic to the Tropics. In J. L. Reid, ed., *Antarctic Oceanology I*. Antarctic Research Series, vol. 15. American Geophysical Union, Washington, D.C., pp. 1–26.

Banfield, L. A. 1994. Seismic Facies investigation of the Late Cenozoic Glacial History of Bransfield Basin, Antarctica. MA Thesis, Rice University, Houston, Texas, 184 pp.

Banfield, L. A., and J. B. Anderson. 1995. Seismic facies investigation of the late Quaternary glacial history of Bransfield Basin, Antarctica. In A. K. Cooper, P. F. Barker, and G. Brancolini, eds., *Geology and Seismic Stratigraphy of the Antarctic Margin*. Antarctic Research Series, vol. 68. American Geophysical Union, Washington, D.C., pp. 123–140.

Bardin, V. I. 1982. Composition of East Antarctic moraines and some problems of Cenozoic history. In C. Craddock, ed., *Antarctic Geoscience*. University of Wisconsin Press, Madison, pp. 1069–1076.

Barker, D. H. N., and J. A. Austin Jr. 1994. Crustal diapirism in Bransfield Strait, West Antarctica: evidence for distributed extension in marginal-basin formation. *Geology*, vol. 22, pp. 657–660.

Barker, P. F. 1976. The tectonic framework of Cenozoic volcanism in the Scotia Sea region – a review. In O. González – Ferrán, ed., *Andean and Antarctic Volcanology Problems*. International Association of Volcanology and Chemistry of the Earth's Interior, pp. 330–346.

1982. The Cenozoic subduction history of the Pacific margin of the Antarctic Peninsula: ridge crest-trench interactions. *J. Geolog. Soc. London*, vol. 139, pp. 787–801.

Barker, P. F., and J. Burrell. 1977. The opening of Drake Passage. *Marine Geol.*, vol. 25, pp. 15–34.

Barker, P. F., A. Camerlenghi, and G. Acton. 1998. Antarctic glacial history and sea-level change-Leg 178 samples Antarctic Peninsula margin sediments. *Joint Ocean. Inst. Deep Earth Sampling*, vol. 24, pp. 7–10.

Barker, P. F., and I. W. D. Dalziel. 1976. The evolution of the southwest Atlantic Ocean basin: Leg 36 data. In P. F. Barker, I. W. D. Dalziel, eds., *Initial Reports of the Deep Sea Drilling Project*. U.S. Government Printing Office, Washington, D.C., pp. 993.

1983. Progress in geodynamics of the Scotia Arc region. In R. Cabré, ed., *Geodynamics of the Eastern Pacific Region, Caribbean and Scotia Arcs*. American Geophysical Union Geodynamic Series, vol. 9, pp. 137–170.

Barker, P. F., I. W. D. Dalziel, and B. C. Storey. 1991. Tectonic development of the Scotia arc region. In R. J. Tingey, ed., *The Geology of Antarctica*. Clarendon Press, Oxford, pp. 215–248.

Barker, P. F., J. P. Kennett, and Scientific Party. 1988a. Weddell Sea palaeoceanography: preliminary results of ODP Leg 113. *Palaeogeog. Palaeoclimatol., Palaeoecol.* vol. 67, pp. 75–102.

Barker, P. F., J. P. Kennett, and the Shipboard Scientific Party. 1988b. Proceedings of the Ocean Drilling Program, Initial Reports, Leg 113. College Station, Texas.

Barkov, N. I., E. S. Korotkevich, and V. N. Petrov. 1991. Stages of Late Pleistocene glaciation of Antarctica. *Problemy arktiki i antarktiki*, vol. 66, pp. 289–296.

Barnes, P. W. 1987. Morphologic studies of the Wilkes Land continental shelf, Antarctica – glacial and iceberg effects. In S. L. Eittreim and M. A Hampton, eds., *The Antarctic Continental Margin: Geology and Geophysics of Offshore Wilkes Land*. CPCEMR Earth Science Series, vol. 5A. Circum-Pacific Council for Energy and Mineral Resources, Houston, Texas, pp. 175–194.

Barnes, P. W., and R. Lien. 1988. Icebergs rework shelf sediments to 500 m off Antarctica. *Geology*, vol. 16, pp. 1130–1133.

Baroni, C., and G. Orombelli. 1991. Holocene raised beaches at Terra Nova Bay, Victoria Land, Antarctica. *Quat. Res.*, vol. 36, pp. 157–177.

Barrera, E., and B. T. Huber. 1993. Eocene to Oligocene oceanography and temperatures in the Antarctic Indian Ocean. In J. P. Kennett and D. A. Warnke, eds., *The Antarctic Paleoenvironment; A Perspective on Global Change II*. Antarctic Research Series, vol. 60. American Geophysical Union, Washington, D.C., pp. 49–65.

Barrett, P. J. 1975. Textural characteristics of Cenozoic preglacial and glacial sediments at Site 270, Ross Sea, Antarctica. In D. E. Hayes and L. A. Frakes, eds., *Initial Reports of the Deep Sea Drilling Project*, vol. 28. U.S. Government Printing Office, Washington, D.C., pp. 757–767.

1981. History of the Ross Sea region during the deposition of the Beacon Supergroup 400–180 million years ago. *J. Royal Soc. N. Z.*, vol. 11, pp. 447–459.

(ed.). 1986a. Antarctic Cenozoic History from the MSSTS-1 Drillhole, McMurdo Sound. DSIR Bulletin 237. Science Information Publishing Centre, Wellington, New Zealand, 174 pp.

1986b. Sediment texture. In P. J. Barrett, ed., *Antarctic Cenozoic History from the MSSTS-1 Drillhole, McMurdo Sound*. DSIR Bulletin 237. Science Information Publishing Centre, Wellington, New Zealand, pp. 53–68.

(ed.). 1989. *Antarctic Cenozoic History from the CIROS-1 Drillhole, McMurdo, Sound*. DSIR Bulletin 245. Science Information Publishing Center, Wellington, New Zealand, 254 pp.

1991. Antarctica and global climate change: a geological perspective. In C. Harris and B. Stonehouse, eds., *Antarctica and Global Climate Change*. Belhaven Press, London, pp. 35–50.

1997. Antarctic paleoenvironment through Cenozoic times: a review. *Terra Antarctica*, vol. 3, pp. 103–119.

Barrett, P. J., C. J. Adams, W. C. McIntosh, C. C. Swisher III, and G. S. Wilson. 1992. Geochronological evidence supporting Antarctic deglaciation three million years ago. *Nature*, vol. 359, pp. 816–818.

Barrett, P. J., D. P. Elston, D. M. Harwood, B. C. McKelvey, and P. N. Webb. 1987. Mid-Cenozoic record of glaciation and sea-level change on the margin of the Victoria Land basin, Antarctica. *Geology*, vol. 15, pp. 634–637.

Barrett, P. J., M. J. Hambrey, D. M. Harwood, A. R. Pyne, and P. N. Webb. 1989. Synthesis. In P. J. Barrett, ed., *Antarctic Cenozoic History from the CIROS-1 Drillhole, McMurdo Sound*. DSIR Bull. 245. DSIR Publishing, Wellington, New Zealand, pp. 241–251.

Barrett, P. J., S. A. Henrys, L. R. Bartek, G. Brancolini, M. Busetti, F. J. Davey, M. J. Hannah, and A. R. Pyne. 1995. Geology of the margin of the Victoria Land Basin off Cape Roberts, southwest Ross Sea. In A. K. Cooper, P. F. Barker, and G. Brancolini, eds., *Geology and Seismic Stratigraphy of the Antarctic Margin*. Antarctic Research Series vol. 68. American Geophysical Union, Washington, D.C., pp. 183–208.

Barrett, P. J., and B. C. McKelvey. 1986. Stratigraphy. In P. J. Barrett, ed., *Antarctic Cenozoic History from the MSSTS-1 Drillhole, McMurdo Sound*. DSIR Bulletin 237. Science Information Publishing Centre, Wellington, New Zealand, pp. 9–52.

Barrett, P. J., A. R. Pyne, and B. L. Ward. 1983. Modern sedimentation in McMurdo Sound, Antarctica. In R. L. Oliver, P. R. James, and J. B. Jago, eds., *Antarctic Earth Science*. Cambridge University Press, New York, pp. 550–554.

Barrie, C. Q., and D. J. W. Piper. 1982. Late Quaternary marine geology of Makkovik Bay, Labrador. Geological Survey of Canada Paper 81–17, 37 pp.

Barron, E. J. 1987. Cretaceous plate tectonic reconstructions. *Palaeogeog. Palaeoclimatol. Palaeoecol.*, vol. 59, pp. 3–29.

Barron, E. J., and C. G. A. Harrison. 1979. Reconstructions of the Campbell Plateau and the Lord Howe Rise. *Earth Plan. Sci. Lett.*, vol. 45, pp. 87–92.

Barron, E. J., C. G. A. Harrison, and W. W. Hay. 1978. A revised reconstruction of the southern continents. EOS (Transactions of the American Geophysical Union), vol. 59, pp. 436–449.

Barron, J. A. 1996. Diatom constraints on the position of the Antarctic Polar Front in the middle part of the Pliocene. *Marine Micropaleontol.*, vol. 27, pp. 195–213.

Barron, J., B. Larsen, and J. G. Baldauf. 1991. Evidence for late Eocene to early Oligocene Antarctic glaciation and observations on late Neogene glacial history of Antarctica: results from Leg 119. In J. Barron and B. Larsen, eds., Proceedings of the Ocean Drilling Program, Scientific Results, vol. 119. College Station, Texas, pp. 869–891.

Barron, J., B. Larsen, and Shipboard Scientific Party (eds.). 1989. Proceedings of the Ocean Drilling Program, Initial Reports, vol. 119. College Station, Texas, 942 pp.

Bart, P. J. 1993. Glacial and Tectonic Development of the Antarctic Peninsula. MA Thesis, Rice University, Houston, Texas, 152 pp.

1998. Seismic-Stratigraphic Analysis of Shelf-margin Delta/Slope Fans and Basin Floor Fans on High-Latitude and Middle-Latitude Margins (Ross Sea and Alabama/West Florida Shelf): Paleoclimatic and Eustatic Implications.

PhD Dissertation, Rice University, Houston, Texas, 236 pp.

Bart, P. J., and J. B. Anderson. 1995. Seismic record of glacial events affecting the Pacific margin of the northwestern Antarctic Peninsula. In A. K Cooper, P. F. Barker, and G. Brancolini, eds., *Geology and Seismic Stratigraphy of the Antarctic Margin*. Antarctic Research Series, vol. 68. American Geophysical Union, Washington, D.C., pp. 75–95.

——— 1996. Seismic expression of depositional sequences associated with expansion and contraction of ice sheets on the northwestern Antarctic Peninsula continental shelf. In M. De Batist and P. Jacobs, eds., *Geology of Siliclastic Shelf Seas*. Geological Society Special Publication No. 117, pp. 171–186.

Bart, P. J., M. De Batist, and H. Miller. 1994. Neogene collapse of glacially deposited, shelf-edge deltas in the Weddell Sea: relationships between deposition during glacial periods and sub-marine fan development. In A. K. Cooper, P. F. Barker, P. N. Webb, and G. Brancolini, eds., The Antarctic Continental Margin. Geophysical and Geological Stratigraphic Records of Cenozoic Glaciation, Paleoenvironments and Sea-Level Change. *Terra Antarctica*, vol. 1, pp. 317–3180.

Bartek, L. R., and J. B. Anderson. 1991. Facies distribution resulting from sedimentation under polar interglacial climatic conditions within a high-latitude marginal basin, McMurdo Sound, Antarctica. In J. B. Anderson and G. M. Ashley, eds., *Glacial Marine Sedimentation; Paleoclimatic Significance*. Geological Society of America Special Paper 261, Boulder, Colorado, pp. 27–49.

Bartek, L. R., J. L. R. Andersen, T. A. Oneacre. 1998. Substrate control on distribution of subglacial and glaciomarine seismic facies based on stochastic models of glacial seismic facies deposition on the Ross Sea continental margin, Antarctica. *Marine Geol.*, vol. 143, pp. 223–262.

Bartek, L. R. J. B. Anderson, and S. A. Henrys. 1994. Contributions of high resolution single-channel data to understanding sequence architecture of a glaciated continental margin, Ross Sea, Antarctica. In A. K. Cooper, P. F. Barker, P. N. Webb, and G. Brancolini, eds., The Antarctic Continental Margin. Geophysical and Geological Stratigraphic Records of Cenozoic Glaciation, Paleoenvironments and Sea-Level Change. *Terra Antarctica*, vol. 1, no. 2, pp. 329–332.

Bartek, L. R., S. A. Henrys, J. B. Anderson, and P. J. Barrett. 1996. Seismic stratigraphy of McMurdo Sound, Antarctica: implications for glacially influenced early Cenozoic eustatic change? *Marine Geol.*, vol. 130, pp. 79–98.

Bartek, L. R., L. C. Sloan, J. B. Anderson, and M. I. Ross. 1992. Evidence from the Antarctic continental margin of late Paleogene ice sheets: a manifestation of plate reorganization and synchronous changes in atmospheric circulation over the emerging Southern Ocean. In D. R. Prothero and W. A. Berggren, eds., *Eocene-Oligocene Climatic and Biotic Evolution*. Princeton University Press, Princeton, N.J., pp. 131–159.

Bartek, L. R., P. R. Vail, J. B. Anderson, P. A. Emmet, and S. Wu. 1991. Effect of Cenozoic ice sheet fluctuations in Antarctica on the stratigraphic signature of the Neogene. *J. Geophys. Res.*, vol. 96, pp. 6753–6778.

Barton, C. M. 1965. The Geology of the South Shetland Islands. III. The Stratigraphy of King George Island. British Antarctic Survey Scientific Reports, no. 44, 37 pp.

Behrendt, J. C. 1962. Geophysical and glaciological studies in the Filchner ice shelf area of Antarctica. *J. Geophys. Res.*, vol. 67, pp. 221–234.

——— 1990. Recent geophysical and geological research in Antarctica related to the assessment of petroleum resources and potential environmental hazards to their development. In J. F. Splettstoesser and G. A. M. Dreschhoff, eds., *Mineral Resources Potential of Antarctica*. Antarctic Research Series vol. 51. American Geophysical Union, Washington, D.C., pp. 163–174.

Behrendt, J. C., and A. K. Cooper. 1991. Evidence of rapid Cenozoic uplift of the shoulder escarpment of the West Antarctic rift system and a speculation on possible climate forcing. *Geology*, vol. 19, pp. 315–319.

Behrendt, J. C., J. R. Henderson, L. Meister, and W. L. Rambo. 1974. Geophysical investigations of the Pensacola Mountains and adjacent glacierized areas of Antarctica. U.S. Geological Survey Professional Paper 844, 28 pp.

Behrendt, J. C., W. LeMasurier, and A. K. Cooper. 1992. The West Antarctic rift system – a propagating rift "captured" by a mantle plume? In Y. Yoshida, K. Kaminuma, and K. Shiraishi, eds., *Recent Progress in Antarctic Earth Science*. Terra Scientific Publishing Co., Tokyo, Japan, pp. 315–322.

Belknap, D. F., and R. C. Shipp. 1991. Seismic stratigraphy of glacial marine units, Maine inner shelf. In J. B. Anderson, and G. M. Ashley, eds., *Glacial Marine Sedimentation; Paleoclimatic Significance*. Geological Society of America Special Paper 261, pp. 137–158.

Bell, C. M. 1975. Structural geology of parts of Alexander Island. *Br. Ant. Surv. Bull.*, no. 41/42, pp. 43–58.

Bell, R. E., J. M. Brozena, W. F. Haxby, and J. L. LaBrecque. 1990. *Continental Margins of the Western Weddell Sea; Insights from Airborne Gravity and GEOSAT-Derived Gravity*. Antarctic Research Series, vol. 50. American Geophysical Union, Washington, D.C., pp. 91–102.

Bennett, M. R., and N. F. Glasser. 1996. Glacial geology, ice sheets and landforms. John Wiley and Sons, New York, 364 pp.

Bentley, C. R. 1964. Ice thickness. In C. R. Bentley, R. L. Cameron, C. Bull, K. Kojima, and A. J. Gow, eds., *Physical Characteristics of the Antarctic Ice Sheet*. Antarctic Map Folio Series – Folio. 2. American Geographical Society, New York, pp. 3–4.

——— 1972. Subglacial topography. In B. C. Heezen, M. Tharp, and C. R. Bentley, eds., *Morphology of the Earth in the Antarctic and Subantarctic*. Antarctic Map Folio Series – Folio 16. American Geographical Society, New York, plate 7.

——— 1973. Crustal structure of Antarctica. *Tectonophysics*, vol. 20, pp. 229–240.

——— 1983. Crustal structure of Antarctica from geophysical evidence – a review. In R. L. Oliver, P. R. James, and J. B. Jago, eds., *Antarctic Earth Science*. Cambridge University Press, New York, pp. 491–497.

——— 1991. Configuration and structure of the subglacial crust. In R. J. Tingey, ed., *The Geology of Antarctica*. Clarendon Press, Oxford, pp. 335–364.

Bentley, C. R. and J. W. Clough. 1972. Antarctic subglacial

structure from seismic refraction measurements. In R. J. Adie, ed., *Antarctic Geology and Geophysics*. International Union of Geological Sciences, Series B, vol. 1. Universitetsforlaget, Oslo, Norway, pp. 683–691.

Bentley, C. R., and N. A. Ostenso. 1961. Glacial and subglacial topography of West Antarctica. *J. Glaciol.*, vol. 3, pp. 882–911.

Bentley, M. J., and J. B. Anderson. 1998. Glacial and marine geological evidence for the ice sheet configuration in the Weddell Sea-Antarctic Peninsula region during the last glacial maximum. *Ant. Sci.*, vol. 10, pp. 309–325.

Berg, J. H., R. A. Hank, and R. I. Kalamarides. 1985. Petrology and geochemistry of inclusions of lower crustal basic granulites from the Erebus Volcanic Province, Antarctica. *Ant. J. U.S.*, vol. 20, pp. 22–23.

Berryhill, H. L., Fr., J. R. Suter, and N. S. Hardin (eds.). 1986. *Late Quaternary Facies and Structure, Northern Gulf of Mexico: Interpretations from Seismic Data (atlas)*. American Association of Petroleum Geologists Studies in Geology, no. 23, Tulsa, Oklahoma, 289 pp.

Bialas, J., and GRAPE-Team. 1990. Seismic investigations along the Antarctic Peninsula. In A. K. Cooper and P. N. Webbs, eds., *International Workshop on Antarctic Offshore Seismic Stratigraphy (ANTOSTRAT)*. Overview and Extended Abstracts, pp. 72–78.

Bibby, J. S. 1966. The Stratigraphy of Part of North-East Graham Land and the James Ross Island Group. British Antarctic Survey Scientific Reports, no. 53, 37 pp.

Biernat, G., K. Birkenmajer, and E. Popiel-Barczyk. 1985. Tertiary brachiopods from the Moby Dick Group of King George Island (South Shetland Islands, Antarctica). *Studia Geologica Polonica*, vol. 81, pp. 109–141.

Bigg, G. R. 1996. *The Oceans and Climate*. Cambridge University Press, Cambridge, 266 pp.

Bindschadler, R. A., S. N. Stephenson, D. R. MacAyeal, and S. Shabtaie. 1987. Ice dynamics at the mouth of Ice Stream B, Antarctica. *J. Geophys. Res.*, vol. 92, pp. 8885–8894.

Birkenmajer, K. 1980. Report on geological investigations of King George Island, South Shetland Islands (West Antarctica) in 1978/79. *Studia Geologica Polonica*, vol. 64, pp. 89–105.

——— 1981a. Geological relations at Lions Rump, King George Island (South Shetland Islands, Antarctica). *Studia Geologica Polonica*, vol. 72, pp. 75–87.

——— 1981b. Lithostratigraphy of the Point Hennequin Group (Miocene volcanics and sediments) at King George Island (South Shetland Islands, Antarctica). *Studia Geologica Polonica*, vol. 72, pp. 59–73.

——— 1982. Late Cenozoic phases of block-faulting on King George Island (South Shetland Islands, West Antarctica). *Bulletin de L'Académie Polonaise des Sciences, Série des Sciences de la Terre*, vol. 30, pp. 21–32.

——— 1984. Pre-Quaternary glaciations of West Antarctica: evidence from the South Shetland Islands. *Polish Polar Research*, vol. 5, pp. 319–329.

——— 1985. Onset of Tertiary continental glaciation in the Antarctic Peninsula sector (West Antarctica). *Acta Geologica Polonica*, vol. 35, pp. 1–31.

——— 1986. Geochronology and migration of Cretaceous through Tertiary plutonic centres, South Shetland Islands (West Antarctica): subduction and hot spot magmatism. *Bulletin de l'Academie Polonaise des Sciences, Série des Sciences de la Terre*, vol. 34, pp. 243–255.

——— 1987. Oligocene-Miocene glacio-marine sequences of King George Island (South Shetland Islands), Antarctica. *Palaeontologia Polonica*, no. 49, pp. 9–40.

——— 1988. Tertiary glacial and interglacial deposits, South Shetland Islands, Antarctica: geochronology versus biostratigraphy (a progress report). Bulletin of the Polish Academy of Sciences, Earth Sciences, vol. 36, pp. 133–145.

——— 1990. Tertiary basaltic hyaloclastites on King George Island (South Shetland Islands, Antarctica). Bulletin of the Polish Academy of Sciences, Earth Sciences, vol. 38, pp. 111–122.

——— 1991. Tertiary glaciation in the South Shetland Islands, West Antarctica: evaluation of data. In M. R. A. Thomson, J. A. Crame, and J. W. Thomson, eds., *Geological Evolution of Antarctica*. Cambridge University Press, New York, pp. 629–632.

——— 1992. Evolution of the Bransfield Basin and Rift, West Antarctica. In Y. Yoshida, K. Kaminuma, and K. Shiraishi, eds., *Recent Progress in Antarctic Earth Science*. Terra Scientific Publishing Co., Tokyo, Japan, pp. 405–410.

Birkenmajer, K., and T. Butkiewicz. 1988. Petrography and provenance of magmatic and metamorphic erratic blocks from lower Miocene glacio-marine deposits of King George Island (South Shetland Islands, Antarctica). *Studia Geologica Polonica*, vol. 95, pp. 23–51.

Birkenmajer, K., M. C. Delitala, W. Narebski, M. Nicoletti, and C. Petrucciani. 1986. Geochronology of Tertiary islandarc volcanics and glacigenic deposits, King George Island, South Shetland Islands (West Antarctica). Bulletin of the Polish Academy of Sciences, Earth Sciences, vol. 34, pp. 257–273.

Birkenmajer, K., and A. Gazdzicki. 1986. Oligocene age of the *Pecten* conglomerate on King George Island, West Antarctica. Bulletin of the Polish Academy of Sciences, Earth Sciences, vol. 34, pp. 219–226.

Birkenmajer, K., A. Gazdzicki, H. Pugaczewska, and R. Wrona. 1987. Recycled Cretaceous belemnites in lower Miocene glacio-marine sediments (Cape Melville Formation) of King George Island, West Antarctica. *Palaeontologia Polonica*, no. 49, pp. 49–65.

Birkenmajer, K., and E. Luczkowska. 1987. Early Miocene foraminiferal zonation, southeast Pacific Basin, Antarctic Peninsula sector. Bulletin of the Polish Academy of Sciences, Earth Sciences, vol. 35, pp. 1–10.

Birkenmajer, K., E. Soliani Jr., and K. Kawashita. 1988. Early Miocene K-Ar age of volcanic basement of the Melville Glaciation deposits, King George Island, West Antarctica. Bulletin of the Polish Academy of Sciences, Earth Sciences, vol. 36, pp. 25–34.

——— 1989. Geochronology of Tertiary glaciations on King George Island, West Antarctica. Bulletin of the Polish Academy of Sciences, Earth Sciences, vol. 37, pp. 27–48.

Bishop, J. F., and J. L. Walton. 1981. Bottom melting under George VI Ice Shelf, Antarctica. *J. Glaciol.*, vol. 27, pp. 429–447.

Black, L. P., P. R. James, and S. L. Harley. 1983. Geochronology and geological evolution of metamorphic rocks in the Field Islands area, East Antarctica. *J. Metamorph. Geol.*, vol. 1, pp. 277–303.

Blackman, D. K., R. P. Von Herzen, and L. A. Lawver. 1987. Heat flow and tectonics in the western Ross Sea, Antarctica. In A. K. Cooper and F. J. Davey, eds., *The Antarctic Continental Margin: Geology and Geophysics of the Western Ross Sea*. CPCEMR Earth Science Series, vol. 5B: Circum-Pacific Council for Energy and Mineral Resources, Houston, Texas, pp. 179–189.

Blank, R. G., and S. V. Margolis. 1975. Pliocene climatic and glacial history of Antarctica as revealed by Southeast Indian Ocean deep-sea cores. *Geolog. Soc. Am. Bull.*, vol. 86, pp. 1058–1066.

Blankenship, D. D., C. R. Bentley, S. T. Rooney, and R. B. Alley. 1987. Till beneath ice stream B. 1. properties derived from seismic travel times. *J. Geophys. Res.*, vol. 92, pp. 8903–8911.

Bloom, A. L, W. S. Broecker, J. Chappell, R. K. Matthews, and K. J. Mesolella. 1974. Quaternary sea level fluctuations on a tectonic coast: new ^{230}Th/^{234}U dates from the Huon Peninsula, New Guinea. *Quat. Res.*, vol. 4, pp. 185–205.

Boeuf, M. G., and H. Doust. 1975. Structure and development of the southern margin of Australia. *Austral. Petrol. Explor. Assoc. J.*, vol. 15, pp. 33–43.

Bohaty, S. M., and D. M. Harwood. 1995. Pliocene silicoflagellate paleotemperature variation from the Kerguelen Plateau. Geological Society of America, Abstracts of Programs, vol. 27, no. 3, pp. 41.

Borg, S. G. 1983. Petrology and geochemistry of the Queen Maud Batholith, central Transantarctic Mountains, with implications for the Ross Orogeny. In R. L. Oliver, P. R. James, and J. B. Jago, eds., *Antarctic Earth Science*. Cambridge University Press, New York, pp. 165–169.

Borg, S. G., and D. J. DePaolo. 1994. Laurentia, Australia, and Antarctica as a Late Proterozoic supercontinent: constraints from isotopic mapping. *Geology*, vol. 22, pp. 307–310.

Borg, S. G., D. J. DePaolo, and B. M. Smith. 1990. Isotopic structure and tectonics of the central Transantarctic Mountains. *J. Geophys. Res.*, vol. 95, pp. 6647–6667.

Borg, S. G., J. W. Goodge, V. C. Bennett, and D. J. DePaolo. 1987. Geochemistry of granites and metamorphic rocks: central Transantarctic Mountains. *Ant. J. U.S.*, vol. 22, pp. 21–23.

Bornhold, B. D. 1983. Ice-rafted debris in sediments from Leg 71, southwest Atlantic Ocean. In W. J. Ludwig, V. A. Krasheninnikov, et al., eds., Initial Reports of the Deep Sea Drilling Project, vol. 71. U.S. Government Printing Office, Washington, D.C., pp. 307–316.

Boulton, G. S. 1970. On the deposition of subglacial and melt-out tills at the margins of certain Svalbard glaciers. *J. Glaciol.*, vol. 9, pp. 231–245.

———. 1978. Boulder shapes and grain-size distributions of debris as indicators of transport paths through a glacier and till genesis. *Sedimentology*, vol. 25, pp. 773–799.

———. 1990. Sedimentary and sea level changes during glacial cycles and their control on glacimarine facies architecture. In J. A. Dowdeswell and J. D. Scourse, eds., *Glacimarine Environments: Processes and Sediments*. Geological Society Special Publication No. 53, London, England, pp. 15–52.

Boulton, G. S., D. L. Dent, and E. M. Morris. 1974. Subglacial shearing and crushing, and the role of water pressures in tills from south-east Iceland. *Geografiska Annaler*, vol. A56, pp. 135–145.

Boulton, G. S., and A. S. Jones. 1979. Stability of temperate ice caps and ice sheets resting on beds of deformable sediments. *J. Glaciol.*, vol. 24, pp. 29–43.

Boulton, G. S., and M. A. Paul. 1976. The influence of genetic processes on some geotechnical properties of glacial tills. *Q. J. Engin. Geol.*, vol. 9, pp. 159–194.

Boyd, R., D. B. Scott, and M. Douma. 1988. Glacial tunnel valleys and Quaternary history of the outer Scotian Shelf. *Nature*, vol. 333, pp. 61–64.

Bradshaw, M. A. 1987. Additional field interpretation of the Jurassic sequence at Carapace Nunatak and Coombs Hills, South Victoria Land, Antarctica. *N.Z. J. Geol. Geophys.*, vol. 30, pp. 37–49.

Brady, H. T. 1982. Late Cenozoic history of Taylor and Wright Valleys and McMurdo Sound inferred from diatoms in Dry Valley Drilling Project cores. In C. Craddock, ed., *Antarctic Geoscience*. University of Wisconsin Press, Madison, pp. 1123–1131.

Brake, C. F. 1982. Sedimentology of the North Victoria Land Continental Margin, Antarctica. MA Thesis, Rice University, Houston, Texas, 175 pp.

Brake, C. F., and J. B. Anderson. 1983. The bathymetry of the North Victoria Land continental margin. *Marine Geod.*, vol. 6, pp. 139–147.

Brancolini, G., A. K. Cooper, and F. Coren. 1995. Seismic facies and glacial history in western Ross Sea (Antarctica). In A. K. Cooper, P. F. Barker, and G. Brancolini, eds., *Geologic and Seismic Stratigraphy of the Antarctic Margin*. Antarctic Research Series vol. 68. American Geophysical Union, Washington, D.C., pp. 209–233.

Brennecke, W. 1921. Die ozeanographischen arbeiten der Deutschen Antarktischen Expedition 1911–1912 (The oceanographic reports of the German Antarctic Expedition 1911–1912). *Archiv der Deutschen Seewarte*, no. 1, 214 pp.

Bromwich, D. H., and D. D. Kurtz. 1984. Katabatic wind forcing of the Terra Nova Bay polynya. *J. Geophys. Res.*, vol. 89, pp. 3561–3572.

Bromwich, D. H., T. R. Parish, A. Pellegrini, C. R. Stearns, and G. A. Weidner. 1993. Spatial and temporal characteristics of the intense katabatic winds at Terra Nova Bay, Antarctica. In D. H. Bromwich and C. R. Stearns, eds., *Antarctic Meteorology and Climatology; Studies Based on Automatic Weather Stations*. Antarctic Research Series, vol. 61. American Geophysical Union, Washington, D.C., pp. 47–68.

Bromwich, D. H., T. R. Parish, and C. A. Zorman. 1990. The confluence zone of the intense katabatic winds at Terra Nova Bay, Antarctica, as derived from airborne sastrugi surveys and mesoscale numerical modeling. *J. Geophys. Res.*, vol. 95, pp. 5495–5509.

Brotchie, J. F., and R. Silvester. 1969. On crustal flexure. *J. Geophys. Res.*, vol. 74, pp. 5240–5252.

Brown, D. A., K. S. W. Campbell, and K. A. W. Crook. 1968. *The Geological Evolution of Australia and New Zealand*. Pergamon Press, Oxford, 409 pp.

Bruno, L. A., H. Baur, T. Graf, C. Schluchter, P. Signer, and R. Wieler. 1997. Dating of Sirius Group tillites in the

Antarctic Dry Valleys with cosmogenic ^{3}He and ^{21}Ne. *Earth Planet. Sci. Lett.*, vol. 147, pp. 37–54.

Budd, W. F. 1991. Antarctica and global change. *Climatic Change*, vol. 18, pp. 271–299.

Budd, W. F., M. J. Corry, and T. H. Jacka. 1982. Results from the Amery Ice Shelf Project. *Ann. Glaciol.*, vol. 3, pp. 36–41.

Budd, W. F., D. Jenssen, E. Mavrakis, and B. Coutts. 1994. Modelling the Antarctic ice-sheet changes through time. *Ann. Glaciol.*, vol. 20, pp. 291–297.

Burckle, L. H. 1984. Diatom distribution and paleoceanographic reconstruction in the Southern Ocean; present and last glacial maximum. *Marine Micropaleontol.*, vol. 9, pp. 241–261.

——— 1993. Is there direct evidence for late Quaternary collapse of the West Antarctic ice sheet? *J. Glaciol.*, vol. 39, pp. 491–494.

Burckle, L. H., and N. Abrams. 1986. Diatom species diachrony in late Neogene sediments of the Southern Ocean. *S. Afr. J. Sci.*, vol. 82, pp. 505–506.

Burckle, L. H., and J. Cirilli. 1987. Origin of diatom ooze belt in the Southern Ocean; implications for late Quaternary paleoceanography. *Micropaleontology*, vol. 33, pp. 82–86.

Burckle, L. H., and E. M. Pokras. 1991. Implications of a Pliocene stand of *Nothofagus* (southern beech) within 500 kilometres of the South Pole. *Ant. Sci.*, vol. 3, pp. 389–403.

Burckle, L. H., and N. Potter Jr. 1995. Pliocene-Pleistocene diatoms in Paleozoic and Mesozoic sedimentary and igneous rocks from Antarctica: a Sirius problem solved. *Geology*, vol. 24, pp. 235–238.

Burckle, L. H., D. Robinson, and D. Cooke. 1982. Reappraisal of sea-ice distribution in Atlantic and Pacific sectors of the Southern Ocean at 18,000 yr BP. *Nature*, vol. 299, pp. 435–437.

Burckle, L. H., A. P. Stroeven, C. Bronge, U. Miller, and A. Wasell. 1996. Deficiencies in the diatom evidence for a Pliocene reduction of the East Antarctic Ice Sheet. *Paleoceanography*, vol. 11, pp. 379–389.

Burgess, J. S., A. P. Spate, and J. Shevlin. 1994. Onset of deglaciation in the Larseman Hills, eastern Antarctica. *Ant. Sci.*, vol. 6, pp. 491–495.

Busetti, M., and A. K. Cooper. 1994. Possible ages and origins of unconformity U6. In A. K. Cooper, P. F. Barker, P. N. Webb, and G. Brancolini, eds., The Antarctic Continental Margin. Geophysical and Geological Stratigraphic Records of Cenozoic Glaciation, Paleoenvironments and Sea Level Change. *Terra Antarctica*, vol. 1, no. 2, pp. 341–344.

Calkin, P. E. 1964. Glacial geology of the Mount Gran area, southern Victoria Land, Antarctica. *Geolog. Soc. Am. Bull.*, vol. 75, pp. 1031–1036.

Cameron, P. J. 1981. The petroleum potential of Antarctica and its continental margin. *Austral. Petrol. Explor. Assoc. J.*, vol. 21, pp. 99–111.

Canals, M., M. de Batist, J. Baraza, J. Acosta and the GEBRA Team. 1994. New reflection seismic data from Bransfield Strait, Preliminary Results of the GEBRA-93 Survey. In A. K. Cooper, P. F. Barker, P. N. Webb, and G. Brancolini, eds., The Antarctic Continental Margin: Geophysical and Geological Stratigraphic Records of Cenozoic Glaciation,

Paleoenvironments and Sea Level Change. *Terra Antarctica*, vol. 1, no. 2, pp. 291–292.

Cande, S. C., and J. C. Mutter. 1982. A revised identification of the oldest sea-floor spreading anomalies between Australia and Antarctica. *Earth Planet. Sci. Lett.*, vol. 58, pp. 151–160.

Caputo, M. V. 1985. Late Devonian glaciation in South America. *Palaeogeog. Palaeoclimatol. Palaeoecol.*, vol. 51, pp. 291–317.

Caputo, M. V., and J. C. Crowell. 1985. Migration of glacial centers across Gondwana during Paleozoic Era. *Geolog. Soc. Am. Bull.*, vol. 96, pp. 1020–1036.

Carey, S. W., and N. Ahmad. 1961. Glacial marine sedimentation. *Geol. Arctic*, vol. 2, pp. 865–894.

Carleton, A. M. 1979. A synoptic climatology of satellite-observed extratropical cyclonic activity for the Southern Hemisphere winter. *Archiv Meteorol., Geophys., Bioklimatol.*, vol. A27, pp. 265–279.

Carlquist, S. 1987. Pliocene Nothofagus wood from the Transantarctic Mountains. *Aliso*, vol. 11, pp. 571–583.

Carmack, E. C., and T. D. Foster. 1975. On the flow of water out of the Weddell Sea. *Deep-Sea Res.*, vol. 22, pp. 711–724.

Carrara, P. 1981. Evidence for a former large ice sheet in the Orville Coast-Ronne Ice Shelf area, Antarctica. *J. Glaciol.*, vol. 27, pp. 487–491.

Carrasco, J. F., and D. H. Bromwich. 1993. Satellite and automatic weather station analyses of katabatic surges across the Ross Ice Shelf. In D. H. Bromwich and C. R. Stearns, eds., *Antarctic Meteorology and Climatology: Studies Based on Automatic Weather Stations.* Antarctic Research Series, vol. 61. American Geophysical Union, Washington, D.C., pp. 93–108.

Chase, T. E., B. A. Seekins, J. D. Young, and S. L. Eittreim. 1987. Marine topography of offshore Antarctica. In S. L. Eittreim and M. A. Hampton, eds., *The Antarctic Continental Margin: Geology and Geophysics of Offshore Wilkes Land.* CPCEMR Earth Science Series, vol. 5A. Circum-Pacific Council for Energy and Mineral Resources, Houston, Texas, pp. 147–150.

Chiu, L. S. 1983. Variation of Antarctic sea ice: an update. *Month. Weather Rev.*, vol. 111, pp. 578–580.

Chriss, T., and L. A. Frakes. 1972. Glacial marine sedimentation in the Ross Sea. In R. J. Adie, ed., *Antarctic Geology and Geophysics.* International Union of Geological Sciences, Series B, vol. 1. Universitetsforlaget, Oslo, Norway, pp. 747–762.

Christoffel, D. A., and R. K. H. Falconer. 1972. Marine magnetic measurements in the southwest Pacific Ocean and the identification of new tectonic features. In D. E. Hayes, ed., *Antarctic Oceanology. II. The Australian-New Zealand Sector.* Antarctic Research Series, vol. 19. American Geophysical Union, Washington, D.C., pp. 197–209.

Ciais, P., J. R. Petit, J. Jouzel, C. Lorius, N. I. Barkov, V. Lipenkov, V. Nicolaiev, and V. Nikolayev. 1992. Evidence for an early Holocene climatic optimum in the Antarctic deep ice-core record. *Clim. Dynam.*, vol. 6, pp. 169–177.

Ciesielski, P. F. 1975. Biostratigraphy and paleoecology of Neogene and Oligocene silicoflagellates from cores recovered during Antarctic Leg 28, Deep Sea Project. In D. E. Hayes,

L. A. Frakes, et al., eds., *Initial Reports of the Deep Sea Drilling Project*, vol. 28. U.S. Government Printing Office, Washington, D.C., pp. 625–664.

Ciesielski, P. F., M. G. Dinkelman, M. T. Ledbetter, and B. B. Ellwood. 1978. New evidence on the timing of late Miocene–middle Pliocene glacial fluctuations from the southwest Atlantic sector of the Southern Ocean. Geological Society of America – Abstracts of Programs, vol. 10, p. 380.

Ciesielski, P. F., and G. P. Grimstead. 1986. Pliocene variations in the position of the Antarctic convergence in the southwest Atlantic. *Paleoceanography*, vol. 1, pp. 197–232.

Ciesielski, P. F., M. T. Ledbetter, and B. B. Ellwood. 1982. The development of Antarctic glaciation and the Neogene paleoenvironment of the Maurice Ewing Bank. *Marine Geol.*, vol. 46, pp. 1–51.

Ciesielski, P. F., and F. M. Weaver. 1974. Early Pliocene temperature changes in the Antarctic seas. *Geology*, vol. 2, pp. 511–515.

Ciesielski, P. F., and S. W. Wise Jr. 1977. Geologic history of the Maurice Ewing Bank of the Falkland Plateau (Southwest Atlantic sector of the Southern Ocean) based on piston and drill cores. *Marine Geol.*, vol. 25, pp. 175–207.

Clapperton, C. M., and D. E. Sugden. 1982. Late Quaternary glacial history of George VI Sound area, West Antarctica. *Quat. Res.*, vol. 18, pp. 243–267.

1990. Late Cenozoic glacial history of the Ross Embayment, Antarctica. *Quat. Sci. Rev.*, vol. 9, pp. 253–272.

Clark, C. D. 1993. Mega-scale glacial lineations and cross-cutting ice-flow landforms. *Earth Surface Processes and Landforms*, vol. 18, pp. 1–29.

Clark, J. A., and C. S. Lingle. 1977. Future sea-level changes due to West Antarctic Ice Sheet fluctuations. *Nature*, vol. 269, pp. 206–209.

Clarkson, P. D., and M. Brook. 1977. Age and position of the Ellsworth Mountains crustal fragment, Antarctica. *Nature*, vol. 265, pp. 615–616.

CLIMAP Project Members. 1976. The surface of ice-age Earth. *Science*, vol. 191, pp. 1131–1137.

Clough, J. W., and B. L. Hansen. 1979. The Ross Ice Shelf Project. *Science*, vol. 203, pp. 433–434.

Colhoun, E. A. 1991. Geological evidence for changes in the East Antarctic Ice Sheet (60 degrees–120 degrees E) during the last glaciation. *Polar Rec.*, vol. 27, pp. 345–355.

Colhoun, E. A., M. C. G. Mabin, D. A. Adamson, and R. M. Kirk. 1992. Antarctic ice volume and contribution to sea-level fall at 20,000 yr BP from raised beaches. *Nature*, vol. 358, pp. 316–319.

Collinson, J. W., and N. R. Kemp. 1983. Permian-Triassic sedimentary sequence in northern Victoria Land, Antarctica. In R. L. Oliver, P. R. James, and J. B. Jago, eds., *Antarctic Earth Science*. Cambridge University Press, New York, pp. 221–225.

Collinson, J. W., N. R. Kemp, and J. T. Eggert. 1987. Comparison of the Triassic Gondwana sequences in the Transantarctic Mountains and Tasmania. In G. D. McKenzie, ed., Gondwana Six: Stratigraphy, Sedimentology, and Paleontology. Geophysical Monograph 41. American Geophysical Union, Washington, D.C., pp. 51–61.

Comisco, J. C., and A. L. Gordon. 1998. Interannual variability in summer sea ice minimum, coastal polynyas and bottom water formation in the Weddell Sea. *Ant. Res. Series*, vol. 74, pp. 293–315.

Conolly, J. R., and M. Ewing. 1965. Pleistocene glacial-marine zones in North Atlantic deep-sea sediments. *Nature*, vol. 208, pp. 135–139.

Cooke, D. W. 1978. Variations in the seasonal extent of sea ice in the Antarctic during the last 140,000 years. PhD Dissertation, Columbia University, New York, 302 pp.

Cooke, D. W., and J. D. Hays. 1982. Estimates of Antarctic Ocean seasonal sea-ice cover during glacial intervals. In C. Craddock, ed., *Antarctic Geoscience*. University of Wisconsin Press, Madison, pp. 1017–1025.

Cooper, A. K. 1989. Crustal structure of the Ross embayment. In H. T. Huh, B.-K. Park, and S.-H. Lee, eds., *Antarctic Science: Geology and Biology*. Seo Deung Co. Ltd, Seoul, Korea, pp. 57–72.

Cooper, A. K., P. F. Barker, and G. Brancolini (eds.). 1995. Geology and Seismic Stratigraphy of the Antarctic Margin. Antarctic Research Series, vol. 68. American Geophysical Union, Washington, D.C., 303 pp.

Cooper, A. K., P. J. Barrett, K. Hinz, V. Traube, G. Leitchenkov, and H. M. J. Stagg. 1991a. Cenozoic prograding sequences of the Antarctic continental margin: a record of glacio-eustatic and tectonic events. *Marine Geol.*, vol. 102, pp. 175–213.

Cooper, A. K., P. J. Barrett, V. V. Traube, K. Hinz, and H. Stagg. 1990. Prograding Cenozoic sedimentary sequences of the Antarctic continental margin. In A. K. Cooper and P. N. Webbs, eds., *International Workshop on Antarctic Offshore Acoustic Stratigraphy (ANTOSTRAT)*. Overview and Extended Abstracts. U.S. Geological Survey Open-File Report 90-309, pp. 109–118.

Cooper, A. K., and J. C. Behrendt. 1991. Rift grabens of the Ross Sea; a record of Mesozoic and younger extension between East and West Antarctica. Geological Society of America – Abstracts of Programs, vol. 23, p. 305.

Cooper, A. K., G. Brancolini, J. C. Behrendt, F. J. Davey, P. J. Barrett, and ANTOSTRAT Ross Sea Regional Working Group. 1994. A record of Cenozoic tectonism throughout the Ross Sea and possible controls on the glacial record. In A. K. Cooper, P. F. Barker, P. N. Webb, and G. Brancolini, eds., The Antarctic Continental Margin: Geophysical Stratigraphic Records of Cenozoic Glaciation, Paleoenvironments and Sea Level Change. *Terra Antarctica*, vol. 1, pp. 353–355.

Cooper, A. K., F. J. Davey, and J. C. Behrendt. 1987a. Seismic stratigraphy and structure of the Victoria Land basin, western Ross Sea, Antarctica. In A. K. Cooper and F. J. Davey, eds., *The Antarctic Continental Margin: Geology and Geophysics of the Western Ross Sea*. CPCEMR Earth Science Series, vol. 5B. Circum-Pacific Council for Energy and Mineral Resources, Houston, Texas, pp. 27–76.

Cooper, A. K., F. J. Davey, and G. R. Cochrane. 1987b. Structure of extensionally rifted crust beneath the western Ross Sea and Iselin Bank, Antarctica, from sonobuoy seismic data. In A. K. Cooper and F. J. Davey, eds., *The Antarctic Continental Margin: Geology and Geophysics of the Western Ross Sea*. CPCEMR Earth Science Series, vol. 5B: Circum-Pacific Council for Energy and Mineral Resources, Houston, Texas, pp. 93–118.

Cooper, A. K., F. J. Davey, and K. Hinz. 1988. Antarctica-1, Ross Sea – geology, hydrocarbon potential. *Oil Gas J.*, vol. 86, pp. 54–58.

Cooper, A. K., F. J. Davey, and K. Hinz. 1991b. Crustal extension and origin of sedimentary basins beneath the Ross Sea and Ross Ice Shelf, Antarctica. In M. R. A. Thomson, J. A. Crame, and J. W. Thomson, eds., *Geological Evolution of Antarctica*. Cambridge University Press, New York, pp. 285–291.

Cooper, A. K., S. Eittreim, U. ten Brink, and I. Zayata. 1993. Cenozoic glacial sequences of the Antarctic continental margin as recorders of Antarctic ice sheet fluctuations. In J. P. Kennett and D. A. Warnke, eds., *The Antarctic Paleoenvironmant: A Perspective on Global Change II*. Antarctic Research Series, vol. 60. American Geophysical Union, Washington, D.C., pp. 75–90.

Cooper, R. A., C. A. Landis, W. E. LeMasurier, and E. G. Speden. 1982. Geologic history and regional patterns in New Zealand and West Antarctica – their paleotectonic and paleogeographic significance. In C. Craddock, ed., *Antarctic Geoscience*: University of Wisconsin Press, Madison, pp. 43–53.

Cooper, A. K., H. Stagg, and E. Geist. 1991c. Seismic stratigraphy and structure of Prydz Bay, Antarctic: implications from Leg 119 drilling. In J. Barron, B. Larsen, et al., eds., Proceeding of the Ocean Drilling Program, Scientific Results, vol. 119. College Station, Texas, pp. 5–25.

Corliss, B. H., D. G. Martinson, and T. Keffer. 1986. Late Quaternary deep-ocean circulation. *Geolog. Soc. Am. Bull.*, vol. 97, pp. 1106–1121.

Cox, K. G. 1978. Flood basalts, subduction and the break-up of Gondwanaland. *Nature*, vol. 274, pp. 47–49.

Crabtree, R. D., B. C. Storey, and C. S. M. Doake. 1985. The structural evolution of George VI Sound, Antarctic Peninsula. *Tectonophysics*, vol. 114, pp. 431–442.

Craddock, C. 1972. Antarctic tectonics. In R. J. Adie, ed., *Antarctic Geology and Geophysics*. International Union of Geological Sciences, Series B, vol. 1. Universitetsforlaget, Oslo, Norway, pp. 449–455.

1982. Antarctica and Gondwanaland – review paper. In C. Craddock, ed., *Antarctic Geoscience*. University of Wisconsin Press, Madison, pp. 3–13.

1983. The East Antarctica-West Antarctica boundary between the ice shelves: a review. In R. L. Oliver, P. R. James, and J. B. Jago, eds., *Antarctic Earth Science*. Cambridge University Press, New York, pp. 94–97.

Craddock, C., J. J. Anderson, and G. F. Webers. 1964. Geologic outline of the Ellsworth Mountains. In R. J. Adie, ed., *Antarctic Geology*. North Holland, Amsterdam, pp. 155–170.

Crawford, A. R., and R. L. Oliver. 1969. The Precambrian Geochronology of Ceylon. Geological Society of Australia Special Publication, vol. 2, pp. 283–306.

Crépon, M., M. N. Houssais, and B. Saint Guily. 1988. The drift of icebergs under wind action. *J. Geophys. Res.*, vol. 93, pp. 3608–3612.

Crowell, J. C., and L. A. Frakes. 1971. Late Paleozoic glaciation of Australia. *J. Geolog. Soc. Austral.*, vol. 17, pp. 115–155.

Curl, J. E. 1980. A glacial history of the South Shetland Islands, Antarctica. Institute of Polar Studies Report No. 63. Ohio State University, Columbus, 129 pp.

Curtis, R. 1966. The Petrology of the Graham Coast, Graham Land. British Antarctic Survey Scientific Reports, no. 50, 51 pp.

Czajkowski, S., and O. Rosler. 1986. Fossil plants from the Fildes Peninsula, King George Island: morphology of leaf impressions. Anais Academia Brasileriade Ciencias, vol. 58, pp. 99–110.

Dagel, M. A. 1985. Stratigraphy and chronology of Stage 6 and 2 glacial deposits, Marshall Valley, Antarctica. MA Thesis, University of Maine, Orono, 102 pp.

Dalziel, I. W. D. 1974. Evolution of the margins of the Scotia Sea. In C. A. Burke and C. L. Drake, eds., *The Geology of Continental Margins*. Springer-Verlag, New York, pp. 567–580.

1980. Comment on 'Mesozoic evolution of the Antarctic Peninsula and the southern Andes'. *Geology*, vol. 8, pp. 260–261.

1981. Back-arc extension in the southern Andes: a review and critical reappraisal. Philosophical Transactions of the Royal Society of London, vol. 300, pp. 319–335.

1982. The early (pre-Middle Jurassic) history of the Scotia arc region: a review and progress report. In C. Craddock, ed., *Antarctic Geoscience*. University of Wisconsin Press, Madison, pp. 111–126.

1991. Pacific margins of Laurentia and East Antarctica-Australia as a conjugate rift pair: evidence and implications for an Eocambrian supercontinent. *Geology*, vol. 19, pp. 598–601.

Dalziel, I. W. D., M. J. de Wit, and K. F. Palmer. 1974. Fossil marginal basin in the southern Andes. *Nature*, vol. 250, pp. 291–294.

Dalziel, I. W. D., and D. H. Elliot. 1973. The Scotia Arc and Antarctic Margin. In A. E. M. Nairn and F. G. Stehli, eds., *The Ocean Basins and Margins. 1. The South Atlantic*. Plenum Press, New York, pp. 171–246.

1982. West Antarctica: problem child of Gondwanaland. *Tectonics*, vol. 1, pp. 3–19.

Dalziel, I. W. D., and R. J. Pankhurst. 1987. Tectonic development of West Antarctica and its relation to East Antarctica: joint U.S./U.K. program 1986–1987. *Ant. J. U.S.*, vol. 22, pp. 50–51.

Dalziel, I. W. D., B. C. Storey, S. W. Garrett, A. M. Grunow, L. D. B. Herrod, and R. J. Pankhurst. 1987. Extensional tectonics and the fragmentation of Gondwanaland. In M. P. Coward, J. F. Dewey, and P. L. Hancock, eds., *Continental Extensional Tectonics*. Geological Society of London Special Publication, vol. 28, pp. 433–441.

Dangeard, L., J. R. Vanney, and G. L. Johnson. 1977. Affeurements, coucants et facies dan la zone Antarctique du Pacifique Oriental (Mers de Bellingshausen et d'Amundsen). *Ann. Inst. Oceanogr.*, Paris, France, vol. 53, pp. 105–124.

DaSilva, J. L. 1995. Seismic Facies Investigation and Late Quaternary Glacial History of the Chilean Shelf and Fjords and Antarctic Peninsula Fjords. MA Thesis, Rice University, Houston, Texas, 273 pp.

DaSilva, J. L., J. B. Anderson, and J. Stravers. 1997. Seismic facies changes along a nearly continuous 24 degree latitudinal transect: the fjords of Chile and the northern Antarctic Peninsula. *Marine Geol.*, vol. 103, pp. 103–123.

Davey, F. J. 1972. Marine gravity measurements in Bransfield Strait and adjacent areas. In R. J. Adie, ed., *Antarctic Geology and Geophysics*. Universitetsforlaget, Oslo, Norway, pp. 39–46.

Davey, F. J. 1981. Geophysical studies in the Ross Sea region. *J. Royal Soc. N.Z.*, vol. 11, pp. 465–479.

——— 1987. Geology and structure of the Ross Sea region. In A. K. Cooper, and F. J. Davey, eds., *The Antarctic Continental Margin: Geology and Geophysics of the Western Ross Sea*. CPCEMR Earth Science Series, vol. 5B. Circum-Pacific Council for Energy and Mineral Resources, Houston, Texas, pp. 1–16.

Davey, F. J., D. J. Bennett, and R. E. Houtz. 1982. Sedimentary basins of the Ross Sea, Antarctica. *N.Z. J. Geol. Geophys.*, vol. 25, pp. 245–255.

Davey, F. J., and G. Brancolini. 1995. The late Mesozoic and Cenozoic structural setting of the Ross Sea region. In A. K. Cooper, P. F. Barker, and G. Brancolini, eds., *Geology and Seismic Stratigraphy of the Antarctic Margin*. Antarctic Research Series, vol. 68. American Geophysical Union, Washington, D.C., pp. 167–182.

Davey, F. J., and D. A. Christoffel. 1978. Magnetic anomalies across Campbell Plateau, New Zealand. *Earth Planet. Sci. Lett.*, vol. 41, pp. 14–20.

Davey, F. J., and A. K. Cooper. 1987. Gravity studies of the Victoria Land basin and Iselin Bank. In A. K. Cooper and F. J. Davey, eds., *The Antarctic Continental Margin: Geology and Geophysics of the Western Ross Sea*. CPCEMR Earth Science Series, vol. 5B. Circum-Pacific Council for Energy and Mineral Resources, Houston, Texas, pp. 119–138.

David, T. W. E., and G. E. Priestley. 1914. Glaciology, physiography, stratigraphy, and tectonic geology of South Victoria Land: British Antarctic Expedition, 1907–1909, Reports of the Scientific Investigations, Geology.

Deacon, G. E. R. 1937. The hydrography of the Southern Ocean. 'Discovery' Reports, vol. 19, 124 pp.

de Batist, M., P. Bart, B. Kuvaas, A. Moons, and H. Miller. 1994. Detailed seismic stratigraphy of the Crary Fan, southeastern Weddell Sea. In A. K. Cooper, P. F. Barker, P. N. Webb, and G. Brancolini, eds., The Antarctic Continental Margin: Geophysical and Geological Stratigraphic Records of Cenozoic Glaciation, Paleoenvironments and Sea-Level Change. *Terra Antartica*, vol. 1, pp. 321–323.

Debenham, F. 1919. A new mode of transportation by ice: the raised marine muds of South Victoria Land. *Q. J. Geolog. Soc. London*, vol. 75, pp. 51–76.

DeFelice, D. R., and S. W. Wise, Jr. 1981. Surface lithofacies, biofacies, and diatom diversification patterns and models for delineation of climatic change in the southeast Atlantic Ocean. *Marine Micropaleontol.*, vol. 6, pp. 29–70.

Deighton, I., D. A. Falvey, and D. J. Taylor. 1976. Depositional environments and geotectonic framework: southern Australia continental margin. *Austral. Petrol. Explor. Assoc. J.*, vol. 16, pp. 25–36.

DeLong, S. E., W. M. Schwartz, and R. M. Anderson. 1978. Subduction of the Kula Ridge at the Aleutian Trench. *Geolog. Soc. Am. Bull.*, vol. 89, pp. 83–95.

del Valle, R. A., D. H. Elliot, and D. I. M. Macdonald. 1992. Sedimentary basins on the east flank of the Antarctic Peninsula: proposed nomenclature. *Ant. Sci.*, vol. 4, pp. 477–478.

del Valle, R. A., N. H. Fourcade, and F. A. Medina. 1982. The stratigraphy of Cape Lamb and The Naze, Vega and James Ross islands, Antarctica. In C. Craddock, ed., *Antarctic Geoscience*. University of Wisconsin Press, Madison, pp. 275–280.

del Valle, R. A., F. Medina, and N. H. Fourcade. 1986. La formacion flora en Bahia Botany (Botanica) Peninsula Antartica (Flora Formation in Botany Bay, Antarctic Peninsula). *Instituto Antártico Argentino Contribución*, vol. 289, 30 pp.

del Valle, R. A., P. A. Rial, and C. A. Rinaldi. 1987. Description e interpretacion de una seccion del Cretacico Superior de la Isla Marambio, Antartida (Description and interpretation of a Cretaceous section of Seymour Island, Antarctica). *Instituto Antártico Argentino Contribución*, vol. 345, 39 pp.

DeMaster, D. J., 1981, The supply and accumulation of silica in the marine environment. *Geochim. Cosmochim. Acta*, vol. 45, p. 1715–1732.

DeMaster, D. J., R. B. Dunbar, L. I. Gordon, A. R. Leventer, J. M. Morrison, D. M. Nelson, C. A. Nittrouer, and W. O. Smith Jr. 1992. Cycling and accumulation of biogenic silica and organic matter in high-latitude environments: the Ross Sea. *Oceanography*, vol. 5, pp. 146–153.

DeMaster, D. J., T. M. Nelson, S. L. Harden, and C. A. Nittrouer. 1991. The cycling and accumulation of biogenic silica and organic carbon in Antarctic deep-sea and continental margin environments. *Marine Chem.*, vol. 35, pp. 489–502.

Denton, G. H., R. P. Ackert, M. L. Prentice, and N. Potter Jr. 1984. Ice-sheet overriding of the ice-free valleys of southern Victoria Land. *Ant. J. U.S.*, vol. 19, pp. 47–48.

Denton, G. H., B. G. Andersen, and H. W. Conway. 1986a. Late Quaternary surface fluctuations of Beardmore Glacier, Antarctica. *Ant. J. U.S.*, vol. 21, pp. 90–92.

Denton, G. H., R. L. Armstrong, and M. Stuiver. 1971. *The Late Cenozoic Glacial History of Antarctica: The Late Cenozoic Glacial ages*. Yale University Press, New Haven, pp. 267–306.

Denton, G. H., Bockheim, J. G., Rutford, R. H., and Anderson, B. G. 1992. Glacial history of the Ellsworth Mountains, West Antarctica. In G. F. Webers, C. Craddock, and J. F. Splettstoesser, eds., *Geology and Paleontology of the Ellsworth Mountains, West Antarctica*. Geological Society of America Memoir 170, Boulder. Colorado, pp. 403–432.

Denton, G. H., J. G. Bockheim, S. C. Wilson, J. E. Leide, and B. G. Anderson. 1989. Late Quaternary ice-surface fluctuations of Beardmore Glacier, Transantarctic Mountains. *Quat. Res.*, vol. 31, pp. 183–209.

Denton, G. H., J. G. Bockheim, S. C. Wilson, and C. Schlüchter. 1986b. Late Cenozoic history of Rennick Glacier and Talos Dome, northern Victoria Land, Antarctica. In E. Stump, ed., *Geological Investigations in Northern Victoria Land*. Antarctic Research Series, vol. 46. American Geophysical Union, Washington, D.C., pp. 339–375.

Denton, G. H., and T. J. Hughes. 1981. The Arctic Ice Sheet: an outrageous hypothesis. In G. H. Denton, and T. J. Hughes, eds., *The Last Great Ice Sheets*. John Wiley & Sons, New York, pp. 437–467.

——— 1986. Potential influence of floating ice shelves on the climate of an ice age. *S. Afr. J. Sci.*, vol. 82, pp. 509–513.

Denton, G. H., M. L. Prentice, and L. H. Burckle, 1991. Cainozoic history of the Antarctic Ice Sheet. In R. J. Tingey, ed., *The Geology of Antarctica*. Clarendon Press, Oxford, pp. 365–433.

Denton, G. H., D. E. Sugden, D. R. Marchant, B. L. Hall, and T. I. Wilch. 1993. East Antarctica Ice Sheet sensitivity to Pliocene climate change from a Dry Valleys perspective. *Geografiska Annaler*, vol. 75A, pp. 155–204.

Department of the Navy. 1985. *Sea Ice Climatic Atlas. 1, Antarctic*: NAVAIR 50-1C-540, 131 pp.

De Santis, L., J. B. Anderson, G. Brancolini, and I. Zayatz. 1995. Seismic record of late Oligocene through Miocene glaciation on the central and eastern continental shelf of the Ross Sea. In A. K. Cooper, P. F. Barker, and G. Brancolini, eds., *Geology and Seismic Stratigraphy of the Antarctic Margin*. Antarctic Research Series, vol. 68. American Geophysical Union, Washington, D.C., pp. 235–260.

Dewar, G. J. 1970. The Geology of Adelaide Island. British Antarctic Survey Scientific Reports, no. 57, 66 pp.

Dewart, G. 1972. Magnetic profile in the Gerlache Strait, Antarctica. Ohio State University Research Foundation Project 2264, no. 1, 8 pp.

de Wit, M. J. 1977. The evolution of the Scotia Arc as a key to the reconstruction of southwestern Gondwanaland. *Tectonophysics*, vol. 37, pp. 53–81.

de Wit, J. J., M. Jeffery, H. Bergh, and L. Nicolaysen. 1988. Geological map of Sectors of Gondwana. American Association of Petroleum Geologists, Tulsa, Oklahoma, 2 sheets.

Diester-Haass, L., C. Robert, and H. Chamley. 1993. Paleoceanographic and paleoclimatic evolution in the Weddell Sea (Antarctica) during the middle Eocene-late Oligocene, from a coarse sediment fraction and clay mineral data (ODP Site 689). *Marine Geol*, vol. 114, pp. 233–250.

Dietz, R. S., and J. C. Holden. 1970. Reconstruction of Pangea: breakup and dispersion of continents, Permian to Present. *J. Geophys. Res.*, vol. 75, pp. 4939–4956.

Dietz, R. S., and W. P. Sproll. 1970. Fit between Africa and Antarctica: a continental drift reconstruction. *Science*, vol. 167, pp. 1612–1614.

Dingle, R. V., and R. A. Scrutton. 1974. Continental breakup and the development of post-Paleozoic sedimentary basins around southern Africa. *Geolog. Soc. Am. Bull.*, vol. 85, pp. 1467–1474.

Ditchfield, P. W., J. D. Marshall, and D. Pirrie. 1994. High latitude paleotemperature variation; new data from the Tithonian to Eocene of James Ross Island, Antarctica. *Palaeogeog. Palaeoclimatol. Palaeoecol.*, vol. 107, pp. 79–102.

Doake, C. S. M. 1982. State of balance of the ice sheet in the Antarctic Peninsula. *Ann. Glaciol.*, vol. 3, pp. 77–82.

Doake, C. S. M. 1985. Antarctic mass balance: glaciological evidence from Antarctic Peninsula and Weddell Sea sector. In *Glaciers, Ice Sheets, and Sea Level: Effect of a CO₂-induced Climatic Change*. U.S. National Research Council, Washington, D.C., pp. 197–209.

Doake, C. S. M., R. D. Crabtree, and I. W. D. Dalziel. 1983. Subglacial morphology between Ellsworth Mountains and Antarctic Peninsula: new data and tectonic significance. In R. L. Oliver, P. R. James, and J. B. Jago, eds., *Antarctic Earth Science*. Cambridge University Press, New York, pp. 270–273.

Doake, C. S. M., R. M. Frolich, D. R. Mantripp, A. M. Smith,

and D. G. Vaughan. 1987. Glaciological studies on Rutford Ice Stream, Antarctica. *J. Geophys. Res.*, vol. 92, pp. 8951–8960.

Doake, C. S. M., and D. G. Vaughan. 1991. Rapid disintegration of the Wordie Ice Shelf in response to atmospheric warming. *Nature*, vol. 350, pp. 328–330.

Doktor, M., A. Gazdzicki, S. A. Marenssi, S. J. Porebski, S. N. Santillana, and A. V. Vrba. 1988. Argentine-Polish geological investigations on Seymour (Marambio) Island, Antarctica, 1988. *Polish Polar Res.*, vol. 9, pp. 521–541.

Domack, E. W. 1982. Sedimentology of glacial and glacial marine deposits on the George V–Adelie continental shelf, East Antarctica. *Boreas*, vol. 11, pp. 79–97.

———. 1987. Preliminary stratigraphy for a portion of the Wilkes Land continental shelf, Antarctica: evidence from till provenance. In S. L. Eittreim and M. A. Hampton, eds., *The Antarctic Continental Margin: Geology and Geophysics of Offshore Wilkes Land*. CPCEMR Earth Science Series, vol. 5A. Circum-Pacific Council for Energy and Mineral Resources, Houston, Texas, pp. 195–203.

———. 1988. Biogenic facies in the Antarctic glacimarine environment: basis for a polar glacimarine summary. *Palaeogeog. Palaeoclimatol. Palaeoecol.*, vol. 63, pp. 357–372.

———. 1990. Laminated terrigenous sediments from the Antarctic Peninsula: the role of subglacial and marine processes. In J. A. Dowdeswell and J. D. Scourse, eds., *Glacimarine Environments: Processes and Sediments*. Geological Society Special Publication No. 53, London, England, pp. 91–103.

Domack, E. W., and J. B. Anderson. 1983. Marine geology of the George V continental margin: combined results of Deep Freeze 79 and the 1911–14 Australasian expedition. In R. L. Oliver, P. R. James, and J. B. Jago, eds., *Antarctic Earth Science*. Cambridge University Press, New York, pp. 402–406.

Domack, E. W., W. W. Fairchild, and J. B. Anderson. 1980. Lower Cretaceous sediment from the East Antarctic continental shelf. *Nature*, vol. 287, pp. 625–626.

Domack, E. W., and S. Ishman. 1993. Oceanographic and physiographic controls on modern sedimentation within Antarctic fjords. *Geolog. Soc. Am. Bull.*, vol. 105, pp. 1175–1189.

———. 1994. Oceanographic controls on benthic foraminifers from the Bellingshausen margin of the Antarctic Peninsula. *Marine Micropaleontol.*, vol. 24, pp. 119–155.

Domack, E. W., S. E. Ishman, A. B. Stein, C. E. McClennen, and A. J. T. Jull. 1995. Late Holocene advance of the Müller Ice Shelf, Antarctic Peninsula: sedimentological, geochemical and palaeontological evidence. *Ant. Sci.*, vol. 7, pp. 159–170.

Domack, E. W., A. J. T. Jull, J. B. Anderson, T. W. Linick, and C. R. Williams. 1989b. Application of tandem accelerator mass–spectrometer dating to Late Pleistocene-Holocene sediments of the East Antarctic continental shelf. *Quat. Res.*, vol. 31, pp. 277–287.

Domack, E. W., A. J. T. Jull, and D. J. Donahue. 1991. Holocene chronology for the unconsolidated sediments at Hole 740A: Prydz Bay, East Antarctica. In J. Barron, B. Larsen, et al., eds., Proceedings of the Ocean Drilling Program, Scientific Results, vol. 119. College Station, Texas, pp. 747–750.

Domack, E. W., T. A. Mashiotta, L. A. Burghley, and S. E. Ishman. 1993. 33-year cyclicity in organic matter preservation in Antarctic fjord sediments. In J. P. Kennett and D. A. Warnke, eds., *The Antarctic Paleoenvironment; a Perspective on Global Change II.* Antarctic Research Series, vol. 60. American Geophysical Union, Washington, D.C., pp. 265–272.

Domack, E. W., and C. E. McClennen. 1996. Accumulation of glacial marine sediments in fjords of the Antarctic Peninsula and their use as late Holocene paleoenvironmental indicators. In R. Ross, E. E. Hofmann, and L. B. Quentin, eds., *Foundations for Ecological Research West of the Antarctic Peninsula.* Antarctic Research Series, vol. 70. American Geophysical Union, Washington, D.C., pp. 135–154.

Domack, E. W., P. O'Brien, P. Harris, F. Taylor, P. G. Quilty, L. DeSantis, and B. Raker. 1998. Late Quaternary sediment facies in Prydz Bay, East Antarctica and their relationship to glacial advance on to the continental shelf. *Ant. Sci.,* vol. 10, pp. 234–244.

Domack, E. W., and C. R. Williams. 1990. Fine structure and suspended sediment transport in three Antarctic fjords. In D. H. Elliot, ed., *Contributions to Antarctic Research I: Antarctic Research Series,* vol. 50. American Geophysical Union, Washington, D.C., pp. 71–89.

Dott, R. H., Jr., R. D. Winn Jr., and C. H. L. Smith. 1982. Relationship of late Mesozoic and early Cenozoic sedimentation to the tectonic evolution of the southernmost Andes and Scotia arc. In C. Craddock, ed., *Antarctic Geoscience.* University of Wisconsin Press, Madison, pp. 193–202.

Douglas, R. G., and S. M. Savin. 1973. Oxygen and carbon isotope analyses of Cretaceous and Tertiary foraminifera from the central North Pacific. In E. L. Winterer, J. I. Ewing, eds., *Initial Reports of the Deep Sea Drilling Project,* vol. 17. U.S. Government Printing Office, Washington, D.C., pp. 591–605.

Dowsett, H. J., J. A. Barron, R. Z. Poore. 1996. Middle Pliocene seasurface temperatures: a global reconstruction. *Marine Micropaleontol.,* vol. 27, pp. 13–25.

Dowsett, H. J., and T. M. Cronin. 1990. High eustatic sea level during the Middle Pliocene; evidence from the southeastern U.S. Atlantic coastal plain with Suppl. Data 90-13. *Geology,* vol. 18, pp. 435–438.

Drewry, D. J. 1975. Radio echo sounding map of Antarctica (~90°E – 180°). *Polar Rec.,* vol. 17, pp. 359–374.

— 1979. Late Wisconsin reconstuction for the Ross sea region, Antarctica. *J. Glaciol.,* vol. 24, pp. 231–244.

— 1981. The record of late Cenozoic glacial events in East Antarctica (60 degrees–171 degrees E). In M. J. Hambrey and W. B. Harland, eds., *Earth's pre-Pleistocene glacial record: IGCP Project No. 38.* Cambridge University Press, Cambridge, pp. 212–216.

— 1983. Antarctica: Glaciological and Geophysical Folio. Scott Polar Research Institute, University of Cambridge, Cambridge, England, 9 sheets.

— 1986. *Glacial Geologic Processes.* Edward Arnold, London, England, 276 pp.

Drewry, D. J., and A. P. R. Cooper. 1981. Processes and models of Antarctic glaciomarine sedimentation. *Ann. Glaciol.,* vol. 2, pp. 117–122.

Drewry, D. J., and E. M. Morris. 1993. The response of large ice sheets to climatic change. In D. J. Drewry, R. M. Laws, and J. A. Pyle, eds., *Antarctica and Environmental Change.* Clarendon Press, Oxford, pp. 35–42.

Dudeney, J. R. 1987. The Antarctic atmosphere. In D. W. H. Walton, ed., *Antarctic Science.* Cambridge University Press, Cambridge, pp. 190–247.

Dunbar, R. B., J. B. Anderson, E. W. Domack, and S. S. Jacobs. 1985. Oceanographic influences on sedimentation along the Antarctic continental shelf. In S. S. Jacobs, ed., *Oceanology of the Antarctic Continental Shelf.* Antarctic Research Series, vol. 43. American Geophysical Union, Washington, D.C., pp. 291–312.

Dunbar, R. B., A. R. Leventer, and W. L. Stockton. 1989. Biogenic sedimentation in McMurdo Sound, Antarctica. *Marine Geol.,* vol. 85, pp. 155–179.

Duphorn, K. 1981. Physiographical and glaciogeological observations in North Victoria Land, Antarctica. *Geologisches Jahrbuch,* vol. B41, pp. 89–109.

Du Toit, A. L. 1937. Our Wandering Continents: A Hypothesis of Continental Drifting. Oliver and Boyd, Edinburgh, Scotland, 366 pp.

Edwards, B. D., H. J. Lee, H. A. Karl, E. Reimnitz, and L. A. Timothy. 1987. Geology and physical properties of Ross Sea, Antarctica, continental shelf sediment. In A. K. Cooper and F. J. Davey, eds., *The Antarctic Continental Margin: Geology and Geophysics of the Western Ross Sea.* CPCEMR Earth Science Series, vol. 5B. Circum-Pacific Council for Energy and Mineral Resources, Houston, Texas, pp. 191–216.

Edwards, C. W. 1982. New paleontological evidence of Triassic sedimentation in West Antarctica. In C. Craddock, ed., *Antarctic Geoscience.* University of Wisconsin Press, Madison, pp. 325–330.

Edwards, D. S. 1968. The detrital mineralogy of surface sediments of the ocean floor in the area of the Antarctic Peninsula, Antarctica. MA Thesis, Florida State University, Tallahassee, Florida, 143 pp.

Ehrmann, W. U. 1991. Implications of sediment composition on the southern Kerguelen Plateau for paleoclimate and depositional environment. In J. Barron, B. Larsen, et al., eds., Proceedings of the Ocean Drilling Program, Scientific Results, vol. 119. College Station, Texas, pp. 185–210.

Ehrmann, W. U. 1995. Clay mineral assemblages in Cenozoic sediments of Ross Sea, Antarctica. Proceedings of the VII ISAES 1995, Siena, Italy, pp. 118.

Ehrmann, W. U., M. Melles, G. Kuhn, and H. Grobe. 1992. Significance of clay mineral assemblages in the Antarctic Ocean. *Marine Geol.,* vol. 107, pp. 249–273.

Eittreim, S. L., A. K. Cooper, and J. Wannesson. 1995. Seismic stratigraphic evidence of ice-sheet advances on the Wilkes Land margin of Antarctica. *Sediment. Geol.,* vol. 96, pp. 131–156.

Eittreim, S. L., and M. A. Hampton (eds.). 1987. *The Antarctic Continental Margin: Geology and Geophysics of Offshore Wilkes Land.* CPCEMR Earth Science Series, vol. 5A. Circum-Pacific Council for Energy and Mineral Resources, Houston, Texas, 221 pp.

Eittreim, S. L., and G. L. Smith. 1987. Seismic sequences and their distribution on the Wilkes Land margin. In S. L. Eittreim and M. A. Hampton, eds., *The Antarctic Con-*

tinental Margin: Geology and Geophysics of Offshore Wilkes Land. CPCEMR Earth Science Series, vol. 5A. Circum-Pacific Council for Energy and Mineral Resources, Houston, Texas, pp. 15–43.

Elliot, C. G., and D. R. Gray. 1992. Correlations between Tasmania and the Tasman-Transantarctic Orogen: evidence for easterly derivation of Tasmania relative to mainland Australia. *Geology*, vol. 20, pp. 621–624.

Elliot, D. H. 1965. Geology of north-west Trinity Peninsula, Graham Land. *Br. Ant. Surv. Bull.*, no. 7, pp. 1–24.

——— 1966. Geology of the Nordenskjöld Coast and a comparison with north-west Trinity Peninsula, Graham Land. *Br. Ant. Surv. Bull.*, no. 10, pp. 1–43.

——— 1975a. Gondwana basins of Antarctica. In K. S. W. Campbell, ed., *Gondwana Geology.* Australian National University Press, Canberra, Australia, pp. 493–536.

——— 1975b. Tectonics of Antarctica: a review. *Am. J. Sci.*, vol. 275-A, pp. 45–106.

——— 1983. The Mid-Mesozoic to Mid-Cenozoic active plate margin of the Antarctic Peninsula. In R. L. Oliver, P. R. James, and J. B. Jago, eds., *Antarctic Earth Science.* Cambridge University Press, New York, pp. 347–351.

——— 1988. Tectonic setting and evolution of the James Ross Basin, northern Antarctic Peninsula. In R. M. Feldman and M. O. Woodburne, eds., *Geology and Paleontology of Seymour Island, Antarctic Peninsula.* Geological Society of America Memoir 169, pp. 541–555.

——— 1992. Jurassic magmatism and tectonism associated with Gondwanaland breakup: an Antarctic perspective. In B. C. Storey, T. Alabaster, and R. J. Pankhurst, eds., *Magmatism and the causes of continental break-up.* Geological Society of America Special Paper, pp. 165–184.

Elliot, D. H., and K. A. Folan. 1986. Potassium-argon age determinations of the Kirkpatrick Basalt, Mesa Range. In E. Stump, ed., *Geological Investigations in Northern Victoria Land.* Antarctic Research Series, vol. 46. American Geophysical Union, Washington, D.C., pp. 279–288.

Elliot, D. H., and T. M. Gracanin. 1983. Conglomeratic strata of Mesozoic age at Hope Bay, northern Antarctic Peninsula. In R. L. Oliver, P. R. James, and J. B. Jago, eds., *Antarctic Earth Science.* Cambridge University Press, New York, pp. 303–307.

Elliot, D. H., S. M. Hoffman, and D. E. Rieske. 1992. Provenance of Paleocene strata, Seymour Island. In Y. Yoshida, K. Kaminuma, and K. Shiraishi, eds., *Recent Progress in Antarctic Earth Science.* Terra Scientific Publishing Co., Tokyo, Japan, pp. 347–355.

Elliot, D. H., P. R. Kyle, and R. J. Pankhurst. 1989. Jurassic basic magmatism and crustal anatexis in Antarctica. Abstracts of the 28th International Geological Congress, vol. 28, pp. 1.446–1.447.

Elliot, D. H., and T. A. Trautman. 1982. Lower Tertiary strata on Seymour Island, Antarctic Peninsula. In C. Craddock, ed., *Antarctic Geoscience.* University of Wisconsin Press, Madison, pp. 287–297.

Ellis, D. J., J. W. Sheraton, R. N. England, and W. B. Dallwitz. 1980. Osumilite-sapphirine-quartz granulites from Enderby Land Antarctica – mineral assemblages and reactions. *Contrib. Mineral. Petrol.*, vol. 72, pp. 123–143.

Elverhøi, A. 1981. Evidence for a late Wisconsin glaciation of the Weddell Sea. *Nature*, vol. 293, pp. 641–642.

——— 1984. Glacigenic and associated marine sediments in the Weddell Sea, fjords of Spitsbergen, and the Barents Sea: a review. *Marine Geol.*, vol. 57, pp. 53–88.

Elverhøi, A., and G. Maisey. 1983. Glacial erosion and morphology of the eastern and southeastern Weddell Sea shelf. In R. L. Oliver, P. R. James, and J. B. Jago, eds., *Antarctic Earth Science.* Cambridge University Press, New York, pp. 483–487.

Elverhøi, A., and E. Roaldset. 1983. Glaciomarine sediments and suspended particulate matter, Weddell Sea Shelf, Antarctica. *Polar Res.*, vol. 1, pp. 1–21.

Emiliani, C. 1969. Interglacial high sea levels and the control of Greenland ice by the precession of the equinoxes. *Science*, vol. 166, pp. 1503–1504.

Engelhardt, H., N. Humphrey, B. Kamb, and M. Fahnestock. 1990. Physical conditions at the base of a fast moving Antarctic ice stream. *Science*, vol. 248, pp. 57–59.

Escutia, C., S. L. Eittreim, and A. K. Cooper. In press. Cenozoic glaciomarine sequences on the Wilkes Land continental rise, Antarctica. Proceedings Volume – VII International Symposium on Antarctic Earth Science.

Exon, N. F., P. J. Hill, and J. Y. Royer. 1995. New maps of crust off Tasmania expand research possibilities. EOS (Transactions of the American Geophysical Union), vol. 76, pp. 201, 206–207.

Fairbanks, R. G. 1989. A 17,000-year glacio-eustatic sea level record: influence of glacial melting rates on the Younger Dryas event and deep-ocean circulation. *Nature*, vol. 342, pp. 637–642.

Fairbanks, R. G., and R. K. Matthews. 1978. The marine oxygen isotope record in Pleistocene coral, Barbados, West Indies. *Quat. Res.*, vol. 10, pp. 181–196.

Falvey, D. A. 1974. The development of continental margins in plate tectonic theory. *Austral. Petrol. Explor. Assoc. J.*, vol. 14, pp. 95–106.

Falvey, D. A., and J. C. Mutter. 1981. Regional plate tectonics and evolution of Australia's passive continental margins. *J. Austral. Geol. Geophy.*, vol. 6, pp. 1–29.

Farman, J. C., B. G. Gardiner, and J. D. Shanklin. 1985. Large losses of total ozone in Antarctica reveal seasonal ClO_x/NO_x interaction. *Nature*, vol. 315, pp. 207–210.

Farquharson, G. W. 1982. Late Mesozoic sedimentation in the northern Antarctic Peninsula and its relationship to the southern Andes. *J. Geolog. Soc. London*, vol. 139, pp. 721–727.

——— 1983. Evolution of late Mesozoic sedimentary basins in the northern Antarctic Peninsula. In R. L. Oliver, P. R. James, and J. B. Jago, eds., *Antarctic Earth Science.* Cambridge University Press, New York, pp. 323–327.

Faure, G., R. Eastin, P. T. Ray, D. McLelland, and C. H. Schultz. 1979. Geochronology of igneous and metamorphic rocks, central Transantarctic Mountains. In B. Laskar and C. S. Raja Rao, eds., *Fourth International Gondwana Symposium: Papers*, vol. II. Hindustan Publishing Corporation, Delhi, India, pp. 805–813.

Fechner, N., and W. Jokat. 1995. A seismic transect across the Ronne shelf. VII International Symposium on Antarctic Earth Sciences, Abstracts. Tipografia Senese, Siena, Italy, pp. 25.

Fedorov, L. V., and J. Hofmann. 1982. Structural development of Precambrian rocks in the mountain fringe of Lambert Glacier and the southern part of the Prince Charles Mountains, East Antarctica (abstract). In C. Craddock, ed., *Antarctic Geoscience*. University of Wisconsin Press, Madison, p. 522.

Fedorov, L. V., M. G. Ravich, and J. Hofmann. 1982. Geologic comparison of southeastern peninsular India and Sri Lanka with a part of East Antarctica (Enderby Land, MacRobertson Land, and Princess Elizabeth Land). In C. Craddock, ed., *Antarctic Geoscience* . University of Wisconsin Press, Madison, pp. 73–78.

Feldmann, R. M. 1984. Decapod crustaceans from the Late Cretaceous and the Eocene of Seymour Island, Antarctic Peninsula. *Ant. J. U.S.*, vol. 19, pp. 4–5.

Fillon, R. H. 1972. Evidence from the Ross Sea for widespread submarine erosion. *Nature*, vol. 238, pp. 40–42.

Fisco, M. P. P. 1982. Sedimentation on the Weddell Sea Continental Margin and Abyssal Plain, Antarctica. MA Thesis, Rice University, Houston, Texas, 170 pp.

Fitzgerald, P. G. 1992. The Transantarctic Mountains of southern Victoria Land; the application of apatite fission track analysis to a rift shoulder uplift. *Tectonics*, vol. 11, pp. 634–662.

——— 1994. Thermochronologic constraints on post-Paleozoic tectonic evolution of the central Transantarctic Mountains, Antarctica. *Tectonics*, vol. 13, pp. 818–836.

Fitzgerald, P. G., and S. Baldwin. 1996. Detachment fault model for the evolution of the Ross Embayment, Antarctica. EOS (Transactions of the American Geophysical Union), vol. 77, 703 pp.

Fitzgerald, P. G., M. Sandiford, P. J. Barrett, and A. J. W. Gleadow. 1986. Asymmetric extension associated with uplift and subsidence in the Transantarctic Mountains and Ross Embayment. *Earth Planet. Sci. Lett.*, vol. 81, pp. 67–78.

Fitzgerald, P. G., and E. Stump. 1991. Early Cretaceous uplift in the Ellsworth Mountains of West Antarctica. *Science*, vol. 254, pp. 92–94.

——— 1997. Cretaceous and Cenozoic episodic denudation of the Transantarctic Mountains, Antarctica: new constraints from apatite fission track thermochronology in the Scott Glacier region. *J. Geophys. Res.*, vol. 102, pp. 7747–7765.

Fitzsimons, S. J. 1990. Ice-marginal depositional processes in a polar maritime environment, Vestfold Hills, Antarctica. *J. Glaciol.*, vol. 36, pp. 279–286.

Flemming, R. F., and J. A. Barron. 1996. Evidence of Pliocene *Nothofagus* in Antarctica from Pliocene marine sedimentary deposits (DSDP site 274). *Marine Micropaleontol.*, vol. 27, pp. 227–236.

Flint, R. F. 1971. *Glacial and Quaternary Geology*: John Wiley & Sons, New York, 892 pp.

Flower, B. P., and J. P. Kennett. 1994. The middle Miocene climatic transition: East Antarctic Ice Sheet development, deep ocean circulation and global carbon cycling. *Palaeogeog. Palaeoclimatol. Palaeoecol.*, vol. 108, pp. 537–555.

Foldvik, A., and T. Gammelsrød. 1988. Notes on southern ocean hydrography, sea-ice and bottom water formation. *Palaeogeog. Palaeoclimatol. Palaeoecol.*, vol. 67, pp. 3–17.

Ford, A. B. 1972. Weddell orogeny – latest Permian to early Mesozoic deformation at the Weddell Sea margin of the Transantarctic Mountains. In R. J. Adie, ed., *Antarctic

Geology and Geophysics*. International Union of Geological Sciences, Series B, vol. 1. Universitetsforlaget, Oslo, Norway, pp. 419–425.

Ford, A. B., and P. J. Barrett. 1975. Basement rocks of the south–central Ross Sea, Site 270, DSDP Leg 28. In D. E. Hayes, L. A. Frakes, et al., eds., Initial Reports of the Deep Sea Drilling Project, vol. 28. U.S. Government Printing Office, Washington, D.C., pp. 861–868.

Ford, A. B., and G. R. Himmelberg. 1991. Geology and crystallization of the Dufek intrusion. In R. J. Tingey, ed., *The Geology of Antarctica*. Clarendon Press, Oxford, pp. 175–214.

Forsyth, D. W. 1975. Fault plane solutions and tectonics of the South Atlantic and Scotia Sea. *J. Geophys. Res.*, vol. 80, pp. 1429–1443.

Foster, T. D., and E. C. Carmack. 1976. Frontal zone mixing and Antarctic Bottom Water formation in the southern Weddell Sea. *Deep-Sea Res.*, vol. 23, pp. 301–317.

Frakes, L. A. 1983. Problems in Antarctic marine geology: a review. In R. L. Oliver, P. R. James, and J. B. Jago, eds., *Antarctic Earth Sciences*. Cambridge University Press, New York, pp. 375–378.

Frakes, L. A., and J. C. Crowell. 1967. Facies and paleogeography of Late Paleozoic diamictite, Falkland Islands. *Geolog. Soc. Am. Bull.*, vol. 78, pp. 37–58.

Frakes, L. A., and J. E. Francis. 1988. A guide to Phanerozoic cold polar climates from high-latitude ice-rafting in the Cretaceous. *Nature*, vol. 333, pp. 547–549.

Francis, J. E. 1995. 381 growth forms of fossil trees and climatic implications for the Pliocene Sirius Group, Dominion Range, Transantarctic Mountains. VII International Symposium on Antarctic Earth Sciences, Abstracts, Tipografia Senese, Siena, Italy, p. 142.

Frezzotti, M. 1992. Fluctuations of ice tongues and ice shelves derived from satellite images in Terra Nova Bay area, Victoria Land, Antarctica. In Y. Yoshida, K. Kaminuma, and K. Shiraishi, eds., *Recent Progress in Antarctic Earth Science*. Terra Scientific Publishing Co., Tokyo, Japan, pp. 733–739.

Fütterer, D. K., and M. Melles. 1990. Sediment patterns in the southern Weddell Sea: Filchner Shelf and Filchner Depression. In U. Bleil and J. Thiede, eds., *Geologic History of the Polar Oceans: Arctic versus Antarctic*. Kluwer Academic Publishers, Boston, pp. 381–401.

Gambôa, L. A. P., and P. R. Maldonado. 1990. Geophysical investigations in the Bransfield Strait and in the Bellingshausen Sea – Antarctica. In B. St. John, ed., *Antarctica as an Exploration Frontier – Hydrocarbon Potential, Geology, and Hazards*. American Association of Petroleum Geologists Studies in Geology, no. 31, pp. 127–141.

Garrett, S. W., L. D. B. Herrod, and D. R. Mantripp. 1987. Crustal structure of the area around Haag Nunataks, West Antarctica: new aeromagnetic and bedrock elevation data. In G. D. McKenzie, ed., *Gondwana Six: Structure, Tectonics, and Geophysics*. Geophysical Monograph, vol. 28. American Geophysical Union, Washington, D.C., pp. 109–115.

Garrett, S. W., and B. C. Storey. 1987. Lithospheric extension on the Antarctic Peninsula during Cenozoic subduction. In M. P. Coward, J. F. Dewey, and P. L. Hancock, eds.,

Continental Extensional Tectonics. American Geological Society Special Publication, vol. 28: Blackwell Scientific Publishers, London, pp. 419–431.

Gazdzicki, A. 1987. Paleontological studies on King George Island, West Antarctica, 1986. *Polish Polar Research*, vol. 8, pp. 85–92.

Gazdzicki, A., and R. Wrona. 1986. Polskie badania paleontogiczna w Antarktyce zachodniej (Polish paleontological investigation in West Antarctica in 1986 – summary). *Przeglad Geologiczny*, vol. 11, pp. 609–617.

Gazdzicka, E., and A. Gazdzicki. 1985a. Kokkolity zlepienca pektenowego Wyspy King George, Antarktyka Zachodnia (Coccoliths in the Pecten Conglomerate from King George Island, West Antarctica – summary). *Prezeglad Geologiczny*, vol. 11, pp. 543–547.

1985b. Oligocene coccoliths of the Pecten Conglomerate, West Antarctica. *Neues Jahrbuch für Geologie und Paläontologie Monatshefte*, vol. 12, pp. 727–735.

Gilbert, R. 1982. Contemporary sedimentary environments on Baffin Island, N.W.T., Canada: glaciomarine processes in fiords of eastern Cumberland Peninsula. *J. Arctic Alp. Res.*, vol. 14, pp. 1–12.

Gill, A. E. 1973. Circulation and bottom water production in the Weddell Sea. *Deep-Sea Res.*, vol. 20, pp. 111–140.

Glasby, G. P. 1976. Manganese nodules in the South Pacific: a review. *N. Z. J. Geol. Geophys.*, vol. 19, pp. 707–736.

Glasby, G. P., P. J. Barrett, J. C. McDougall, and D. G. McKnight. 1975. Localized variations in sedimentation characteristics in the Ross Sea and McMurdo Sound regions, Antarctica. *N.Z. J. Geol. Geophy.*, vol. 18, pp. 605–621.

Gleadow, A. J. W., and P. G. Fitzgerald. 1987. Uplift history and structure of the Transantarctic Mountains: new evidence from fission track dating of basement apatites in the Dry Valleys area, southern Victoria Land. *Earth Planet. Sci. Lett.*, vol. 82, pp. 1–14.

Gledhill, A., D. C. Rex, and P. W. G. Tanner. 1982. Rb-Sr and K-Ar geochronology of rocks from the Antarctic Peninsula between Anvers Island and Marguerite Bay. In C. Craddock, ed., *Antarctic Geoscience*. University of Wisconsin Press, Madison, pp. 315–323.

Goldring, D.C. 1962. The Geology of the Loubet Coast, Graham Land, British Antarctic Survey Scientific Reports, no. 36, 60 pp.

Goldstein, R. M., H. Engelhardt, B. Kamb, and R. M. Frolich. 1993. Satellite radar interferometry for monitoring ice sheet motion: application to an Antarctic ice stream. *Science*, vol. 262, pp. 1525–1530.

Goldstrand, P. M., P. G. Fitzgerald, T. F. Redfield, E. Stump, and C. Hobbs. 1994. Stratigraphic evidence for the Ross Orogeny in the Ellsworth Mountains, West Antarctica: implication for the evolution of the paleo-Pacific margin of Gondwana. *Geology*, vol. 22, pp. 427–430.

González-Ferrán, O. 1991. The Bransfield rift and its active volcanism. In M. R. A. Thomson, J. A. Crame and J. W. Thomson, eds., *Geological evolution of Antarctica*. Cambridge University Press, New York, pp. 505–509.

1983. The Larsen Rift: an active extension fracture in West Antarctica. In R. L. Oliver, P. R. James, and J. B. Jago, eds., *Antarctic Earth Science*. Cambridge University Press, New York, pp. 344–346.

1985. Volcanic and tectonic evolution of the northern Antarctic Peninsula – Late Cenozoic to Recent. *Tectonophysics*, vol. 114, pp. 389–409.

Goodell, H. G. 1973. The sediments. In H. G. Goodell, R. Houtz, M. Ewing, et al., eds., *Marine Sediments of the Southern Oceans*. Antarctic Map Folio Series – Folio 17. American Geographical Society, New York, pp. 1–9.

Goodell, H. G., R. Houtz, M. Ewing, D. Hayes, B. Naini, R. J. Echols, J. P. Kennett, and J. G. Donahue. 1973. *Marine Sediments of the Southern Oceans*. Antarctic Map Folio Series – Folio 17. American Geographical Society, New York, 18 pp.

Goodell, H. G., M. A. Meylan, and B. Grant. 1971. Ferromanganese deposits of the south Pacific Ocean, Drake Passage, and Scotia Sea. In J. L. Reid, ed., *Antarctic Oceanology I*. Antarctic Research Series, vol. 15. American Geophysical Union, Washington, D.C., pp. 27–92.

Goodell, H. G., N. D. Watkins, T. T. Mather, and S. Koster. 1968. The Antarctic glacial history recorded in sediments of the Southern Ocean. *Palaeogeog. Palaeoclimatol. Palaeoecol.*, vol. 5, pp. 41–62.

Goodwin, I. D. 1993. Holocene deglaciation, sea-level change, and the emergence of the Windmill Islands, Budd Coast, Antarctica. *Quat. Res.*, vol. 40, pp. 70–80.

1996. Mid to late Holocene readvance of the Law Dome ice margin, Budd Coast, East Antarctica. *Ant. Sci.*, vol. 8, pp. 395–406.

Gordon, A. L. 1967. *Structure of Antarctic Waters Between 20°W and 170°W*. Antarctic Map Folio Series – Folio 6. American Geographical Society, New York, 10 pp.

1971. Oceanography of Antarctic waters. In J. L. Reid, ed., *Antarctic Oceanology I*. Antarctic Research Series, vol. 15. American Geophysical Union, Washington, D.C., pp. 169–203.

Gordon, A. L., and R. D. Goldberg. 1970. *Circumpolar Characteristics of Antarctic Waters*. Antarctic Map Folio Series – Folio 13. American Geographical Society, New York, 5 pp.

Gordon, A. L., and E. J. Molinelli. 1982. Thermohaline and chemical distributions and the atlas data set. In A. L. Gordon, E. J. Molinelli, and T. N. Baker, eds., *Southern Ocean Atlas*: Columbia University Press, New York, pp. 3–11.

Gordon, A. L., and P. Tchernia. 1972. Waters of the continental margin off Adélie Coast, Antarctica. In D. E. Hayes, ed., *Antarctic Oceanology II*: Antarctic Research Series, vol. 19. American Geophysical Union, Washington, D.C., pp. 59–69.

Grácia, E., M. Canals, M. L. Farran, M. J. Prieto, J. Sorribas, and GEBRA Team. 1996. Morphostructure and evolution of the central and eastern Bransfield Basin (northwestern Antarctic Peninsula). *Marine Geophys. Res.*, vol. 18, pp. 429–448.

Greischar, L. L., and C. R. Bentley. 1980. Implications for the Late Wisconsin/Holocene extent of the West Antarctic Ice Sheet from a regional isostatic gravity map of the Ross Embayment (abstract). American Quaternary Association Sixth Biennial Meeting Abstract and Program, no. 6, p. 87.

Grew, E. S. 1982. Geology of the southern Prince Charles Mountains, East Antarctica. In C. Craddock, ed., *Antarc-*

tic Geoscience. University of Wisconsin Press, Madison, pp. 473–478.

Grew, E. S., and W. I. Manton. 1986. A new correlation of sapphirine granulites in the Indo-Antarctic metamorphic terrain: Late Proterozoic dates from the Eastern Ghats Province of India. *Precambrian Res.*, vol. 33, pp. 123–137.

Griffith, T. W. 1988. A Geological and Geophysical Investigation of Sedimentation and Recent Glacial History in the Gerlache Strait Region, Graham Land, Antarctica. MA Thesis, Rice University, Houston, Texas, 449 pp.

Griffith, T. W., and J. B. Anderson. 1989. Climatic control of sedimentation in bays and fjords of the northern Antarctic Peninsula. *Marine Geol.*, vol. 85, pp. 181–204.

Griffiths, J. R. 1971. Reconstruction of the south-west Pacific margin of Gondwanaland. *Nature*, vol. 234, pp. 203–207.

Grikurov, G. E. 1982. Structure of Antarctica and outline of its evolution. In C. Craddock, ed., *Antarctic Geoscience.* University of Wisconsin Press, Madison, pp. 791–804.

Grindley, G. W., and F. J. Davey. 1982. The reconstruction of New Zealand, Australia, and Antarctica. In C. Craddock, ed., *Antarctic Geoscience.* University of Wisconsin Press, Madison, pp. 15–29.

Grindley, G. W., and I. McDougall. 1969. Age and correlation of the Nimrod Group and other Precambrian rock units in the central Transantarctic Mountains, Antarctica. *N.Z. J. Geol. Geophys.*, vol. 12, pp. 391–411.

Grindley, G. W., and P. J. Oliver. 1983. Post-Ross Orogeny cratonisation of northern Victoria Land. In R. L. Oliver, P. R. James, and J. B. Jago, eds., *Antarctic Earth Science.* Cambridge University Press, New York, pp. 133–139.

Grindley, G. W., and G. Warren. 1964. Stratigraphic nomenclature and correlation in the western Ross Sea region. Proceedings of the First International Symposium on Antarctic Geology. North-Holland, Amsterdam, pp. 314–333.

Grinnell, D. V. 1971. Physiography of the continental margin of Antarctica from 125°E to 150°E. *Ant. J. U.S.*, vol. 6, pp. 164–165.

Grobe, H., D. K. Fütterer, and V. Spiess. 1990. Oligocene to Quaternary sedimentation processes on the Antarctic continental margin, ODP Leg 113, Site 693. In P. F. Barker, J. P. Kennett, et al., eds., Proceedings of the Ocean Drilling Program, Scientific Results, vol. 113. College Station, Texas, pp. 121–131.

Grobe, H., P. Huybrechts, and D. K. Fütterer. 1993. Late Quaternary record of sea-level changes in the Antarctic. *Geologische Rundschau*, vol. 82, pp. 263–275.

Grobe, H., and A. Mackensen. 1992. Late Quaternary climate cycles as recorded in sediments from the Antarctic continental margin. In J. P. Kennett and N. B. Watkins, eds., *The Antarctic Paleoenvironment.* American Geophysical Union, Antarctic Research Series, Washington, D.C., vol. 56, pp. 349–376.

Grunow, A. M. 1993. Creation and destruction of the Weddell Sea floor in the Jurassic. *Geology*, vol. 21, pp. 647–650.

Grunow, A. M., I. W. D. Dalziel, T. M. Harrison, and M. T. Heizler. 1992. Structural geology and geochronology of subduction complexes along the margin of Gondwanaland: new data from the Antarctic Peninsula and southernmost Andes. *Geolog. Soc. Am. Bull.*, vol. 104, pp. 1497–1514.

Grunow, A. M., I. W. D. Dalziel, and D. V. Kent. 1991. *An overview of Paleomagnetic Results from West Antarctica: Geological Evolution of Antarctica.* Cambridge, Cambridge University Press, 56 pp.

Grunow, A. M., D. V. Kent, and I. W. D. Dalziel. 1987b. Mesozoic evolution of West Antarctica and the Weddell Sea Basin: new paleomagnetic constraints. *Earth Planet. Sci. Lett.*, vol. 86, pp. 16–26.

1991c. New paleomagnetic data from Thurston Island and their implications for the tectonics of West Antarctica. *J. Geophys. Res.*, vol. 96, pp. 17,935–17,954.

Gunn, B. M., and G. Warren. 1962. Geology of Victoria Land between the Mawson and Mulock Glaciers, Antarctica. *N.Z. Geolog. Surv. Bull.*, vol. 71, 157 pp.

Gunner, J. 1976. Isotopic and Geochemical Studies of the Pre-Devonian Basement Complex, Beardmore Glacier Region, Antarctica. Institute of Polar Studies Report, no. 41. Ohio State University, Columbus, 126 pp.

Guterch, A., M. Grad, T. Janik, and E. Perchuc. 1991. Tectonophysical models of the crust between the Antarctic Peninsula and the South Shetland trench. In M. R. A. Thomson, J. A. Crame, and J. W. Thomson, eds., *Geological Evolution of Antarctica.* Cambridge University Press, New York, pp. 499–504.

Guterch, A., M. Grad, T. Janik, E. Perchuc, and J. Pajchel. 1985. Seismic studies of the crustal structure in West Antarctica 1979–1980 – preliminary results. *Tectonophysics*, vol. 114, pp. 411–429.

Haase, G. M. 1986. Glaciomarine sediments along the Filchner/Rønne Ice Shelf, southern Weddell Sea – first results of the 1983/84 Antarktis-II/4 Expedition. *Marine Geol.*, vol. 72, pp. 241–258.

Hall, K. 1983. A reconstruction of the Quaternary ice cover on Marion Island. In R. L. Oliver, P. R. James, and J. B. Jago, eds., *Antarctic Earth Science.* Cambridge University Press, New York, pp. 461–464.

1984. Evidence in favour of an extensive ice cover on subantarctic Kerguelen Island during the last glacial. *Palaeogeog. Palaeoclimatol. Palaeoecol.*, vol. 47, pp. 225–232.

Hall, B. L., G. H. Denton, D. R. Lux, and C. Schluchter. 1997. Pliocene paleoenvironment and Antarctic ice sheet behavior: evidence from Wright Valley. *J. Geol.*, vol. 105, pp. 285–294.

Halpern, M. 1968. Ages of Antarctic and Argentine rocks bearing on continental drift. *Earth Planet. Sci. Lett.*, vol. 5, pp. 159–167.

Halpern, M., and G. E. Grikurov. 1975. Rubidium-strontium data from the southern Prince Charles Mountains. *Ant. J. U.S.*, vol. 10, pp. 9–15.

Hambrey, M. J. 1993. Cenozoic sedimentary and climatic record, Ross Sea region, Antarctica. In J. P. Kennett and D. A. Warnke, eds., *The Antarctic Paleoenvironment: A Perspective on Global Change II.* Antarctic Research Series, vol. 60. American Geophysical Union, Washington, D.C., pp. 91–124.

Hambrey, M. J., W. U. Ehrmann, and B. Larsen. 1991. Cenozoic glacial record of the Prydz Bay continental shelf, East Antarctica. In J. Barron, B. Larsen, et al., eds., Proceedings of the Ocean Drilling Program, Scientific Results, vol. 119. College Station, Texas, pp. 77–132.

Hambrey, M. J., and W. B. Harland (eds.). 1981. *Earth's Pre-*

Pleistocene Glacial Record. Cambridge University Press, New York, 1004 pp.

Hambrey, M. J., B. Larsen, W. U. Ehrmann, and ODP Leg 119 Shipboard Scientific Party. 1989. Forty million years of Antarctic glacial history yielded by Leg 119 of the Ocean Drilling Program. *Polar Rec.*, vol. 25, pp. 99–106.

Hampton, M. A. 1972. The role of subaqueous debris flows in generating turbidity currents. *J. Sediment. Petrol.*, vol. 42, pp. 775–793.

Hampton, M. A., S. L. Eittreim, and B. M. Richmond. 1987. Post-breakup sedimentation on the Wilkes Land margin, Antarctica. In S. L. Eittreim and M. A. Hampton, eds., *The Antarctic Continental Margin: Geology and Geophysics of Offshore Wilkes Land.* CPCEMR Earth Science Series, vol. 5A. Circum-Pacific Council for Energy and Mineral Resources, Houston, Texas, pp. 75–88.

Han, M. W., and E. Suess. 1987. Lateral migration of pore fluids through sediments of an active back-arc basin: Bransfield Strait, Antarctica (abstract): EOS (Transactions of the American Geophysical Union), vol. 68, p. 1796.

Hannah, M. J. 1994. Eocene dinoflagellates from CIROS-1 drillhole, McMurdo Sound, Antarctica. In A. K. Cooper, P. F. Barker, P. N. Webb, and G. Brancolini, eds., The Antarctic Continental Margin: Geophysical and Geological Stratigraphic Records of Cenozoic Glaciation, Paleoenvironments and Sea-Level Change. *Terra Antarctica*, vol. 1, p. 371.

Hansom, J. D. 1983. Ice-formed intertidal boulder pavements in the sub-Antarctic. *J. Sediment. Petrol.*, vol. 53, pp. 135–145.

Haq, B. U. 1980. Biogeographic history of Miocene calcareous nannoplankton and paleoceanography of the Atlantic Ocean. *Micropaleontology*, vol. 26, pp. 414–443.

Haq, B. U., J. Hardenbol, and P. R. Vail. 1987. Chronology of fluctuating sea levels since the Triassic. *Science*, vol. 235, pp. 1156–1167.

Haq, B. U., G. P. Lohmann, and S. W. J. Wise. 1977. Calcareous nannoplankton biogeography and its paleoclimatic implications; Cenozoic of the Falkland Plateau (DSDP Leg 36) and the Miocene of the Atlantic. In P. F. Barker, I. W. D. Dalziel, et al., eds., *Initial Reports of the Deep Sea Drilling Project*, vol. 36. U.S. Government Printing Office, Washington, D.C., pp. 745–760.

Harden, S. L., D. J. DeMaster, and C. A. Nittrouer. 1992. Developing sediment geochronologies for high-latitude continental shelf deposits: a radiochemical approach. *Marine Geol.*, vol. 103, pp. 69–97.

Hardenbol, J., J. Thierry, M. B. Farley, T. Jacquin, P.-C. De Graciansky, and P. R. Vail. In press. Mesozoic-Cenozoic sequence chronostratigraphic of European basins. In P.-C. De Graciansky, J. Hardenbol, T. Jacquin, P. R. Vail, and M. B. Farley, eds., *Sequence Stratigraphy of European Basins.* Society of Economic Paleontologists and Mineralogists Special Publication, vol. 57, Tulsa, Oklahoma.

Harris, P. T., P. E. O'Brien, P. G. Quilty, F. Taylor, E. Domack, L. DeSantis, and B. Raker. 1997. Vincennes Bay, Prydz Bay and MacRobertson Shelf, Post-cruise report AGSO Cruise 186, Anare Voyage 5, 1996/97, AGSO. Canberra City, Australia, 75 pp.

Hart, G. F. 1964. Where was the Lower Karoo Sea? Scientific Society of Africa, vol. 1, pp. 289–290.

Harwood, D. M. 1983. Diatoms from the Sirius Formation, Transantarctic Mountains. *An. J. U.S.*, vol. 18, pp. 98–100.

1986. Diatom biostratigraphy and paleoecology with a Cenozoic history of Antarctic Ice Sheets. PhD Dissertation, Ohio State University, Columbus, 592 pp.

1991. Cenozoic diatom biogeography in the southern high latitudes: inferred biogeographic barriers and progressive endemism. In M. R. A. Thomson, J. A. Crame, and J. W. Thomson, eds., *Geological Evolution of Antarctica.* Cambridge University Press, New York, pp. 667–673.

Harwood, D. M., P. Webb. 1998. Glacial transport of diatoms in the Antarctic sirius group: Pliocene refrigerator. *GSA Today*, vol. 8, pp. 4–5.

Haughton, S. H. 1969. Geological History of Southern Africa. Geological Society of South Africa, Capetown, 535 pp.

Haugland, K. 1982. Seismic reconnaissance survey in the Weddell Sea. In C. Craddock, ed., *Antarctic Geoscience*: University of Wisconsin Press, Madison, pp. 405–413.

Haugland, K., Y. Kristoffersen, and A. Velde. 1985. Seismic investigations in the Weddell Sea embayment. *Tectonophysics*, vol. 114, pp. 293–313.

Hawkes, D. D. 1981. Tectonic segmentation of the northern Antarctic Peninsula. *Geology*, vol. 9, pp. 220–224.

Hayes, D. E. (ed.). 1991. *Marine Geological and Geophysical Atlas of the Circum-Antarctic to 30 degrees S.* Antarctic Research Series, vol. 54. American Geophysical Union, Washington, D.C., 56 pp.

Hayes, D. E., and F. J. Davey. 1975. A geophysical study of the Ross Sea, Antarctica. In D. E. Hayes and L. A. Frakes, eds., *Initial Reports of the Deep Sea Drilling Project*, vol. 28. U.S. Government Printing Office, Washington, D.C., pp. 887–907.

Hayes, D. E., and L. A. Frakes. 1975. General Synthesis, Deep Sea Drilling Project 28. In D. E. Hayes and L. A. Frakes, eds., *Initial Reports of Deep Sea Drilling Project*, vol. 28. U.S. Government Printing Office, Washington, D.C., pp. 919–942.

Hays, J. D., and N. D. Opdyke. 1967. Antarctic radiolaria, magnetic reversals, and climatic change. *Science*, vol. 158, pp. 1001–1011.

Heezen, B. C. 1972. Soundings and earthquakes. In B. C. Heezen, M. Tharp, and C. R. Bentley, eds., *Morphology of the Earth in the Antarctic and Subantarctic.* Antarctic Map Folio Series – Folio16. American Geographical Society, New York, plate 2.

Heezen, B. C., and C. D. Hollister. 1971. *The Face of the Deep.* Oxford University Press, New York, 659 pp.

Heezen, B. C., M. Tharp, and C. R. Bentley. 1972. *Morphology of the Earth in the Antarctic and Subantarctic.* Antarctic Map Folio Series – Folio 16. American Geographical Society, New York, 16 pp.

Helfer, M., and C. Schluchter. 1995. The Antarctic Sirius Formation: geological analysis of two sites. VII International Symposium on Antarctic Earth Sciences, Abstracts, Tipografia, Senese, Siena, Italy, p. 189.

Henrys, S. A., L. R. Bartek, J. B. Anderson, and P. J. Barrett. 1994. Seismic stratigraphy in McMurdo Sound: correlation of high resolution data sets. In A. K. Cooper, P. F. Barker, P. N. Webb, and G. Brancolini, eds., The Antarctic Continental Margin: Geophysical and Geological Stratigraphic Records of Cenozoic Glaciation, Paleoenvironments and Sea-Level Change. *Terra Antarctica*, vol. 1, pp. 373–374.

Herron, E. M., and B. E. Tucholke. 1976. Sea-floor magnetic patterns and basement structure in the southeastern Pacific. In C. D. Hollister and C. Craddock, eds., *Initial Reports of the Deep Sea Drilling Project*, vol. 35. U.S. Government Printing Office, Washington, D.C., pp. 263–278.

Herron, M. J. 1988. Marine Geology and Geophysics of the Western South Orkney Plateau, Antarctica: Implications for Quaternary Glacial History Tectonics and Paleoceanography. MA Thesis, Rice University, Houston, Texas, 256 pp.

Herron, M. J., and J. B. Anderson. 1990. Late Quaternary glacial history of the South Orkney Plateau, Antarctica. *Quat. Res.*, vol. 33, pp. 265–275.

Hill, R. S., and E. M. Truswell. 1993. *Nothofagus* fossils in the Sirius Group, Transantarctic Mountains; leaves and pollen and their climatic implications. In J. P. Kennett and D. A. Warnke, eds., *The Antarctic Paleoenvironment: A Perspective on Global Change II*. Antarctic Research Series, vol. 60. American Geophysical Union, Washington, D.C., pp. 67–73.

Hinz, K., and M. Block. 1984. Results of geophysical investigations in the Weddell Sea and in the Ross Sea, Antarctica. In Proceedings of the 11th World Petroleum Congress, London 1983, vol. 2. John Wiley & Sons, New York, pp. 79–91.

Hinz, K., and W. Krause. 1982. The continental margin of Queen Maud Land/Antarctica: seismic sequences, structural elements and geological development. *Geologisches Jahrbuch*, vol. E23, pp. 17–41.

Hinz, K., and Y. Kristoffersen. 1987. Antarctica – recent advances in the understanding of the continental shelf. *Geologisches Jahrbuch*, vol. E37, pp. 3–54.

Hirakawa, K., and K. Moriwaki. 1990. Former ice sheet based on the newly observed glacial landforms and erratics in the central Sør Rondane Mountains, East Antarctica. In Proceedings of the NIPR Symposium on Antarctic Geosciences, vol. 4. National Institute of Polar Research, Tokyo, Japan, pp. 41–54.

Hirvas, H., P. Lintinen, and K. Nenonen. 1994, Moreenin hienoaineksen ominaisuuksista Suomessa ja Etelamantereella (Properties of till fines in Finland and Antarctica). In M. Perttunen, ed., *Uudet menetelmat ja miiden sovellukset kvartaaritutkimuksessa (New Methods and Their Application to Quaternary Studies)*. Acta Universitatis Ouluensis, Series A, Scientiae Rerum Naturalium, vol. 251. Oulun Yliopisto, Finland, pp. 9–23.

Hjort, C., O. Ingólfsson, and S. Björck. 1992. The last major deglaciation in the Antarctic peninsula region-a review of recent Swedish quaternary research. In Y. Yoshida, ed., *Recent Progress in Antarctic Earth Science*. Terra Scientific Publishing, Tokyo, Japan, pp. 741–743.

Hjort, C., O. Ingólfsson, P. Möller, and J. M. Lirio. 1997. Holocene glacial history and sea-level changes on James Ross Island, Antarctic peninsula. *J. Quat. Sci.*, vol. 12, pp. 259–273.

Hobday, D. K. 1982. The southeast African margin, in A. E. M. Nairn and F. G. Stehli, eds., *The Ocean Basins and Mar-*

gins. 6. *The Indian Ocean*. Plenum Press, New York, pp. 149–183.

Hofmann, J. 1978. Tektonische Beobachtungen im hoch – und schwachmetamorphen Praekambrium der Gebirgsumrandung des Lambert-Gletschers (Ostantarktis); (Tectonic observations in high-grade and low-grade Precambrian metamorphism of the mountainous region of the Lambert Glacier, eastern Antarctica). *Mineralogie-Geochemie*, vol. C335, pp. 1–111.

Hollin, J. T. 1962. On the glacial history of Antarctica. *J. Glaciol.*, vol. 4, pp. 173–195.

——— 1964. Origin of ice ages: an ice shelf theory for Pleistocene glaciation. *Nature*, vol. 202, pp. 1099–1100.

Hollister, C. D., and R. B. Elder. 1969. Contour currents in the Weddell Sea. *Deep-Sea Res.*, vol. 16, pp. 99–101.

Holtedahl, O. 1929. On the geology and physiography of some Antarctic and sub-Antarctic islands. Scientific Results of the Norwegian Antarctic Expeditions, 1927–1928 and 1928–1929, 172 pp.

——— 1970. On the morphology of the West Greenland Shelf, with general remarks on the "marginal channel" problem. *Marine Geol.*, vol. 8, pp. 155–172.

Hooper, P. R. 1962. The petrology of Anvers Island and adjacent islands. Falkland Islands Dependencies Survey Scientific Report, no. 34, 69 pp.

Houtz, R. E., and F. J. Davey. 1973. Seismic profiler and sonobuoy measurements in the Ross Sea, Antarctica. *J. Geophys. Res.*, vol. 78, pp. 3448–3468.

Houtz, R. E., and R. Meijer. 1970. Structure of the Ross Sea shelf from profiler data. *J. Geophys. Res.*, vol. 75, pp. 6592–6597.

Huang, T. C., and N. D. Watkins. 1977. Contrasts between the Brunhes and Matuyama sedimentary records of bottom water activity in the South Pacific. *Marine Geol.*, vol. 23, pp. 113–132.

Hughes, T. 1973. Is the West Antarctic Ice Sheet disintegrating? *J. Geophys. Res.*, vol. 78, pp. 7884–7910.

——— 1977. West Antarctic ice streams. *Rev. Geophy. Space Phys.*, vol. 15, pp. 1–46.

——— 1981. Numerical reconstruction of paleo-ice sheets. In G. H. Denton and T. J. Hughes, eds., *The Last Great Ice Sheets*. John Wiley & Sons, New York, pp. 221–261.

——— 1987. Deluge II and the continent of doom: rising sea level and collapsing Antarctic ice. *Boreas*, vol. 16, pp. 89–100.

Hunter, D. R., A. C. Johnson, and N. D. Aleshkova. 1996. Aeromagnetic data from the southern Weddell Sea embayment and adjacent areas: synthesis and interpretation. In B. C. Storey, E. C. King, and R. A. Livermore, eds., *Weddell Sea Tectonics and Gondwana Break-Up*. Geological Society Special Publication No. 108, pp. 143–154.

Hunter, D. R., and D. A. Pretorius. 1981. Structural framework (of Southern Africa). In D. R. Hunter, ed., *Precambrian of the Southern Hemisphere*. Elsevier, Amsterdam, pp. 397–422.

Huybrechts, P. 1990. The Antarctic ice sheet during the last glacial-interglacial cycle: a three-dimensional experiment. *Ann. Glaciol.*, vol. 14, pp. 115–119.

——— 1992. The Antarctic Ice Sheet and Environmental Change: A Three-dimensional Modelling Study. Unpublished PhD Dissertation, Bremerhaven, Germany.

——— 1993. Glaciological modelling of the late Cenozoic East

Antarctic ice sheet: stability or dynamism? *Geografiska Annaler*, vol. 75A, pp. 221–238.

Igarashi, A., N. Harada, and K. Moriwaki. 1995. Marine fossils of 30–40 ka raised beach deposits and the late Pleistocene glacial history around Lützow-Holm Bay, East Antarctica. NIPR Symposium on Antarctic Geosciences, Proceedings: National Institute of Polar Research, pp. 219–229.

Ineson, J. R., J. A. Crame, and M. R. A. Thomson. 1986. Lithostratigraphy of the Cretaceous strata of west James Ross Island, Antarctica. *Cretaceous Res.*, vol. 7, pp. 141–159.

Ivanov, V. L. 1983. Sedimentary basins of Antarctica and their preliminary structural and morphological classification. In R. L. Oliver, P. R. James, and J. B. Jago, eds., *Antarctic Earth Science*. Cambridge University Press, New York, pp. 539–541.

Ivy-Ochs, S., C. Schlüchter, P. W. Kubik, B. Dittrich-Hannen, and J. Beer. 1995. Minimum ^{10}Be exposure ages of early Pliocene for the Table Mountain plateau and the Sirius Group at Mount Fleming, Dry Valleys, Antarctica. *Geology*, vol. 23, pp. 1007–1010.

Jackson, M. P. A. 1979. A major charnockite-granulite province in southwestern Africa. *Geology*, vol. 7, pp. 22–26.

Jacobel, R. W., and R. Bindschadler. 1993. Radar studies at the mouths of Ice Streams D and E, Antarctica. *Ann. Glaciol.*, vol. 17, pp. 262–268.

Jacobs, S. S. 1992. Is the Antarctic ice sheet growing? *Nature*, vol. 360, pp. 29–33.

Jacobs, S. S., A. F. Amos, and P. M. Bruchhausen. 1970. Ross Sea oceanography and Antarctic Bottom Water formation. *Deep-Sea Res.*, vol. 17, pp. 935–962.

Jacobs, S. S., R. G. Fairbanks, and Y. Horibe. 1985. Origin and evolution of water masses near the Antarctic continental margin: evidence from $H_2^{18}O/H_2^{16}O$ ratios in seawater. In S. S. Jacobs, ed., *Oceanology of the Antarctic Continental Shelf*. Antarctic Research Series, vol. 43. American Geophysical Union, Washington, D.C., pp. 59–85.

Jacobs, S. S., A. L. Gordon, and A. F. Amos. 1979. Effect of glacial ice melting on the Antarctic Surface Water. *Nature*, vol. 277, pp. 469–471.

Jacobs, S. S., and W. E. Haines. 1982. Ross Ice Shelf Project, Oceanographic Stations, 1976–1979: Technical Report, no. 1, vol. 82. Lamont-Doherty Geological Observatory, Palisades, New York, 505 pp.

Jacobs, S. S., H. H. Hellmer, C. S. M. Doake, A. Jenkins, and R. M. Frolich. 1992. Melting of ice shelves and the mass balance of Antarctica. *J. Glaciol.*, vol. 38, pp. 375–387.

Jacobs, S. S., H. H. Hellmer, and A. Jenkins. 1996. Antarctic ice sheet melting in the southeast Pacific. *Geophys. Res. Lett.*, vol. 23, pp. 957–960.

Jahns, E. 1994. Evidence for a fluidized till deposit on the Ross Sea continental shelf. *Ant. J. U.S.*, vol. 29, pp. 139–141.

—— 1995. Evidence for a fluidized till deposit on the Ross Sea continental shelf. *Ant. J. U. S.*, vol. 29, pp. 139–141.

Jankowski, E. J., and D. J. Drewry. 1981. The structure of West Antarctica from geophysical studies. *Nature*, vol. 291, pp. 17–21.

Jankowski, E. J., D. J. Drewry, and J. C. Behrendt. 1983. Magnetic studies of upper crustal structure in West Antarctica

and the boundary with East Antarctica. In R. L. Oliver, P. R. James, and J. B. Jago, eds., *Antarctic Earth Science*. Cambridge University Press, New York, pp. 197–203.

Jeffers, J. D. 1988. Tectonics and Sedimentary Evolution of the Bransfield Basin, Antarctica. MA Thesis, Rice University, Houston, Texas, 142 pp.

Jeffers, J. D., and J. B. Anderson. 1990. Sequence stratigraphy of the Bransfield Basin, Antarctica: implications for tectonic history and hydrocarbon potential. In B. St. John, ed., *Antarctica as an Exploration Frontier – Hydrocarbon Potential, Geology, and Hazards*. American Association of Petroleum Geologists, Studies in Geology, no. 31, pp. 13–30.

Jeffers, J. D., J. B. Anderson, and L. A. Lawver. 1991. Evolution of the Bransfield Basin, Antarctic Peninsula. In M. R. A. Thomson, J. A. Crame, and J. W. Thomson, eds., *Geological Evolution of Antarctica*. Cambridge University Press, New York, pp. 481–485.

Jenkins, A., and C. S. M. Doake. 1991. Ice-ocean interaction on Rønne Ice Shelf, Antarctica. *J. Geophys. Res.*, vol. 96, pp. 791–813.

Jenkins, A., D. G. Vaughan, S. S. Jacobs, H. H. Hellmer, and J. R. Keys. 1996. Glaciological and oceanographic evidence of high melt rates beneath Pine Island Glacier, West Antarctica. *J. Glaciol.*, vol. 43, pp. 114–121.

John, B. S. and D. E. Sugden. 1975. Coastal geomorphology of high latitudes. *Progr. Geogr.*, vol. 7, pp. 53–112.

Johnson, A. C., and A. M. Smith. 1992. New aeromagnetic map of West Antarctica (Weddell Sea sector): introduction to important features. In Y. Yoshida, K. Kaminuma, and K. Shiraishi, eds., *Recent Progress in Antarctic Earth Science*. Terra Scientific Publishing Co., Tokyo, Japan pp. 555–562.

Johnson, B. D., C. M. Powell, and J. J. Veevers. 1976. Spreading history of the eastern Indian Ocean and Greater India's northward flight from Antarctica and Australia. *Geolog. Soc. Am. Bull.*, vol. 87, pp. 1560–1566.

Johnson, G. L., J.-R. Vanney, A. Elverhøi, and J. L. LaBrecque, 1981, Morphology of the Weddell Sea and southwest Indian Ocean. *Deutsche Hydrographische Zeitschrift*, vol. 34, pp. 263–272.

Johnson, G. L., J. R. Vanney, and D. Hayes. 1982. The Antarctic continental shelf – review paper. In C. Craddock, ed., *Antarctic Geoscience*. University of Wisconsin Press, Madison, pp. 995–1002.

Jokat, W., N. Fechner, and M. Studinger. 1995. Filchner-Rønne Shelf: new seismic data. Abstracts of the VII International Symposium on Antarctic Earth Sciences, Siena, Italy, p. 214.

Jouzel, J., C. Lorius, J. R. Petit, C. Genthon, N. I. Barkov, V. M. Kotlyakov, and V. M. Petrov. 1987. Vostok ice core: a continuous isotope temperature record over the last climatic cycle (160,000 years). *Nature*, vol. 329, pp. 403–408.

Kagami, H., H. Kuramochi, and Y. Shima. 1991. Submarine canyons in the Bellingshausen and Riiser-Larsen Seas around Antarctica. In Proceedings of the NIPR Symposium on Antarctic Geosciences, no. 5, pp. 84–98.

Kaharoeddin, F. A., M. R. Eggers, E. H. Goldstein, R. S. Graves, D. K. Watkins, J. A. Bergen, S. C. Jones, and D. S. Cassidy. 1980. ARA *Islas Orcadas* cruise 1578; sediment descriptions. Sedimentology Research Laboratory

Contribution, vol. 48. Florida State University, Tallahassee, 162 pp.

Kaharoeddin, F. A., M. R. Eggers, R. S. Graves, E. H. Goldstein, J. G. Hattner, S. C. Jones, and P. F. Ciesielski. 1979. ARA *Islas Orcadas* Cruise 1277 Sediment Descriptions. Sedimentology Research Laboratory Contribution, vol. 47. Florida State University, Tallahassee, 108 pp.

Kalamarides, R. I., J. H. Berg, and R. A. Hank. 1987. Lateral isotopic discontinuity in the lower crust: an example from Antarctica. *Science*, vol. 237, pp. 1192–1195.

Kamenev, E. N. 1972. Geological structure of Enderby Land. In R. J. Adie, ed., *Antarctic Geology and Geophysics*. International Union of Geological Sciences, Series B, vol. 1. Universitetsforlaget, Oslo, Norway, pp. 579–583.

———. 1982. Antarctica's oldest metamorphic rocks in the Fyfe Hills, Enderby Land. In C. Craddock, ed., *Antarctic Geoscience*. University of Wisconsin Press, Madison, pp. 505–510.

Kamenev, E. N., and V. L. Ivanov. 1983. Structure and outline of geologic history of the southern Weddell Sea Basin. In R. L. Oliver, P. R. James, and J. B. Jago, eds., *Antarctic Earth Science*. Cambridge University Press, New York, pp. 194–196.

Katz, H. R. 1973. Contrasts in tectonic evolution of orogenic belts in the south-east Pacific. *J. Royal Soc. N.Z.*, vol. 3, pp. 333–362.

Katz, M. B. 1978. Sri Lanka in Gondwanaland and the evolution of the Indian Ocean. *Geolog. Mag.*, vol. 115, pp. 237–244.

Kaul, N. 1992. High resolution seismics and stratigraphy off Kapp Norvegia, Antarctica. *Zeitschrift für Geomorphologie*, pp. 105–112.

Keany, J. 1978. Paleoclimatic trends in Early and Middle Pliocene deep-sea sediments of the Antarctic. *Marine Micropaleontol.*, vol. 3, pp. 35–49.

Keller, R. A., and M. R. Fisk. 1987. Magmatism associated with the initial stages of back-arc rifting, Bransfield Strait, Antarctica (abstract). EOS (Transactions of the American Geophysical Union), vol. 68, pp. 1533.

Keller, R. A., M. R. Fisk, and W. M. White. 1991. Rifting of an island arc, Bransfield Strait and South Shetland Islands, Antarctica. International Symposium on Antarctic Earth Sciences, vol. 6, pp. 298–303.

Kellogg, D. E., and T. B. Kellogg. 1987. Microfossil distributions in modern Amundsen Sea sediments. *Marine Micropaleontol.*, vol. 12, pp. 203–222.

———. 1996. Diatoms in South Pole ice: implications for eolian contamination of Sirius Group deposits. *Geology*, vol. 24, pp. 115–118.

Kellogg, T. B. 1987. Glacial-Interglacial changes in global deepwater circulation. *Paleoceanography*, vol. 2, pp. 259–271.

Kellogg, T. B., T. Hughes, and D. E. Kellogg. 1996. Late Pleistocene interactions of East and West Antarctic ice-flow regimes: evidence from the McMurdo Ice Shelf. *J. Glaciol.*, vol. 42, pp. 486–500.

Kellogg, T. B., and D. E. Kellogg. 1988. Antarctic cryogenic sediments: biotic and inorganic facies of ice shelf and marine-based ice sheet environments. *Palaeogeog. Palaeoclimatol. Palaeoecol.*, vol. 67, pp. 51–74.

Kellogg, T. B., D. E. Kellogg, and M. Stuiver. 1990. Late Quaternary history of the southwestern Ross Sea: evidence from debris bands on the McMurdo Ice Shelf, Antarctica. In D. H. Elliot, ed., *Contributions to Antarctic Research I*. Antarctic Research Series, vol. 50. American Geophysical Union, Washington, D.C., pp. 25–56.

Kellogg, T. B., R. S. Truesdale, and L. E. Osterman. 1979. Late Quaternary extent of the West Antarctic Ice Sheet: new evidence from Ross Sea cores. *Geology*, vol. 7, pp. 249–253.

Kemp, E. M., L. A. Frakes, and D. E. Hayes. 1975. Paleoclimatic significance of diachronous biogenic facies, Leg 28. In D. E. Hayes and L. A. Frakes, et al., eds., *Initial Reports of the Deep Sea Drilling Project*, vol. 28: U.S. Government Printing Office, Washington, D.C., pp. 909–917.

Kennedy, D. S. 1988. Modern Sedimentary Dynamics and Quaternary Glacial History of Marguerite Bay, Antarctic Peninsula. MA Thesis, Rice University, Houston, Texas, 203 pp.

Kennedy, D. S., and J. B. Anderson. 1989. Glacial-marine sedimentation and Quaternary glacial history of Marguerite Bay, Antarctic Peninsula. *Quat. Res.*, vol. 31, pp. 255–276.

Kennett, J. P. 1966. Foraminiferal evidence of a shallow calcium carbonate solution boundary, Ross Sea, Antarctica. *Science*, vol. 153, pp. 191–193.

———. 1977. Cenozoic evolution of Antarctic glaciation, the circum-Antarctic ocean, and their impact on global paleoceanography. *J. Geophys. Res.*, vol. 82, pp. 3843–3860.

———. 1978. Cenozoic microfossil datums in Antarctic-subAntarctic deep-sea sedimentary sequences and the evolution of Southern Ocean planktonic biogeography. In N. Ikebe, ed., Correlation of Tropical Through High Latitude Marine Neogene Deposits of the Pacific Basin. Stanford University Publications, *Geol. Sci.*, vol. 14, pp. 30–31.

Kennett, J. P., and P. F. Barker. 1990. Latest Cretaceous to Cenozoic climate and oceanographic developments in the Weddell Sea, Antarctica: an ocean-drilling perspective. In P. F. Barker, J. P. Kennett, et al., eds., Proceedings of the Ocean Drilling Program, Scientific Results, vol. 113, pp. 937–960.

Kennett, J. P., and D. A. Hodell. 1993. Evidence for relative climatic stability of Antarctica during the early Pliocene: a marine perspective. *Geografiska Annaler*, vol. 75A, pp. 205–220.

———. 1995. Stability or instability of Antarctic ice sheets during warm climates of the Pliocene? *GSA Today*, vol. 5, pp. 1–22.

Kennett, J. P., and N. J. Shackleton. 1976. Oxygen isotopic evidence for the development of the psychrosphere 38 Myr ago. *Nature*, vol. 260, pp. 513–515.

Kennett, J. P., and L. D. Stott. 1990. Proteus and Protooceanus: ancestral Paleogene oceans as revealed from Antarctic stable isotopic results. In P. F. Barker, J. P. Kennett, et al., eds., Proceedings of the Ocean Drilling Program, Scientific Results, vol. 113, pp. 865–880.

Kennett, J. P., and P. Vella. 1975. Late Cenozoic planktonic foraminifera and paleoceanography at DSDP Site 284 in the cool subtropical South Pacific. In J. P. Kennett, R. E. Houtz, et al., eds., *Initial Reports of the Deep Sea Drilling Project*, vol. 29. U.S. Government Printing Office, Washington, D.C., pp. 769–799.

Kennett, J. P., N. D. Watkins, and P. Vella. 1971. Paleomagnetic chronology of Pliocene-Early Pleistocene climates and the Plio-Pleistocene boundary in New Zealand. *Science*, vol. 171, pp. 276–279.

Keys, J. R. 1990. Ice. In G. P. Glasby, ed., *Antarctic Sector of the Pacific*. Elsevier Oceanography Series, vol. 51, New York, pp. 95–123.

Kidd, R. B., and T. A. Davies. 1978. Indian Ocean sediment distribution since the Late Jurassic. *Marine Geol.*, vol. 26, pp. 49–70.

Killingley, J. S. 1983. Effects of diagenetic recrystallization on 180/160 values of deep-sea sediments. *Nature*, vol. 301, pp. 594–597.

Kim, Y., T. W. Chung, and S. H. Nam. 1992. Marine magnetic anomalies in Bransfield Strait, Antarctica. In Y. Yoshida, K. Kaminuma, and K. Shiraishi, eds., *Recent Progress in Antarctic Earth Science*. Terra Scientific Publishing Co., Tokyo, Japan, pp. 431–437.

Kim, Y., H. S. Kim, R. D. Larter, A. Camerlenghi, L. A. P. Gambôa, and S. Rudowski. 1994. Tectonic deformation in the upper crust and sediments at the South Shetland Trench. In A. K. Cooper, P. F. Barker, P. N. Webb, and G. Brancolini, eds., The Antarctic Continental Margin: Geophysical and Geological Stratigraphic Records of Cenozoic Glaciation, Paleoenvironments, and Sea-Level Change. *Terra Antarctica*, vol. 1, no. 2, pp. 299–301.

Kim, Y., L. D. McGinnis, and R. H. Bowen. 1986. The Victoria Land Basin: part of an extended crustal complex between East and West Antarctica. In M. Barazangi and L. Brown, eds., *Reflection Seismology: the Continental Crust*. Geodynamic Series, vol. 14. American Geophysical Union, Washington, D.C., pp. 323–330.

Kimura, K. 1982. Geological and geophysical survey in the Bellingshausen Basin, off Antarctica. *Ant. Rec.*, vol. 75, pp. 12–24.

King, E. C., and A. C. Bell. 1996. New seismic data from the Ronne Ice Shelf, Antarctica. In B. C. Storey, E. C. King, and R. A. Livermore, eds., *Weddell Sea Tectonics and Gondwana Break-Up*. Geological Society Publication No. 108, Geological Society of London, UK, pp. 213–226.

King, L. H. 1993. Till in the marine environment. *J. Quat. Sci.*, vol. 8, pp. 347–358.

King, L. H., and G. B. J. Fader. 1986. Wisconsinian glaciation of the Atlantic continental shelf of southeast Canada. *Geolog. Surv. Can. Bull.*, vol. 363, 72 pp.

King, L. H., K. Rokoengen, G. B. J. Fader, and T. Gunleiksrud. 1991. Till-tongue stratigraphy. *Geolog. Soc. Am. Bull.*, vol. 103, pp. 637–659.

Kleinschmidt, G., S. Matzer, F. Henjes-Kunst, and G. Fenn. 1992. New field data from Surgeon Island, north Victoria Land, Antarctica. *Polarforschung*, vol. 60, pp. 133–134.

Klepikov, V. V. 1963. Weddell Sea hydrology; works of the Soviet Antarctic Expedition. *Hydrol. Ant. Waters*, vol. 16, pp. 45–93.

Klepikov, V. V., and Y. A. Grigor'yev. 1966. Water circulation in the Ross Sea. Soviet Antarctic Expedition Reports, vol. 6, pp. 52–54.

Koerner, R. M. 1964. Glaciological observations in Trinity Peninsula and the islands of the Prince Gustav Channel, Graham Land, 1958–1960. British Antarctic Survey Scientific Reports, no. 42, 45 pp.

Kolobov, D. D., and L. M. Savatyugin. 1982. Bottom sediments under the Novolazarevskiy Ice Shelf. *Polar Geog. Geol.*, vol. 6, pp. 267–271.

Kothe, J., F. Tessensohn, W. Thonhauser, and R. Wendebourg. 1981. The expedition and its logistics. *Geologisches Jahrbuch*, vol. B41, pp. 3–30.

Krantz, D. E. 1991. A chronology of Pliocene sea-level fluctuations; the U.S. middle Atlantic Coastal Plain record. *Quat. Sci. Rev.*, vol. 10, pp. 163–174.

Kristoffersen, Y., and K. Hinz. 1991. Evolution of the Gondwana plate boundary in the Weddell Sea area. In M. R. A. Thomson, J. A. Crame, and J. W. Thomson, eds., *Geologic Evolution of Antarctica*. Cambridge University Press, New York, pp. 225–230.

Krumbein, W. C. 1941. Measurement and geological significance of shape and roundness of sedimentary particles. *J. Sediment. Petrol.*, vol. 11, pp. 64–72.

Kuhn, G., M. Melles, W. U. Ehrmann, M. J. Hambrey, and G. Schmiedl. 1993. Character of clasts in glaciomarine sediments as an indicator of transport and depositional processes, Weddell and Lazarev seas, Antarctica. *J. Sediment. Petrol.*, vol. 63, pp. 477–487.

Kuhn, G., and M. E. Weber. 1993. Acoustical characterization of sediments by Parasound and 3.5 kHz systems; related sedimentary processes on the southeastern Weddell continental slope, Antarctica. *Marine Geol.*, vol. 113, pp. 201–217.

Kurinin, R. G., and G. E. Grikurov. 1982. Crustal structure of part of East Antarctica from geophysical data. In C. Craddock, ed., *Antarctic Geoscience*: University of Wisconsin Press, Madison, pp. 895–901.

Kurtz, D. D., and J. B. Anderson. 1979. Recognition and sedimentologic description of recent debris flow deposits from the Ross and Weddell Seas, Antarctica. *J. Sediment. Petrol.*, vol. 49, pp. 1159–1170.

Kurtz, D. D., J. B. Anderson, K. M. Balshaw, and M. L. Cole. 1979. Glacial marine sedimentation: relationship to the distribution of sediment geotechnical properties on high latitude continental margins. Proceedings of the 11th Annual Offshore Technology Conference, OTC 3437, no. 11, vol. 1, pp. 683–688.

Kuvaas, B., and Y. Kristoffersen. 1991. The Crary Fan: a trough-mouth fan on the Weddell Sea continental margin, Antarctica. *Marine Geol.*, vol. 97, pp. 345–362.

Kuvaas, B. and G. Leitchenkov. 1992. Glaciomarine turbidite and current controlled deposits Prydz Bay, Antarctica. *Marine Geol.*, vol. 108, pp. 365–381.

Kvasov, D. D., and M. Y. Verbitsky. 1981. Causes of Antarctic glaciation in the Cenozoic. *Quat. Res.*, vol. 15, pp. 1–17.

Kyle, P. R. 1981. Glacial history of the McMurdo Sound area as indicated by the distribution and nature of McMurdo Volcanic Group rocks. In L. D. McGinnis, ed., *Dry Valley Drilling Project*. Antarctic Research Series, vol. 33. American Geophysical Union, Washington, D.C., pp. 403–412.

Kyle, P. R., D. H. Elliot, and J. F. Sutter. 1981. Jurassic Ferrar Supergroup tholeiites from the Transantarctic Mountains, Antarctica, and their relationship to the initial fragmentation of Gondwana. In M. M. Cresswell and P. Vella, eds., *Gondwana Five*. A.A. Balkema, Rotterdam, pp. 283–287.

LaBrecque, J., S. Cande, R. Bell, C. Raymond, J. Brozena,

M. Keller, J. C. Parra, and G. Yanez. 1986. Aerogeophysical survey yields new data in the Weddell Sea. *Ant. J. U.S.*, vol. 21, pp. 69–70.

Laird, M. G. 1981. Lower Palaeozoic rocks of Antarctica. In C. H. Holland, ed., *Lower Palaeozoic of the Middle East, Eastern and Southern Africa, and Antarctica.* John Wiley & Sons, New York, pp. 257–314.

— 1991. The Late Proterozoic-middle Palaeozoic rocks of Antarctica. In R. J. Tingey, ed., *The Geology of Antarctica.* Clarendon Press, Oxford, pp. 74–119.

Laird, M. G., G. D. Mansergh, and J. M. A. Chappell. 1971. Geology of the central Nimrod Glacier area, Antarctica. *N.Z. J. Geol. Geophys.*, vol. 14, pp. 427–468.

Larter, R. D., and P. F. Barker. 1989. Seismic stratigraphy of the Antarctic Peninsula Pacific margin: a record of Pliocene-Pleistocene ice volume and paleoclimate. *Geology*, vol. 17, pp. 731–734.

— 1991a. Effects of ridge crest-trench interaction on Antarctic-Phoenix spreading: forces on a young subducting plate. *J. Geophys. Res.*, vol. 96, pp. 19583–19607.

— 1991b. Neogene interaction of tectonic and glacial processes at the Pacific margin of the Antarctic Peninsula. In D. I. M. MacDonald, ed., *Sedimentation, Tectonics, and Eustasy: sea level changes at active margins.* International Association of Sedimentologists Special Publication, vol. 12, pp. 165–186.

Larter, R. D., and A. P. Cunningham. 1993. The depositional pattern and distribution of glacial-interglacial sequences on the Antarctic Peninsula Pacific margin. *Marine Geol.*, vol. 109, pp. 203–219.

Larter, R. D., M. Rebesco, L. E. Vanneste, and P. F. Barker. 1997. Cenozoic Tectonic, sedimentary and glacial history of the continental shelf west of Graham Land, Antarctic Peninsula. In P. F. Barker and A. D. Cooper, eds., *Geology and Seismic Statigraphy of the Antarctic Margin.* Antarctic Research Series, vol. 71. American Geophysical Union, Washington, D.C., pp. 1–27.

Larter, R. D., and L. E. Vanneste. 1995. Relict subglacial deltas on the Antarctic Peninsula outer shelf. *Geology*, vol. 23, pp. 33–36.

Lawver, L. A. 1984. Problems with reconstructions of Gondwana. *Ant. J. U.S.*, vol. 19, pp. 1–2.

Lawver, L. A., I. W. D. Dalziel, and D. T. Sandwell. 1993. Antarctic plate: tectonics from a gravity anomaly and infrared satellite image. *GSA Today*, vol. 3, pp. 117–122.

Lawver, L. A., and L. M. Gahagan. 1991. Constraints on Mesozoic transtension in Antarctica. International Symposium on Antarctic Earth Sciences, vol. 6. Cambridge University Press, New York, pp. 343–344.

— 1994. Simplified Cenozoic Antarctic-Australian-New Zealand tectonics. *EOS Transactions*, vol. 75, pp. 609–610.

Lawver, L. A., L. M. Gahagan, and M. F. Coffin. 1992. The development of paleoseaways around Antarctica. In J. P. Kennett and D. A. Warnke, eds., *The Antarctic Paleoenvironment: A Perspective on Global Change 1.* Antarctic Research Series, vol. 56. American Geophysical Union, Washington, D.C., pp. 7–30.

Lawver, L. A., and J. W. Hawkins. 1978. Diffuse magnetic anomalies in marginal basins: their possible tectonic and petrologic significance. *Tectonophysics*, vol. 45, pp. 323–339.

Lawver, L. A., J.-Y. Royer, D. T. Sandwell, and C. R. Scotese. 1991. Evolution of the Antarctic continental margins. In M. R. A. Thomson, J. A. Crame, and J. W. Thomson, eds., *Geological Evolution of Antarctica.* Cambridge University Press, New York, pp. 533–539.

Lawver, L. A., J. G. Sclater, and L. Meinke. 1985. Mesozoic and Cenozoic reconstructions of the South Atlantic. *Tectonophysics*, vol. 114, pp. 233–254.

Lawver, L. A., and C. R. Scotese. 1987. A revised reconstruction of Gondwanaland. In G. D. McKenzie, ed., *Gondwana Six: Structure, Tectonics, and Geophysics.* Geophysical Monograph 40. American Geophysical Union, Washington, D.C., pp. 17–23.

Lawver, L. A., B. J. Sloan, D. H. N. Barker, M. Ghidella, R. P. Von Herzen, R. A. Keller, G. P. Klinkhammer, and C. S. Chin. 1996. Distributed, active extension in Bransfield Basin, Antarctic Peninsula, evidence from multibeam bathymetry. *GSA Today*, vol. 6, pp. 1–6.

Lazarus, D., and J. P. Caulet. 1993. Cenozoic Southern Ocean reconstructions from sedimentologic, radiolarian, and other microfossil data. In J. P. Kennett and D. A. Warnke, eds., *The Antarctic Paleoenvironment: A Perspective on Global Change II*: Antarctic Research Series, vol. 60. American Geophysical Union, Washington, D.C., pp. 145–174.

Lazarus, D., A. Pallant, and J. D. Hays. 1987. Data-base of Antarctic pre-Pleistocene sediment cores, Woods Hole, Woods Hole Oceanographic Institution, p. 42.

Leckie, R. M., and P. N. Webb. 1983. Late Oligocene–early Miocene glacial record of the Ross Sea, Antarctica: evidence from DSDP Site 270. *Geology*, vol. 11, pp. 578–582.

Ledbetter, M. T. 1979. Fluctuations of Antarctic Bottom Water velocity in the Vema Channel during the last 160,000 years. *Marine Geol.*, vol. 33, pp. 71–89.

— 1981. Palaeooceanographic significance of bottom-current fluctuations in the Southern Ocean. *Nature*, vol. 294, pp. 554–556.

— 1986. Bottom-current pathways in the Argentine Basin revealed by mean silt particle size. *Nature*, vol. 321, pp. 423–425.

Ledbetter, M. T., and P. F. Ciesielski. 1982. Bottom-current erosion along a traverse in the South Atlantic sector of the Southern Ocean. *Marine Geol.*, vol. 46, pp. 329–341.

— 1986. Post-Miocene disconformities and paleoceanography in the Atlantic sector of the Southern Ocean. *Palaeogeog. Palaeoclimatol. Palaeoecol.*, vol. 52, pp. 185–214.

Ledbetter, M. T., P. F. Ciesielski, N. I. Osborn, and E. T. Allison. 1983. Bottom-current erosion in the southeast Indian and southwest Pacific oceans during the last 5.4 million years. In R. L. Oliver, P. R. James, and J. B. Jago, eds., *Antarctic Earth Science.* Cambridge University Press, New York, pp. 379–383.

Ledbetter, M. T., and B. B. Ellwood. 1980. Spatial and temporal changes in bottom-water velocity and direction from analysis of particle size and alignment in deep-sea sediment. *Marine Geol.*, vol. 38, pp. 245–261.

— 1982. Variations in particle alignment and size in sediments of the Vema Channel record Antarctic bottom-water velocity changes during the last 400,000 years. In C. Craddock, ed., *Antarctic Geoscience.* University of Wisconsin Press, Madison, pp. 1033–1038.

Ledbetter, M. T., and T. C. Huang. 1978. Manganese micron-odules in the South Pacific: an index of bottom current activity during the last 3 million years. Geological Society of America – Abstracts of Programs, vol. 10, p. 442.

Ledbetter, M. T., and D. A. Johnson. 1976. Increased transport of Antarctic Bottom Water in the Vema Channel during the last ice age. *Science*, vol. 194, pp. 837–839.

Ledbetter, M. T., and N. D. Watkins. 1978. Separation of pri-mary ice-rafted debris from lag deposits, utilizing man-ganese micronodule accumulation rates in abyssal sedi-ments of the Southern Ocean. *Geolog. Soc. Am. Bull.*, vol. 89, pp. 1619–1629.

Ledford-Hoffman, P. A., D. J. DeMaster, and C. A. Nittrouer. 1986. Biogenic-silica accumulation in the Ross Sea and the importance of Antarctic continental-shelf deposits in the marine silica budget. *Geochimica et Cosmochimica Acta*, vol. 50, pp. 2099–2110.

Leitchenkov, G. L., H. Miller, and E. N. Zatzepin. 1996. Struc-ture and Mesozoic evolution of the eastern Weddell Sea, Antarctica: history of early Gondwana break-up. In B. C. Storey, E. C. King, and R. A. Livermore, eds., *Weddell Sea Tectonics and Gondwana Break-Up*. Geological So-ciety Special Publications No. 108, Geological Society of London, UK, pp. 175–190.

Leitchenkov, G., F. Shelestov, V. Gandjuhin, and V. But-senko. 1990. Outline of structure and evolution of the Cooperation Sea sedimentary basin. In A. K. Cooper and P. N. Webb, eds., International Workshop on Antarctic Offshore Acoustic Stratigraphy (Antostrat); overview and extended abstracts. U.S. Geological Survey Open-File Re-port, Denver, Colorado, pp. 202–211.

LeMasurier, W. E., O. Melander, G. W. Grindley, and W. C. McIntosh. 1979. Tillite, glacial striae, and hyaloclastite associations on Hobbs Coast, Marie Byrd Land. *Ant. J. U.S.*, vol. 14, pp. 48–50.

LeMasurier, W. E., and D. C. Rex. 1982. Volcanic record of Cenozoic glacial history in Marie Byrd Land and west-ern Ellsworth Land: revised chronology and evaluation of tectonic factors. In C. Craddock, ed., *Antarctic Geosci-ence*: University of Wisconsin Press, Madison, pp. 725–734.

1983. Rates of uplift and the scale of ice level instabilities recorded by volcanic rocks in Marie Byrd Land, West Antarctica. In R. L. Oliver, P. R. James, and J. B. Jago, eds., *Antarctic Earth Science*. Cambridge University Press, New York, pp. 663–670.

Lemke, P. 1986. Stochastic description of atmosphere-sea-ice-ocean interaction. In N. Untersteiner, ed., *The Geophysics of Sea Ice*. NATO ASI Series B, Physics, vol. 146. Plenum Press, New York, pp. 785–823.

Lemke, P., E. W. Trinkl, and K. Hasselman. 1981. Stochas-tic dynamic analysis of polar sea ice variability. *J. Phys. Oceanog.*, vol. 10, pp. 2100–2120.

Lepley, L. K. 1966. Submarine geomorphology of eastern Ross Sea and Sulzberger Bay, Antarctica. Ocean Surveys Divi-sion Technical Report, no. 172, U.S. Naval Oceanographic Office, Washington, D.C., 34 pp.

Leventer, A., E. W. Domack, S. E. Ishman, S. Brachfeld, C. E. McClennen, and P. Manly. 1996. Productivity cycles of 200–300 years in the Antarctic Peninsula region: un-derstanding linkages among the sun, atmosphere, oceans,

sea ice, and biota. *Geolog. Soc. Am. Bull.*, vol. 108, pp. 1626–1644.

Leventer, A., R. B. Dunbar, and D. J. DeMaster. 1993. Diatom evidence for late Holocene climatic events in Granite Har-bor, Antarctica. *Paleoceanography*, vol. 8, pp. 373–386.

Licht, K. J., N. W. Dunbar, J. T. Andrews, and A. E. Jennings. In press. Distinguishing subglacial till and glacial marine diamictons in the western Ross Sea, Antarctica: implica-tions for last glacial maximum grounding line. *Geo. Soc. Am. Bull.*, in press.

Licht, K. J., N. W. Dunbar, J. T. Andrews, and A. E. Jennings. In press. Distinguishing subglacial till and glacial marine diamictons in the western Ross Sea, Antarctica: implica-tions for last glacial maximum grounding line. *Geo. Soc. Am. Bull.*

Licht, K. J., A. E. Jennings, J. T. Andrews, and K. M. Williams. 1996. Chronology of late Wisconsin ice retreat from the western Ross Sea, Antarctica. *Geology*, vol. 24, pp. 223–226.

Lien, R. 1981. Sea bed features in the Blaaenga area, Wed-dell Sea, Antarctica. Proceedings of the Sixth International Conference on Port and Ocean Engineering under Arctic Conditions, no. 6. Université Laval, Québec, pp. 706–716.

Lindsay, J. F. 1970. Depositional environment of Paleozoic glacial rocks in the central Transantarctic Mountains. *Ge-olog. Soc. Am. Bull.*, vol. 81, pp. 1149–1171.

Lindstrom, D., and D. Tyler. 1984. Preliminary results of Pine Island and Thwaites Glaciers study. *Ant. J. U.S.*, vol. 19, pp. 53–55.

Lingle, C. S., and J. A. Clark. 1979. Antarctic ice-sheet volume at 18,000 years B.P. and Holocene sea-level changes at the West Antarctic margin. *J. Glaciol.*, vol. 24, pp. 213–230.

Lisitsin, A. P. 1962. Bottom sediments of the Antarctic. In H. Wexler, ed., *Antarctic Research*. American Geophys-ical Union Monograph No. 7, pp. 81–88.

1972. Sedimentation in the World Ocean; with emphasis on the nature, distribution and behavior of marine suspen-sions. Soc. Economic Paleontologists and Mineralogists Special Publication, vol. 17, 218 pp.

Livermore, R. A., and R. W. Woollett. 1993. Seafloor spreading in the Weddell Sea and southwest Atlantic since the Late Cretaceous. *Earth Planet. Sci. Lett.*, vol. 117, pp. 475–495.

Lodolo, E., and F. Coren. 1994. The westernmost Pacific-Antarctic plate boundary in the vicinity of the Macquarie triple junction. In C. A. Ricci, ed., *Terra Antartica*, vol. 1, no. 1, pp. 158–161.

Lopatin, B. G., and V. S. Semenov. 1982. Amphibolite facies rocks of the southern Prince Charles Mountains, East Antarctica. In C. Craddock, ed., *Antarctic Geoscience*. University of Wisconsin Press, Madison, pp. 465–471.

Lorius, C., J. Jouzel, and D. Raynaud. 1993. Glacials-interglacials in Vostok: climate and greenhouse gases. *Glob. Planet. Change*, vol. 7, pp. 131–143.

Lorius, C., J. Jouzel, C. Ritz, L. Merlivat, N. I. Barkov, Y. S. Korotkevich, and V. M. Kotlyakov. 1985. A 150,000-year climatic record from Antarctic ice. *Nature*, vol. 316, pp. 591–596.

Lorius, C., D. Raynaud, J.-R. Petit, J. Jouzel, and L. Merlivat. 1984. Late-glacial maximum-Holocene atmospheric and ice-thickness changes from Antarctic ice-core studies. *Ann. Glaciol.*, vol. 5, pp. 88–94.

Lozano, J. A. and J. D. Hayes. 1976. Relationship of radiolarian assemblages to sediment types and physical oceanography in the Atlantic and western Indian Ocean sectors of the Antarctic Ocean. Geological Society of America Memoir, no. 145, pp. 303–336.

Lyra, C. S. 1986. Tertiary sediment palynology at Fildes Peninsula, King George Island, South Shetland Islands, and some paleoenvironmental considerations. Anais Academia Brasileira de Ciencias, vol. 58, pp. 137–147.

Mabin, M. C. G. 1992. Late Quaternary ice-surface fluctuations of the Lambert Glacier. In Y. Yoshida, K. Kaminuma, and K. Shiraishi, eds., *Recent Progress in Antarctic Earth Science*. Terra Scientific Publishing Co., Tokyo, Japan, pp. 683–687.

MacAyeal, D. R. 1992. Irregular oscillations of the West Antarctic Ice Sheet. *Nature*, vol. 359, pp. 29–32.

Macdonald, D. I. M. 1986. Proximal to distal sedimentological variation in a linear turbidite trough: implications for the fan model. *Sedimentology*, vol. 33, pp. 243–259.

Macdonald, D. I. M., and P. J. Butterworth. 1990. The stratigraphy, setting, and hydrocarbon potential of the Mesozoic sedimentary basins of the Antarctic Peninsula. In B. St. John, ed., *Antarctica as an Exploration Frontier – Hydrocarbon Potential, Geology, and Hazards*. American Association of Petroleum Geologists, Studies in Geology, no. 31, pp. 101–126.

MacDonald, S. E., and J. B. Anderson. 1986. Paleoceanographic implications of terrigenous deposits on the Maurice Ewing Bank, southwest Atlantic Ocean. *Marine Geol.*, vol. 71, pp. 259–287.

MacDonald, T. R., J. G. Ferrigno, R. S. Williams Jr., and B. K. Lucchitta. 1989. Velocities of Antarctic outlet glaciers determined from sequential Landsat images. *Ant. J. U.S.*, vol. 24, pp. 105–106.

Macellari, C. E. 1984. Late Cretaceous stratigraphy, sedimentology, and macropaleontology of Seymour Island, Antarctic Peninsula. PhD Dissertation, Ohio State University, Columbus, 599 pp.

——— 1988. Stratigraphy, sedimentology, and paleoecology of Upper Cretaceous/Paleocene shelf-deltaic sediments of Seymour Island. In R. M. Feldman and M. O. Woodburne, eds., *Geology and Paleontology of Seymour Island, Antarctic Peninsula*. Geological Society of America Memoir 169, pp. 25–53.

Mackensen, A., H. Grobe, G. Kuhn, and D. K. Fütterer. 1989. Stabel isotope stratigraphy from the Antarctic continental margin during the last one million years. *Marine Geology*, vol. 87, pp. 315–321.

Maldonado, A., F. Aldaya, J. C. Balanya, J. Galindo-Zaldivar, A. Jabaloy, R. D. Larter, J. Rodriguez-Fernandez, M. Roussanov, and C. Sanz de Galdeano. 1994. Cenozoic continental margin growth patterns in the northern Antarctic Peninsula. In A. K. Cooper, P. F. Barker, P. N. Webb, and G. Brancolini, eds., The Antarctic Continental Margin: Geophysical and Geological Stratigraphic Records of Cenozoic Glaciation, Paleoenvironments, and Sea-Level Change. *Terra Antarctica*, vol. 1, no. 2, pp. 311–314.

Marchant, D. R., and G. H. Denton. 1996. Miocene and Pliocene paleoclimate of the Dry Valleys region, southern Victoria Land; a geomorphological approach. *Marine Micropaleontol.*, vol. 27, pp. 253–271.

Marchant, D. R., G. H. Denton, D. E. Sugden, and C. C. Swisher. 1993. Miocene-Pleistocene glacial history of Arena Valley, Quartermain Mountains, Antarctica. *Geografiska Annaler*, vol. A75, pp. 269–302.

Marchant, D. R., G. H. Denton, C. C. Swisher III, and N. Potter Jr. 1996. Late Cenozoic Antarctic paleoclimate reconstructed from volcanic ashes in the Dry Valleys region of southern Victoria Land. *Geol. Soc. Am. Bull.*, vol. 108, pp. 181–194.

Margolis, S. V., and R. G. Burns. 1976. Pacific deep-sea manganese nodules; their distribution, composition, and origin. *Ann. Rev. Earth Planet. Sci.*, vol. 4, pp. 229–263.

Margolis, S. V., and J. P. Kennett. 1971. Cenozoic paleoglacial history of Antarctica recorded in subantarctic deep-sea cores. *Am. J. Sci.*, vol. 271, pp. 1–36.

Martin, P. J., and D. A. Peel. 1978. The spatial distribution of 10 m temperatures in the Antarctic Peninsula. *J. Glaciol.*, vol. 20, pp. 311–317.

Martinson, D. G., and R. A. Iannuzzi. 1998. Antarctic ocean-ice interaction: implications from ocean bulk propterty distributions in the Weddell Gyre. *Ant. Res. Series*, vol. 74, pp. 243–271.

Martinson, D. G., N. G. Pisias, J. D. Hays, J. Imbrie, T. C. Moore, Jr., and N. J. Shackleton. 1987. Age dating and the orbital theory of the Ice Ages: Development of a high-resolution 0 to 300,000-year chronostratigraphy. *Quat. Res.*, vol. 27, pp. 1–29.

Maslanyj, M. P. 1987. Seismic bedrock depth measurements and the origin of George VI Sound, Antarctic Peninsula. *British Antarctic Survey Bulletin*, no. 75, pp. 51–65.

Matsch, C. L., and R. W. Ojakangas. 1991. Comparisons in depositional style of "polar" and "temperate" glacial ice; Late Paleozoic Whiteout Conglomerate (West Antarctica) and late Proterozoic Mineral Fork Formation (Utah). In J. B. Anderson and G. M. Ashley, eds., *Glacial Marine Sedimentation; Paleoclimatic Significance*. Geological Society of America Special Paper 261, Boulder, Colorado, pp. 191–206.

Matthews, R. K., and R. Z. Poore. 1980. Tertiary $\delta^{18}O$ record and glacio-eustatic sea-level fluctuations. *Geology*, vol. 8, pp. 501–504.

Mawson, D. 1942. Geographical narrative and cartography, Australian Antarctic Expedition 1911–1914. Scientific report, Series A, vol. 1. Government Printing Office, Sydney, Australia, pp. 1–364.

Mayewski, P. A. 1975. Glacial geology and late Cenozoic history of the Transantarctic Mountains, Antarctica. Ohio State University Institute of Polar Studies Report, 168 pp.

Mayewski, P. A., M. S. Twickler, S. I. Whitlow, L. D. Meeker, Q. Yang, J. Thomas, K. Kreutz, P. M. Grootes, D. L. Morse, E. J. Steig, E. D. Waddington, E. S. Saltzman, P.-Y. Whung, and K. C. Taylor. 1996. Climate change during the last deglaciation in Antarctica. *Science*, vol. 272, pp. 1636–1638.

Mazzullo, J., and J. B. Anderson. 1987. Grain shape and surface texture analysis of till and glacial-marine sand grains from the Weddell and Ross seas, Antarctica. In J. R. Marshall, ed., *Clastic Particles: Scanning Electron Microscopy and Shape Analysis of Sedimentary and Volcanic Clasts.* Van Nostrand Reinhold Co., New York, pp. 314–327.

McCoy, F. W. 1991. Southern ocean sediments; Circum-Antarctic to 30 degrees S. In D. E. Hayes, ed., *Marine Geological and Geophysical Atlas of the Circum-Antarctic to 30 degrees S.* Antarctic Research Series, vol. 54. American Geophysical Union, Washington, D.C., pp. 37–46.

McDougall, I. 1961. Determination of the age of a basic igneous intrusion by the potassium-argon method. *Nature,* vol. 190, pp. 1184–1186.

McGinnis, J. P., and D. E. Hayes. 1994. Sediment drift formation along the Antarctic Peninsula. In A. K. Cooper, P. F. Barker, P. N. Webb, and G. Brancolini, eds., The Antarctic Continental Margin: Geophysical and Geological Stratigraphic Records of Cenozoic Glaciation, Paleoenvironments, and Sea-Level Change. *Terra Antarctica,* vol. 1, no. 2, pp. 275–276.

———— 1995. The roles of downslope and along-slope depositional processes: southern Antarctic Peninsula continental rise. In A. K. Cooper, P. F. Barker, and G. Brancolini, eds., *Geology and Seismic Stratigraphy of the Antarctic Margin.* Antarctic Research Series, vol. 68. American Geophysical Union, Washington, D.C., pp. 141–156.

McGinnis, L. D., and Y. Kim. 1985. Deep seismic soundings along the boundary between East and West Antarctica. *Ant. J. U.S.,* vol. 20, pp. 21–22.

McKelvey, B. C. 1991. The Cainozoic glacial record in south Victoria Land: a geological evaluation of the McMurdo Sound drilling projects. In R. J. Tingey, ed., *The Geology of Antarctica.* Clarendon Press, Oxford, pp. 434–454.

McKelvey, B. C., and N. C. N. Stephenson. 1990. A geological reconnaissance of the Radok Lake area, Amery Oasis, Prince Charles Mountains. *Ant. Sci.,* vol. 2, pp. 53–66.

McKelvey, B.C., P.-N. Webb, D. M. Harwood, and M. C. G. Mabin. 1991. The Dominion Range Sirius Group: a record of the late Pliocene-early Pleistocene Beardmore Glacier. In M. R. A. Thomson, J. A. Crame, and J. W. Thomson, eds., *Geological Evolution of Antarctica.* Cambridge University Press, New York, pp. 675–682.

McKenzie, D. P. 1978. Some remarks on the development of sedimentary basins. *Earth Planet. Sci. Lett.,* vol. 40, pp. 25–32.

McLachlan, I. R. 1973. Problematic microfossils from the Lower Karoo Beds in South Africa. *Paleontologia Africana,* vol. 15, pp. 1–11.

Melles, M., and G. Kuhn. 1993. Sub-bottom profiling and sedimentological studies in the southern Weddell Sea, Antarctica; evidence for large-scale erosional/depositional processes. *Deep-Sea Res.* I, vol. 40, pp. 739–760.

Meneilly, A. W., S. M. Harrison, B. A. Piercy, and B. C. Storey. 1987. Structural evolution of the magmatic arc in northern Palmer Land, Antarctic Peninsula. In G. D. McKenzie, ed., *Gondwana Six: Structure, Tectonics, and Geophysics.* Geophysical Monograph 40. American Geophysical Union, Washington, D.C., pp. 209–219.

Mensing, T. M. 1987. Geology and Petrogenesis of the Kirkpatrick Basalt, Pain Mesa and Solo Nunatak, Northern Victoria Land, Antarctica. PhD Dissertation, Ohio State University, Columbus, 391 pp.

Menzies, J. (ed.). 1995. *Modern Glacial Environments: Processes, Dynamics and Sediments.* Butterworth-Heinemann, Oxford, 621 pp.

Mercer, J. H. 1967. Glaciers of the Antarctic: Antarctic Map Folio Series – Folio 7. American Geographical Society, New York, 10 pp.

———— 1968. Antarctic ice and Sangamon sea level. Commission of Snow and Ice Reports and Discussions, Publication no. 79, De L'Association Internationale D'Hydrologie Scientifique, Gentbrugge, Belgium, pp. 217–225.

———— 1972. Some observations on the glacial geology of the Beardmore Glacier area. In R. J. Adie, ed., *Antarctic Geology and Geophysics.* International Union of Geological Sciences, Series B, vol. 1. Universitetsforlaget, Oslo, Norway, pp. 427–433.

———— 1978. Glacial development and temperature trends in the Antarctic and in South America. In E. M. van Zinderen Bakker, ed., *Antarctic Glacial History and World Paleoenvironments*: A. A. Balkema, Rotterdam, Netherlands pp. 73–93.

———— 1981. Tertiary terrestrial deposits of the Ross Ice Shelf area, Antarctica. In M. J. Hambrey and W. B. Harland, eds., *Earth's Pre-Pleistocene Glacial Records.* Cambridge University Press, Cambridge, pp. 204–207.

———— 1983. Cenozoic glaciation in the Southern Hemisphere. *Annu. Rev. Earth Planet. Sci.,* vol. 11, pp. 99–132.

Mercer, J. H., and J. F. Sutter. 1982. Late Miocene-earliest Pliocene glaciation in southern Argentina; implications for global ice-sheet history. *Palaeogeog. Palaeoclimatol. Palaeoecol.,* vol. 38, pp. 185–206.

Milam, R. W., and J. B. Anderson. 1981. Distribution and ecology of recent benthonic foraminifera of the Adélie-George V continental shelf and slope, Antarctica. *Marine Micropaleontol.,* vol. 6, pp. 297–325.

Mildenhall, D. 1980. New Zealand late Cretaceous and Cenozoic plant biogeography; a contribution. *Palaeogeog. Palaeoclimatol. Palaeoecol.,* vol. 31, pp. 197–233.

———— 1989. Terrestrial palynology. In P. Barrett, ed., *Antarctic Cenozoic History from the CIROS-1 Drillhole, McMurdo Sound.* Deptartment of Scientific and Industrial Research Bulletin, vol. 245, Wellington, pp. 119–127.

Miller, D. J. 1953. Late Cenozoic marine glacial sediments and marine terraces of Middleton Island, Alaska. *J. Geol.,* vol. 61, pp. 17–40.

Miller, H., M. De Batist, W. Jokat, N. Kaul, S. Steinmetz, G. Uenzelmann-Neben, and W. Versteeg. 1990a. Revised interpretation of tectonic features in the southern Weddell Sea, Antarctica, from new seismic data. *Polarforschung,* vol. 60, pp. 33–38.

Miller, H., J. P. Henriet, N. Kaul, and A. Moons. 1990b. A fine-scale seismic stratigraphy of the eastern margin of the Weddell Sea. In U. Bleil and J. Thiede, eds., *Geological History of the Polar Oceans: Arctic Versus Antarctic*: Kluwer Academic Publishers, Boston, pp. 131–161.

Miller, H., W. Loske, and U. Kramm. 1987a. Zircon provenance and Gondwana reconstruction: U-Pb data of detrital zircons from Triassic Trinity Peninsula Formation metasandstones. *Polarforschung,* vol. 57, pp. 59–69.

Miller, J. M. G. 1989. Glacial advance and retreat sequences in a Permo-Carboniferous section, central Transantarctic Mountains. *Sedimentology,* vol. 36, pp. 419–430.

Miller, K. G., R. G. Fairbanks, and G. S. Mountain. 1987b.

Tertiary oxygen isotope synthesis, sea level history, and continental margin erosion. *Paleoceanography*, vol. 2, pp. 1–19.

Miller, K. G., G. S. Mountain, the Leg 150 Shipboard Party, and Members of the New Jersey Coastal Plain Drilling Project. 1996. Drilling and dating New Jersey Oligocene-Miocene sequences: ice volume, global sea level, and Exxon records. *Science*, vol. 271, pp. 1092–1095.

Miller, K. G., J. D. Wright, and R. G. Fairbanks. 1991. Unlocking the ice house: Oligocene-Miocene oxygen isotopes, eustasy, and margin erosion. *J. Geophys. Res.*, vol. 96, pp. 6829–6848.

Mitchum, R. M. Jr., J. B. Sangree, P. R. Vail, and W. W. Wornardt. 1994. Recognizing sequences and systems tracts from well logs, seismic data, and biostratigraphy; examples from the late Cenozoic of the Gulf of Mexico. American Association of Petroleum Geologists Memoir 58, pp. 163–197.

Molnar, P., T. Atwater, J. Mammerickx, and S. M. Smith. 1975. Magnetic anomalies, bathymetry and the tectonic evolution of the South Pacific since the Late Cretaceous. *Geophys. J. Royal Astronomical Soc.*, vol. 40, pp. 383–420.

Moons, A., M. De Batist, J. P. Henriet, and H. Miller. 1992. Sequence stratigraphy of the Crary Fan, southeastern Weddell Sea. In Y. Yoshida, K. Kaminuma, and K. Shiraishi, eds., *Recent Progress in Antarctic Earth Science*. Terra Scientific Publishing Co., Tokyo, Japan, pp. 613–618.

Moore, T. C. J., T. H. Van-Andel, C. Sancetta, and N. G. Pisias. 1978. Cenozoic hiatuses in pelagic sediments. *Micropaleontology*, vol. 24, pp. 113–138.

Moores, E.M. 1991. Southwest U.S.-East Antarctic (SWEAT) connection: a hypothesis. *Geology*, vol. 19, pp. 425–428.

Morgan, V. I., and W. F. Budd. 1978. The distribution, movement and melt rate of Antarctic icebergs. In A. A. Husseiny, ed., *Iceberg Utilization*. Proceedings of the First International Conference and Workshops on Iceberg Utilization: Pergamon Press, New York, pp. 220–228.

Moriwaki, K. 1979. Submarine topography of the central part of Lützow-Holm Bay and around Ongul Islands, Antarctica. In T. Nagata, ed., Memoirs of National Institute of Polar Research, Spec. Issue, no. 14, National Institute of Polar Research, Tokyo, pp. 194–209.

Moriwaki, K., Y. Yoshida, and D. M. Harwood. 1992. Cenozoic glacial history of Antarctica – a correlative synthesis. In Y. Yoshida, K. Kaminuma, and K. Shiraishi, eds., *Recent Progress in Antarctic Earth Science*. Terra Scientific Publishing Co., Tokyo, pp. 773–780.

Mortlock, R. A., C. D. Charles, P. N. Froelich, M. A. Zibello, J. Saltzman, J. D. Hays, and L. H. Buckle. 1991. Evidence for lower productivity in the Antarctic Ocean during the last glaciation. *Nature*, vol. 351, pp. 220–223.

Mosley-Thompson, E. 1992. Paleoenvironmental conditions in Antarctica since A. D. 1500; ice core evidence. In R. S. Bradley and P. D. Jones, eds., *Climate Since A. D. 1500*. University of Massachusetts Press, Amherst, pp. 572–591.

1996. Holocene climate changes recorded in an East Antarctic ice core. In P. D. Jones, R. S. Bradley, and J. Jouzel, eds., *Climate Variations and Forcing Mechanisms of the Last 2000 Years*. NATO ASI Series, vol. 41. Springer Verlag, Berlin, pp. 243–262.

Moyes, A. B., J. M. Barton Jr., and P. B. Groenewald. 1993.

Late Proterozoic to early Palaeozoic tectonism in Dronning Maud Land, Antarctica: supercontinental fragmentation and amalgamation. *J. Geolog. Soc. London*, vol. 150, pp. 833–842.

Mullan, A. B., and J. S. Hickman. 1990. Meteorology. In G. P. Glasby, ed., *Antarctic Sector of the Pacific*. Elsevier Oceanographic Series, vol. 51, New York, pp. 21–54.

Muller, E. H. 1983. Till genesis and the glacier sole. In E. B. Evanson, C. Schlüchter, and J. Rabassa, eds., *Tills and Related Deposits*. A. A. Balkema, Rotterdam, Netherlands, pp. 19–22.

Mutter, J. C., J. A. Heggarty, S. C. Cande, and J. K. Weissel. 1985. Breakup between Australia and Antarctica, a brief review in light of new data. *Tectonophysics*, vol. 114, pp. 255–279.

Mutterlose, J. 1986. Upper Jurassic belemnites from the Orville Coast, western Antarctica, and their palaeobiogeographical significance. *British Antarctic Survey Bulletin*, no. 70, pp. 1–22.

Myers, N. C. 1982a. Marine geology of the western Ross Sea: implications for Antarctic glacial history. MA Thesis, Rice University, Houston, Texas, 233 pp.

1982b. Petrology of Ross Sea basal tills: implications for Antarctic glacial history. *Ant. J. U.S.*, vol. 17, pp. 123–124.

Nagihara, S., and L. A. Lawver. 1989. Heat-flow measurements in the King George Basin, Bransfield Strait. *Ant. J. U.S.*, vol. 24, pp. 123–125.

Nakawo, M. 1989. Deterioration of Antarctic ice sheet and sea-level rise. *J. Geog.*, vol. 98, pp. 32–45.

National Geographic Society. 1975. Antarctica (map) (1:8,841,000). National Geographic Society, Washington, D.C.

Nayudu, Y. R. 1971. Lithology and chemistry of surface sediments in subAntarctic regions of the Pacific Ocean. In J. L. Reid, ed., *Antarctic Oceanology I*. Antarctic Research Series, vol. 15. American Geophysical Union, Washington, D.C., pp. 247–282.

Neethling, D. C. 1972. Age and correlation of the Ritscher Supergroup and other Precambrian rock units, Dronning Maud Land. In R. J. Adie, ed., *Antarctic Geology and Geophysics*. International Union of Geological Sciences, Series B, vol. 1. Universitetsforlaget, Oslo, Norway, pp. 547–556.

Nicols, R. L. 1964. Present status of Antarctic glacial geology. First Symposium on Antarctic Geology and Geophysics, Amsterdam, North Holland, pp. 123–137.

Nitsche, R. D., K. Gohl, K. Vanneste, and H. Miller. 1997. Seismic expression of glacially deposited sequences in the Bellingshausen and Amundsen Seas, West Antarctica. In P. F. Barker and A. K. Cooper, eds., *Geology and Seismic Stratigraphy of the Antarctic Margin*. Antarctic Research Series, vol. 71. American Geophysical Union, Washington, D.C., pp. 95–108.

Norton, I. O., and J. G. Sclater. 1979. A model for the evolution of the Indian Ocean and the breakup of Gondwanaland. *J. Geophys. Res.*, vol. 84, pp. 6803–6830.

O'Brien, P. E. 1994. Morphology and Late glacial history of Prydz Bay, Antarctica, based on echo sounder data. *Terra Antarctica*, vol. 1, pp. 403–405.

O'Brien, P. E., and P. T. Harris. 1996. Patterns of glacial erosion and deposition in Prydz Bay and the past behavior of the Lambert Glacier. Papers and Proceedings of the Royal Society of Tasmania, vol. 130, no. 2, pp. 79–85.

O'Brien, P. E., P. T. Harris, P. G. Quilty, F. Taylor, and P. Wells. 1995. Post-cruise report, Antarctic CRC marine geoscience, Prydz Bay, MacRobertson Shelf, and Kerguelen Plateau. Australian Geological Survey Record, 1995/29.

O'Brien, P. E., and G. Leitchenkov. 1997. Deglaciation of Prydz Bay, East Antarctica, based on echo sounding and topographic features. In P. F. Barker and A. K. Cooper, eds., Geology and Seismic Stratigraphy of the Antarctic Margin. Antarctic Research Series, vol. 71. American Geophysical Union, Washington D.C., pp. 109–125.

O'Brien, P. E., E. M. Truswell, and T. Burton. 1994. Morphology and late glacial history of Prydz Bay, Antarctica based on echo sounder data. In A. K. Cooper, P. F. Barker, P. N. Webb, and G. Brancolini, eds., The Antarctic Continental Margin: Geophysical and Geological Stratigraphic Records of Cenozoic Glaciation, Paleoenvironments and Sea Level Change. Terra Antarctica, vol. 1, no. 2, pp. 403–405.

Oerlemans, J. 1982. A model of the Antarctic ice sheet. Nature, vol. 297, pp. 550–553.

Oeschger, H., and U. Siegenthaler. 1988. How has the atmospheric concentration of CO_2 changed? In F. S. Rowland and I. S. A. Isaken, eds., The Changing Atmosphere. Physical, Chemical, and Earth Sciences Research Reports, vol. 7, pp. 5–23.

Ojakangas, R. W., and C. L. Matsch. 1981. The Late Palaeozoic Whiteout Conglomerate: a glacial and glaciomarine sequence in the Ellsworth Mountains, West Antarctica. In M. J. Hambrey and W. B. Harland, eds., Earth's Pre-Pleistocene Glacial Record. Cambridge University Press, New York, pp. 241–244.

Okuda, Y., T. Yamazaki, S. Sato, T. Saki, and N. Oikawa. 1983. Framework of the Weddell Basin infered from the New Geophysical and geological data. In T. Nagata, ed., Proceedings of the Third Symposium on Antarctic Geosciences, Memoirs of the National Institute of Polar Research, Special Publication 28. Nat. Inst. Polar Res., Tokyo, Japan, pp. 93–114.

Olivero, E., R. A. Scasso, and C. A. Rinaldi. 1986. Revision of the Marambio Group, James Ross Island, Antarctica. Instituto Antártico Argentino Contribución, no. 331, 29 pp.

Orheim, O. 1980. Physical characteristics and life expectancy of tabular Antarctic icebergs. Ann. Glaciol., vol. 1, pp. 11–18.

Orheim, O., and L. S. Govorukha. 1982. Present-day glaciation in the South Shetland Islands. Ann. Glaciol., vol. 3, pp. 233–238.

Orombelli, G., C. Baroni, and G. H. Denton. 1991. Late Cenozoic glacial history of the Terra Nova Bay region, northern Victoria Land, Antarctica. Geografia Fisica e Dinamica Quaternaria, vol. 13, pp. 139–163.

Pankhurst, R. J. 1982. Sr-isotope and trace element geochemistry of Cenozoic volcanic rocks from the Scotia arc and the northern Antarctic Peninsula. In C. Craddock, ed., Antarctic Geoscience. University of Wisconsin Press, Madison, pp. 229–234.

Parada, M. A., J.-B. Orsini, and R. Ardila. 1992. Transverse variations in the Gerlache Strait plutonic rocks: effects of

the Aluk ridge-trench collision in the northern Antarctica Peninsula. In Y. Yoshida, K. Kaminuma, and K. Shiraishi, eds., Recent Progress in Antarctic Earth Science. Terra Scientific Publishing Co., Tokyo, Japan, pp. 395–403.

Parish, T. R. 1981. The katabatic winds of Cape Denison and Port Martin. Polar Rec., vol. 20, pp. 525–532.

Parish, T. R., and D. H. Bromwich. 1987. The surface windfield over the Antarctic ice sheets. Nature, vol. 328, pp. 51–54. 1989. Instrumented aircraft observations of the katabatic wind regime near Terra Nova Bay. Month. Weather Rev., vol. 117, pp. 1570–1585.

Park, B.-K., M.-S. Lee, H.-I. Yoon, and S.-H. Nam. 1989. Marine geology and petrochemistry in the Maxwell Bay area, South Shetland Islands. In H. T. Huh, B.-K. Park, and S.-H. Lee, eds., Antarctic Science: Geology and Biology. Seo Deung Co. Ltd., Seoul, Korea, pp. 85–120.

Parkinson, C. L. 1998. Length of the sea ice season in the Southern Ocean, 1988–1994. Ant. Res. Series, vol. 74, pp. 173–186.

Patterson, S. L., and T. Whitworth. 1990. Physical oceanography. In G. P. Glasby, ed., Antarctic Sector of the Pacific. Elsevier Oceanographic Series, vol. 51, New York, pp. 55–93.

Pattyn, F., H. Decleir, and P. Huybrechts. 1992. Glaciation of the central part of the Sør Rondane, Antarctica: glaciological evidence. In Y. Yoshida, K. Kaminuma, and K. Shiraishi, eds., Recent Progress in Antarctic Earth Science. Terra Scientific Publishing Co., Tokyo, Japan, pp. 669–678.

Payne, R. F., J. R. Conolly, and W. H. Abbott. 1972. Turbidite muds within diatom ooze off Antarctica: Pleistocene sediment variaton defined by closely spaced piston cores. Geolog. Soc. Am. Bull., vol. 83, pp. 481–486.

Pedly, M., J. G. Paren, and J. R. Potter. 1988. Localized basal freezing within George VI Ice Shelf, Antarctica. J. Glaciol., vol. 34, pp. 71–77.

Pelayo, A. M., and D. A. Wiens. 1989. Seismotectonics and relative plate motions in the Scotia Sea region. J. Geophys. Res., vol. 94, pp. 7293–7320.

Pennington, D. C., and J. W. Collinson. 1984. Sedimentary petrology of Permian and Triassic Beacon sandstones, northern Victoria Land. Ant. J. U.S., vol. 19, pp. 10–11.

Petit, J. R., I. Basile, A. Leruyuet, D. Raynaud, C. Lorius, J. Jouzel, M. Stievenard, V. Y. Lipenkov, N. I. Barkov, B. B. Kudryashov, M. Davis, E. Saltzman, V. Kotlyakov. 1997. Four climate cycles in Vostok ice core. Nature, vol. 387, p. 359.

Pewé, T. L. 1960. Multiple glaciation in the McMurdo Sound region. Antarctica – a progress report. J. Glaciol., vol. 68, pp. 498–514.

Pezzetti, T. F., and L. A. Krissek. 1986. Re-evaluation of the Eocene La Meseta Formation of Seymour Island, Antarctic Peninsula. Ant. J. U.S., vol. 21, p. 75.

Pickard, J., D. A. Adamson, D. M. Harwood, G. H. Miller, P. G. Quilty, and R. K. Dell. 1988. Early Pliocene marine sediments, coastline, and climate of East Antarctica. Geology, vol. 16, pp. 158–161.

Pillsbury, R. D., and S. S. Jacobs. 1985. Preliminary observations from long-term current meter moorings near the Ross Ice Shelf. In S. S. Jacobs, ed., Oceanology of the Antarctic Continental Shelf. Antarctic Research Series, vol. 43. American Geophysical Union, Washington, D.C., pp. 87–107.

Piotrowski, J. A. 1994. Tunnel-valley formation in northwest Germany – geology, mechanisms of formation and subglacial bed conditions for the Bornhöved tunnel valley. *Sediment. Geol.*, vol. 89, pp. 107–141.

Piper, D. J. W., and C. D. Brisco. 1975. Deep-water continental margin sedimentation, DSDP Leg 28, Antarctica. In E. D. Hayes, L. A. Frakes, eds., *Initial Reports of the Deep Sea Drilling Project*, vol. 28. U.S. Government Printing Office, Washington, D.C., pp. 359–400.

Piper, D. J. W., T. R. Swint, L. G. Sullivan, and F. W. McCoy. 1985. Manganese nodules, seafloor sediment, and sedimentation rates of the Circum-Pacific region. Circum-Pacific Council for Energy and Mineral Resources. American Association of Petroleum Geologists, Tulsa, Oklahoma (chart).

Piper, D. J. W., and R. J. Iuliucci. 1978. Reconnaissance of the marine geology of Makkovik Bay, Labrador. Canadian Geological Survey Paper 78-1a, pp. 333–336.

Pirrie, D., J. A. Crame, J. B. Riding, A. R. Butcher, P. D. Taylor. 1997. Miocene glaciomarine sedimentation of the northern Antarctic Peninsula region: the stratigraphy and sedimentology of the Hobbs Glacier Formation, James Ross Island. *Geological Magazine*, vol. 134, pp. 745–762.

Pirrie, D., A. G. Whitham, and J. R. Ineson. 1991. The role of tectonics and eustasy in the evolution of a marginal basin: Cretaceous-Tertiary Larsen Basin, Antarctica. In D. I. M. MacDonald, ed., *Sedimentation, Tectonics, and Eustasy: Sea Level Changes at Active Margins*. International Association of Sedimentologists Special Publication, vol. 12, pp. 293–305.

Plumb, K. A. 1979. The tectonic evolution of Australia. *Earth Sci. Rev.*, vol. 14, pp. 205–249.

Poore, R. Z., and R. K. Matthews. 1984. Oxygen isotope ranking of late Eocene and Oligocene planktonic foraminifers; implications for Oligocene sea surface temperatures and global ice volumes. *Marine Micropaleontol.*, vol. 9, pp. 111–134.

Pope, P. G., and J. B. Anderson. 1992. Late Quaternary glacial history of the northern Antarctic Peninsula's western continental shelf: evidence from the marine record. In D. H. Elliot, ed., *Contributions to Antarctic Research III*. Antarctic Research Series, vol. 57. American Geophysical Union, Washington, D.C., pp. 63–91.

Porebski, S. J., D. Meischner, and K. Görlich. 1991. Quaternary mud turbidites from the South Shetland trench (West Antarctica): recognition and implications for turbidite facies models. *Sedimentology*, vol. 38, pp. 691–715.

Potter, J. R., and J. G. Paren. 1985. Interaction between ice shelf and ocean in George VI Sound, Antarctica. In S. S. Jacobs, ed., *Oceanology of the Antarctic Continental Shelf*. Antarctic Research Series, vol. 43. American Geophysical Union, Washington, D.C., pp. 35–58.

Powell, C. M., B. D. Johnson, and J. J. Veevers. 1980. A revised fit of East and West Gondwanaland. *Tectonophysics*, vol. 63, pp. 13–29.

Powell, R. D. 1983. Glacial-marine sedimentation processes and lithofacies of temperate tidewater glaciers, Glacier Bay, Alaska. In B. F. Molnia, ed., *Glacial-Marine Sedimentation*. Plenum Press, New York, pp. 185–232.

Powell, R. D., and B. F. Molnia. 1989. Glacimarine sedimentary processes, facies and morphology of the south-southeast Alaska shelf and fjords. *Marine Geol.*, vol. 85, pp. 359–390.

Prell, W. L., J. Imbrie, D. G. Martinson, J. J. Morley, N. G. Pisias, N. J. Shackleton, and H. F. Streeter. 1986. Graphic correlation of oxygen isotope stratigraphy application to the late Quaternary. *Paleoceanography*, vol. 1, pp. 137–162.

Prentice, M. L. 1985. Peleus glaciation of Wright Valley, Antarctica. *S. Afr. J. Sci.*, vol. 81, pp. 241–243.

Prentice, M. L., G. H. Denton, L. H. Burckle, and D. A. Hodell. 1987. Evidence from Wright Valley for the response of the Antarctic ice sheet to climate warming. *Ant. J. U.S.*, vol. 22, pp. 56–58.

Prentice, M. L., and J. L. Fastook. 1990. Late Neogene Antarctic ice sheet dynamics; ice model-data convergence. EOS (Transactions of the American Geophysical Union), vol. 71, p. 1378.

Prentice, M. L., and R. K. Matthews. 1988. Cenozoic ice-volume history: development of a composite oxygen isotope record. *Geology*, vol. 17, pp. 963–966.

Prieto, M. J., M. Canals, G. Ercilla, and M. De Batist. In press. Structure and geodynamic evolution of Central Bransfield Basin (NW Antarctica) from seismic reflection data. *Marine Geol.*, in press.

Prieto, M. J., E. Grácia, M. Canals, G. Ercilla, and M. de Batist. 1997. Sedimentary history of the central Bransfield Basin (NW Antarctic Peninsula). In C. A. Ricci, ed., *The Antarctic Region*. Geological Evolution and Processes, Special Publication, Issue VII, ISAES.

Pudsey, C. J. 1992. Late Quaternary changes in Antarctic Bottom Water velocity inferred from sediment grain size in the northern Weddell Sea. *Marine Geol.*, vol. 107, pp. 9–33.

Pudsey, C. J., P. F. Barker, and N. Hamilton. 1988. Weddell Sea abyssal sediments; a record of Antarctic bottom water flow. *Marine Geol.*, vol. 81, pp. 289–314.

Pudsey, C. J., P. F. Barker, and R. D. Larter. 1994. Ice sheets retreat from the Antarctic Peninsula shelf. *Cont. Shelf Res.*, vol. 14, pp. 1647–1675.

Quilty, P. G. 1991. The geology of Marine Plain, Vestfold Hills, East Antarctica. In M. R. A. Thomson, J. A. Crame, and J. W. Thomson, eds., *Geological Evolution of Antarctica*. Cambridge University Press, New York, pp. 683–686.

Rabinowitz, P. D., and J. LaBrecque. 1979. The Mesozoic South Atlantic Ocean and evolution of its continental margins. *J. Geophys. Res.*, vol. 84, pp. 5973–6002.

Rathburn, A. E., J. J. Pichon, M. A. Ayress, and P. Deckker. 1997. Microfossil and stable-isotope evidence for changes in Late Holocene paleo-productivity and paleoceanographic conditions in the Prydz Bay region of Antarctica. *Palaeogeog. Palaeoclimatol. Palaeoecol.*, vol. 131, pp. 485–510.

Ravich, M. G. 1982a. Comparison of the folded complexes of the Adelaide (Australia) and Ross (Antarctica) geosyncline areas. In C. Craddock, ed., *Antarctic Geoscience*. University of Wisconsin Press, Madison, pp. 65–71.

1982b. The lower Precambrian of Antarctica – review paper. In C. Craddock, ed., *Antarctic Geoscience*. University of Wisconsin Press, Madison, pp. 421–427.

Ravich, M. G., and L. V. Fedorov. 1982. Geologic structure of

MacRobertson Land and Princess Elizabeth Land, East Antarctica. In C. Craddock, ed., *Antarctic Geoscience*. University of Wisconsin Press, Madison, pp. 499–504.

Raynaud, D., and B. Lebel. 1979. Total gas content and surface elevation of polar ice sheets. *Nature*, vol. 281, pp. 289–291.

Raynaud, D., and I. M. Whillans. 1982. Air content of the Byrd core and past changes in the West Antarctic ice sheet. *Ann. Glaciol.*, vol. 3, pp. 269–273.

Rebesco, M., R. D. Larter, P. F. Barker, A. Camerlenghi, and L. E. Vanneste. 1994. A history of sedimentation on the continental rise west of the Antarctic Peninsula. In A. K. Cooper, P. F. Barker, P. N. Webb, and G. Brancolini, eds., The Antarctic Continental Margin. Geophysical and Geological Stratigraphic Records of Cenozoic Glaciation, Paleoenvironments and Sea Level Change. *Terra Antarctica*, vol. 1, no. 2, pp. 277–280.

Rebesco, M., R. D. Larter, P. R. Barker, A. Camerlenghi, and L. E. Vanneste. 1997. History of sedimentation on the continental rise west of the Antarctic Peninsula. In P. F. Barker and A. K. Cooper, eds., *Geology and Seismic Stratigraphy of the Antarctic Margin*. Antarctic Research Series, vol. 71. American Geophysical Union, Washington D.C., pp. 29–49.

Reece, A. 1950. The ice of Crown Prince Gustav Channel, Graham Land, Antarctica. *J. Glaciol.*, vol. 1, pp. 404–408.

Reid, D. E., and J. B. Anderson. 1990. Hazards to Antarctic exploration and production. In B. St. John, ed., *Antarctic as an Exploration Frontier – Hydrocarbon Potential, Geology, and Hazards*. American Association of Petroleum Geologists Studies in Geology, no. 31, pp. 31–46.

Renner, R. G. B., L. J. S. Sturgeon, and S. W. Garrett. 1985. Reconnaissance gravity and aeromagnetic surveys of the Antarctic Peninsula. British Antarctic Survey Scientific Reports, no. 110, 50 pp.

Rex, R. W., S. V. Margolis, and B. Murray. 1970. Possible interglacial dune sands from 300 meters water depth in the Weddell Sea, Antarctica. *Geolog. Soc. Am. Bull.*, vol. 81, pp. 3465–3472.

Reynolds, J. M. 1981. Distribution of mean annual air temperatures in the Antarctic Peninsula. *British Antarctic Survey Bulletin*, no. 54, pp. 123–133.

Rinaldi, C. A. 1982. The Upper Cretaceous in the James Ross Island Group. In C. Craddock, ed., *Antarctic Geoscience*. University of Wisconsin Press, Madison, pp. 281–286

Rinaldi, C. A., A. Massabie, J. Morelli, H. L. Rosenman, and R. del Valle. 1978. Geologia de la Islas Vicecomodoro Marambio (Geology of Seymour Island). Instituto Antártico Argentino Contribución, no. 217, 54 pp.

Roach, P. J. 1978. The nature of back-arc extension in Bransfield Strait (abstract). *Royal Astronom. Soc. Geophys. J.*, vol. 53, pp. 165.

Robert, C., and J. P. Kennett. 1994. Antarctic subtropical humid episode at the Paleocene-Eocene boundary; clay-mineral evidence. *Geology*, vol. 22, pp. 211–214.

Robert, C. M., and H. Maillot. 1990. Paleoenvironments in the Weddell Sea area and Antarctic climates, as deduced from clay mineral associations and geochemical data, ODP Leg 113. In P. F. Barker, J. P. Kennett, et al., eds., Proceedings of the Ocean Drilling Program, Scientific Results, vol. 113. College Station, Texas, pp. 51–70.

Robin, G. d. Q. 1979. Formation, flow, and disintegration of ice shelves. *J. Glaciol.*, vol. 24, pp. 259–271.

——— 1983. Climate Record in Polar Ice Sheets; A Study of Isotopic Temperature Profiles in Polar Ice Sheets Based on a Workshop Held in the Scott Polar Research Institute, Cambridge. Cambridge University Press, Cambridge, 220 pp.

——— 1988. The Antarctic Ice Sheet, its history and response to sea level and climatic changes over the past 100 million years. *Palaeogeog. Palaeoclimatol. Palaeoecol.*, vol. 67, pp. 31–50.

Robin, G. d. Q., and R. J. Adie. 1964. The ice cover. In R. Priestley, G. d. Q. Robin, and R. J. Adie, eds., *Antarctic Research: A Review of British Scientific Achievement in Antarctica*. Butterworths, London, pp. 10–117.

Robinson, E. S. 1964. Geological structure of the Transantarctic Mountains and adjacent ice covered area, Antarctica. PhD Dissertation, University of Wisconsin, Madison, 319 pp.

Robinson, P. H., A. R. Pyne, M. J. Hambrey, K. J. Hall, and P. J. Barrett. 1987. Core log, photographs and grain size analyses from the CIROS-1 drillhole, western McMurdo Sound, Antarctica. Antarctic Data Series, vol. 14, Antarctic Research Centre Publication, Research School of Earth Sciences, Victoria University of Wellington, Wellington, New Zealand, 241 pp.

Rollinson, H. R., B. F. Windley, and M. Ramakrishnan. 1981. Contrasting high and intermediate pressures of metamorphism in the Archean Sarqur Schists of southern India. *Contrib. Mineral. Petrol.*, vol. 76, pp. 420–429.

Romanov, A. A. 1984. Ice of the Southern Ocean (in navigable conditions). Gidrometeoizdat, Leningrad, 150 pp.

Rooney, S. T., D. D. Blankenship, R. B. Alley, and C. R. Bentley. 1987. Till beneath ice stream B, 2. Structure and continuity. *J. Geophys. Res.*, vol. 92, pp. 8913–8920.

Ropelewski, C. F. 1983. Spatial and temporal variations in Antarctic sea-ice (1973–82). *J. Clim. App. Meteorol.*, vol. 22, pp. 470–473.

Roqueplo-Brouillet, C. 1982. Seismic Stratigraphy of the Eastern Continental Shelf of the Weddell Sea, Antarctica. MA Thesis, Rice University, Houston, Texas, 132 pp.

Rose, K. E. 1979. Characteristics of ice flow in Marie Byrd Land, Antarctica. *J. Glaciol.*, vol. 24, pp. 63–75.

Rott, H., P. Skvarca, and T. Nagler. 1996. Rapid collapse of northern Larsen Ice Shelf, Antarctica. *Science*, vol. 271, pp. 788–792.

Rowell, A. J., M. N. Rees, and P. Braddock. 1987. Silurian marine fauna not confirmed from Antarctica: *Alcheringa*, vol. 11, p. 137.

Rowley, P. D., K. S. Kellogg, and W. R. Vennum. 1985. Geologic studies in the English Coast, eastern Ellsworth Land, Antarctica. *Ant. J. U.S.*, vol. 20, pp. 34–36.

Rowley, P. D., W. R. Vennum, K. S. Kellogg, T. S. Laudon, P. E. Carrara, J. M. Boyles, and M. R. A. Thomson. 1983. Geology and plate tectonic setting of the Orville Coast and eastern Ellsworth Land, Antarctica. In R. L. Oliver, P. R. James, and J. B. Jago, eds., *Antarctic Earth Science*. Cambridge University Press, New York, pp. 245–250.

Rowley, P. D., and P. L. Williams. 1982. Geology of the northern Lassiter Coast and southern Black Coast, Antarctic

Peninsula. In C. Craddock, ed., *Antarctic Geoscience.* University of Wisconsin Press, Madison, pp. 339–348.

Rundle, A. S. 1974. Glaciology of the Marr Ice Piedmont, Anvers Island, Antarctica. Institute of Polar Studies Report No. 47. The Ohio State University Research Foundation, Columbus, 216 pp.

Rust, I. C. 1975. Tectonic and sedimentary framework of Gondwana basins in South Africa. In K. S. W. Campbell, ed., *Gondwana Geology.* Australia National University Press, Canberra, pp. 537–564.

Rutford, R. H., C. Craddock, and T. W. Bastien. 1968. Late Tertiary glaciation and sea-level changes in Antarctica. *Paleogeog. Paleoclimatol. Paleoecol.,* vol. 5, pp. 15–39.

Rutford, R. H., C. Craddock, C. M. White, and R. L. Armstrong. 1972. Tertiary glaciation in the Jones Mountains. In R. J. Adie, ed., *Antarctic Geology and Geophysics.* International Union of Geological Sciences, Series B, vol. 1. Universitetsforlaget, Oslo, Norway, pp. 239–243.

Rutland, R. W. R. 1981. Structural framework of the Australian Precambrian. In D. R. Hunter, ed., *Precambrian of the Southern Hemisphere.* Elsevier, New York, pp. 1–32.

Sadler, I. 1968. Observations on the ice caps of Galindez and Skua islands, Argentine Islands, 1960–66. *British Antarctic Survey Bulletin,* no. 17, pp. 21–49.

——— 1988. Geometry and stratification of uppermost Cretaceous and Paleogene units on Seymour Island, northern Antarctic Peninsula. In R. M. Feldmann and M. O. Woodburne, eds., *Geology and Paleontology of Seymour Island, Antarctic Peninsula.* Geological Society of America Memoir 169, pp. 303–320.

Sahni, A. 1982. The structure, sedimentation, and evolution of the Indian continental margins. In A. E. M. Nairn and F. G. Stehli, eds., *The Ocean Margins and Basins 6. The Indian Ocean.* Plenum Press, New York, pp. 353–398.

Salvini, F., G. Brancolini, M. Bussetti, F. Stroti, F. Mazzarini, and F. Coren. 1997. Cenozoic geodynamics of the Ross Sea region, Antarctica: crustal extension, interplate strike–slip faulting and tectonic inheritance. *J. Geophys. Res.,* vol. 102, pp. 669–696.

Sandwell, D. T. 1992. Antarctic marine gravity field from high-density satellite altimetry. *Geophys. J. Internat.,* vol. 109, pp. 437–448.

Sato, S., N. Asakura, T. Saki, N. Oikawa, and Y. Kaneda. 1984. Preliminary results of geological and geophysical surveys in the Ross Sea and in the Dumont D'Urville Sea, off Antarctica. Memoirs of National Institute of Polar Research Special Issue No. 33, Tokyo, Japan, pp. 66–92.

Saunders, A. D., and J. Tarney. 1982. Igneous activity in the southern Andes and northern Antarctic Peninsula: a review. *J. Geolog. Soc. London,* vol. 139, pp. 691–700.

Savage, M. L., and P. F. Ciesielski. 1983. A revised history of glacial sedimentation in the Ross Sea region. In R. L. Oliver, P. R. James, and J. B. Jago, eds., *Antarctic Earth Science.* Cambridge University Press, New York, pp. 555–559.

Savin, S. M., R. G. Douglas, and F. G. Stehli. 1975. Tertiary marine paleotemperatures. *Geolog. Soc. Am. Bull.,* vol. 86, pp. 1499–1510.

Scherer, R. 1991. Quaternary and Tertiary microfossils from beneath Ice Stream B: evidence for a dynamic West Antarctic Ice Sheet. *Palaeogeog. Palaeoclimatol. Palaeoecol.,* vol. 90, pp. 395–412.

Scherer, R. P., A. Aldahan, S. Tulaczyk, G. Possnert, H. Engelhardt, and B. Kamb. 1998. Pleistocene collapse of the West Antarctic Ice Sheet. *Science,* vol. 281, pp. 82–85.

Schermerhorn, L. J. G. 1974. Late Precambrian mixtites: glacial and/or nonglacial? *Am. J. Sci.,* vol. 274, pp. 673–824.

Schmidt, D. L., and P. D. Rowley. 1986. Continental rifting and transform faulting along the Jurassic Transantarctic rift, Antarctica. *Tectonics,* vol. 5, pp. 279–291.

Schnitker, D. 1980. Global paleoceanography and its deep water linkage to the Antarctic glaciation. *Earth Sci. Rev.,* vol. 16, pp. 1–20.

Schopf, T. J. M. 1969. Ellsworth Mountains: position in West Antarctica due to sea-floor spreading. *Science,* vol. 164, pp. 63–66.

Schwerdtfeger, W. 1984. *Weather and Climate of the Antarctic, Developments in Atmospheric Sciences.* Elsevier, New York, 261 pp.

Schwerdtfeger, W., and S. Kachelhoffer. 1973. The frequency of cyclonic vortices over the Southern Ocean in relation to the extension of the pack ice belt. *Ant. J. U. S.,* vol. 8, p. 234.

Scotese, C. R., R. K. Bambach, C. Barton, R. Van der Voo, and A. M. Ziegler. 1979. Paleozoic base maps. *J. Geol.,* vol. 87, pp. 217–277.

Scott, D. P. D. 1981. Cartographie du fond des oceans: la carte generale bathymetrique des oceans (GEBCO) (Cartography of ocean floors; the general bathymetric map of the oceans (GEBCO). Canadian Hydrographic Service. *Nature et Resources,* vol. 17, pp. 6–10.

Scott, K. M. 1965. Geology of the southern Gerlache Strait region, Antarctica. *J. Geol.,* vol. 73, pp. 518–527.

Scott, R. F. 1905. *The Voyage of the Discovery.* Charles Scribner's Sons, New York, 508 pp.

Seabrook, J. M., G. L. Hufford, and R. B. Elder. 1971. Formation of Antarctic Bottom Water in the Weddell Sea. *J. Geophys. Res.,* vol. 76, pp. 2164–2178.

Shabtaie, S., and C. R. Bentley. 1982. Tabular icebergs: implications from geophysical studies of ice shelves. *J. Glaciol.,* vol. 28, pp. 413–430.

——— 1987. West Antarctic ice streams draining into the Ross Ice Shelf: configuration and mass balance: *J. Geophys. Res.,* vol. 92, pp. 1311–1336.

Shackleton, N. J. 1986. Paleogene stable isotopes events: *Palaeogeog. Palaeoclimatol. Palaeoecol.,* vol. 57, pp. 91–102.

Shackleton, N. J., and A. Boersma. 1981. The climate of the Eocene ocean. *J. Geolog. Soc. London,* vol. 138, pp. 153–157.

Shackleton, N. J., M. A. Hall and D. Pate. 1995. Pliocene stable isotope stratigraphy of site 846. In N. G. Psias, L. A. Mayer, T. R. Janacek, A. Palmer-Julson, and T. H. van Andel, eds., Proceedings of the Ocean Drilling Program, Scientific Results, vol. 138, pp. 337–355.

Shackleton, N. J., and J. P. Kennett. 1975. Paleotemperature history of the Cenozoic and the initiation of Antarctic glaciation: oxygen and carbon isotope analysis in DSDP Sites, 277, 279, and 281. *Initial Reports of the Deep Sea Drilling Project,* vol. 29. U.S. Government Printing Office, Washington, D.C., pp. 743–755.

Shackleton, N. J., and N. D. Opdyke. 1976. Oxygen-isotope and paleomagnetic stratigraphy of Pacific Core V28-239, Late Pliocene to latest Pleistocene. In R. M. Cline and J. D. Hays, Investigation of late Quaternary paleoceanography and paleoclimatology. Geological Society of America Memoir 145, pp. 449–464.

Shaw, J. 1988. Subglacial erosional marks, Wilton Creek, Ontario. *Can. J. Earth Sci.*, vol. 25, pp. 1256–1267.

Shepard, F. P. 1931. Glacial troughs of the continental shelves. *J. Geol.*, vol. 39, pp. 345–360.

Shepard, F. P. 1973. *Submarine Geology.* Harper & Row, New York, 517 pp.

1973. *Submarine Geology.* Harper & Row, New York, 517 pp.

Sheraton, J. W., and L. P. Black. 1981. Geochemistry and geochronology of Proterozoic tholeiite dykes of East Antarctica: evidence for mantle metasomatism. *Contrib. Mineral. Petrol.*, vol. 78, pp. 305–317.

Shergold, J. H., J. B. Jago, R. A. Cooper, and J. R. Laurie. 1985. *The Cambrian system in Australia, Antarctica and New Zealand.* International Union of Geological Sciences Publication, no. 19, Ottawa, Canada, 94 pp.

Shevenell, A. E., E. W. Domack, and G. M. Kerman. 1996. Record of Holocene paleoclimate change along the Antarctic Peninsula: evidence from glacial marine sediments, Lallemand Fjord. Papers and Proceedings of the Royal Society of Tasmania, vol. 130, pp. 55–64.

Shevenell, A. E., M. H. Lopiccolo, B. T. Straten, E. W. Domack. 1996. Holocene paleoenvironmental studies within Antarctic fjords along the western side of the Antarctic Peninsula; understanding hypsithermal and Neoglacial fluctuations. Geological Society of America – Abstrats of Programs, vol. 28, 99 pp.

Shil'nikov,V. I. 1969. Icebergs. In Atlas Antarctiki (Antarctic Atlas), vols. 1 and 2. Gidrometeoizdat, Leningrad.

Shipp, S., and J. B. Anderson. 1994. High-resolution seismic survey of the Ross Sea continental shelf: implications for ice-sheet retreat behavior. *Ant. J. U.S.*, vol. 29, pp. 137–138.

1995. Retreat history of the West Antarctic ice sheet; ties to the steps in the Holocene sea level curve. In A. C. Hine and R. B. Halley, eds., *Linked Earth Systems*; congress program and abstracts, vol. 1. Society of Economic Paleontologists and Mineralogists, Tulsa, Oklahoma, p. 113.

1997 (a). Lineations of the Ross Sea continental shelf, Antarctica. In T. A. Davies, H. Josenhans, L. Polyak, A. Solheim, M. S. Stoker, and J. A. Stravers, eds., *Glaciated Continental Margins, An Atlas of Acoustic Images.* Chapman and Hall, London, pp. 54–55.

1997 (b). Grounding zone wedges on the Antarctic continental shelf, Ross Sea. In T. A. Davies, H. Josenhans, L. Polyak, A. Solheim, M. S. Stoker, and J. A. Stravers, eds., *Glaciated Continental Margins, An Atlas of Acoustic Images.* Chapman and Hall, London, pp. 104–105.

1997 (c). Drumlin field on the Ross Sea continental shelf, Antarctica. In T. A. Davies, H. Josenhans, L. Polyak, A. Solheim, M. S. Stoker, and J. A. Stravers, eds., *Glaciated Continental Margins, An Atlas of Acoustic Images.* Chapman and Hall, London, pp. 52–53.

1997 (d). Paleo-ice streams and ice stream boundaries, Ross Sea, Antarctica. In T. A. Davies, H. Josenhans,

L. Polyak, A. Solheim, M. S. Stoker, and J. A. Stravers, eds., *Glaciated Continental Margins, An Atlas of Acoustic Images.* Chapman and Hall, London, pp. 106–109.

1997 (e). Till sheets on the Ross Sea continental shelf, Antarctica. In T. A. Davies, H. Josenhans, L. Polyak, A. Solheim, M. S. Stoker, and J. A. Stravers, eds., *Glaciated Continental Margins, An Atlas of Acoustic Images.* Chapman and Hall, London, pp. 235–237.

Shipp, S., J. Anderson, L. DeSantis, L. R. Bartek, B. Alonso, and I. Zayatz. 1994. High- to intermediate resolution seismic stratigraphic analysis of mid-late-Miocene to Pleistocene strata in eastern Ross Sea: implications for changing glacial/climate regime. In A. K. Cooper, P. F. Barker, P. N. Webb, and G. Brancolini, eds., The Antarctic Continental Margin: Geophysical and Geological Stratigraphic Records of Cenozoic Glaciation, Paleoenvironments and Sea-Level Change. *Terra Antarctica*, vol. 1, pp. 381–384.

Shipp, S., J. B. Anderson, and E. W. Domack. In press. High-resolution seismic signature of the late Pleistocene fluctuation of the West Antarctic Ice Sheet system in Ross Sea: Part I-Geophysical Results. *Geo. Soc. Am. Bull.*, in press.

Shiraishi, K., D. J. Ellis, Y. Hiroi, C. M. Fanning, Y. Motoyoshi, and Y. Nakai. 1994. Cambrian orogenic belt in East Antarctica and Sri Lanka: implications for Gondwana assembly. *J. Geol.*, vol. 102, pp. 47–65.

Singer, J. K. 1987. Terrigenous, Biogenic, and Volcaniclastic Sedimentation Patterns of the Bransfield Strait and Bays of the Northern Antarctic Peninsula: Implications for the Quaternary Glacial History. PhD Dissertation, Rice University, Houston, Texas, 342 pp.

Singer, J. K. and J. B. Anderson. 1984. Use of total grain-size distributions to define bed erosion and transport for poorly sorted sediment undergoing simulated bioturbation. *Marine Geol.*, vol. 57, pp. 335–359.

Sivaprakash, C. 1981. Petrology of calc-silicate rocks from Koduru, Andhra Pradesh, India. *Contrib. Mineral. Petrol.*, vol. 77, pp. 121–128.

Skornyakova, N. S., and V. P. Petelin. 1967. Osadki tsentral'nogo rayona yuzhnoy chasti Tikhogo Okeana (Sediments in the central part of the South Pacific). Okeanol. (Akad. Nauk SSSR), vol. 7, pp. 1005–1019.

Sloan, B. J., L. A. Lawver, and J. B. Anderson. 1995. Seismic stratigraphy of the Palmer Basin. In A. K. Cooper, P. F. Barker, and G. Brancolini, eds., *Geology and seismic stratigraphy of the Antarctic Margin.* American Geophysical Union, Antarctic Research Series, Washington, D.C., vol. 68, pp. 235–260.

Smellie, J. L. 1979. The geology of Low Island, South Shetland Islands, and Austin Rocks. *British Antarctic Survey Bulletin*, no. 49, pp. 239–257.

1981. A complete arc-trench system recognized in Gondwana sequences of the Antarctic Peninsula region. *Geolog. Mag.*, vol. 118, pp. 139–159.

Smith, A. G., and D. J. Drewry. 1984. Delayed phase change due to hot asthenosphere causes Transantarctic uplift? *Nature*, vol. 309, pp. 536–538.

Smith, A. G., and A. Hallam. 1970. The fit of the southern continents. *Nature*, vol. 225, pp. 139–144.

Smith, M. J. 1985. Marine geology of the Northwestern Weddell Sea and Adjacent Coastal Fjords and Bays: Implications for Glacial History. MA Thesis, Rice University, Houston, Texas, 157 pp.

Smithson, S. B. 1972. Gravity interpretation in the Transantarctic Mountains near McMurdo Sound, Antarctica. *Geolog. Soc. Am. Bull.*, vol. 83, pp. 3437–3442.

Sobotovich, E. V., E. N. Kamenev, A. A. Komaristyy, and V. A. Rudnik. 1976. The oldest rocks in Antarctica (Enderby Land). *Internat. Geol. Rev.*, vol. 18, pp. 371–388.

Solheim, A., C. F. Forsberg, and A. Pittenger. 1991. Stepwise consolidation of glacigenic sediments related to the glacial history of Prydz Bay, East Antarctica. In J. Barron, B. Larsen, et al., eds., Proceeding of the Ocean Drilling Program, Scientific Results, vol. 119. College Station, Texas, pp. 169–182.

Solheim, A., L. Russwurm, A. Elverhøi, and M. N. Berg. 1990. Glacial geomorphic features in the northern Barents Sea: direct evidence for grounded ice and implications for the pattern of deglaciation and late glacial sedimentation. In J. A. Dowdeswell, and J. D. Scourse, eds., *Glacimarine Environments: Processes and Sediments*. Geological Society of America Special Publication, no. 53, pp. 253–268.

Solomon, S., R. R. Garcia, F. S. Rowland, and D. J. Wuebbles. 1986. On the depletion of Antarctic Ozone. *Nature*, vol. 321, pp. 755–758.

Sprigg, R. C. 1967. A short geological history of Australia. *Austral. Petrol. Explor. Assoc. J.*, vol. 7, pp. 59–82.

Stagg, H. M. J. 1985. Structure and origin of Prydz Bay and MacRobertson Shelf, East Antarctica. *Tectonophysics*, vol. 114, pp. 315–340.

Stagg, H. M. J., D. C. Ramsay, and R. Whitworth. 1983. Preliminary report of a marine geophysical survey between Davis and Mawson Stations, 1982. In R. L. Oliver, P. R. James, and J. B. Jago, eds., *Antarctic Earth Science*. Cambridge University Press, New York, pp. 527–532.

Stagg, H. M. J., and J. B. Willcox. 1992. A case for Australia-Antarctica separation in the Neocomian (ca. 125 Ma). *Tectonophysics*, vol. 210, pp. 21–32.

Stearns, C. R., L. M. Keller, G. A. Weidner, and M. Sievers. 1993. Monthly mean climatic data for Antarctic automatic weather stations. In D. H. Bromwich and C. R. Stearns, eds., *Antarctic Meteorology and Climatology: Studies Based on Automatic Weather Stations*. Antarctic Research Series, vol. 61. American Geophysical Union, Washington, D.C., pp. 1–22.

Stern, T. A., and U. S. ten Brink. 1989. Flexural uplift of the Transantarctic Mountains. *J. Geophys. Res.*, vol. 94, pp. 10,315–10,330.

Stetson, H. C., and J. E. Upson. 1937. Bottom deposits of the Ross Sea [Antarctic]. *J. Sediment. Petrol.*, vol. 7, pp. 55–66.

Stock, J., and P. Molnar. 1987. Revised history of early Tertiary plate motion in the south-west Pacific. *Nature*, vol. 325, pp. 495–499.

Stoker, M. S. 1990. Glacially-influenced sedimentation of the Hebridean slope, northwestern United Kingdom continental margin. Geological Society Special Publications, vol. 53, pp. 349–362.

Storey, B. C. 1993. The changing face of the late Precambrian and early Palaeozoic reconstructions. *J. Geolog. Soc. London*, vol. 150, pp. 665–668.

Storey, B. C., I. W. D. Dalziel, S. W. Garrett, A. M. Grunow, R. J. Pankhurst, and W. R. Vennum. 1988. West Antarctica in Gondwanaland: crustal blocks, reconstruction and breakup processes. *Tectonophysics*, vol. 155, pp. 381–390.

Storey, B. C., and S. W. Garrett. 1985. Crustal growth of the Antarctic Peninsula by accretion, magmatism and extension. *Geolog. Mag.*, vol. 122, pp. 5–14.

Storey, B. C., M. R. A. Thomson, and A. W. Meneilly. 1987. The Gondwanian orogeny within the Antarctic Peninsula: a discussion. In G. D. McKenzie, ed., *Gondwana Six: Structure, Tectonics, and Geophysics*. Geophysical Monograph 40. American Geophysical Union, Washington, D.C., pp. 191–198.

Stott, L. D., J. P. Kennett, N. J. Shackleton, and R. M. Corfield. 1990. The evolution of Antarctic surface waters during the Paleogene: inferences from the stable isotopic composition of planktonic foraminifers, ODP Leg 113. In P. F. Barker, J. P. Kennett, et al., eds., Proceedings of the Ocean Drilling Program, Scientific Results, vol. 113. Ocean Drilling Program, College Station, Texas, pp. 849–863.

Streten, N. A., and D. J. Pike. 1980. Characteristics of the broadscale Antarctic sea ice extent and the associated atmospheric circulation 1972–1977. *Archiv für Meteorologie, Geophysik und Bioklimatologie*, vol. A29, pp. 279–299.

Streten, N. A., and A. J. Troup. 1973. A synoptic climatology of satellite observed cloud vortices over the Southern Hemisphere. *Quart. J. Royal Meteorolog. Soc.*, vol. 99, pp. 56–72.

Stroeven, A. P., and M. L. Prentice. 1997. A case for Sirius Group alpine glaciation at Mount Fleming, South Victoria Land, Antarctica: A case against Pliocene ice sheet reduction. *Geo. Soc. Am. Bull.*, vol. 109, pp. 825–840.

Stroeven, A. P., M. L. Prentice, and J. Kleman. 1996. On microfossil transport and pathways in Antarctica during the Neogene: evidence from the Sirius Group at Mount Fleming. *Geology*, vol. 24, pp. 727–730.

Stuchlik, L. 1981. Tertiary pollen spectra from the Ezcurra Inlet Group of Admiralty Bay, King George Island (South Shetland Islands, Antarctica). *Studia Geologica Polonica*, vol. 72, pp. 109–137.

Stuiver, M., G. H. Denton, T. J. Hughes, and J. L. Fastook. 1981. History of the marine ice sheet in West Antarctica during the last glaciation: a working hypothesis. In G. H. Denton and T. J. Hughes, eds., *The Last Great Ice Sheets*. John Wiley & Sons, New York, pp. 319–436.

Stump, E. 1973. Earth evolution in the Transantarctic Mountains and West Antarctica. In D. H. Tarling and S. K. Runcorn, eds., *Implications of Continental Drift to the Earth Sciences*, vol. 2. Academic Press, New York, pp. 909–924.

——— 1980. Two episodes of deformation at Mt. Madison, Antarctica. *Ant. J. U.S.*, vol. 15, pp. 13–14.

——— 1992. The Ross orogen of the Transantarctic Mountains in light of the Laurentia-Gondwana split. *GSA Today*, vol. 2, pp. 25–31.

——— 1995. *The Ross orogen of the Transantarctic Mountains*. Cambridge University Press, New York, 284 pp.

Stump, E., M. F. Sheridan, S. G. Borg, and J. F. Sutter. 1980. Early Miocene subglacial basalts, the East Antarctic ice sheet, and uplift of the Transantarctic Mountains. *Science*, vol. 207, pp. 757–759.

Suárez, M. 1976. Plate-tectonic model for southern Antarctic Peninsula and its relation to southern Andes. *Geology*, vol. 4, pp. 211–214.

Sugden, D. E., and C. M. Clapperton. 1977. The maximum ice sheet extent on island groups in the Scotia Sea, Antarctica. *Quat. Res.*, vol. 7, pp. 268–282.

——— 1981. An ice-shelf moraine, George VI Sound, Antarctica. *Ann. Glaciol.*, vol. 2, pp. 135–141.

Sugden, D. E., and B. S. John. 1976. *Glaciers and Landscape – A Geomorphological Approach*. Edward Arnold, New York, 376 pp.

Sugden, D. E., D. R. Marchant, and G. H. Denton. 1993. The case for a stable East Antarctic ice sheet; the background. *Geografiska Annaler*, vol. 75A, pp. 151–155.

Sverdrup, H. U., M. W. Johnson, and R. H. Fleming. 1942. *The Oceans, Their Physics, Chemistry and General Biology*. Prentice-Hall, New York, 1087 pp.

Swithinbank, C. 1988. Antarctica – Satellite Image Atlas of Glaciers of the World. U.S. Geological Survey Professional Paper 1386-B, 278 pp.

Swithinbank, C., P. McClain, and P. Little. 1977. Drift tracks of Antarctic icebergs. *Polar Rec.*, vol. 18, pp. 495–501.

Syvitski, J. P. M. 1989. On the deposition of sediment within glacier-influenced fjords: oceanographic controls. *Marine Geol.*, vol. 85, pp. 301–329.

Syvitski, J. P. M. and C. P. Blakeney. 1983. Sedimentology of Arctic fjords experiment. Hu 82-031 Data Report, vol. 1, Open File 960, Geological Survey of Canada.

Taljaard, J. J. 1972. Synoptic meteorology of the Southern Hemisphere. In C. W. Newton, ed., *Meteorology of the Southern Hemisphere: Meteorological Monographs*, vol. 13. American Meteorological Society, Boston, Massachusetts, pp. 139–213.

Tanahashi, M., T. Saki, N. Oikawa, and S. Sato. 1987. An interpretation of the multichannel seismic reflection profiles across the continental margin of the Dumont d'Urville Sea, off Wilkes Land, East Antarctica. In S. L. Eittreim and M. A. Hampton, eds., *The Antarctic Continental Margin: Geology and Geophysics of Offshore Wilkes Land*. CPCEMR Earth Science Series, vol. 5A. Circum-Pacific Council for Energy and Mineral Resources, Houston, Texas, pp. 1–14.

Tankard, A. J., M. P. A. Jackson, K. A. Eriksson, D. K. Hobday, D. R. Hunter, and W. E. L. Minter. 1982. *Crustal Evolution of Southern Africa: 3.8 Billion Years of Earth History*. Springer Verlag, New York, 523 pp.

Tanner, P. W. G., and D. I. M. Macdonald. 1982. Models for the deposition and simple shear deformation of a turbidite sequence in the South Georgia portion of the southern Andes back-arc basin. *J. Geolog. Soc. London*, vol. 139, pp. 739–754.

Tauber, G. M. 1960. Characteristics of Antarctic katabatic winds. In Proceedings of the Symposium on Antarctic Meteorology. Pergamon Press, New York, pp. 52–64.

Taviani, M., D. E. Reid, and J. B. Anderson. 1993. Skeletal and isotopic composition and paleoclimatic significance of Late Pleistocene carbonates, Ross Sea, Antarctica. *J. Sediment. Petrol.*, vol. 63, pp. 84–90.

Taylor, B. J., M. R. A. Thomson, and L. E. Willey. 1979. The Geology of the Upper Jurassic–Lower Cretaceous succession between Ablation Point and Keystone Cliffs, eastern Alexander Island. In B. J. Taylor, M. R. A. Thomson, and L. E. Willey, eds., *The Geology of the Ablation Point; Keystone Cliffs Area, Alexander Island*. British Antarctic Survey Scientific Reports, no. 82, p. 62.

Taylor, T. N., E. L. Taylor, and M. J. Farabee. 1988. Palynostratigraphy of the Falla Formation (Upper Triassic), Beardmore Glacier region. *Ant. J. U.S.*, vol. 23, pp. 8–9.

Tchernia, P., and P. F. Jeannin. 1984. Circulation in Antarctic waters as revealed by iceberg tracks 1972–1983. *Polar Rec.*, vol. 22, pp. 263–269.

ten Brink, N. M. 1974. Glacio-isostasy: new data from West Greenland and geophysical implications. *Geolog. Soc. Am. Bull.*, vol. 85, pp. 219–228.

ten Brink, U. S., S. Bannister, B. C. Beaudoin, and T. A. Stern. 1993. Geophysical investigations of the tectonic boundary between East and West Antarctica. Science, vol. 261, pp. 45–50.

ten Brink, U. S., and A. K. Cooper. 1992. Modeling the bathymetry of the Antarctic continental shelf. In Y. Yoshida, K. Kaminuma, and K. Shiraishi, eds., *Recent Progress in Antarctic Earth Science*. Terra Scientific Publishing Co, Tokyo, Japan, pp. 763–771.

ten Brink, U. S., C. Schneider, and A. H. Johnson. 1995. Morphology and stratal geometry of the Antarctic continental shelf: insights from models. In A. K. Cooper, P. F. Barker, and G. Brancolini, eds., *Geology and Seismic Stratigraphy of the Antarctic Margin*. Antarctic Research Series, vol. 68. American Geophysical Union, Washington, D.C., pp. 1–24.

Tessensohn, T. 1994. LIRA; Lithosphere investigations in the Ross Sea area. *Terra Antarctica*, vol. 1, pp. 226–228.

Tessensohn, F., K. Duphorn, H. Jordan, G. Kleinschmidt, D. N. B. Skinner, U. Vetter, T. O. Wright, and D. Wyborn. 1981. Geological comparison of basement units in North Victoria Land, Antarctica. *Geologisches Jahrbuch*, vol. B41, pp. 31–88.

Thomas, M. A., and J. B. Anderson. 1994. Sea-level controls on the facies architecture of the Trinity/Sabine incised-valley system, Texas continental shelf. In R. W. Dalrymple, R. Boyd, and B. A. Zaitlin, eds., *Incised-Valley Systems: Origin and Sedimentary Sequences*. Society of Economic Petrologists and Mineralogists Special Publication no. 51, pp. 63–82.

Thomas, R. H. 1979. The dynamics of marine ice sheets. *J. Glaciol.*, vol. 24, pp. 167–177.

Thomas, R. H., and C. R. Bentley. 1978. A model for Holocene retreat of the West Antarctic Ice Sheet. *Quat. Res.*, vol. 10, pp. 150–170.

Thomson, M. R. A. 1982. Mesozoic paleogeography of West Antarctica. In C. Craddock, ed., *Antarctic Geoscience*: University of Wisconsin Press, Madison, pp. 331–337.

——— 1983. Late Jurassic ammonites from the Orville Coast, Antarctica. In R. L. Oliver, P. R. James, and J. B. Jago, eds., *Antarctic Earth Science*. Cambridge University Press, New York, pp. 315–319.

Thomson, M. R. A., R. J. Pankhurst, and P. D. Clarkson. 1983. The Antarctic Peninsula – a late Mesozoic-Cenozoic arc (review). In R. L. Oliver, P. R. James, and J. B. Jago, eds., *Antarctic Earth Science*. Cambridge University Press, New York, pp. 289–294.

Thomson, M. R. A., and T. H. Tranter. 1986. Early Jurassic

fossils from central Alexander Island and their geological setting. *British Antarctic Survey Bulletin*, no. 70, pp. 23–39.

Tingey, R. J. 1982. The geologic evolution of the Prince Charles Mountains – an Antarctic Archean cratonic block. In C. Craddock, ed., *Antarctic Geoscience*. University of Wisconsin Press, Madison, pp. 455–464.

——— 1991. The regional geology of Archaean and Proterozoic rocks in Antarctica. In R. J. Tingey, ed., *The Geology of Antarctica*. Clarendon Press, Oxford, pp. 1–73.

Tokarski, A. K. 1987. Structural events in the South Shetland Islands (Antarctica), IV. Structural evolution of King George Island and regional implications. *Studia Geologica Polonica*, vol. 93, pp. 63–112.

Tomlinson, J. S., C. J. Pudsey, R. A. Livermore, R. D. Larter, and P. F. Barker. 1992. Long-range sidescan sonar (GLORIA) survey of the Antarctic Peninsula Pacific margin. In Y. Yoshida, K. Kaminuma, and K. Shiraishi, eds., *Recent Progress in Antarctic Earth Science*. Terra Scientific Publishing Co., Tokyo, Japan, pp. 423–429.

Tranter, T. H. 1987. The structural history of the LeMay Group of central Alexander Island, Antarctic Peninsula. *British Antarctic Survey Bulletin*, no. 77, pp. 61–80.

Traube, V., and A. Rybnikov. 1990. Preliminary results of the first marine geophysical investigations in the north-western Weddell Sea. International Workshop on Antarctic Offshore Seismic Stratigraphy (ANTOSTRAT). Overview and Extended Abstracts. pp. 278–280.

Treshnikov, A. F. 1964. Surface water circulation in the Antarctic Ocean. *Soviet Antarctic Expedition Information Bulletin*, No. 45, pp. 5–8 (English translation: 1965, *Soviet Antarctic Expedition Information Bulletin*, vol. 5. American Geophysical Union, Washington, D.C., pp. 81–83).

Truswell, E. M. 1982. Palynology of seafloor samples collected by the 1911–14 Australiasian Antarctic Expedition: implications for the geology of coastal East Antarctica. *J. Geolog. Soc. of Austral.*, vol. 29, pp. 343–356.

——— 1983. Geological implications of recycled palynomorphs in continental shelf sediments around Antarctica. In R. L. Oliver, P. R. James, and J. B. Jago, eds., *Antarctic Earth Science*. Cambridge University Press, New York, pp. 394–399.

Truswell, E. M., and J. B. Anderson. 1984. Recycled palynomorphs and the age of sedimentary sequences in the eastern Weddell Sea. *Ant. J. U.S.*, vol. 19, pp. 90–92.

Truswell, E. M., and D. J. Drewry. 1984. Distribution and provenance of recycled palynomorphs in surficial sediments of the Ross Sea, Antarctica. *Marine Geol.*, vol. 59, pp. 187–214.

Tucholke, B. E. 1977. Sedimentation processes and acoustic stratigraphy in the Bellingshausen Basin. *Marine Geol.*, vol. 25, pp. 209–230.

Tucholke, B. E., C. D. Hollister, F. M. Weaver, and W. R. Vennum. 1976. Continental rise and abyssal plain sedimentation in the southeast Pacific Basin – Leg 35 Deep Sea Drilling Project. In C. D. Hollister, C. Craddock, eds., *Initial Reports of the Deep Sea Drilling Project*, vol. 35. U.S. Government Printing Office, Washington, D.C., pp. 359–400.

Tucholke, B. E., and R. E. Houtz. 1976. Sedimentary framework of the Bellingshausen Basin from seismic profiler data. In C. D. Hollister, C. Craddock, eds., *Initial Reports of the Deep Sea Drilling Project*, vol. 35. U.S. Government Printing Office, Washington, D.C., pp. 197–228.

Turner, B. R., and D. Padley. 1991. Lower Cretaceous coal-bearing sediments from Prydz Bay, East Antarctica. In J. Barron, J. Anderson, J. Baldauf, and B. Larsen, eds., Proceedings of the Ocean Drilling Program, Scientific Results, vol. 119, p. 57–60.

U. S. Naval Oceanographic Office. 1960. Sailing Directions for Antarctica, H.O. Publication. U.S. Government Printing Office, Washington, D.C., p. 431.

Urien, C. M., and J. J. Zambarrano. 1973. *Geology of the basins of the Argentine continental margin and Malvinas Plateau: The South Atlantic*. New York, Plenum Press, pp. 135–169.

Vail, P. R., and J. Hardenbol. 1979. Sea level changes during the Tertiary. *Oceanus*, vol. 22, pp. 71–79.

Vail, P. R., R. M. Mitchum Jr., R. G. Todd, J. M. Widmier, S. Thompson III, J. B. Sangree, J. N. Bubb, and W. G. Hatlelid. 1977. Seismic stratigraphy and global changes of sea level. In C. E. Payton, ed., *Seismic Stratigraphy – Applications to Hydrocarbon Exploration*. American Association of Petroleum Geologists Memoir 26, pp. 49–205.

Valenzuela, E., and F. Hervé. 1972. Geology of Byers Peninsula, Livingston Island, South Shetland Islands. In R. J. Adie, ed., *Antarctic Geology and Geophysics*. International Union of Geological Sciences, Series B, vol. 1. Universitetsforlaget, Oslo, Norway, pp. 83–89.

Van der Wateren, F. M., and A. L. L. M. Verbers. 1992. Cenozoic glacial geology and mountain uplift in northern Victoria Land, Antarctica. In Y. Yoshida, K. Kaminuma, and K. Shiraishi, eds., *Recent Progress in Antarctica Earth Science*. Terra Scientific Publishing Co., Tokyo, Japan, pp. 707–714.

Vanneste, L. E., and R. D. Larter. 1995. Deep-tow boomer survey on the Antarctic Peninsula Pacific margin: an investigation of the morphology and acoustic characteristics of late Quaternary sedimentary deposits on the outer continental shelf and upper slope. In A. K. Cooper, B. F. Barker, and B. Brancolini, eds., *Geology and Seismic Stratigraphy of the Antarctic Margin*. Antarctic Research Series, vol. 68. American Geophysical Union, Washington, D.C., pp. 97–121.

Vanney, J. R., R. K. H. Falconer, and G. L. Johnson. 1981. Geomorphology of the Ross Sea and adjacent oceanic provinces. *Marine Geol.*, vol. 41, pp. 73–102.

Vanney, J. R., and G. L. Johnson. 1976. Geomorphology of the Pacific continental margin of the Antarctic Peninsula. In C. D. Hollister and C. Craddock, eds., *Initial Reports of the Deep Sea Drilling Project*, vol. 35. U.S. Government Printing Office, Washington, D.C., pp. 279–289.

——— 1979. Wilkes Land continental margin physiography, East Antarctica. *Polarforschung*, vol. 49, pp. 20–29.

Veevers, J. J. 1986. Breakup of Australia and Antarctica estimated as mid-Cretaceous (95 ± 5 Ma) from magnetic and seismic data at the continental margin. *Earth Planet. Sci. Lett.*, vol. 77, pp. 91–99.

Veevers, J. J. 1987a. The conjugate continental margins of Antarctica and Australia. In S. L. Eittreim and M. A. Hampton, eds., *The Antarctic Continental Margin: Geology and Geophysics of Offshore Wilkes Land*. CPCEMR

Earth Science Series, vol. 5A. Circum-Pacific Council for Energy and Mineral Resources, Houston, pp. 45–74.

1987b. Earth history of the southeast Indian Ocean and the conjugate margins of Australia and Antarctica. *J. Proc. Royal Soc. New South Wales*, vol. 120, pp. 57–70.

1990. Antarctica-Australia fit resolved by satellite mapping of oceanic fracture zones. *Austral. J. Earth Sci.*, vol. 37, pp. 123–126.

Veevers, J. J., and P. R. Evans. 1975. Late Paleozoic and Mesozoic History of Australia. In K. S. W. Campbell, ed., *Gondwana Geology*. Australia National University Press, Canberra, pp. 579–607.

Visser, J. N. J. 1991. The paleoclimate setting of the late Paleozoic marine ice sheet in the Karoo Basin of Southern Africa in J. B. Anderson and G. M. Ashley, eds., *Glacial Marine Sedimentation; Paleoclimatic Significance*. Geological Society of America Special Paper 261, Boulder, Colorado, pp. 181–189.

Vorren, T. O., M. Hald, M. Edvarsen, and O. Lind-Hansen. 1983. Glacigenic sediments and sedimentary environments on continental shelves: general principles with a case study from the Norwegian Shelf. In J. Ehlers, ed., *Glacial Deposits of Northwest Europe*. A. A. Balkema, Rotterdam, Netherlands, pp. 61–73.

Vorren, T. O., E. Lebesbye, K. Andreassen, and K. B. Larsen. 1989. Glacigenic sediments on a passive continental margin as exemplified by the Barents Sea. *Marine Geol.*, vol. 85, pp. 251–272.

Wade, F. A. 1972. Geologic survey of Marie Byrd Land. *Ant. J. U.S.*, vol. 7, pp. 144–145.

Wade, F. A., and J. R. Wilbanks. 1972. Geology of Marie Byrd and Ellsworth Lands. In R. J. Adie, ed., *Antarctic Geology and Geophysics*. International Union of Geological Sciences, Series B, vol. 1. Universitetsforlaget, Oslo, Norway, pp. 207–214.

Wadhams, P. 1986. The seasonal ice zone. In N. Untersteiner, ed., *The Geophysics of Sea Ice*. Plenum, New York, pp. 825–991.

1988. Winter observations of iceberg frequencies and sizes in the South Atlantic Ocean. *J. Geophys. Res.*, vol. 93, pp. 3583–3590.

Wager, A. C., and A. W. Jamieson. 1983. Glaciological characteristics of Spartan Glacier, Alexander Island. *British Antarctic Survey Bulletin*, no. 52, pp. 221–228.

Waitt, R. B. 1983. Thicker West Antarctic ice sheet and peninsula icecap in late-Wisconsin time – sparse evidence from northern Lassiter Coast. *Ant. J. U.S.*, vol. 18, pp. 91–93.

Walcott, R. I. 1970. Isostatic response to loading of the crust in *Canada*. *Canad. J. Earth Sci.*, vol. 7, pp. 716–727.

Wannesson, J. 1990. Geology and petroleum potential of the Adelie Coast margin, East Antarctica. In B. St. John, ed., *Antarctica as an Exploration Frontier – Hydrocarbon Potential, Geology, and Hazards*. American Association of Petroleum Geologists Studies in Geology, no. 31, pp. 77–88.

Wannesson, J., M. Pelras, B. Petitperrin, M. Perret, and J. Segoufin. 1985. A geophysical transect of the Adélie Margin, East Antarctica. *Marine Petroleum Geol.*, vol. 2, pp. 192–201.

Watkins, N. D., J. Keany, M. T. Ledbetter, and T. Huang. 1974. Antarctic glacial history from analyses of ice-rafted

deposits in marine sediments: new model and initial tests. *Science*, vol. 186, pp. 533–536.

Watkins, N. D., and J. P. Kennett. 1972. Regional sedimentary disconformities and Upper Cenozoic changes in bottom water velocities between Australasia and Antarctica. In D. E. Hayes, ed., *Antarctic Oceanology II; The Australian-New Zealand Sector*. American Geophysical Union, Washington, D.C., pp. 273–294.

1977. Erosion of deep-sea sediments in the Southern Ocean between longitudes 70E and 190E and contrasts in manganese nodule development. *Marine Geol.*, vol. 23, pp. 103–111.

Watkins, N. D., M. T. Ledbetter, and T. C. Huang. 1982. Antarctic glacial history using spatial and temporal variations of ice-rafted debris in abyssal sediments of the Southern Ocean. In C. Craddock, ed., *Antarctic Geoscience*. University of Wisconsin Press, Madison, pp. 1013–1016.

Weaver, S. D., A. D. Saunders, R. J. Pankhurst, and J. Tarney. 1979. A geochemical study of magmatism associated with the initial stages of back-arc spreading: the Quaternary volcanics of Bransfield Strait, from South Shetland Islands. *Contrib. Mineral. Petrol.*, vol. 68, pp. 151–169.

Weaver, S. D., A. D. Saunders, and J. Tarney. 1982. Mesozoic-Cenozoic volcanism in the South Shetland Islands and the Antarctic Peninsula: geochemical nature and plate tectonic significance. In C. Craddock, ed., *Antarctic Geoscience*. University of Wisconsin Press, Madison, pp. 263–273.

Webb, P. N. 1987. Cenozoic sedimentary and paleontological history of Antarctica. Fifth International Symposium on Antarctic Earth Sciences, Abstracts. Cambridge, August 1987, 135.

1990. The Cenozoic history of Antarctica and its global impact. *Ant. Sci.*, vol. 2, pp. 3–21.

1991. A review of the Cenozoic stratigraphy and paleontology of Antarctica. In M. R. A. Thomson, J. A. Crane, and J. W. Thomson, eds., *Geological Evolution of Antarctica*. Cambridge University Press, New York, pp. 599–607.

Webb, P. N., and D. M. Harwood. 1987. Terrestrial flora of the Sirius Formation: its significance for late Cenozoic glacial history. *Ant. J. U.S.*, vol. 22, pp. 7–11.

1991. Late Cenozoic glacial history of the Ross embayment, Antarctica. *Quat. Sci. Rev.*, vol. 10, pp. 215–223.

Webb, P. N., R. M. Leckie, and B. L. Ward. 1986. Foraminifera (late Oligocene): Antarctic Cenozoic history from the MSSTS-1 drillhole, McMurdo Sound. Science Information Publishing Centre, Wellington, New Zealand, pp. 115–125.

Webb, P. N., T. E. Ronan Jr., J. H. Lipps, and T. E. DeLaca. 1979. Miocene glaciomarine sediments from beneath the southern Ross Ice Shelf, Antarctica. *Science*, vol. 203, pp. 435–437.

Webb, P. N., and J. H. Wrenn. 1982. Upper Cenozoic micropaleontology and biostratigraphy of eastern Taylor Valley, Antarctica. In C. Craddock, ed., *Antarctic Geoscience*. University of Wisconsin Press, Madison, pp. 1117–1122.

Weber, M. E., G. Bonani, and K. D. Fütterer. 1994. Sedimentation processes within channel-ridge systems, southeastern Weddell Sea, Antarctica. *Paleoceanography*, vol. 9, pp. 1027–1048.

Weber, W., and J. J. Livschitz. 1985. Beitrag zur Geologie der Hutton Mountains und der Geuttard Range (Palmer Land, Antarktische Halbinsel): contributions to the geology of the Hutton Mountains and Guettard Range, Lassiter Coast/Palmer Land (Antarctica). *Geodaetische und Geophysikalische Veroeffentlichungen*, vol. 12, pp. 57–70.

Weber, W., and K. Rank. 1987. Beitrag zur Geologie der Hutton Mountains und der Guettard Range (Palmer Land, Antarktische Halbinsel): contribution to the geology of the Hutton Mountains and the Guettard Range (Palmer Land, Antarctic Peninsula). *Freiberger Forschungshefte*, vol. 412 C, pp. 51–63.

Webers, G. F., C. Craddock, M. A. Rogers, and J. J. Anderson. 1982. Geology of the Whitmore Mountains. In C. Craddock, ed., *Antarctic Geoscience*. University of Wisconsin Press, Madison, pp. 841–847.

Weertman, J. 1972. The theory of glacier sliding. In C. Embleton, ed., *Glaciers and Glacial Erosion*. Macmillan, London, pp. 244–268.

Wei, W., and S. W. Jr. Wise. 1990. Biogeographic gradients of middle Eocene-Oligocene calcareous nannoplankton in the South Atlantic Ocean. *Palaeogeogr., Palaeoclimatol., Palaeoecol.*, vol. 79, pp. 29–61.

Wei, W., and S. W. Jr. Wise. 1992. Oligocene-Pleistocene calcareous nannofossils from Southern Ocean sites 747, 748, and 751. In S. W. Wise, Jr., R. Schlich, eds., Proceedings of the Ocean Drilling Program, Scientific Results, vol. 120. College Station, Texas, pp. 509–521.

Weissel, J. K., and D. E. Hayes. 1972. Magnetic anomalies in the southeast Indian Ocean. In D. E. Hays, ed., *Antarctic Oceanology II; The Australian-New Zealand Sector*. Antarctic Research Series, vol. 19. American Geophysical Union, Washington, D.C., pp. 165–196.

Wellman, P., and R. J. Tingey. 1976. Gravity evidence for a major crustal fracture in eastern Antarctica. Bureau of Mineral Resources. *J. Austral. Geol. Geophys.*, vol. 1, pp. 105-108.

Whillans, I. M., M. Jackson, and Y.-H. Tseng. 1993. Velocity pattern in a transect across Ice Stream B, Antarctica. *J. Glaciol.*, vol. 39, pp. 562–572.

White, C. M., and C. Craddock. 1987. Compositions of igneous rocks in the Thurston Island area, Antarctica: evidence for a Late Paleozoic-Middle Mesozoic Andinotype continental margin. *J. Geol.*, vol. 95, pp. 699–709.

White, W. M., and M. Cheatham. 1987. Geochemistry of back-arc basin volcanics from Bransfield Strait, Antarctic Peninsula. EOS (Transactions of the American Geophysical Union), vol. 68, p. 1520.

Wilch, T. I. 1997. Volcanic record of the West Antarctic Ice Sheet in Marie Byrd Land. Unpublished PhD Dissertation. New Mexico Institute of Mining and Technology, Socorro, New Mexico.

Wilch, T. I., D. R. Lux, G. H. Denton, and W. C. McIntosh. 1993. Minimal Pliocene-Pleistocene uplift of the dry valleys sector of the Transantarctic Mountains; a key parameter in ice-sheet reconstructions. *Geology*, vol. 21, pp. 841–844.

Wilch, T. I., D. R. Lux, W. C. McIntosh, and G. H. Denton. 1989. Plio–Pleistocene uplift of the McMurdo dry valley sector of the Transantarctic Mountains. *Ant. J. U.S.*, vol. 24, pp. 30–33.

Williams, C. 1996. Lake Vostok – freshwater time capsule. *Geotimes*, vol. 41, pp. 6–7.

Williams, D. F., and B. H. Corliss. 1982. The south Australian continental margin and the Australian-Antarctic sector of the Southern Ocean. In A. E. M. Nairn and F. G. Stehli, eds., *The Ocean Basins and Margins. 6. The Indian Ocean*. Plenum Press, New York, pp. 545–584.

Williams, P. L., D. L. Schmidt, C. C. Plummer, and L. E. Brown. 1972. Geology of the Lassiter Coast area, Antarctic Peninsula: Preliminary Report. In R. J. Adie, ed., *Antarctic Geology and Geophysics*. International Union of Geological Sciences, Series B, vol. 1. Universitetsforlaget, Oslo, Norway, pp. 143–148.

Wilson, G. S. 1995a. The Neogene East Antarctic ice sheet: a dynamic or stable feature? *Quat. Sci. Rev.*, vol. 14, pp. 101–123.

Wilson, G. S., A. P. Roberts, K. L. Verosub, F. Florindo, and L. Sagnotti. 1998. Magnetobiostratigraphic chronology of the Eocene–Oligocene transition in the CIROS–1 core, Victoria Land margin, Antarctica: implications for Antarctic glacial history. *Geo. Soc. Am. Bull.*, vol. 110, pp. 35–47.

Wilson, T. J. 1991. Jurassic faulting and magmatism in the Transantarctic Mountains: implications for Gondwana breakup. In R. H. Findlay, R. Unrug, M. R. Banks, and J. J. Veevers, eds., Eighth Gondwana Symposium, Hobart, Tasmania. Balkema Publishers, Brookfield, Vermont, pp. 563–572.

1995b. Cenozoic transtension along the Transantarctic Mountains–West Antarctic rift boundary, southern Victoria Land: Antarctica. *Tectonics*, vol. 14, pp. 531–545.

Wise, S. W., Jr. A. M. Gombos, and J. P. Muza. 1985. Cenozoic evolution of polar water masses, southwest Atlantic Ocean. In K. J. Hsue and H. J. Weissert, eds., *South Atlantic Paleoceanography*. Cambridge University Press, Cambridge, pp. 283–324.

Wise, S. W., Jr., J. R. Breza, D. M. Harwood, and W. Wei. 1991. Paleogene glacial history of Antarctica. in D. W. Muller, J. A. McKenzie, and H. Weissert, eds., *Controversies in Modern Geology; Evolution of Geological Theories in Sedimentology, Earth History and Tectonics*. Academic Press, London, pp. 133–171.

Wise, S. W. Jr., J. R. Breza, D. M. Harwood, W. Wei, and J. C. Zachos. 1992. Paleogene glacial history of Antarctica in light of Leg 120 drilling results. In S. W. Wise Jr., R. Schlich, et al., eds., Proceedings of the Ocean Drilling Program, Scientific Results, vol. 120. College Station, Texas, pp. 1001–1030.

Wodzicki, A., and R. Robert Jr. 1986. Geology of the Bowers Supergroup, central Bowers Mountains, northern Victoria Land. In E. Stump, ed., *Geological Investigations in Northern Victoria Land*. Antarctic Research Series, vol. 46. American Geophysical Union, Washington, D.C., pp. 39–68.

Wong, H. K., and D. A. Christoffel. 1981. A reconnaissance seismic survey of McMurdo Sound and Terra Nova Bay, Ross Sea. In L. D. McGinnis, ed., *Dry Valley Drilling Project*. Antarctic Research Series, vol. 33. American Geophysical Union, Washington, D.C., pp. 37–62.

Woodruff, F., S. M. Savin, and R. G. Douglas. 1981. Miocene stable isotope record: a detailed deep Pacific Ocean study and its paleoclimatic implications. *Science*, vol. 212, pp. 665–668.

Wordie, J. M. 1921. Depths and deposits of the Weddell Sea, Shackleton Antarctic Expedition 1914–1917. Transactions of the Royal Society of Edinburgh, vol. 52, pp. 781–793.

Wrenn, J. H., and P. N. Webb. 1982. Physiographic analysis and interpretation of the Ferrar Glacier-Victoria Valley area, Antarctica. In C. Craddock, ed., *Antarctic Geoscience*. University of Wisconsin Press, Madison, pp. 1091–1099.

Wright, J. D., and K. G. Miller. 1992. Miocene stable isotope stratigraphy, site 747, Kerguelen Plateau. In S. W. Wise, Jr., R. Schlich, et al., eds., Proceedings of the Ocean Drilling Program, Scientific Results, vol. 120. College Station, Texas, pp. 855–866.

1993. Southern Ocean influences on late Eocene to Miocene deepwater circulation. In J. P. Kennett and D. A. Warnke, eds., *The Antarctic Paleoenvironment; A Perspective on Global Change II*. Antarctic Research Series, vol. 60. American Geophysical Union, Washington, D.C., pp. 1–25.

Wright, R. 1980. Sediment Gravity Transport on the Weddell Sea Continental Margin. MA Thesis, Rice University, Houston, Texas, 96 pp.

Wright, R., and J. B. Anderson. 1982. The importance of sediment gravity flow to sediment transport and sorting in a glacial marine environment: Eastern Weddell Sea, Antarctica. *Geolog. Soc. Am. Bull.*, vol. 93, pp. 951–963.

Wright, R., J. B. Anderson, and P. P. Fisco. 1983. Distribution and association of sediment gravity flow deposits and glacial/glacial-marine sediments around the continental margin of Antarctica. In B. F. Molnia, ed., *Glacial-Marine Sedimentation*. Plenum Press, New York, pp. 265–300.

Yamaguchi, K., Y. Tamura, I. Mizukoshi, and T. Tsuru. 1988. Preliminary report of geophysical and geological surveys in the Amundsen Sea, West Antarctica. In Proceedings of the NIPR Symposium on Antarctic Geosciences, vol. 2. National Institute of Polar Research, Tokyo, Japan, pp. 55–67.

Yang, S., and D. M. Harwood. 1995. Late Quaternary environmental fluctuations from Yanou Lake, King George Island, Fildes Peninsula, Antarctica. Abstracts, VII International Symposium on Antarctic Earth Sciences. Tipgrafia Senese, Siena, Italy, 414 pp.

Yoon, H. I., M. W. Han, B. K. Park, J. K. Oh, and S. K. Chang. 1994. Depositional environment of near-surface sediments, King George Basin, Bransfield Strait, Antarctica. *Geo-Marine Lett.*, vol. 14, pp. 1–9.

Yoshida, Y. 1983. Physiography of the Prince Olav and Prince Harald Coasts, East Antarctica. Memoirs of the NIPR Research, Series C, Earth Sciences, vol. 13, 83 pp.

1989. Tectonic and/or structural land forms in eastern Queen Maud Land. Proceedings of the NIPR Symposium on Antarctic Geosciences, vol. 3, p. 144.

Young, N. W., D. Raynaud, M. A. de Angelis, J. R. Petit, and C. Lorius. 1984. Past changes of the Antarctic ice sheet in Terre Adelie as deducted from ice-core data and ice modelling. *Ann. of Glaciol.*, vol. 5, 239 pp.

Zachos, J. C., J. R. Breza, and S. W. Wise. 1992. Early Oligocene ice-sheet expansion on Antarctica: stable isotope and sedimentological evidence from Kerguelen Plateau, southern Indian Ocean. *Geology*, vol. 20, pp. 569–573.

Zachos, J. C., L. D. Stott, and K. C. Lohmann. 1994. Evolution of early Cenozoic marine temperatures. *Paleoceanography*, vol. 9, pp. 353–387.

Zemmels, I. 1978. A Study of the Sediment Composition and Sedimentary Geochemical Processes in the Vicinity of the Pacific-Antarctic Ridge. PhD Dissertation, Florida State University, Tallahassee, Florida, 349 pp.

Zhivago, A. V., and A. P. Lisitsin. 1957, new data on the bottom and submarine deposits in the eastern Antarctic. *Izvestiya Akademii Nauk SSSR*, vol. 1, pp. 19–35.

Zinsmeister, W. J. 1982. Review of the Upper Cretaceous-Lower Tertiary sequence on Seymour Island, Antarctica. *J. Geolog. Soc. London*, vol. 139, pp. 779–785.

Zotikov, I. A., V. S. Zagorodnov, and J. V. Raidovsky. 1980. Core drilling through the Ross Ice Shelf (Antarctica) confirmed basal freezing. *Science*, vol. 207, pp. 1463–1465.

Index